Ground Water Manual

A Water Resources Technical Publication

A Guide for the Investigation, Development, and Management of Ground-Water Resources

First Edition 1977

Revised Reprint 1981

U.S. Department of the Interior
Water and Power Resources Service

A Wiley-Interscience Publication

JOHN WILEY & SONS

New York Chichester Brisbane Toronto Singapore

Selected Government Publications. Part of a series reprinted by John Wiley & Sons, Inc. to facilitate worldwide distribution.

Printed in the United States of America

ISBN 0-471-80008-2

PREFACE

This manual has been prepared as a guide to field personnel in the more practical aspects and commonly encountered problems of ground-water investigations, development, and management.

Information is presented concerning such aspects as ground-water occurence and flow, well-aquifer relationships, ground-water investigations, aquifer test analysis, estimating aquifer yield, data collection, and geophysical investigations. In addition, permeability tests, well design, dewatering systems, well specifications and drilling, well sterilization, pumps, and other aspects have been discussed. An extensive bibliography has also been included.

The manual has been developed over a period of years, and its many contributors have diversified technical backgrounds. Contributors include personnel from the Bureau of Reclamation Engineering and Research Center and field offices, other agencies, foreign governments, and many individual scientists and engineers.

Principal contributors and authors include T. P. Ahrens, H. H. Ham, and W. N. Tapp. Mathematical treatment and development are those of R. E. Glover, W. T. Moody, R. W. Ribbens, D. Jarvis, C. N. Zanger, and J. W. Christopher. Other technical contributors include W. A. Pennington, D. Wantland, H. R. McDonald, L. A. Johnson, A. C. Barlow, C. R. Maierhofer, and R. J. Winger, Jr. Major contributors to the overall conception, preparation, and editing of the manual were made by T. P. Ahrens, R. J. Winger, Jr., H. H. Ham, R. D. Mohr, and W. E. Foote. The manual also includes material from papers by C. V. Theis, M. I. Rorabaugh, W. C. Walton, C. E. Jacob, R. W. Stallman, M. S. Hantush, S. W. Lohman, and other scientists and engineers who have been appropriately cited.

There are occasional references to proprietary materials or products in this publication. These must not be construed in any way as an endorsement, as the Bureau cannot endorse proprietary products or processes of manufacturers or the services of commercial firms for advertising, publicity, sales, or other purposes.

CONTENTS

CHAPTER II. THEORY OF GROUND-WATER FLOW, AQUIFER PROPERTIES, AND DEFINITIONS

CHAPTER III. WELL AND AQUIFER RELATIONSHIPS

CHAPTER IV. PLANNING GROUND–WATER INVESTIGATIONS AND PRESENTATION OF RESULTS

CHAPTER V. ANALYSIS OF DISCHARGING WELL AND OTHER TEST DATA

CHAPTER VI. ESTIMATES OF AQUIFER YIELD, HYDROLOGIC BUDGETS, AND INVENTORIES

CHAPTER VII. INITIAL OPERATIONS AND COLLECTION OF CORRELATIVE DATA

CHAPTER VIII. GEOPHYSICAL INVESTIGATIONS, BORE HOLE LOGS, AND SURVEYS

CHAPTER IX. METHODS OF DETERMINING AQUIFER CHARACTERISTICS

CHAPTER X. PERMEABILITY TESTS IN INDIVIDUAL DRILL HOLES

CHAPTER XI. COMPONENTS OF A WELL AND PARTICULARS OF DESIGN

CHAPTER XII. INFILTRATION GALLERIES

CHAPTER XIII. DEWATERING SYSTEMS

CHAPTER XIV. CORROSION AND INCRUSTATION

CHAPTER XV. WELL SPECIFICATIONS

CHAPTER XVI. WATER WELL DRILLING

CHAPTER XVII. WATER WELL DEVELOPMENT

CHAPTER XVIII. WELL STERILIZATION

CHAPTER XIX. VERTICAL TURBINE PUMPS

CHAPTER XX. WELL AND PUMP COST FACTORS, OPERATION AND MAINTENANCE

CHAPTER XXI. WELL REHABILITATION

FIGURES

TABLES

INDEX

«Chapter I

GROUND-WATER OCCURRENCE, PROPERTIES, AND CONTROLS

1-1. Introduction.—Ground-water engineering is the art and science of investigating, developing, and managing ground water for the benefit of man. The technology involves specialized fields of soil science, hydraulics, hydrology, drainage, geophysics, geology, mathematics, agronomy, metallurgy, bacteriology, and electrical, mechanical, and chemical engineering.

With the ever-increasing demand for water, ground-water engineering can be expected to become increasingly important.

In addition to the solution of problems on ground-water recovery for water supply, ground-water engineering is important in problems concerning seepage from surface reservoirs and canals, the effects of bank storage, stability of slopes, recharging of ground-water reservoirs, controlling of saltwater intrusion, dewatering of excavations, subsurface drainage, and construction, land subsidence, waste disposal, and contamination control.

Ground-water engineering involves the determination of aquifer properties and characteristics and the application of hydraulic principles to ground-water behavior for the solution of engineering problems. Determination of aquifer characteristics and the application of those data by appropriate mathematical and other methods are essential to the solution of complex problems in which ground water is a factor. The extent to which the determination of aquifer properties and characteristics must be made depends upon the complexity of the problem involved. The required investigation may range from cursory to detailed. It may entail study or consideration of all or only one or two aquifer properties and hydraulic principles. Conditions often may be so complex as to preclude the determination of finite values and the application of available theory to the solution of some problems. Under such circumstances, solutions may be largely subjective and their reliability dependent upon the experience and judgment of the ground-water technical specialist.

1-2. History of Use.—The first use of ground water as a source of supply is lost in antiquity. For centuries [8, 10] [1] its use was limited by the difficulties of its development and by the absence of clear understanding of its origin and occurrence.

---

[1] Numbers in brackets refer to items in the bibliography, section 1-9.

Shallow, hand-dug wells and crude water-lifting devices marked the early exploitation of ground water. The introduction of well-drilling machinery and motor-driven pumps made possible the recovery of ground water in large amounts and at increased depths. Expanded knowledge of ground-water hydrology and other sciences added to man's ability to understand and utilize this resource.

As technology has improved, the benefits of ground-water development have become increasingly important. The use of water for domestic purposes (human and animal consumption) usually has the highest priority, followed by industrial requirements, and then agricultural usage (irrigation). Development of the ground-water resources of the United States has been increasing in recent years as development of surface water sources approaches the point of full potential.

1-3. Origin.—(a) *The Hydrologic Cycle.*—Precipitation, storage, runoff, and evaporation of the earth's water follow an unending sequence known as the hydrologic cycle [4,7,10]. During this cycle the total amount of water in the atmosphere and in or on the earth remains the same; however, its form may change. Although minor quantities of magmatic water or water from other deep-seated sources may find its way to the surface, all water is assumed to be part of the hydrologic cycle.

The movement of water within the hydrologic cycle is shown on figure 1-1. Water vapor in the atmosphere is condensed into ice crystals or water droplets that fall to the earth as rain or snow. A portion evaporates and returns to the atmosphere. Another portion flows across the ground surface until it reaches a stream, thence flows to the ocean. The remaining portion infiltrates directly into the ground and seeps downward. Some of this portion may be transpired by the roots of plants or moved back to the ground surface by capillarity and evaporated. The remainder seeps downward to join the ground-water body.

Ground water returns to the ground surface through springs and seepage to streams where it is subject to evaporation or is directly evaporated from the ground surface or transpired by vegetation. Thus, the hydrologic cycle is completed. When sufficient water vapor again gathers in the atmosphere, the cycle repeats.

The elements of the hydrologic cycle for any area can be quantified in an equation. For ground-water investigations, the equation can be expressed in terms of ground-water components. However, it may be necessary to evaluate the broad hydrologic picture to quantify

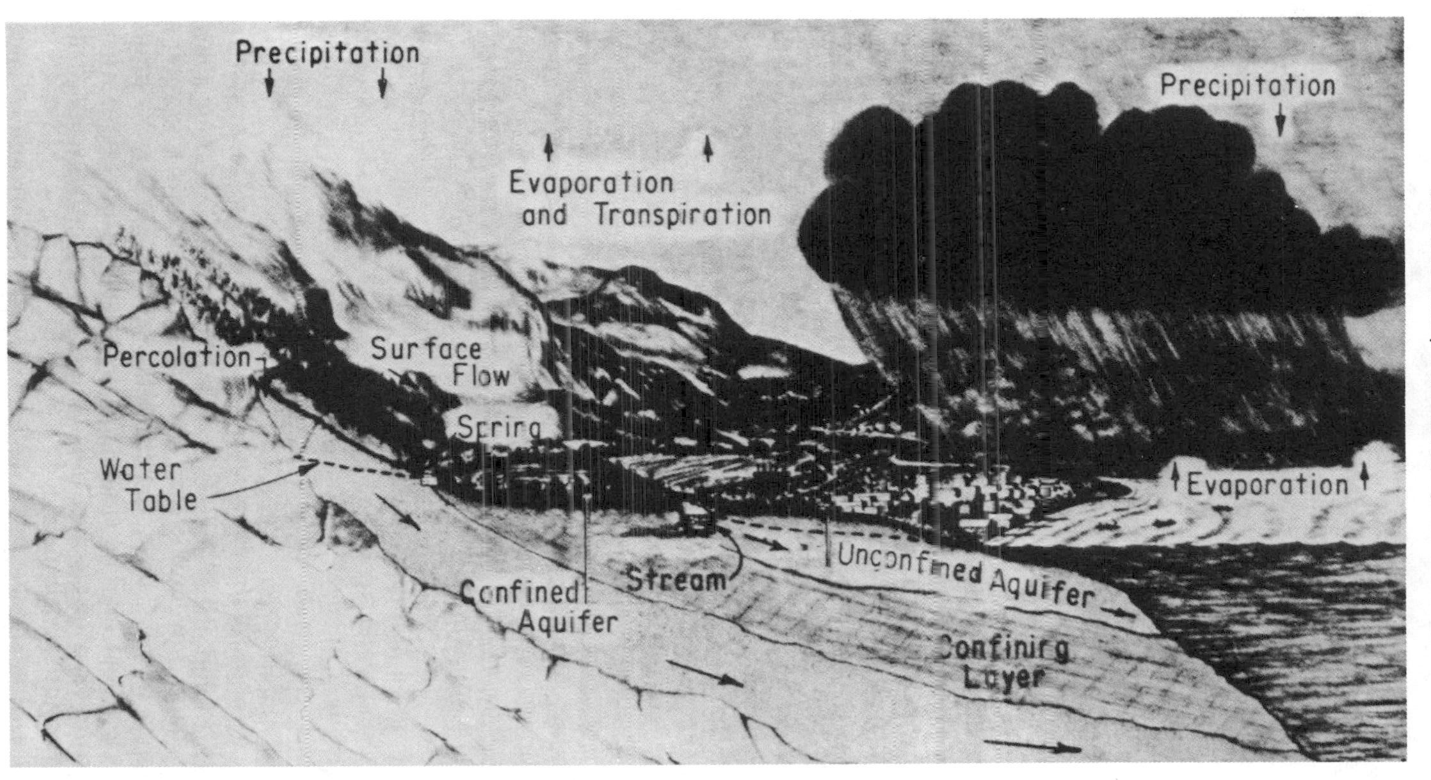

FIGURE 1-1.—The hydrologic cycle. 103–D–1400

the ground-water components. Determination of components in the ground-water equation is tedious and time consuming, and the results, at best, are only approximate. Therefore, an analysis should be made to determine that such an evaluation is necessary and justified before it is undertaken.

(b) *Ground-water Equation.*—A basic ground-water equation [4,7] which will permit an approach to a quantitative estimate of ground-water availability can be established for an area to account for those factors of the hydrologic cycle which have a direct effect on flow and storage of ground water. The equation can be stated as:

$$\Delta S_{gw} = \text{recharge} - \text{discharge}$$

where ΔS_{gw} is the change in ground-water storage during the period of study. Theoretically, under natural conditions and over a long period of time, which includes both wet and dry cycles, ΔS_{gw} will be zero and inflow (recharge) will equal outflow (discharge).

The natural recharge to the ground-water body includes deep percolation from precipitation, seepage from streams and lakes, and subsurface underflow. Artificial recharge includes deep percolation from irrigation and water spreading, seepage from canals and reservoirs, and recharge from recharge wells. The natural discharge or outflow from the ground-water body consists of seepage to streams, flow from springs, subsurface underflow, transpiration, and evaporation. Artificial discharge occurs by wells or drains. If ground-water storage in an area is less at the end of the selected period of time than at the beginning, discharge is indicated as having exceeded recharge. Conversely, recharge may exceed discharge.

(c) *Recharge to and Discharge from Aquifers.*—Recharge from natural sources include the following:

- *Deep percolation from precipitation.*—Deep percolation of precipitation is one of the most important sources of ground-water recharge. The amount of recharge in a particular area is influenced by vegetative cover, topography, and nature of soils; as well as the type, intensity, and frequency of precipitation.
- *Seepage from streams and lakes.*—Seepage from streams, lakes, and other water bodies is another important source of recharge. In humid and subhumid areas where ground-water levels may be high, the influence of seepage may be limited in extent and may be seasonal. However, in arid regions where the entire flow of streams may be lost to an aquifer, seepage may be of major significance.
- *Underflow from another aquifer.*—An aquifer may be recharged by underflow from a nearby, hydraulically connected aquifer.

The amount of this recharge depends on the head differential, the nature of the connection, and the hydraulic properties of the aquifers.

• *Artificial recharge.*—Artificial recharge to the ground water may be achieved through planned systems or may be unforeseen or unintentional. Planned major contributions to the ground-water reservoir may be through spreading grounds, infiltration ponds, and recharge wells. Irrigation applications, sewage effluent spreading grounds, septic tank seepage fields, and other activities have a similar, but usually unintentional effect. Seepage from reservoirs, canals, drainage ditches, ponds, and similar water impounding and conveyance structures may be local sources of major ground-water recharge. Recharge from such sources can completely change the ground-water regimen over a considerable area.

(d) *Ground-Water Discharge.*—Losses from the ground-water reservoir occur in the following four ways:

(1) *Seepage to streams.*—In certain reaches of streams and in certain seasons of the year, ground water may discharge into streams and maintain their base flows. This condition is more prevalent in humid areas than in arid or semiarid areas.

(2) *Flow from springs and seeps.*—Springs and seeps exist where the water table intersects the land surface or a confined aquifer outlets to the surface.

(3) *Evaporation and transpiration.*—Ground water may be lost by evaporation if the water table is near enough to the land surface to maintain flow by capillary rise. Also, plants may transpire ground water from the capillary fringe or the saturated zone.

(4) *Artificial discharge.*—Wells and drains are imposed artificial withdrawals on ground-water storage and in some areas are responsible for the major depletion.

Ground water moves in response to a hydraulic gradient in the same manner as water flowing in an open channel or pipe. However, the flow of ground water is appreciably restricted by friction with the porous medium through which it flows. This results in low velocities and high head losses as compared to open channel or pipe flow.

1–4. Occurrence of Ground Water.—(a) *General.*—Webster defines an aquifer as "a water-bearing bed or stratum of earth, gravel, or porous stone." Some strata are good aquifers, whereas others are poor. The most important requirement is that the stratum must have interconnected openings or pores through which water can move. The nature of each aquifer depends on the material of which it is composed, its origin, the relationship of the constituent grains or particles and associated pores, its relative position in the Earth's surface, its exposure to a recharge source, and other factors.

Rocks—used here to denote all material of the earth, whether consolidated and firm or unconsolidated and loose or soft—are generally classified as sedimentary, igneous, and metamorphic. The geologic structure, lithology, and stratigraphy of rocks in an area provide general knowledge of their potentials as aquifers.

(b) *Sedimentary Rocks.*—In general, the best aquifers are the coarse-grained, saturated portions of the unconsolidated, granular sedimentary mantle [4] which cover the consolidated rocks over much of the surface of the earth. Widespread presence of unconsolidated sediments is more common at lower elevations in proximity to streams. These sediments consist of stream alluvium, glacial outwash, wind-deposited sand, alluvial fans, and similar water- or wind-deposited coarse-grained, granular materials. In addition, some residual materials resulting from the weathering inplace of consolidated rock are good aquifers.

The coarser grained consolidated rocks such as conglomerates and sandstones are also often good aquifers, but are usually found below the unconsolidated granular sedimentary mantle. Their value as aquifers depends to a large extent on the degree of cementation and fracturing to which they have been subjected. In addition, some massive sedimentary rocks such as limestone, dolomite, and gypsum may also be good aquifers. These rocks are relatively soluble and, over the years, solution along fractures or partings may form voids which range in size from a fraction of an inch to several hundred feet. Some of the best known and most productive aquifers are cavernous limestones.

(c) *Igneous and Metamorphic Rocks.*—The value of igneous and metamorphic rocks as aquifers depends greatly on the amount of stress and weathering to which they have been subjected after their initial formation. In general, the crystalline igneous rocks are very poor aquifers if they remain undisturbed. However, mechanical and other stresses cause fractures and faults in these rocks in which ground water may occur. Such openings may range from hairline cracks to voids several inches wide. In general, these openings disappear with depth and do not yield significant quantities of water below depths of several hundred to a thousand feet.

In coarse-grained crystalline igneous rocks, where inplace weathering has occurred, a thin permeable zone may be found in the transition zone between the sound rock and the thoroughly weathered, usually relatively impermeable, overlying residual material. Some lavas, especially those of viscous basaltic composition, may contain good to excellent aquifers in the zones between successive flows. The scoriaceous upper and lower surfaces of flows are usually porous and permeable, and cooling fractures may be present in a zone extending into the flow from the upper and lower surfaces. Furthermore, coarse-grained sedimentary material may also be present between flows.

(d) *Unconfined and Confined Aquifers.*—An unconfined aquifer (fig. 1–2) is one that does not have a confining layer overlying it. It is often referred to as a free or "water table" aquifer or as being under "water-table conditions." Water infiltrating into the ground surface percolates downward through air-filled interstices of the material above the saturated zone and joins the ground-water body. The water table, or upper surface of the saturated ground-water body, is in direct contact with the atmosphere through the open pores of the material above and is everywhere in balance with atmospheric pressure. Movement of the ground water is in direct response to gravity.

A confined or artesian aquifer (fig. 1–2) has an overlying, confining layer of lower permeability than the aquifer and has only an indirect or distant connection with the atmosphere. Water in an artesian aquifer is under pressure and when the aquifer is penetrated by a tightly cased well or piezometer, the water will rise above the bottom of the confining bed to an elevation at which it is in balance with the atmospheric pressure and which reflects the pressure in the aquifer at the point of penetration. If this elevation is greater than that of the land surface at the well, water will flow from the well. The imaginary surface, conforming to the elevations to which water will rise in wells penetrating an artesian aquifer, is known as the potentiometric or piezometric surface.

(e) *Perched Aquifers.*—Beds of clay or silt, unfractured consolidated rock, or other material with relatively lower permeability than the surrounding materials may be present in some areas above the regional water table. Downward percolating water may be intercepted and a saturated zone of limited areal extent formed. This results in a perched aquifer with a perched water table [4]. An unsaturated zone is present between the bottom of the perching bed and the regional water table. A perched aquifer is a special case of an unconfined aquifer. Depending on climatic conditions or overlying land use, a perched water table may be a permanent phenomenon or one which is seasonally intermittent (fig. 1–2).

(f) *Zones of Moisture.*—Water may occur in several recognizable subsurface zones under different conditions, as shown in table 1–1 which was adapted from Meinzer [4].

The thickness of each zone above the zone of rock flowage varies according to the area and with time. During a period of recharge, the

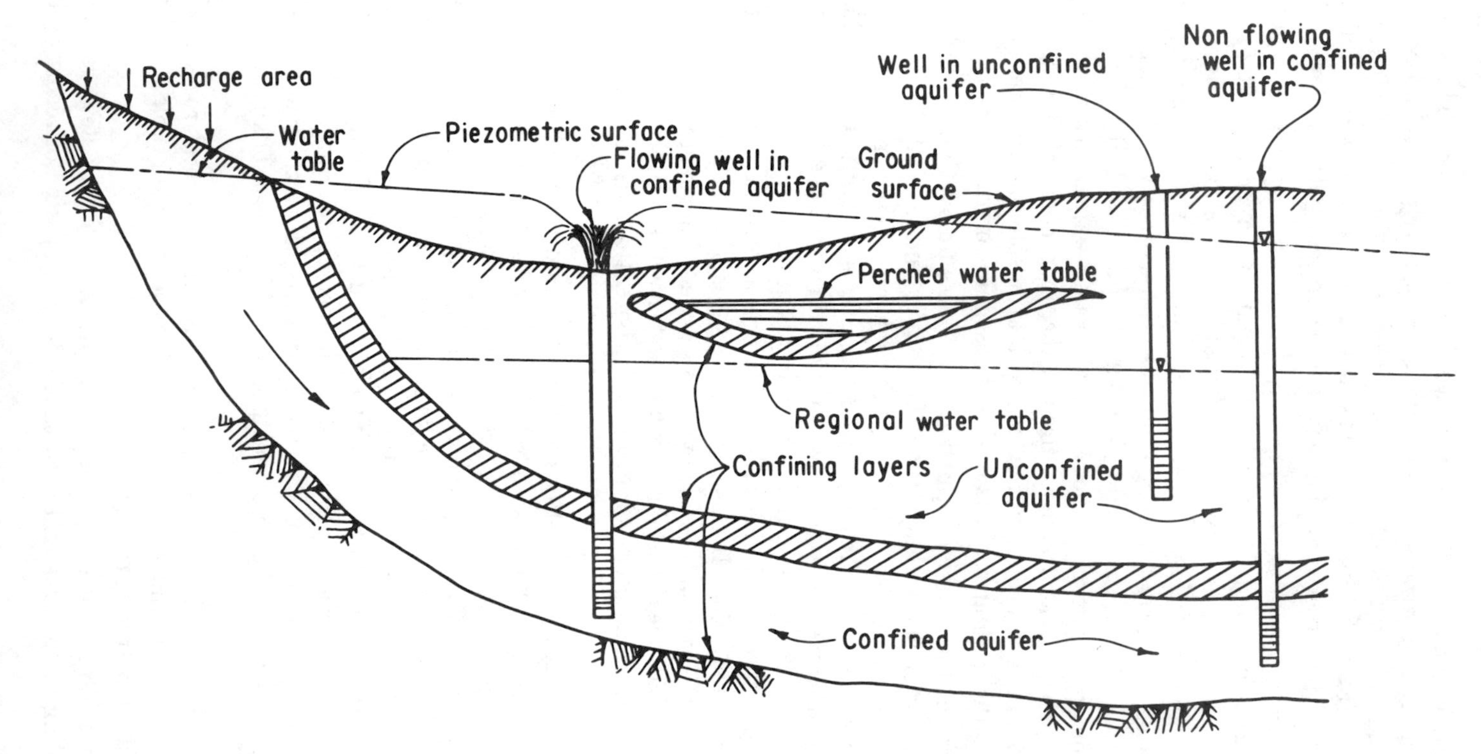

FIGURE 1-2.—Types of aquifers. 103-D-1401.

TABLE 1-1.—*Status of water in various soil zones*

Zone	Horizon	Condition of water	Condition of soil
Aeration (above water table).	Soil water	Under tension	Unsaturated.
	Intermediate	Under tension	Unsaturated.
	Capillary fringe	Under tension	Saturated and unsaturated.
Saturation (below water table).	Unconfined ground water.	Under pressure but upper surface at atmospheric pressure.	Saturated.
	Confined or artesian ground water.	Under pressure but upper surface above atmospheric pressure.	Saturated.
Rock flowage		Combined—no free water.	Dry.

zone of saturation thickens at the expense of the zone of aeration. When discharge exceeds recharge, the zone of saturation thins and the zone of aeration thickens. During periods of recharge, a temporary downward migrating saturated lense may move through the zone of aeration.

The foregoing comments refer to ground water in temperate and tropical areas. However, in the colder areas of the northern and southern hemispheres, permafrost or permanently frozen ground may extend to considerable depths and influence ground-water conditions. The engineering problems associated with such conditions may be unusual and are not considered in this manual.

1-5. Ground-Water Quality.—(a) *General.*—Precipitation usually contains minute amounts of silica and other minerals [1,3,9] and dissolved gases such as carbon dioxide, sulphur dioxide, nitrogen, and oxygen which are present in the air and are entrapped as precipitation occurs. As a result, the pH value of most precipitation is below 7 (acid condition) and the water is corrosive. Upon reaching the earth's surface, the water may pick up organic acids from humus and similar materials which increase its corrosive characteristics. While the water is percolating through rock, minerals are attacked by the acid waters and dissolved, forming salts which are taken into solution. The amount and character of the salts depend upon the chemical composition of the water; the mineralogical and physical structure of the rocks encountered; and the temperature, pressure, and duration of contact.

Nearly all elements may be present in ground water, and its mineral content varies from aquifer to aquifer and from place to place within an

TABLE 1–2.—*Chemical constituents commonly found in ground water*

Cations	Anions
Calcium, Ca	Bicarbonate, HCO_3
Magnesium, Mg	Sulphate, SO_4
Sodium, Na	Chloride, Cl
Potassium, K	Nitrate, NO_3
Iron, Fe	Fluoride, F
	Silica, SiO_2

aquifer. The most commonly encountered elements and compounds are listed in table 1–2.

Less commonly encountered constituents which are nevertheless important because of their known beneficial or deleterious effects in use of water are: boron (B), manganese (Mn), lead (Pb), arsenic (As), selenium (Se), barium (Ba), copper (Cu), zinc (Zn), hydrogen sulfide (H_2S), methane (CH_4), oxygen (O_2), carbon dioxide (CO_2), and nitrite (NO_2).

(b) *Limits for Minerals and Gases.*—The mineral content of water is so variable and the acceptable quality for various uses has such a large range that it is not practical to discuss them other than in generalities.

Quality standards recommended for municipal and domestic water in the United States are published in "The United States Public Health Service 1962 Drinking Water Standards"[12]. These standards contain the acceptable and mandatory limits of the various elements, compounds, and organisms contained in water and should be followed if at all possible. The mineral limitations are reproduced in table 1–3. If water cannot be obtained to meet these standards, the advice of the Public Health Service should be sought to determine whether it would be permissible to exceed the standards.

The American Public Health Association's publication "Standard Methods for the Examination of Water, Sewage, and Industrial Wastes" [1] outlines standard procedures for water analysis for municipal and domestic purposes.

The quality of water required for irrigation purposes is very complex. Many factors such as soil, drainage, climate, and crops must be considered. The U.S. Department of Agriculture Handbook No. 60 [9] is a useful guide on the acceptability of water of a certain chemical composition for irrigation and on the relationship of the chemicals in water and soils. This handbook also contains procedures for laboratory analysis of irrigation water.

TABLE 1–3—*Chemical limits for drinking water* [12]

Dissolved chemicals in water	Recommended maximum amounts, in p/m [1]	Mandatory limits for rejection of water in p/m [1]
Iron	0. 3	
Manganese	0. 05	
Nitrate	45. 0	
Detergent (ABS)	0. 05	
Copper	1. 0	
Arsenic	0. 01	0. 05
Fluoride	[2] 1. 7	[2] 3. 4
Lead		0. 05
Selenium		0. 01
Silver		0. 05
Barium		1. 0
Cadmium		0. 01
Chromium (hexavalent)		0. 05
Cyanide	0. 01	0. 2
Magnesium	50. 0	
Zinc	5. 0	
Chloride	250. 0	
Sulfate	250. 0	
Chloroform—soluble extract	0. 2	
Phenol	0. 001	
Total dissolved solids	500. 0	1, 000. 0

[1] Parts per million.
[2] These values change with the daily air temperature range. Refer to [12] for the different limits.

Water quality is also important because of its influence on the operating efficiency and life of equipment including pumps and wells. Acidic water with a pH value of less than 7 is usually corrosive; whereas alkaline water (pH>7) is usually less corrosive. However, alkaline water is likely to form deposits on well screens and on pipes. Hard water with a pH value over 7 may be corrosive, and may form deposits if it contains relatively large amounts of sulfate, bicarbonate, and chloride radicals. Gases such as hydrogen sulfide, carbon dioxide, methane, and oxygen may be damaging by both corrosion and cavitation.

(c) *Contamination and Pollution.*—Contaminated or polluted water contains organisms and substances that make it unsuitable or unfit for use. Ground water may become contaminated as a result of leakage from septic tanks, sewage effluent spreading grounds, garbage dumps, or similar features for the disposal of vegetable and animal waste. Other sources of contamination which are becoming of increasing concern in many areas are improperly sealed wells, other subsurface structures and excavations, and those areas created by the disposal of oil field brines and industrial wastes through evaporation ponds, spreading fields, and disposal wells.

The distance that organisms may migrate in ground water varies. In general, it should be assumed that in crevassed, fissured, and cavernous rock and in coarse, clean gravels, organically contaminated water may travel as far as several miles. In finer-grained materials, natural filtering action and adsorption may remove such organisms in less than a 100-foot distance. However, chemical contaminants may persist indefinitely in ground water. Accordingly, no water should be considered suitable until both chemical and bacterial analyses have shown it to be so.

(d) *Other Uses of Water Quality Data.*—A study of differences and changes in the chemical content of water may be useful in determining the source or sources of recharge, direction of flow, and presence of boundaries [3,7]. The age of water determined by tritium content, carbon 14 dating, and similar analysis may be useful in estimating the age, recharge conditions, or potential direction of flow of ground water or other aspects of ground-water hydrology. Also, quality data may be essential in determining the compatibility of water intended for artificial recharge.

1–6. Ground-Water and Surface-Water Relationships.—(a) *Humid Area Relationships.*—Ground water in humid areas maintains the base flow of streams by seepage into stream channels. However, the headwater reaches of some streams may be above the water table, and therefore are dry during seasons of low precipitation. In such reaches, seepage from the streambed may charge an underlying aquifer. Consequently, some reaches of a stream may be replenished by while others lose water to the ground-water reservoir.

(b) *Arid Area Relationships.*—In many arid drainage basins, the perennial master streams receive seepage from the ground-water reservoir; whereas other streams may be above the water table and streamflow occurs only during periods of high surface runoff. Where the water table is below the streambed, practically all the streamflow may be lost by seepage to the ground-water reservoir. Beneath many such streambeds, considerable underflow may be present in the channel fill, although the channel is dry.

It is in the semiarid to arid areas, where irrigation is usually practiced, that water losses from canals and deep percolation from irrigation applications frequently alter natural ground-water conditions. Such alterations include water table rise and waterlogging and salination of soils. Artificial drainage by open or buried pipe drains, wells, or other means is often required to lower the water table, maintain a salt balance, and permit the continued production of crops.

(c) *Artificial Ground-Water Recharge.*—In recent years, much interest has developed in recharging ground-water reservoirs with excess surface water [5,6]. Such recharge is intended to maintain ground-water levels, store water for use during droughts, control salt water intrusion, dispose of treated sewage effluent, or for other purposes. In addition, pollutants such as oil field brines and toxic and radioactive industrial wastes are often disposed of by storing them in deep isolated aquifers.

(d) *Ground-Water Reservoirs.*—Suitable surface water reservoir sites are becoming scarce. Consequently, interest has increased in the underground storage of water. While underground reservoirs are not as obvious or as readily delineated as surface reservoirs, they offer a possible alternative in many areas where conventional storage would be costly or otherwise undesirable. As is true of all alternative solutions, each type of reservoir offers advantages and disadvantages. To assist in the evaluation of the alternatives, table 1–4 lists the major advantages of each type of reservoir.

1–7. Ground-Water Rights.—(a) *General.*—In the United States, doctrines of law and statutes relating to the ownership and use of water [11] are the responsibility of the courts and legislative bodies of the several States. There are no Federal statutes under which a water right can be acquired; that is, a right granted by law to use or take possession of water in a natural source and put it to a beneficial use. As a consequence, both surface and ground-water rights for Reclamation projects are obtained in conformance with the laws of the States in which the project is located. Two entirely different systems for acquiring water rights are followed in the contiguous 48 States. These are the doctrine of riparian rights, recognized in the 31 predominantly Eastern States, and the doctrine of prior appropriation, recognized in the 17 Western States (see fig. 1–3).

(b) *Doctrine of Riparian Rights.*—The doctrine of riparian rights is based on the common law of England and stems from ownership of land contiguous to a natural water source such as a stream or lake. For ground water, ownership of land overlying an aquifer is sufficient to establish a ground-water right. This doctrine is often referred to as the *English rule of unlimited use.*

(c) *Doctrine of Prior Appropriation.*—Under this doctrine, ownership of water is vested in the State, that is, the common property of the people. An appropriator who is first in time to beneficially use a certain water source has a prior right to its use. However,

TABLE 1–4.—*Advantages of surface versus subsurface reservoirs*

Item	Subsurface reservoirs	Surface reservoirs
1	Many large capacity sites available	Few new sites available.
2	Slight to no evaporation loss	High evaporation loss even in humid climate.
3	Require little land area	Require large land area.
4	Slight to no danger of catastrophic structural failure.	Ever present danger of catastrophic failure.
5	Uniform water temperature	Fluctuating water temperature.
6	High biological purity	Easily contaminated.
7	Safe from immediate radioactive fallout.	Easily contaminated by radioactive material.
8	Reservoir serves as conveyance system—canals or pipeline across lands of others unnecessary.	Water must be conveyed.
9	Water must be pumped	Water may be available by gravity flow.
10	Storage and conveyance use only	Multiple use.
11	Water may be mineralized	Water generally of relatively low mineral content.
12	Minor flood control value	Maximum flood control value.
13	Limited flow at any point	Large flows.
14	Power head usually not available	Power head available.
15	Difficult and costly to investigate, evaluate, and manage.	Relatively easy to evaluate, investigate, and manage.
16	Recharge opportunity usually dependent on surplus surface flows.	Recharge dependent on annual precipitation.
17	Recharge water may require expensive treatment.	No treatment required.
18	Continuous expensive treatment of recharge areas or wells.	Little treatment required

water for domestic use usually is not subject to the need for appropriation.

(d) *Prescriptive Rights.*—In some States where the doctrine of prior appropriation is followed, a prescriptive water right can be acquired by taking and putting to beneficial use, for some number of consecutive years, water to which other landowners or prior appropriators have rights.

(e) *Ground-Water Regulations.*—In addition to those providing for water rights, other statutes and rules relating to the administration and control of ground water have been established in some States to protect the public interest and to provide for orderly development of this resource. Some of the more common regulations provide for licensing and bonding of well drillers, obtaining permits

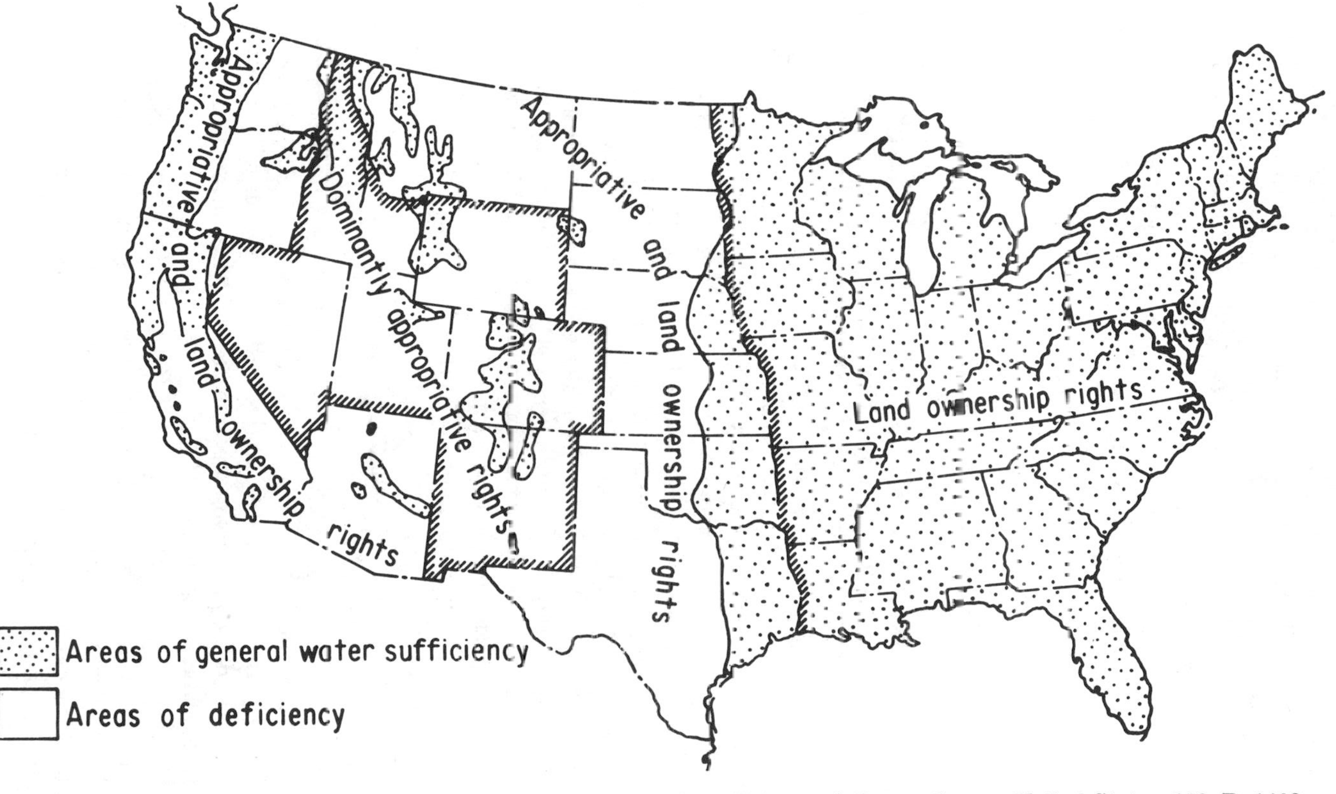

Figure 1-3.—Water rights doctrines by States and water supply sufficiency of the contiguous United States. 103-D-1402.

to drill new wells or to rehabilitate existing wells, filing of geologic logs of new wells, and following construction practices that ensure against contamination. Also, some States have regulations restricting the subsurface disposal of pollutants such as brines and industrial wastes that might contaminate the public ground-water supplies.

(f) *Conjunctive Use of Surface and Ground Water.*—The recognition of the interrelationship of surface and ground water and the development of the philosophy of conjunctive use has caused legal conflicts between ground and surface water users. Optimum use of water resources is often precluded because of legal barriers. Any plan involving conjunctive use should be examined carefully from the aspects of both legal and physical practicability.

1–8. Application of Ground-Water Engineering.—(a) *Water Supply.*—The major application of ground-water engineering has been and probably always will be the provision of a water supply by means of wells and infiltration galleries. Facilities range from isolated individual small wells yielding less than a gallon a minute for domestic and stock purposes to well fields consisting of a number of irrigation, municipal, or industrial water supply wells with individual discharges in excess of 5,000 gallons per minute. The small individual well seldom presents a problem if it is designed according to good engineering practice. The larger installations, particularly those with numerous wells, require evaluation of the aquifer characteristics, estimates of well spacing, drawdowns, quality of water, and possibly recharge-discharge relationships. Wells must be designed and pumps selected for economical, long, and trouble-free operation within the capabilities of the aquifer, with the consideration of any possible corrosion and encrustation problems which may be present.

Proposed development may be further complicated by restrictions imposed by overlying or underlying saline aquifers, salt water intrusion, influences on the discharge of adjacent surface water streams, and land subsidence.

Some aquifers have little measurable recharge or discharge but contain large quantities of water in storage which have accumulated over long periods. Estimates can be made of the desirability of mining the water and the probable economic life of such aquifers under various degrees of development.

(b) *Ground-Water Reservoirs and Artificial Recharge.*—The storage of surface waters in underground reservoirs and the recharge of depleted ground-water reservoirs are other aspects of ground-water engineering of growing importance and interest. Recharge wells, basins, channels, and waste disposal facilities present special problems of aquifer plugging caused by chemical, biological, and physical factors and of contamination of overlying or adjacent potable aquifers.

The maintenance of minimum streamflows by supplementing surface water with pumped ground water during low flow periods and recharging the ground-water reservoir during high runoff is also of growing interest.

(c) *Drainage.*—Drainage may involve the lowering of ground-water levels beneath irrigated lands to permit crop growth; the lowering of water levels or prevention of boils in limited areas to permit excavation and construction activities in the dry; reduction of pressures to maintain stability of slopes, and the reduction of pressures and exit velocities to assure stability of dams and similar structures incident to reservoir and dam construction. Applications of ground-water hydraulics and engineering are involved in all such problems.

1.9. Bibliography.—

[1] "Standard Methods for Examination of Water, Sewage, and Industrial Waste," 10th edition. Published jointly by American Public Health Association, American Water Works Association, and Federation of Sewage and Industrial Wastes Association, New York, 1955.

[2] "Ground Water Basin Management," ASCE Manual of Engineering Practice No. 40, American Society of Civil Engineers, New York, 1961.

[3] Hem, J. D., "Study and Interpretation of the Chemical Characteristics of Natural Water," U.S. Geological Survey Water-Supply Paper 1473, 1959.

[4] Meinzer, O. E. (editor), "Physics of the Earth—IX, Hydrology", Dover Publications, New York, 1949.

[5] Rima, D. R., Chase, E. B., and Myers, B. M., "Subsurface Disposal by Means of Wells—A Selected Annotated Bibliography," U.S. Geological Survey Water-Supply Paper 2020, 1971.

[6] Signor, D. C., Growitz, D. J., and Kam, W., "Annotated Bibliography on Artificial Recharge of Ground Water 1955–67," U.S. Geological Survey Water-Supply Paper 1990, 1970.

[7] Todd, D. K., "Ground Water Hydrology," John Wiley & Sons, New York, 1959.

[8] Tolman, C. F., "Ground Water," McGraw-Hill, New York, 1957.

[9] "Diagnosis and Improvement of Saline and Alkaline Soils," Agricultural Handbook No. 60, U.S. Department of Agriculture, 1954.

[10] "Water," Agricultural Year Book for 1955, U.S. Department of Agriculture, 1956.

[11] Thomas, H. E., "Ground Water Law." Transcript of Lecture Presented at Ground Water Short Course, U.S. Geological Survey and Bureau of Reclamation, Fort Collins, Colo., April 6–17, 1953.

[12] "The United States Public Health Service 1962 Drinking Water Standards," 42CFR, Part 72, Interstate Quarantine, Drinking Water Standards.

Chapter II

THEORY OF GROUND-WATER FLOW, AQUIFER PROPERTIES, AND DEFINITIONS

2-1. Darcy's Law.—The foundation of ground-water hydraulics is Darcy's Law [12,14] [1] which states that the flow rate through a porous medium is proportional to the head loss and inversely proportional to the length of the flow path (see fig. 2-1). The law is applicable when flow is laminar and without turbulence. Formulas expressing the law are given in a number of forms, the most common of which are presented below and are derivations of $Q=AV$:

$$V=Ki$$

$$Q=KiA$$

$$Q=KA\frac{h_1-h_2}{L}$$

$$K=\frac{Q}{iA}=\frac{\frac{L^3}{t}}{L^2}=\frac{L}{t}$$

where:

V=velocity, $\frac{L}{t}$

K=permeability or hydraulic conductivity of the porous medium, $\frac{L}{t}$

i=hydraulic gradient=$\frac{h_1-h_2}{L}$, nondimensional

A=area normal to the direction of flow, L^2

Q=rate of flow, $\frac{L^3}{t}$

h_1 and h_2=the water level or potential at two points on a line parallel to the direction of flow

L=length of flow path between h_1 and h_2

t=time

Rearrangement of Darcy's equation leads to $K=\frac{V}{i}=\frac{Q}{iA}$, where K is a proportionality constant commonly known as the coefficient of permeability, or the hydraulic conductivity [9] when discussing ground water, and has the dimensions $\frac{\frac{L^3}{t}}{L^2}$ which reduces to $\frac{L}{t}$ or a velocity.

[1] Numbers in brackets refer to items in the bibliography, section 2-21.

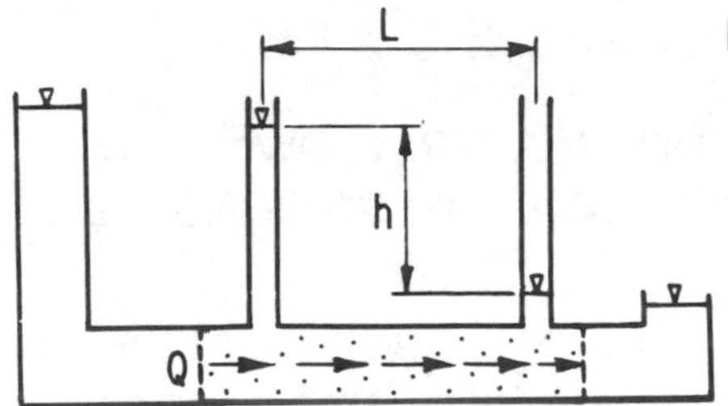

FIGURE 2–1.—Illustration of Darcy's Law. 103–D–1403.

Numerous expressions, some on the basis of $\frac{Q}{t}$, others on $\frac{L}{t}$, and with a variety of consistent and inconsistent units and i values, have been used for expressing K. In Reclamation practice, K is usually expressed for water as $\frac{L}{t}$ under a unit gradient, where L is in feet and where t may be in seconds, or larger units of time (see fig. 2–2). Factors for conversions between the most commonly used units are given in table 2–1.

The value of K varies for different fluids depending upon their density and viscosity as follows,

$$K=\frac{k\gamma}{\mu},$$

where γ is the specific weight and μ is the viscosity of the fluid. In this formula, k is the intrinsic permeability of the medium; that is,

$$V=Ki=\frac{ki\gamma}{\mu}.$$

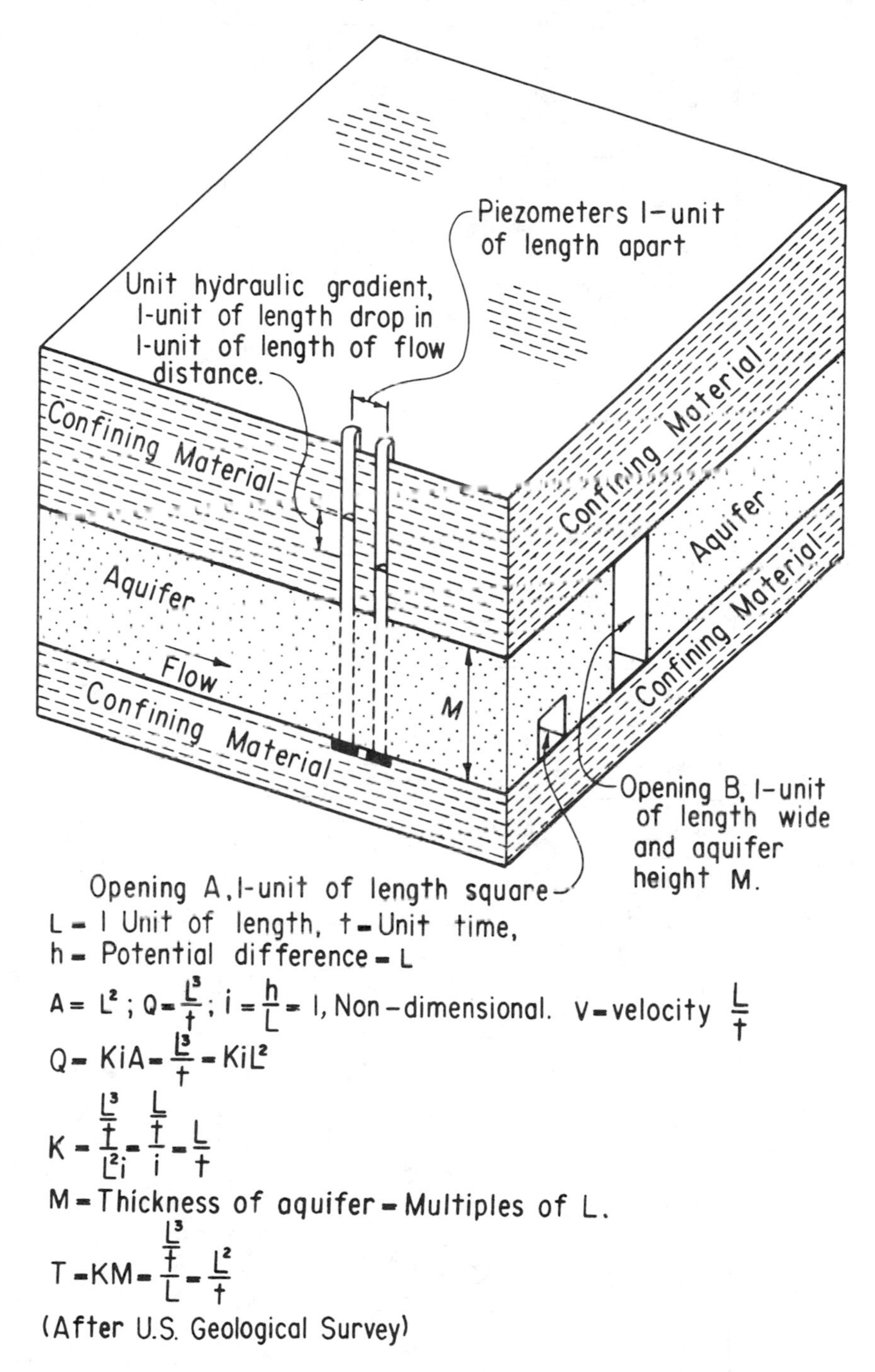

FIGURE 2–2.—Illustration of permeability and transmissivity. 103–D–1404.

TABLE 2-1.—*Conversion factors for various units of hydraulic conductivity.* 103–D–1454.

(1)	(2)	(3)	(4)	(5)	(6)	(7)	(8)	(9)	(10)	(11)
$ft^3/ft^2/yr$	$ft^3/ft^2/day$	$ft^3/ft^2/hr$	$ft^3/ft^2/min$	$ft^3/ft^2/sec$	$in^3/in^2/day$	$in^3/in^2/hr$	$gal/ft^2/day$	$m^3/m^2/day$	$cm^3/cm^2/hr$	Darcy $cm^3/s\text{-}cm^2$ (atm/cm)
1	2.74×10^{-3}	1.141×10^{-4}	1.903×10^{-6}	3.171×10^{-8}	3.287×10^{-2}	1.37×10^{-3}	2.049×10^{-2}	8.35×10^{-4}	3.479×10^{-3}	1.133×10^{-3}
365	1	4.167×10^{-2}	6.945×10^{-4}	1.157×10^{-5}	12	5.0×10^{-1}	7.4805	3.05×10^{-1}	1.270	4.115×10^{-1}
8,760	24	1	1.667×10^{-2}	2.778×10^{-4}	288	12	179.5	7.32	30.48	9.872
525,600	1,440	60	1	1.667×10^{-2}	17,280	720.0	10,772	438.9	1,829	591.7
31,536,000	86,400	3,600	60	1	1,036,800	43,200	646,315	26,335	109,723	35,549
30.42	8.333×10^{-2}	3.472×10^{-3}	5.787×10^{-5}	9.645×10^{-7}	1	4.166×10^{-2}	6.234×10^{-1}	2.54×10^{-2}	1.058×10^{-1}	3.435×10^{-2}
730	2.0	8.334×10^{-2}	1.389×10^{-3}	2.315×10^{-5}	24	1	14.96	6.1×10^{-1}	2.540	8.217×10^{-1}
48.78	1.337×10^{-1}	5.569×10^{-3}	9.282×10^{-5}	1.547×10^{-6}	1.604	6.682×10^{-2}	1	4.07×10^{-2}	1.697×10^{-1}	5.494×10^{-2}
1,198	3.28	1.368×10^{-1}	2.27×10^{-3}	3.78×10^{-5}	39.38	1.64	24.54	1	4.167	1.35
287.4	7.874×10^{-1}	3.281×10^{-2}	5.469×10^{-4}	9.114×10^{-6}	9.449	3.939×10^{-1}	5.890	0.24	1	3.246×10^{-1}
886.96	2.43	10.13×10^{-2}	16.88×10^{-4}	28.13×10^{-6}	29.20	1.217	18.2	7.41×10^{-1}	3.08	1

All factors computed for 68°F with viscosity of 1.0050 centipoises.

Examples

(1) The permeability of a soil has been determined to be 15 $gal/ft^2/day$. What is this in $in^3/in^2/hr$? Find value of 1 in Column (8) and move horizontally to value for $in^3/in^2/hr$ in Column (7) Multiply value in Column (7) (0.0668) by 15 = 1.002 $in^3/in^2/hr$.

(2) The permeability of a soil has been determined to be 4,000 $ft^3/ft^2/yr$. What is this in $gal/ft^2/day$? Find value of 1 in Column (1) and move horizontally to value for $gal/ft^2/day$ in Column (8) Multiply value in Column (8) (0.02049) by 4,000 = 82.0 $gal/ft^2/day$.

In ground-water engineering this refinement is seldom required. In laboratory determinations of K using water, the results are usually expressed as the value obtained at a water temperature of 60° or 68°F. Laboratory results neglecting the slight change in weight with temperature can be compared to field determinations with the expression

$$K_L = \frac{V_F}{V_L} K_F$$

where K_L is the standard or laboratory determination, K_F is the field determination, V_F is the kinematic viscosity of water at field temperature, and V_L is the kinematic viscosity at laboratory temperature. Since ground-water temperatures at depths to 200 feet from the surface seldom vary more than about 2°F from the average annual temperature of the area in which they occur, the above conversion for K is seldom necessary since the determined values will be used in the area where the test was made. Table 2–2 gives the variation of properties of pure water with temperature.

Laboratory determinations of hydraulic conductivity are only representative of a specific sample of aquifer material, whereas field determinations by pumping tests are usually an average value representing an integration of all the permeability variations in all directions and from place to place in an aquifer. For this reason, laboratory determinations even if made using undisturbed samples are not as representative as those found in actual field aquifer tests and may be misleading.

As the term *hydraulic conductivity* fails to describe adequately the flow characteristics of an aquifer, C. V. Theis [13] introduced the term *transmissivity* [10], $T = KM$, which is equal to the average permeability times the saturated thickness of the aquifer to clarify this deficiency. Transmissivity has dimensions of

$$\frac{\frac{L^3}{t}}{L}$$

or

$$\frac{L^2}{t}$$

since it represents flow through a vertical strip of aquifer one unit wide. Where K may be considered as the hydraulic conductivity of a unit cross-sectional area of the aquifer, T may be considered as the hydraulic conductivity of a unit width of the full thickness of the aquifer.

TABLE 2–2.—*Variation of properties of pure water with temperature*

Temperature, °C	Temperature, °F	Density at 1 atmosphere, [1](gm-cm^{-3})	Dynamic viscosity, in centipoises [2](10^{-2} dyne-sec cm^{-2})	Kinematic viscosity, in centistokes [3](10^{-2} cm^2 sec^{-1})	Surface tension against air, [4](dyne cm^{-1})	Vapor pressure, [5](mm Hg)
5	41. 0	0. 999965	1. 5188	1. 5189	74. 92	6. 543
6	42. 8	. 999941	1. 4726	1. 4727	74. 78	7. 013
7	44. 6	. 999902	1. 4288	1. 4289	74. 64	7. 513
8	46. 4	. 999849	1. 3872	1. 3874	74. 50	8. 045
9	48. 2	. 999781	1. 3476	1. 3479	74. 36	8. 609
10	50. 0	. 999700	1. 3097	1. 3101	74. 22	9. 209
11	51. 8	. 999605	1. 2735	1. 2740	74. 07	9. 844
12	53. 6	. 999498	1. 2390	1. 2396	73. 93	10. 518
13	55. 4	. 999377	1. 2061	1. 2069	73. 78	11. 231
14	57. 2	. 999244	1. 1748	1. 1757	73. 64	11. 987
15	59. 0	. 999099	1. 1447	1. 1457	73. 49	12. 788
16	60. 8	. 998943	1. 1156	1. 1168	73. 34	13. 634
17	62. 6	. 998774	1. 0875	1. 0889	73. 19	14. 530
18	64. 4	. 998595	1. 0603	1. 0618	73. 05	15. 477
19	66. 2	. 998405	1. 0340	1. 0357	72. 90	16. 477
20	68. 0	. 998203	1. 0087	1. 0105	72. 75	17. 535
21	69. 8	. 997992	0. 9843	0. 9863	72. 59	18. 650
22	71. 6	. 997770	. 9608	. 9629	72. 44	19. 827
23	73. 4	. 997538	. 9380	. 9403	72. 28	21. 068
24	75. 2	. 997296	. 9161	. 9186	72. 13	22. 377
25	77. 0	. 997044	. 8949	. 8976	71. 97	23. 756
26	78. 8	. 996783	. 8746	. 8774	71. 82	25. 209
27	80. 6	. 996512	. 8551	. 8581	71. 66	26. 739
28	82. 4	. 996232	. 8363	. 8395	71. 50	28. 349
29	84. 2	. 995944	. 8181	. 8214	71. 35	30. 043
30	86. 0	. 995646	. 8004	. 8039	71. 18	31. 824
31	87. 8	. 995340	. 7834	. 7871	[6] 71. 02	33. 695
32	89. 6	. 995025	. 7670	. 7708	[6] 70. 86	35. 663
33	91. 4	. 994702	. 7511	. 7551	[6] 70. 70	37. 729
34	93. 2	. 994371	. 7357	. 7399	[6] 70. 53	39. 898
35	95. 0	. 99403	. 7208	. 7251	70. 38	42. 175
36	96. 8	. 99368	. 7064	. 7109	[6] 70. 21	44. 563
37	98. 6	. 99333	. 6925	. 6971	[6] 70. 05	47. 067
38	100. 4	. 99296	. 6791	. 6839	[6] 69. 88	49. 692

[1] Handbook of Chemistry and Physics, 46th ed., 1965–66: Cleveland, Chemical Rubber Publishing Co., table F–4, computed from the relative values.

[2] International critical tables of numerical data, physics, chemistry, and technology: Natl. Acad. Sciences, vol. 5, p. 10.

[3] Dynamic viscosity divided by density.

[4] International critical tables of numerical data, physics, chemistry, and technology: Natl. Acad. Sciences, vol. 4, p. 447.

[5] Handbook of Chemistry and Physics, 46th ed., 1965–66: Cleveland, Chemical Rubber Publishing Co., table D–94.

[6] Interpolated.

2-2. Storativity.—The terms specific yield, effective porosity, coefficient of storage, and storativity [10] have often been used interchangeably to express the storage capacity of an aquifer. However, some authors have limited the use of specific yield to free aquifers and coefficient of storage to confined aquifers. Since the influence is essentially the same in either case and S is the commonly used symbol to express the value regardless of the nature of the aquifer, the term *storativity* will be used herein to designate both concepts.

Storativity is defined as the volume of water released from or taken into storage per unit surface area of the aquifer per unit change in the component of hydraulic head normal to that surface. In a vertical column with a horizontal cross section of one square unit extending through an aquifer (fig. 2-3), the storativity equals the volume of water released from or gained by the aquifer when the piezometric surface or water table declines or rises one unit. Storativity is expressed as the ratio:

$$S=\frac{V'}{V}$$

where V' equals the volume of water released and V is the volume of material drained in a free aquifer or the volume defined by the change in piezometric head for an artesian aquifer.

Since $\frac{V'}{V}=\frac{L^3}{L^3}$, S is nondimensional.

In equations for unsteady (transient) flow, S must be considered. Release of water from the aquifer in response to a change in head generally is assumed to be instantaneous. In many cases, however, initial release is relatively rapid but decreases in rate with time. In other cases, because of the fine-grained nature of the aquifer, drainage may be so slow that the response to change in head is similar to that of a leaky aquifer for a fairly prolonged period of time (sec. 5-10).[2]

In a free aquifer, S is a function of the size and number of interconnected voids and represents the actual volume of water drained from the aquifer by lowering of the water table. The S value ranges from as low as 1 percent to over 40 percent, but is usually in the range of 10 percent to 30 percent. The less uniform, finer grained and more dense a material, the smaller the S value.

---

[2] Indicates sections of the manual in which the subject is further discussed, usually from a different aspect.

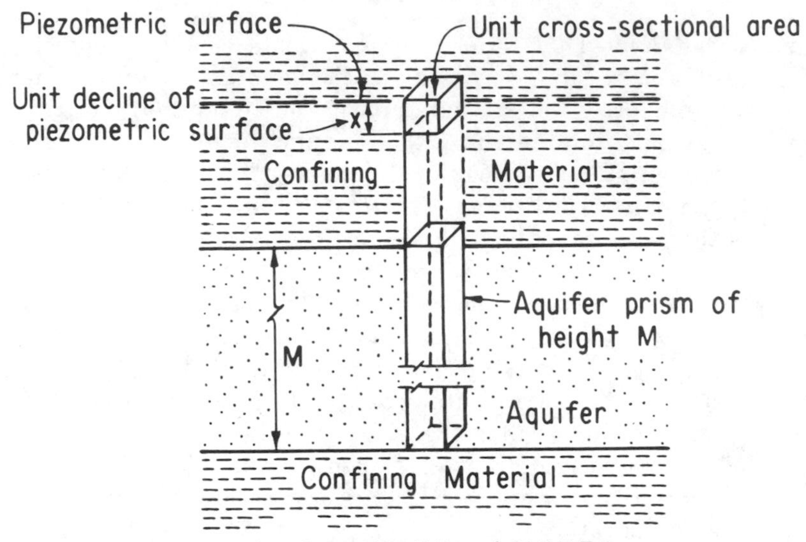

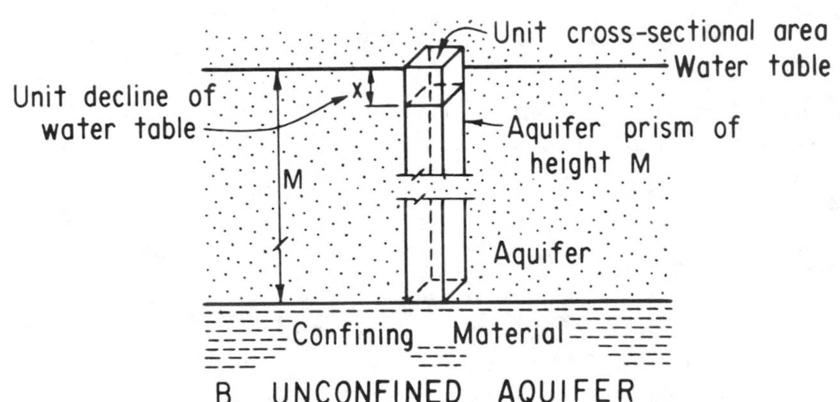

Storativity, $S = \frac{V'}{V}$, is non-dimensional

L = Unit length, V = The volume involved in a unit drop or increase in piezometric surface or water-table elevation within a prism of 1-unit^2 cross section (L^3). V' = The volume of water released by a change in piezometric surface or drained or recharged by a change in water-table elevation within the prism (L^3).

(After U.S. Geological Survey)

FIGURE 2-3.—Illustration of storativity. 103-D-1405.

In a confined aquifer, where the cone of depression is not drawn below the bottom of the upper confining layer, there is no actual drainage of the aquifer. Water released is due to: (1) the small expansion of the water resulting from the reduction in pressure and (2) water being forced out of the aquifer by compaction of the aquifer skeleton because of this reduction in pressure. The value of S in an artesian aquifer may be independent of void content of the aquifer material and ranges from 1/1,000 of 1 percent to 1/10 of 1 percent (0.00001 to 0.001).

2–3. Hydraulic Diffusivity.—The ratio of transmissivity to storativity in transient flow conditions can be expressed in the formula

$$\alpha=\frac{T}{S},$$

where alpha is hydraulic diffusivity and has the dimensions $\frac{L^2}{t}$. In an ideal aquifer, the time of response at a distant location to an imposed stress, such as a discharging well, is inversely proportional to the diffusivity [5].

Transmissivity (or permeability times aquifer thickness) is the overriding factor in well yield in terms of yield versus drawdown or specific capacity, see section 5–17. Under transient conditions, storativity is also a controlling factor.

Figures 2–4 and 2–5 give generalized conversion scales for units of transmissivity and permeability and illustrate general relationships involving well yield and potential and permeability of common aquifer materials.

2–4. Steady One-Directional Flow.—Steady flow of ground water in a confined aquifer of uniform thickness behaves in accordance with Darcy's law; that is, the head decreases linearly in the direction of flow [15]. However, in a free aquifer the water table is also a flow line. The shape of the water table determines the flow distribution, but conversely, the flow distribution determines the shape of the water table. Accordingly, a general analytical solution of the flow is therefore not possible. However, Dupuit [12] in an attempt to simplify analysis of undirectional flow, made the following assumptions: (1) the velocity of flow is proportional to the tangent of the hydraulic gradient rather than the sine as determined by Darcy; and (2) the flow is horizontal and uniform everywhere in a vertical section. Dupuit derived the following equation:

$$Q=\frac{K(h_2^2-h_1^2)}{2L} \tag{1}$$

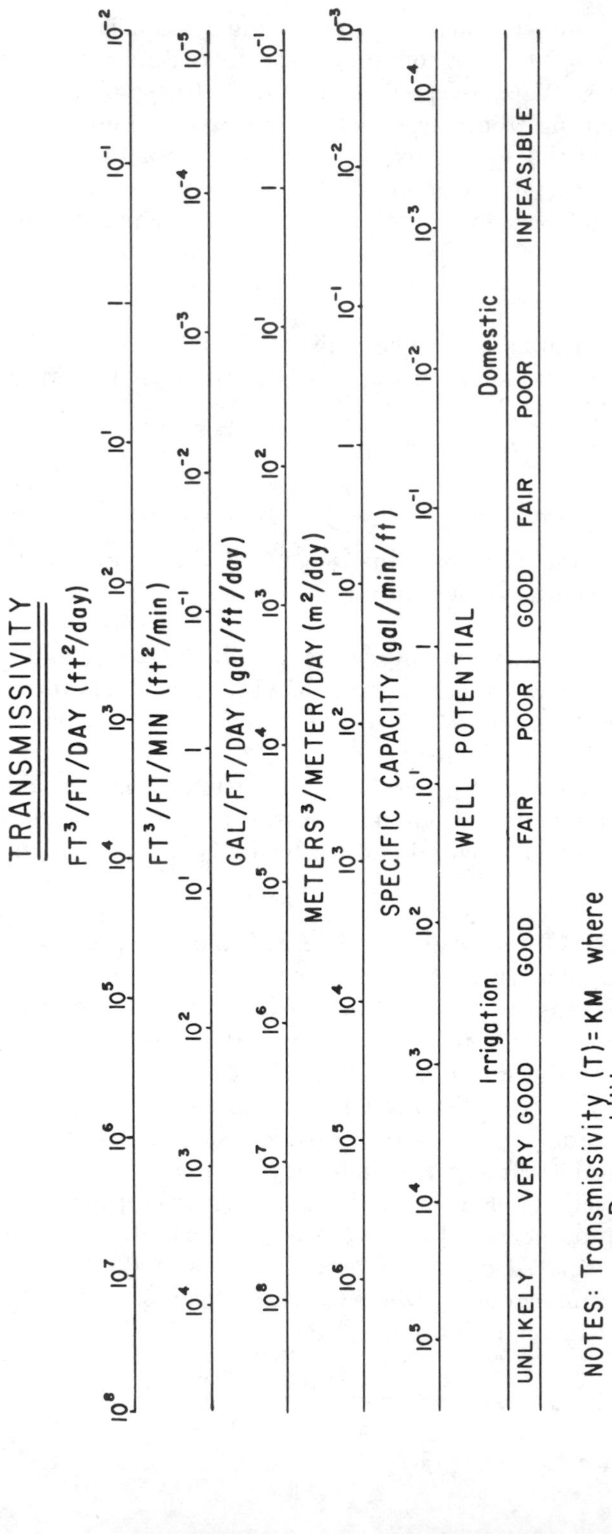

TRANSMISSIVITY

FT³/FT/DAY (ft^2/day)

10^8 10^7 10^6 10^5 10^4 10^3 10^2 10^1 1 10^{-1} 10^{-2}

FT³/FT/MIN (ft^2/min)

10^4 10^3 10^2 10^1 1 10^{-1} 10^{-2} 10^{-3} 10^{-4} 10^{-5}

GAL/FT/DAY (gal/ft/day)

10^8 10^7 10^6 10^5 10^4 10^3 10^2 10^1 1 10^{-1}

METERS³/METER/DAY (m^2/day)

10^6 10^5 10^4 10^3 10^2 10^1 1 10^{-1} 10^{-2} 10^{-3}

SPECIFIC CAPACITY (gal/min/ft)

10^5 10^4 10^3 10^2 10^1 1 10^{-1} 10^{-2} 10^{-3} 10^{-4}

WELL POTENTIAL

Irrigation					Domestic			
UNLIKELY	VERY GOOD	GOOD	FAIR	POOR	GOOD	FAIR	POOR	INFEASIBLE

NOTES: Transmissivity (T) = KM where
K = Permeability
M = Saturated thickness of the aquifer
Specific capacity values based on pumping period of approximately 8-hours but are otherwise generalized.

FIGURE 2-4.—Comparison of transmissivity, specific capacity, and well potential. 103-D-1406.

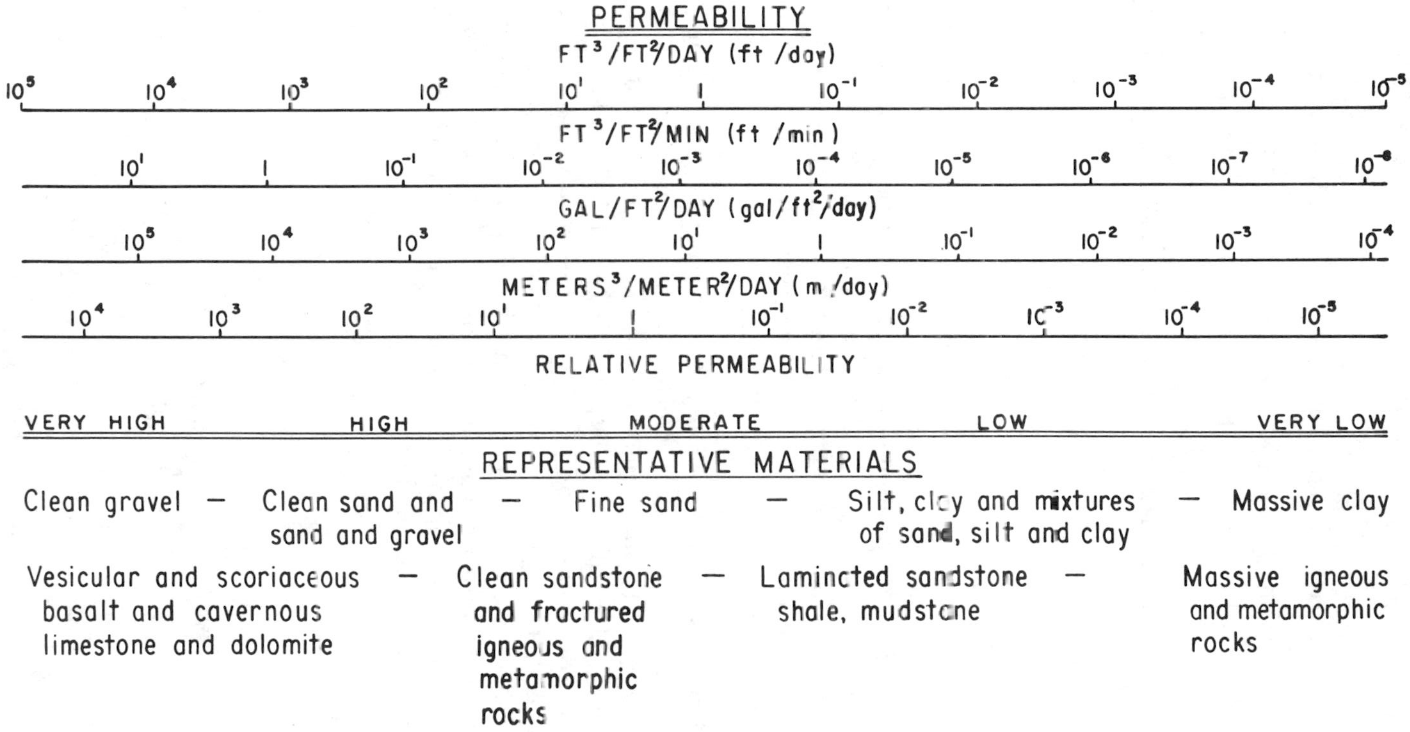

PERMEABILITY

FT³/FT²/DAY (ft/day)

10^5 10^4 10^3 10^2 10^1 1 10^{-1} 10^{-2} 10^{-3} 10^{-4} 10^{-5}

FT³/FT²/MIN (ft/min)

10^1 1 10^{-1} 10^{-2} 10^{-3} 10^{-4} 10^{-5} 10^{-6} 10^{-7} 10^{-8}

GAL/FT²/DAY (gal/ft²/day)

10^5 10^4 10^3 10^2 10^1 1 10^{-1} 10^{-2} 10^{-3} 10^{-4}

METERS³/METER²/DAY (m/day)

10^4 10^3 10^2 10^1 1 10^{-1} 10^{-2} 10^{-3} 10^{-4} 10^{-5}

RELATIVE PERMEABILITY

VERY HIGH — HIGH — MODERATE — LOW — VERY LOW

REPRESENTATIVE MATERIALS

Clean gravel — Clean sand and sand and gravel — Fine sand — Silt, clay and mixtures of sand, silt and clay — Massive clay

Vesicular and scoriaceous basalt and cavernous limestone and dolomite — Clean sandstone and fractured igneous and metamorphic rocks — Laminated sandstone shale, mudstone — Massive igneous and metamorphic rocks

FIGURE 2-5.—Comparison of permeability and representative aquifer materials. 103-D-1407.

where:

Q=flow of water per unit time $\frac{L^3}{t}$ per unit width normal to the direction of flow,

h_1=saturated thickness of the aquifer at one point in the line of flow,

h_2=saturated thickness of the aquifer at a second point in the line of flow from h_1, and

L=distance between the points parallel to the direction of flow, L.

However, because of the assumptions of horizontal flow (no vertical component), the computed (Dupuit) water table deviates more and more from the actual water table in the direction of flow. Nevertheless, despite the simplifying assumptions, the equation closely approximates the water table position where the sine and tangent of the slope of the water table are approximately equal. The equation is applicable under such conditions to determine Q and K but should be used with caution near a point or zone of discharge where the drawdown curve may be accentuated.

2–5. Steady Radial Flow.—Early in this century, C. Theim [12] and P. Forchheimer [4] independently derived equations for steady radial flow [4,12,14] to a fully penetrating well with 100 percent penetration and open hole using Darcy's law and Dupuit's assumptions. The equations, known today as the steady state, Theim, Dupuit-Forchheimer, or Theim-Forchheimer equations, can be used to determine the coefficient of permeability of an aquifer from measurements made during a pumping test using a fully penetrating well with 100 percent open hole and two or more observation wells. The equation for a confined aquifer is:

$$K=\frac{Q\log_e\left(\frac{r_2}{r_1}\right)}{2\pi M(s_1-s_2)} \qquad (2)$$

and for a free aquifer:

$$K=\frac{Q\log_e\left(\frac{r_2}{r_1}\right)}{\pi(h_2^2-h_1^2)} \qquad (3)$$

where:

K=the coefficient of permeability, $\frac{L}{t}$,

Q=the discharge of the well, $\frac{L^3}{T}$,

r_1 and r_2 = horizontal distances from the center of the discharging well to the centers of observation wells located on a line passing through the center of the discharging well; distance increases with the value of the subscript, L,

M = thickness of the aquifer, L,

s_1 and s_2 = drawdown in observation wells r_1 and r_2, respectively, L,

h_1 and h_2 = saturated thickness of the aquifer at r_1 and r_2, respectively, L, and

$\log_e$ = natural log = common log × 2.303.

The steady state equation for a free aquifer ignores vertical components of flow and curvature of the equipotential lines but recognizes decrease in aquifer thickness in the direction of the well. An additional assumption is that the well has infinitesimal diameter compared to a fixed radius of influence. Despite the simplifying assumptions, the equations give relatively reliable determinations of Q and K or T from measurements made of pumping tests of adequate duration with fully penetrating wells in confined aquifers and in free aquifers where the drawdowns in observation wells do not exceed 0.25 M. Piezometers should be used in confined aquifers instead of observation wells to assure reliable drawdown data. The steady state equations do not take into consideration time and the release of water from storage. All water is assumed to originate beyond the radius of influence.

2–6. Unsteady One-Directional Flow.—Moody and Ribbens [11] modified and applied equations derived by Glover [5] for unsteady one-directional flow using much the same basic principles that are involved in the equation for unsteady radial flow (sec. 2–7). The assumptions on which the equations are based are as follows:

- The aquifer is homogeneous and isotropic.
- Hydraulic conductivity K and storativity S are constant with time.
- The aquifer is of infinite horizontal extent.
- The aquifer is confined or, if free, the saturated thickness is large compared to the drawdown.
- Water is withdrawn at a constant rate from a fully penetrating vertical plane sink which is oriented parallel to the source of water. The drawdown is constant at a given time for all points along the sink.

The equation is as follows:

$$s=\frac{Q}{2K}\left[\sqrt{\frac{4Kt}{\pi MS}}\exp\left(-\frac{r^2S}{4Tt}\right)-\frac{r}{M}\left(1-\text{erf}\sqrt{\frac{r^2S}{4Tt}}\right)\right] \quad (4)$$

where:

s = drawdown at any point a distance r from the plane sink on a line normal to the sink, L,

Q = the rate of withdrawal of water per linear foot from the sink, $\frac{L^3}{t}$,

K = hydraulic conductivity of the aquifer, $\frac{L}{t}$,

t = time since pumping or discharge per linear foot from the sink began, t,

M = saturated thickness of the aquifer, L,

S = storativity (nondimensional),

r = distance from the plane sink on a line normal to it, L,

T = transmissivity of the aquifer, $\frac{L^2}{t}$,

exp = the exponential function ($\exp(x)=e^x$), and

erf = the error or probability function.

Tables of the exponential function e^x are found in Applied Mathematics Series 14 and the error or probability function in Applied Mathematics Series 41 of the National Bureau of Standards.

Moody and Ribbens [11] also give a function, equation (5), with which to compute the additional drawdown due to convergence of parallel flow to a horizontal line sink; that is, a drain or ditch instead of a fully penetrating plane sink.

$$s=\frac{Q}{2K}\left\{\frac{r}{M}-\frac{2}{\pi}\log_e\left[\exp\left(\frac{\pi r}{2M}\right)-\exp\left(-\frac{\pi r}{2M}\right)\right]\right\} \quad (5)$$

In equation (5), M = the thickness of the aquifer and the other symbols are as in equation (4). If two line sinks exist in the neighborhood of each other, the drawdowns are additive.

2-7. Unsteady Radial Flow.—The limitations and errors in the steady state well equations resulting from the simplifying assumptions made in their derivations were recognized by early investigators. In 1935, Theis [13] perceived the analogy between the flow of heat and flow of water and adapted the equation for the flow of heat in a conducting solid to the flow of water to a well in a confined aquifer. In 1940, Jacob [7] derived an identical equation from purely hydraulic considerations. The Theis or nonequilibrium equation, which takes into consideration both time and storativity, has the form:

$$s=\frac{Q}{4\pi T}\int_{u}^{\infty}\frac{e^{-u}}{u}\,du \tag{6}$$

where:

s=the drawdown at any point r on the cone of depression, L,

Q=uniform discharge of a well per unit time, $\frac{L^3}{t}$,

$T=KM$, the transmissivity, by definition the coefficient of permeability times the thickness of the aquifer, $\frac{L^2}{t}$,

r=the distance from the center of the discharging well to the point of measurement of s, L,

S=the storativity or coefficient of storage (nondimensional),

t=the time since discharge of the well began, t, and

$u=\frac{r^2S}{4Tt}$ (nondimensional).

The assumptions on which the nonequilibrium equation is based are:

- The aquifer is homogeneous, isotropic, of uniform thickness, and of infinite areal extent.
- The discharging well is of infinitesimal diameter, completely penetrates, and is open to the aquifer.
- Discharge of water from storage is instantaneous with the reduction in pressure due to drawdown.
- Flow to the well is radial and horizontal.

The exponential integral of u is frequently expressed as $W(u)$, the well function of u, and the equation can then be rewritten as:

$$s=\frac{QW(u)}{4\pi T} \tag{7}$$

The above assumptions are rarely all present in actual conditions. Also, equation (7) is theoretically applicable only to confined aquifers. However, the error in the analysis of free aquifers is minor provided the drawdown at the point of observation does not exceed 25 percent of the aquifer thickness. Use of the nonequilibrium equation permits analysis of aquifer conditions and predictions of aquifer behavior that change with time and involve storage. This makes possible many of the modern modeling and computer techniques used in ground-water analyses.

2–8. Specific Retention.—If a unit volume of dry porous material is saturated and then permitted to drain by gravity, the volume of water released is less than that required for saturation [10]. The volume of water retained in the material is held by capillary action and molecular forces against the pull of gravity. The ratio of the volume of the retained water to the volume of the material is the *specific retention*, a nondimensional value expressed as a percentage. The value of specific retention ranges from less than 1 percent to 100 percent. It increases with a decrease in grain and pore sizes of the material.

The specific retention may be expressed as follows:

$$R_s = \frac{V_{wr}}{V} 100 \tag{8}$$

where:

R_s = specific retention,
V_{wr} = volume of water retained, L^3, and
V = volume of material, L^3.

2-9. Boundaries.—The nonequilibrium equations are based on flow through a homogeneous and isotropic aquifer of infinite areal extent. However, all aquifers have boundaries that modify such flow conditions [1].

The confining beds of an artesian aquifer and the water table and the lower confining bed of a free aquifer represent a type of boundary which limits transmissivity.

However, most common boundaries are those limiting the horizontal extent of aquifers. These may be negative (impermeable), or positive (recharge), or both types of boundaries may be present.

An impermeable boundary is one in which there is a significant reduction in transmissivity. Examples would be where the permeable alluvial fill of a valley abuts the buried valley sides that consist of impermeable granite or where a permeable sandstone is faulted against an impermeable shale. An impermeable boundary influences a discharging well by retarding or stopping the expansion of the cone of depression, which results in increased drawdown between the well and the boundary and subsequent removal of water from storage in this area of increased drawdown. In the discharging well the rate of drawdown is increased, the specific capacity is decreased, and the slope of the cone of depression not only decreases in the direction of the boundary but increases on the opposite side of the pumping well (see sec. 5-11).

A recharge boundary is one in which there is significant increase in transmissivity; for example, where a permeable material is in direct connection with a surface body of water or a permeable material is faulted against a more permeable one. A recharge boundary also influences a discharging well by retarding or stopping the expansion of the cone of depression. However, as the boundary provides a source of recharge to replace the normal flow from outside the boundary, the drawdown stabilizes between the well and the boundary, and removal of water from storage is limited. In the discharging well the rate of drawdown is lessened, the specific capacity increased, and the slope of the cone of depression is not only increased in the direction of the boundary, but is decreased on the opposite side of the pumping well (see sec. 5–11).

Boundaries are of concern when predicting the influence and probable yield and drawdown of wells. If an aquifer test is not run long enough for the area of influence to intercept a boundary, a substantial error may be made in estimating well performance. Also, a well may draw as much as 90 percent of its discharge from a stream that is hydraulically connected to the aquifer (see sec. 5–16).

Boundary effects can sometimes be anticipated on the basis of known geological conditions or the conditions may be hidden and revealed only by analyses of a pumping test. In some instances, two or more boundaries may influence well performance to the extent that reliable determinations of aquifer characteristics and boundary locations are precluded. Methods of analyses of boundary effects from pumping tests are discussed in sections 5–11 and 5–12.

2–10. Leaky Aquifers.—A confining bed of a leaky aquifer [5,8,9] may also be considered a type of boundary. When a well discharges from such an aquifer, the reduction in head may promote increased flow through confining beds to the aquifer or reduction in flow from the aquifer. When the area of influence has expanded sufficiently so that the amount of increased seepage into or reduced seepage out of the aquifer equals the pumping rate, the discharge drawdown, relationship, and flow pattern about the well stabilize (see sec. 5–8). A similar response is reflected when a free aquifer overlying a confined aquifer is pumped. The lowering of the water table in the cone of depression reduces the pressure and increases upflow from the artesian aquifer. The influences on aquifer tests of boundary and leaky aquifer conditions may appear similar but usually can be differentiated (see subsec. 5–5(b)).

2–11. Delayed Drainage.—A free aquifer may consist in whole or in part of fine-grained material from which drainage is relatively slow [2]. Pumping tests run under such conditions may yield unusual S-shaped plots of log t versus log s which may be attributed to leaky aquifer influence. Furthermore, in extreme cases, an uneconomically long test might be required to differentiate between the two situations. In the final analysis, judgment based on the knowledge of subsurface and other conditions may be the principal basis for interpretation.

2–12. Anisotropy.—Most aquifers are anisotropic, that is, flow conditions vary with direction [6,15]. In granular material, the particle shape and orientation, and the process and sequence of deposition usually result in vertical permeability being less than horizontal permeability. In nongranular rocks, the size, shape, orientation, and spacing of fractures and other voids may result in anisotropy. Regardless of the nature of the anisotropy, the effects on yield and drawdown by distortion of the distribution of flow are similar. Where anisotropy is the result of differences in vertical and horizontal permeability, the effect is to distort the distribution of drawdown in free aquifers. The distortion is related to the distance to the observation well, the thickness of the aquifer, the ratio of the horizontal and vertical permeabilities, and to the degree of aquifer penetration by the pumping well. Hantush [6] and Weeks [15] derived theoretical methods of determining horizontal and vertical permeability ratios and values from the analysis of pump test data.

Where anisotropy is the result of lower vertical than horizontal permeability, the flow distortion effect is relatively small in flow to a well with a 100 percent open hole in a confined aquifer or in a free aquifer with small drawdown. However, where the ratio of $\frac{K_h}{K_v}$ and the vertical component of flow are large, such as in a partially penetrating well in a confined aquifer or in a free aquifer with large drawdown, the decrease in yield or increase in drawdown compared to the ideal aquifer may be significant (see fig. 2–6) and the bottom of such a well may be considered to be the bottom of the aquifer.

2–13. Porosity.—Porosity [10] is a nondimensional value that expresses the ratio of the volume of pores to the total volume of a porous material and is usually expressed as a percentage:

$$P=\frac{V_p}{V}100 \tag{9}$$

where:

P=porosity,
V_p=volume of pore space, L^3, and
V=volume of material, L^3.

Porosity ranges from less than 1 percent to as much as 80 percent in some recently deposited clays, but in most granular materials it falls between about 5 and 40 percent. In free aquifers the porosity is equal to the specific retention plus the specific yield. Porosity must be considered in recharge and storage analyses since the volume of water recharged into a dry aquifer will be greater than the recoverable volume. Some rocks may have considerable porosity represented by

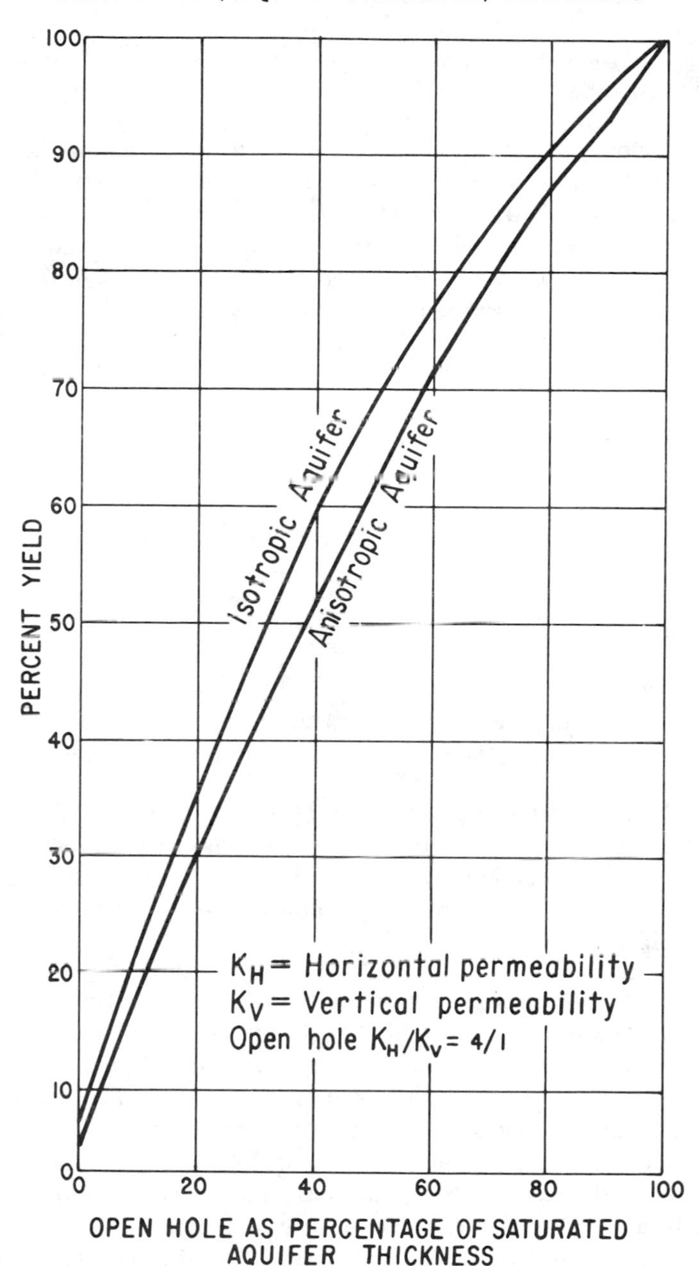

FIGURE 2-6.—Relationship of yield to open hole in isotropic and anisotropic confined aquifers. 103-D-1408.

vugs or holes which are not interconnected. From the standpoint of ground-water flow, such rocks have no porosity.

2–14. Velocity.—The discharge velocity of a porous medium [10], V, is defined as the volume of water that flows per unit time across a unit cross-sectional area normal to the direction of flow. However, in an ideal porous medium, only the void space, which is equal to the porosity, is available for flow. Hence, the actual average interstitial or seepage velocity V_s is the discharge velocity divided by the porosity, or $\frac{V}{P}$, and is expressed as simple velocity $\frac{L}{t}$. Analogously, the velocity under a unit gradient is equal to the hydraulic conductivity divided by the porosity, or $\frac{K}{P}$. Another concept takes into account that water in small crevices and cracks, that is the specific retention, is probably stagnant and that flow occurs only through the area represented by the specific yield, or

$$V_s=\frac{V}{S}.$$

In this case, V_s will be somewhat higher. However, this concept should not be used to estimate the precise rate, distance, or time required for a given molecule of water to move from one place to another.

2–15. Structural Geology and Stratigraphy.—Geologic factors such as stratigraphy, structure, and lithology constitute the skeleton or framework which controls the occurrence and movement of ground water and must be considered in the analysis and solution of ground-water problems.

2–16. Recharge and Discharge Areas.—The location, size, and features of the area within which recharge occurs to an aquifer are pertinent to many ground-water problems. In some free aquifers, recharge occurs over the entire aquifer area, while in others it may be limited by the presence of natural or artificial impermeable materials overlying parts of the area or to aquifers connected with a body of surface water. In confined aquifers the recharge area is limited to a large extent by the exposure of the aquifer at the ground surface or to its subsurface connection with another aquifer or a body of surface water.

Discharge areas are of similar complexity and variation. Primary avenues of natural discharge include evapotranspiration, spring flow, seepage to streams, and leakage to other aquifers. The determination and delineation of recharge and discharge areas are sometimes a complex problem.

2–17. Contributing Areas.—Recharge areas are those within which water enters an aquifer. However, water may enter a recharge area from adjacent and surrounding terrain. The entire area from which water is tributary to a recharge area is the contributing area.

2–18. The Radius of Influence and the Cone of Depression.—The equilibrium equations assume creation of a fixed radius of influence of a well and further assume that all water pumped by the well enters the cone of depression from beyond the radius of influence [3,10]. This can be equated to a well discharging from an aquifer underlying a circular island and which is in hydraulic connection with the surrounding sea. However, the nonequilibrium equation assumes that all water comes from storage within the radius of influence and that this radius increases with time. In an ideal aquifer of infinite areal extent, the radius of influence and the drawdown theoretically increase at a constantly diminishing rate as long as the well is pumped. Under field conditions, however, the rate of change becomes so slow after a sufficiently long period of pumping that it is difficult to measure (see fig. 2–7). In many aquifers the area of influence intercepts the natural aquifer discharge or encounters recharge in sufficient quantity to balance well discharge and true stabilization occurs.

2–19. Well Interference.—If two or more wells discharging from the same aquifer are close enough to each other so that their respective areas of influence overlap, each well interferes with the other and the chord joining the two points of intersection of the areas of influence [1,10] then becomes a divide across which there is no flow. This phenomenon is called well interference and, as a consequence, the rate of drawdown of each well is accelerated.

A recharging well has a similar but opposite effect on a discharging well. The chord joining the two points of intersection of the areas of influence then becomes a line of no drawdown. Flow across it is only in one direction and is equal to the flow originating in the recharging well. Consequently, the rate of drawdown is retarded (see sec. 5–11).

2–20. Principle of Superposition.—If the transmissivity and storativity of an ideal aquifer and the yield and duration of discharge or recharge of two or more wells are known, the combined drawdown or buildup at any point within their interfering area of influence may be estimated by adding algebraically the component of drawdown of each well. This is illustrated on figures 5–17 and 5–18 in section 5–11 which show the impressed heads of a real and image well and the resultant actual drawdown. The effects would be identical for two real wells. The principle of superposition is used to determine desirable spacing

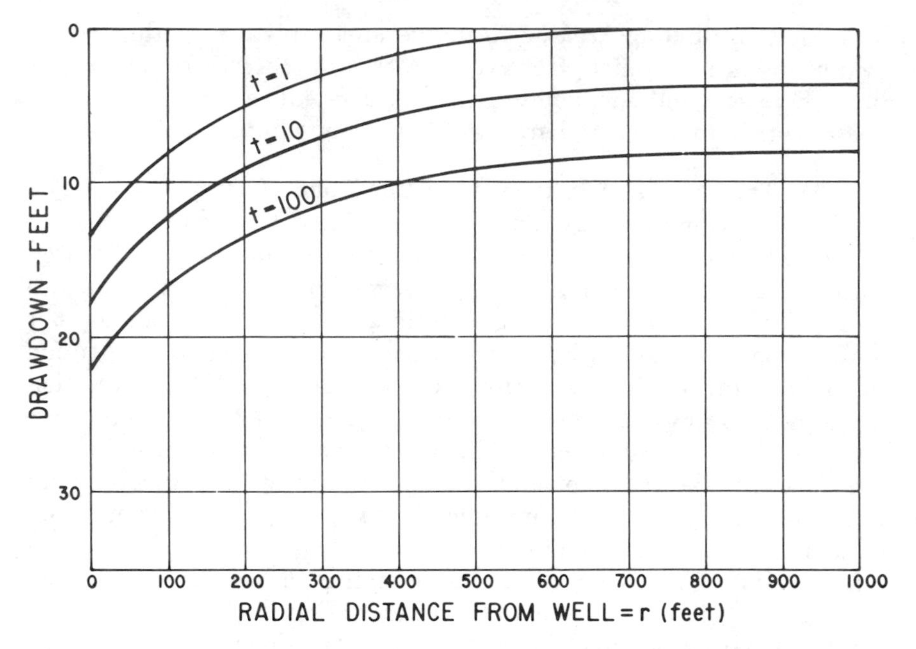

FIGURE 2-7.—Time-drawdown relationship with all factors constant except time, t. 103-D-1409.

of wells in well fields, effects of recharging wells, and in the evaluation of boundary conditions (sec. 5-12).

2-21. Bibliography.—

[1] Bentall, Ray (compiler), "Shortcuts and Special Problems in Aquifer Tests," U.S. Geological Survey Water-Supply Paper 1545-C, 1963.

[2] Boulton, N. S., "Analysis of Data from Non-Equilibrium Pumping Tests Allowing for Delayed Yield from Storage," Proceedings of the Institution of Civil Engineers (London), Paper No. 6693, vol. 26, pp. 469-478, 1963.

[3] Ferris, J. G., Knowles, D. B., Brown, R. H., and Stallman, R. W., "Theory of Aquifer Tests," U.S. Geological Survey Water-Supply Paper 1536-E, pp. 69-174, 1962.

[4] Forchheimer, Philipp, "Hydraulik," third edition, B. G. Teubner Verlagsgesellschaft, Berlin, 1930.

[5] Glover, R. E., "Studies of Ground Water Movement," Bureau of Reclamation Technical Memorandum 657, March 1960.

[6] Hantush, M. S., "Analysis of Data from Pumping Tests in Anisotropic Aquifers," Journal of Geophysical Research, vol. 71, No. 2, pp. 421-425, January 15, 1966.

[7] Jacob, C. E., "On the Flow of Water in an Elastic Artesian Aquifer," Transactions of the American Geophysical Union, vol. 21, part II, pp. 574–586, 1940.

[8] ——— "Radial Flow in a Leaky Artesian Aquifer," Transactions of the American Geophysical Union, vol. 27, No. II, pp. 198–208, April 1946.

[9] Lohman, S. W., et al., "Definitions of Selected Ground-Water Terms—Revisions and Conceptual Refinements," U.S. Geological Survey Water-Supply Paper 1988, January 1972.

[10] Meinzer, O. E. (editor), "The Physics of the Earth—IX, Hydrology," Dover Publications, New York, 1949.

[11] Moody, W. T., and Ribbens, R. W., "Ground Water—Tehama-Colusa Canal Reach No. 3, Sacramento Canals Unit, Central Valley Project," Memorandum to Chief, Canals Branch, Bureau of Reclamation, Office of Chief Engineer, Denver, Colo., December 1965.

[12] Muskat, Morris, "The Flow of Homogeneous Fluids through Porous Media," J. W. Edwards, Ann Arbor, Mich., 1946.

[13] Theis, C. V., "The Relation Between the Lowering of the Piezometric Surface and the Rate and Duration of Discharge of a Well Using Ground Water Storage," Transactions, American Geophysical Union, vol. 16, pp. 519–524, 1935.

[14] Todd, D. K., "Ground Water Hydrology," John Wiley & Sons, New York, 1960.

[15] Weeks, E. P., "Field Methods for Determining Vertical Permeability and Aquifer Anisotropy," U.S. Geological Survey Professional Paper 501-D, pp. 193–198, 1964.

Chapter III

WELL AND AQUIFER RELATIONSHIPS

3–1. Aquifer and Well Hydraulics.—Aquifer characteristics exert primary control over well performance in terms of yield versus drawdown. Accordingly, determination of the effects of well geometry on the flow and head distribution in aquifers and on the yield and drawdown of wells has been the goal of most research on well hydraulics. Mathematical analyses have been made on the basis of steady state flows according to Darcy's law, and Dupuit's assumptions of horizontal radial flow and fixed radii of influence, as well as unsteady state conditions in ideal artesian and free aquifers which are isotropic, homogeneous, of uniform thickness, and infinite areal extent. The conclusions are generally adequate for estimating the performance of wells in artesian aquifers and in free aquifers where the drawdown is a small percentage of the aquifer thickness and the discharging well is fully penetrating. Corrections for partial penetration of the discharging well, large drawdowns in free aquifers, and anisotropy have been derived, but adequate data for application of the corrections are often not readily available. Much research has also been done on analogs and other models of various types, but too often the geometry of the test apparatus has not duplicated field conditions.

However, the nature of the well is also an important factor in well performance. Experience has shown that well design features and construction practices have measurable effects on well performance and operating life and on the economic utilization of the aquifer. Nevertheless, despite this relationship, the engineering and scientific aspects of well hydraulics have received little attention because pumpage of a desired amount of water has been the principal criterion of a successful well. The few laboratory analog analyses which have been made of these relationships have seldom duplicated field conditions. As a result of this and other factors, water well design has been commonly based on experience, observations, and judgment of the designer and driller.

3–2. Flow to Wells.—Theoretically, on initiation of discharge from a well, the water level or head in the well is lowered relative to the undisturbed condition of the potentiometric surface or water table outside the well [4,5,6].[1] The water in the aquifer surrounding the well responds by flowing radially to the lower level in the well.

[1] Numbers in brackets refer to items in the bibliography, section 3–11.

In an artesian aquifer, except for a slight delay caused by inertia, the actual distribution of flow to the well conforms relatively close to the theoretical shortly after pumping is started. However, in a free aquifer the materials in the cone of depression must drain and establish progressively the surface configuration of the cone. Hence, the actual distribution of flow may not conform to the theoretical. Figure 3–1 illustrates schematically the successive stages of development of flow distribution under such conditions by means of equipotential lines and flow lines around a well.

In an ideal artesian aquifer, assume a 100-percent open hole well of radius r_w surrounded by two concentric cylinders of radius r_1 and r_2 with heights equal to the thickness M of the aquifer. The surface area A of the well is $2\pi r_w M$, and of the cylinders $2\pi r_1 M$ and $2\pi r_2 M$, respectively (see fig. 3–2). Under steady state conditions the same quantity of water per unit time must flow through each cylinder and ultimately into the well. According to Darcy's law, $Q=KiA$ or $V=Ki$. If Q and K are constant, the gradient i and the velocity must increase in value as A decreases. Hence, if h_o is the effective head (potentiometric surface) at the radius of influence r_o, and h_2 and h_1 are the heads at radii r_2 and r_1 about the well, the velocity and gradient must increase in the direction of the well. The result is a funnel-shaped area of lowering of the potentiometric surface centered about the well.

Figure 3–3 illustrates schematically the distribution of flow in an artesian aquifer by means of a network of flow lines and equipotential lines lying in a vertical section passing through the axis of a fully penetrating well which has a 100 percent open hole. If the piezometric surface is not drawn down below the bottom of the upper confining bed, the flow lines to the well remain parallel and horizontal and the equipotential lines parallel and vertical. If the drawdown falls below the bottom of upper confining bed, a mixed condition of artesian and free aquifer flow results, which is difficult to assess.

Figure 3–4 illustrates a well that penetrates through the upper confining bed but not into the artesian aquifer. The flow lines and equipotential lines develop hemispherically about the well radius, and a strong vertical component of flow is established out to a distance about equal to the thickness of the aquifer. At a distance about 1.5 times the thickness of the ideal aquifer, the flow lines and equipotential lines assume a relationship similar to that of the fully penetrating 100-percent open hole well.

Figure 3–5 illustrates a well penetrating into and open to about 50 percent of the thickness of an artesian aquifer. The vertical component of flow is less than in the nonpenetrating well but the equipotential lines are still strongly curved although not as great as on figure 3–4. The transition to strictly horizontal flow at a distance from

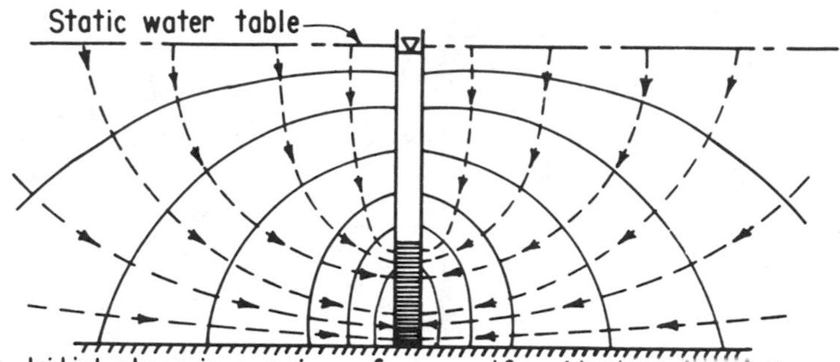

A. Initial stage in pumping a free aquifer. Most water follows a path with a high vertical component from the water table to the screen.

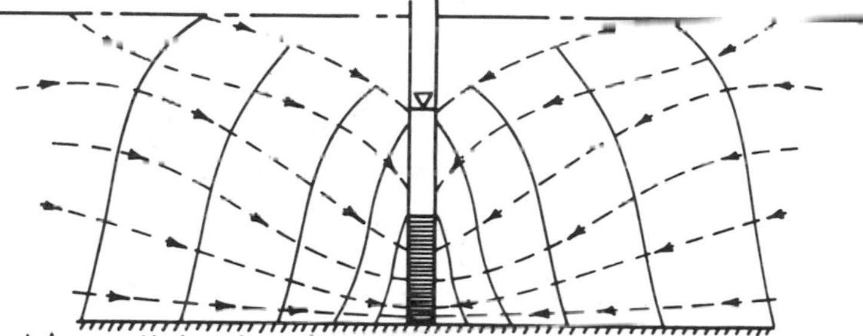

B. Intermediate stage in pumping a free aquifer. Radial component of flow becomes more pronounced but contribution from drawdown cone in immediate vicinity of well is still important.

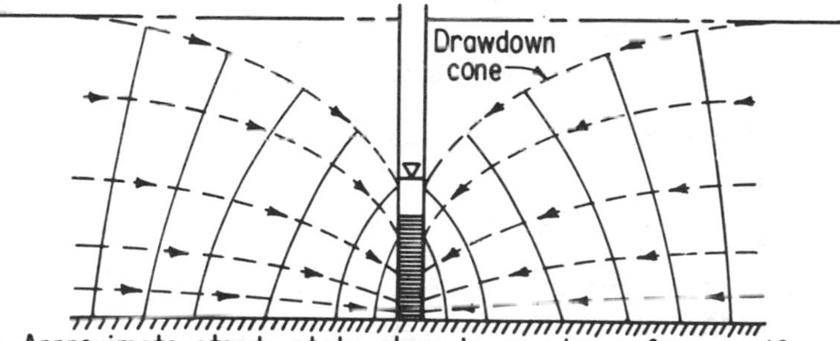

C. Approximate steady state stage in pumping a free aquifer. Profile of cone of depression is established. Nearly all water originating near outer edge of area of influence and stable primarily radial flow pattern established.

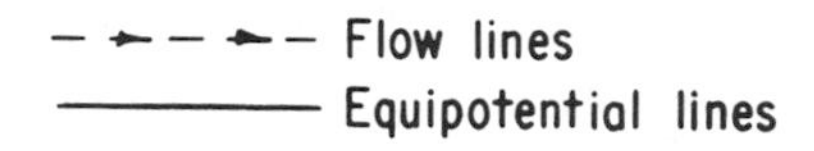

FIGURE 3-1.—Development of flow distribution about a discharging well in a free aquifer—a fully penetrating and 33-percent open hole. 103-D-1410.

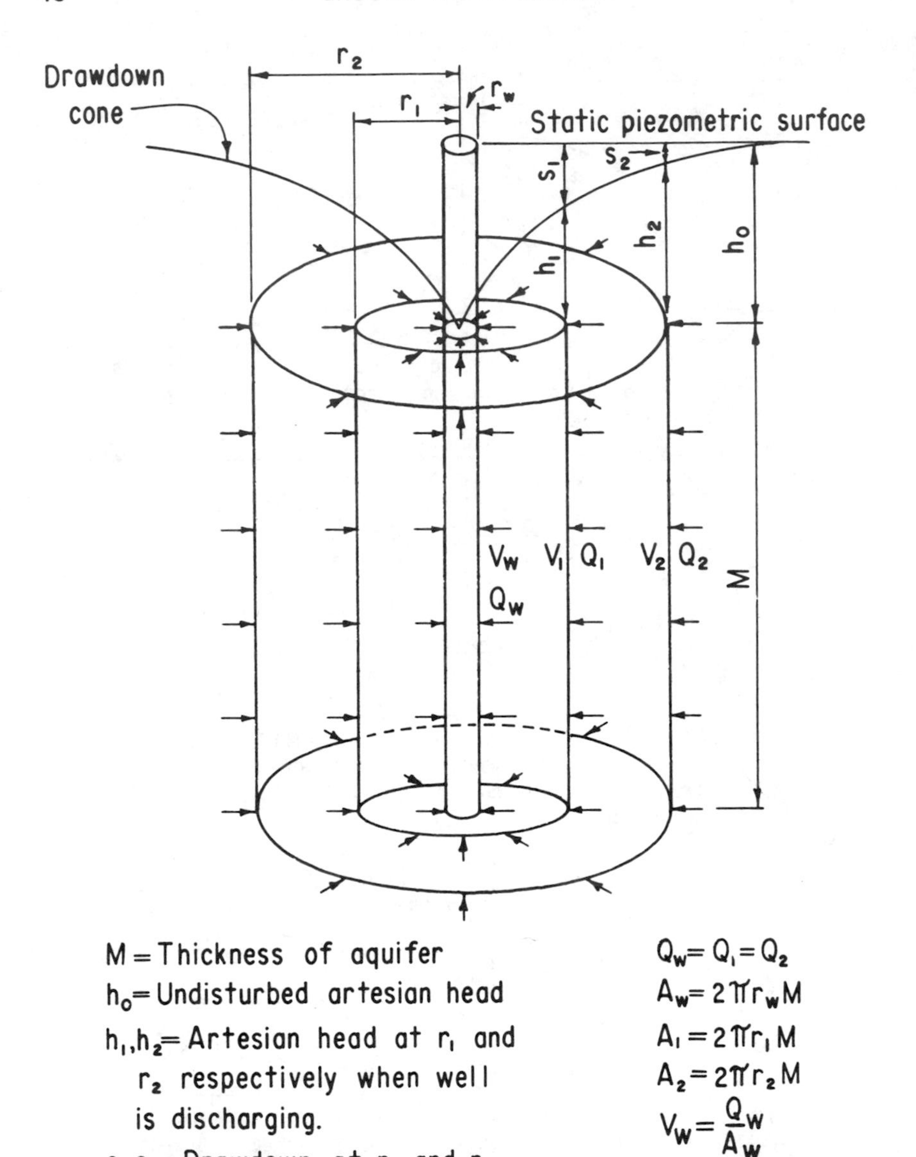

M = Thickness of aquifer
h_o = Undisturbed artesian head
h_1, h_2 = Artesian head at r_1 and r_2 respectively when well is discharging.
s_1, s_2 = Drawdown at r_1 and r_2 respectively when well is discharging

$Q_w = Q_1 = Q_2$
$A_w = 2\pi r_w M$
$A_1 = 2\pi r_1 M$
$A_2 = 2\pi r_2 M$
$V_w = \frac{Q_w}{A_w}$
$V_1 = \frac{Q_1}{A_1}$
$V_2 = \frac{Q_2}{A_2}$

FIGURE 3–2.—Flow distribution to a discharging well in an artesian aquifer—a fully penetrating and 100-percent open hole. 103–D–1411.

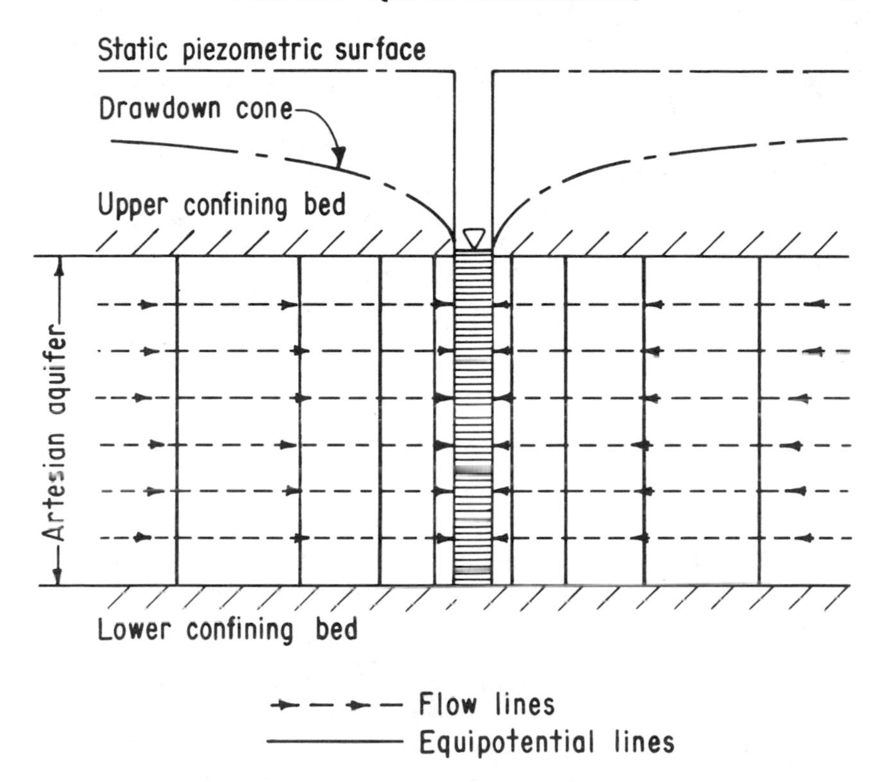

FIGURE 3-3.—Distribution of flow to a discharging well in an artesian aquifer—a fully penetrating and 100-percent open hole. 103-D-1412.

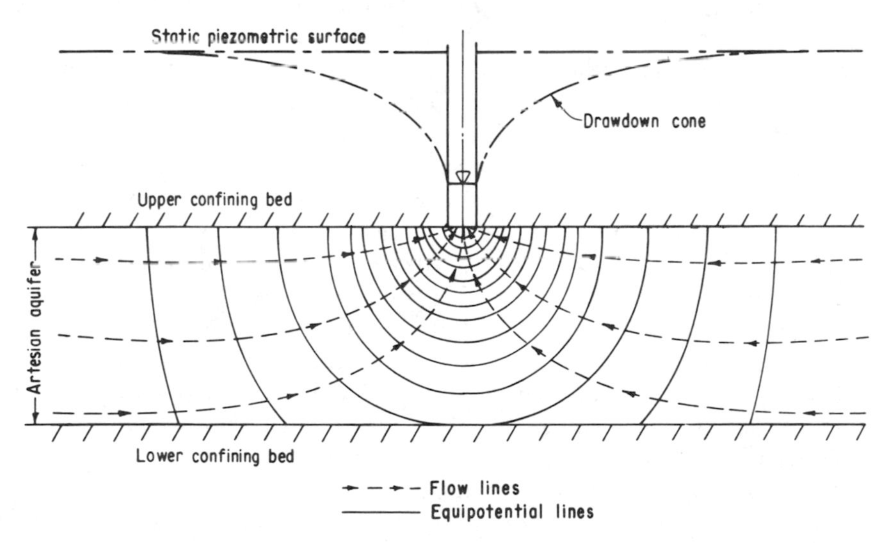

FIGURE 3-4.—Distribution of flow to a discharging well—just penetrating to the top of an artesian aquifer. 103-D-1413.

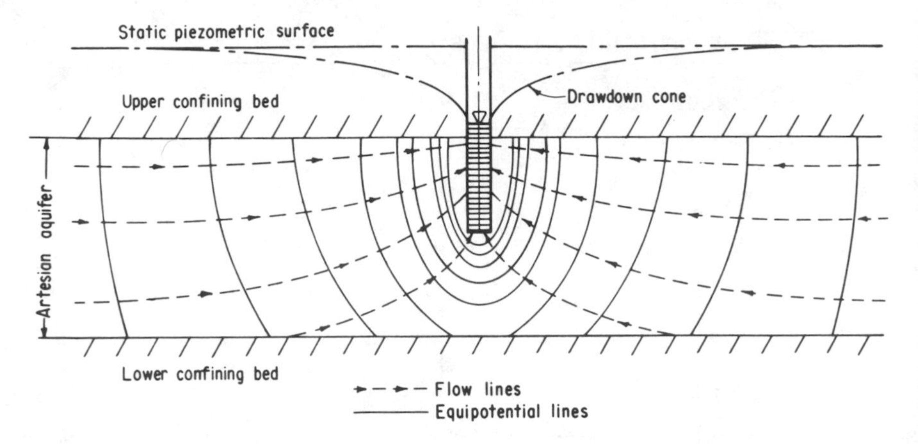

FIGURE 3-5.—Distribution of flow to a well in an artesian aquifer—a 50-percent penetrating and open hole. 103-D-1414.

the well of about 1.5 times the thickness of the aquifers is also apparent. If the open hole was in the lower rather than the upper half of the aquifer, the pattern of flow and equipotential lines would be similar to figure 3-5 if it was inverted.

Vertical convergent flow in wells of less than 100 percent penetration and open hole results in increasing drawdown with a decreasing percentage of open hole.

Figure 3-6 illustrates schematically the distribution of flow and equipotential lines around a fully penetrating and 50-percent open hole well in a free aquifer. The drawdown in the well is about one-half the thickness of the aquifer. Drainage of material above the drawdown cone results in a decline in aquifer thickness, M, and a similar decline in transmissivity. Also, the drawdown in a free aquifer accentuates the vertical components of flow. Thus, the drawdown is accentuated by these two factors.

Reduction in the percentage of open hole in a free aquifer has a similar effect on the drawdown cone and flow lines as in an artesian aquifer. However, the effect is further accentuated because of dewatering of the aquifer in the direction of the well.

Wells often must be drilled in aquifers of large, but unknown, thickness where the cost of full penetration would be prohibitive. Under such circumstances, the usual practice is to compromise on theoretical aspects and drill only to the depth that will furnish the desired supply of water at an acceptable lift.

Kozeny [6] derived an equation for estimating the yield of partially open holes in an ideal artesian aquifer. Figure 3-7 is a graph of this equation for parameters usually encountered or used in well design. The plot shows approximate values which may be used for estimating

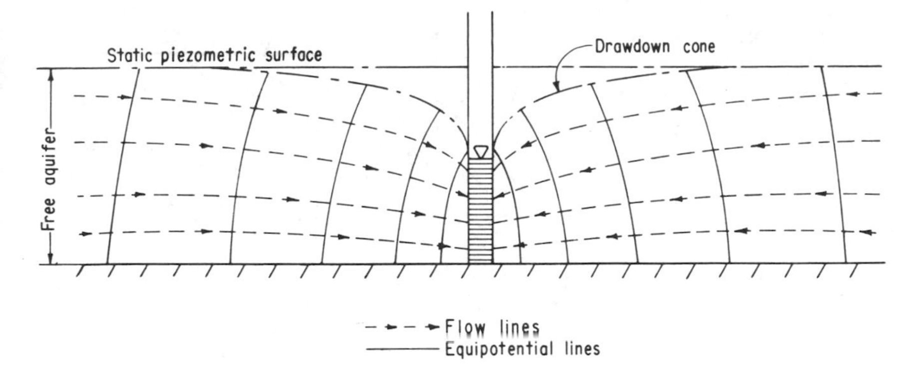

FIGURE 3-6.—Distribution of flow to a discharging well in a free aquifer—a fully penetrating and 50-percent open hole. 103-D-1415.

purposes if the aquifer is fairly uniform and homogeneous and its thickness and characteristics are known. Jacob [1] also derived a method for use in determining aquifer characteristics to correct observed drawdowns resulting from partial penetration to those of the ideal condition if the aquifer thickness was known. Both methods, however, are influenced by other often unknown factors such as aquifer anisotropy and boundaries.

3-3. Discharge and Drawdown Relationships.—The following two steady state equations [2,5,6] can be expressed in approximately the same form as the unsteady state:

$$s_1 - s_2 = \frac{Q \log_e \left(\frac{r_2}{r_1}\right)}{2\pi KM} \text{ and } h_2^2 - h_1^2 = \frac{Q \log_e \left(\frac{r_2}{r_1}\right)}{\pi K}$$

and in the unsteady state $s = \frac{Q}{4T\pi} W(u)$. This relationship facilitates recognition of the following:

(a) The drawdown at any point in the cone of depression is proportional to Q (see fig. 3-8).

(b) At a given discharge, the drawdown at any point on the cone of depression is inversely proportional to the log of r in all equations, and in the unsteady state equation, the drawdown is also inversely proportional to storativity and proportional to log t (see fig. 3-9). At a given Q, drawdown decreases with increased values of transmissivity (see fig. 3-10).

The relationships are applicable for any point in the cone of depression for either fully or partially open hole discharging wells, although

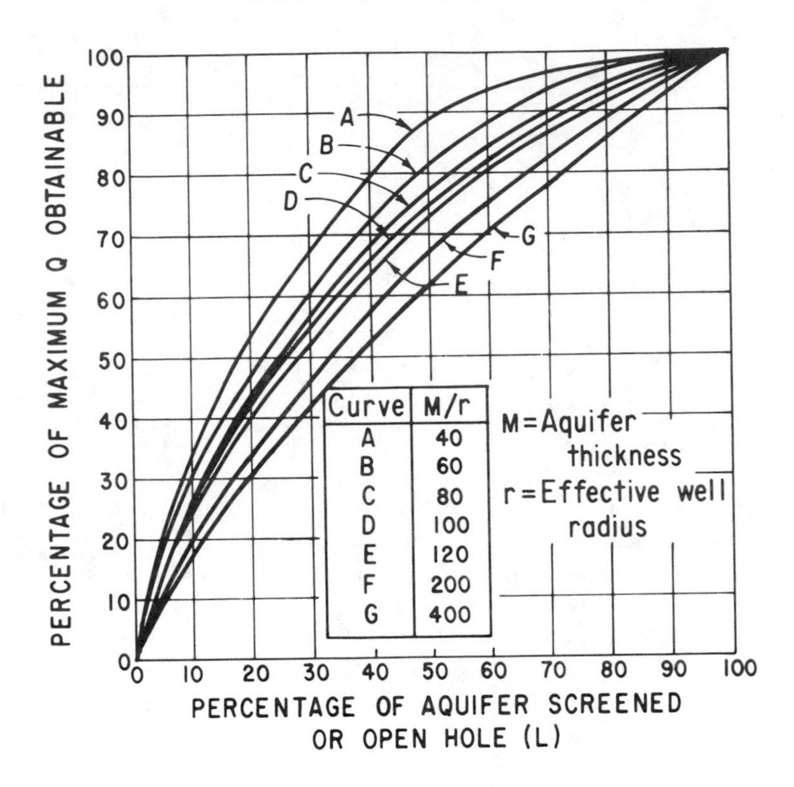

$$\frac{Q_p}{Q} = L\left[1 + 7\left(\frac{r}{2ML}\right)^{1/2} \cos \frac{\pi L}{2}\right]$$

Q_p – Relative yield of a partially open hole.

Q – Yield of a fully open hole.

L – Length of open hole as a fraction of aquifer thickness

r – Well radius

M – Aquifer thickness

FIGURE 3-7.—Graph of Kozeny's equation for relative yield and percentage of open hole in an ideal artesian aquifer. 103–D–1416.

the actual form of the cone of depression will be distorted from the ideal by a partially open hole, anisotropy, or boundaries.

When time is infinite, the nonsteady state equation becomes the same as the steady state equation. Measurements made simultaneously with the drawdowns in the discharging well and one observation well have been used to compute transmissivity or permeability using the equilibrium equation. Theoretically this is possible, but practically

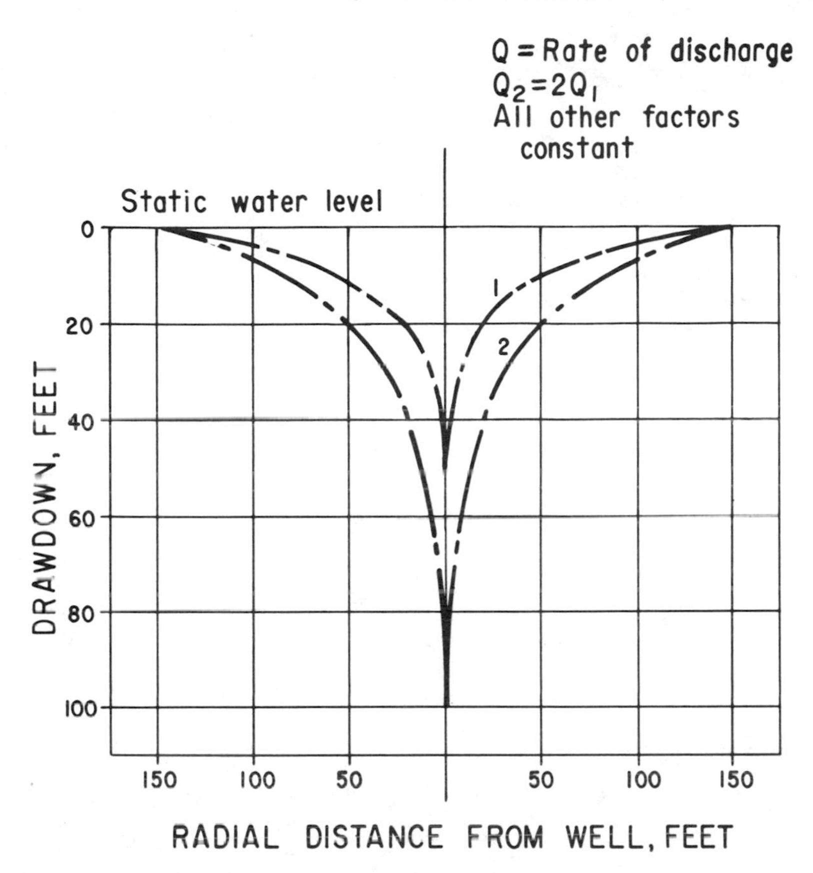

FIGURE 3-8.—Influence of rate of discharge on drawdown in a well. 103-D-1417.

it is not recommended because the water level inside a well generally is lower than outside due to well losses. The result is generally a computed transmissivity or permeability that is too small.

The previous discussion has been primarily concerned with the effect of partial and full penetration on distribution of flow to wells. These are important theoretical concepts, but of equal importance is the effect on well performance. The steepening of the flow lines resulting from a partially open hole in an aquifer results in increased drawdown for the same yield; in other words, a decrease in specific capacity of the well. A well in an artesian aquifer should be open through the entire thickness of the aquifer. A well in a free aquifer should be open in the lower one-third to one-half of the aquifer. However, this is not economically feasible in very deep and thick aquifers. In such aquifers, the usual practice is to penetrate a sufficient thickness of the aquifer to assure the required discharge at an acceptable pumping lift.

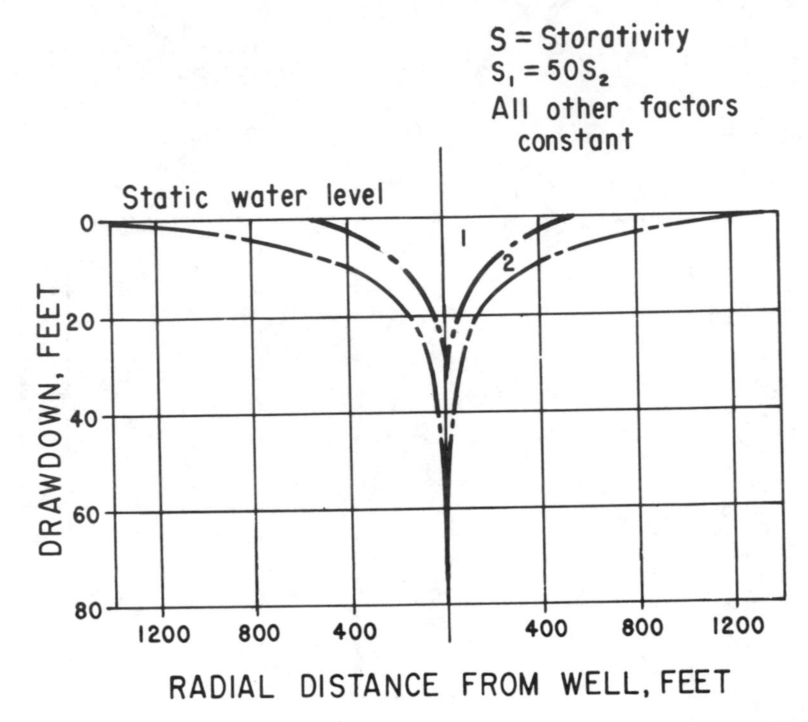

FIGURE 3-9.—Influence of storativity on drawdown in a well. 103-D-1418.

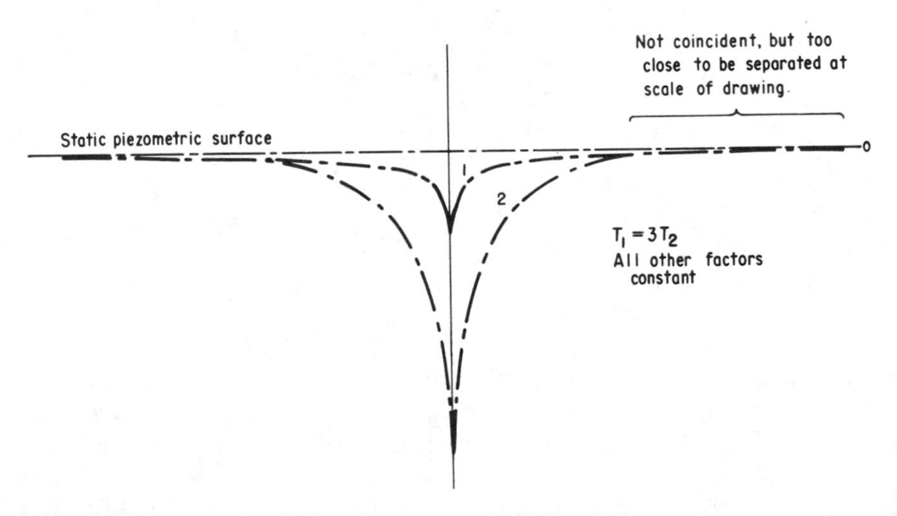

FIGURE 3-10.—Influence of transmissivity on drawdown in a well. 103-D-1419.

In a correctly designed well in an artesian aquifer where turbulent flow is minimal and in which the drawdown is not below the bottom of the upper confining bed, there is practically a linear relationship between yield and drawdown. Hence, specific capacity remains relatively constant regardless of yield. Conditions differ, however, in a free aquifer in which the saturated thickness decreases with an increase in drawdown and vertical components of flow prevail. A number of equations relating yield and drawdown in a free aquifer are available, but all of these are rough approximations. Some show maximum yield with 100 percent drawdown, which would be impracticable because 100 percent drawdown would not leave any well face area available for inflow.

Figure 3–11 is an inexact, composite curve which may serve only as a useful guide to yield-drawdown relationships. Available equations and observations show a nearly linear relationship for yield and drawdown in a free aquifer for drawdown as much as 50 percent of the saturated thickness and acceptable ratios approaching 65 percent. Beyond this, however, the specific capacity begins to fall rapidly, although yield continues to increase. For these reasons, most screened wells are screened in the lower one-third or one-half of the aquifer.

A similar drawing for an artesian aquifer is not included because the yield is approximately proportional to the drawdown in a 100 percent efficient artesian well with 100 percent open hole as long as the drawdown does not lower the hydrostatic head below the bottom of the upper confining bed.

3–4. Well Diameter and Yield.—A common misconception holds that well yield is proportional to well diameter and that doubling the diameter will increase the yield proportionately other things being equal. Well diameter in this case refers to the diameter of the hole that penetrates the aquifer.

The fallacy in this reasoning can be shown by assuming all factors (other than the well diameter) constant in the equilibrium equations and rearranging the equations.

The resultant equation is:

$$Q=\frac{c}{\log_e\left(\frac{r_e}{r_w}\right)}.$$

For an artesian aquifer:

$$c=2\pi KM(s_2-s_1),$$

and for a free aquifer:

$$c=K(h_2^2-h_1^2).$$

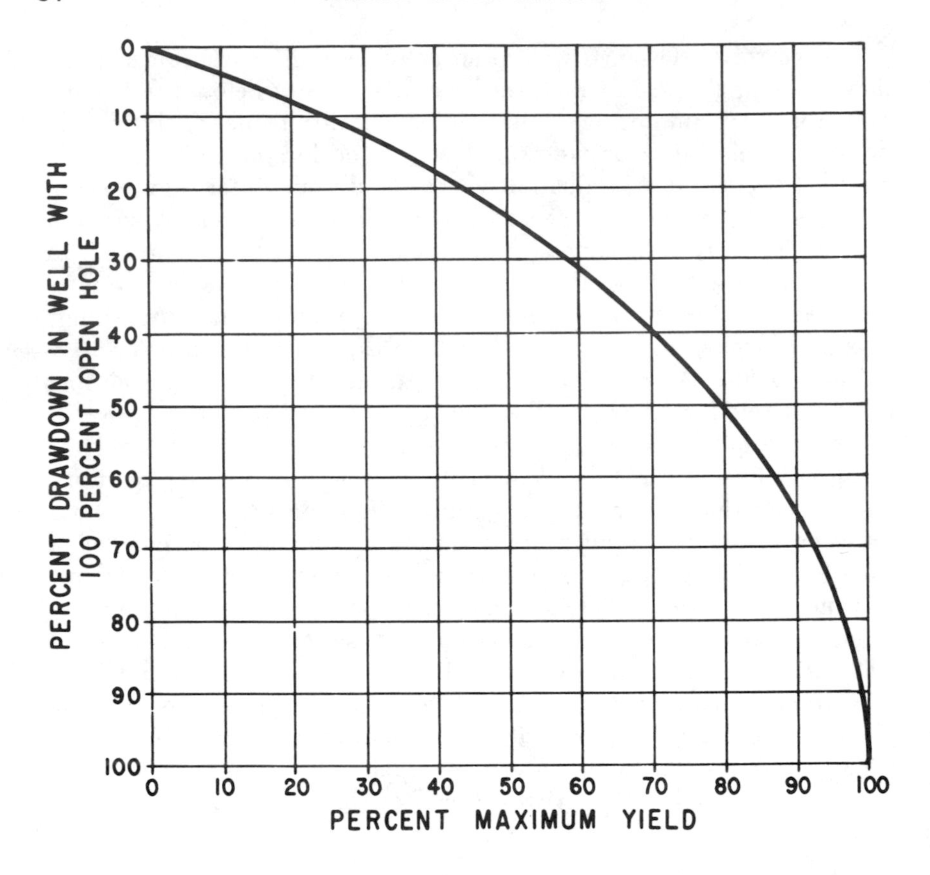

FIGURE 3-11.—Comparison of yield with drawdown in a 100-percent open hole in an ideal aquifer. 103-D-1420.

where:

Q=relative discharge,

c=constant,

r_e=radius of influence,

r_w=radius of well, and

K, M, s_2, s_1, h_2, and h_1 are as previously defined.

By analysis it can be shown that the yield is proportional to the reciprocal of $\log_e \left(\frac{r_e}{r_w}\right)$.

Assuming the effective radii of wells in an artesian aquifer as 8,000 feet and in a free aquifer as 500 feet with well diameters of 1 foot, the equation indicates that in the artesian aquifer doubling the well diameter will increase the yield from 8 to 13 percent and in the free aquifer from 10 to 17 percent (see fig. 3–12). To theoretically double the yield within these parameters would require increasing the diameter of the well in the artesian aquifer about 90 times and in the free aquifer about 45 times. This analysis assumes laminar flow and steady state conditions. Experience has indicated, however, that doubling the diameter of wells will, in some instances, result in an increase in yield of as much as 25 percent. The difference is probably caused by a reduction in turbulence in the aquifer and in the well, increased entrance area, and other factors which reduce well losses.

Zanger and Jarvis [13], on the basis of electric analog studies of recharge wells, determined that one effect of screen diameter in a partially open hole is to reduce the effective diameter of a well to that represented by a hole of the same length but having a surface area equivalent to the open area of the screen. Restated, a 12-inch screen with 25 percent open area would have an effective well diameter of 3 inches in the well and aquifer relationship. It is not known whether this relationship applies to a pumping well of any percentage of open hole, since tests have not been made regarding this feature at this time.

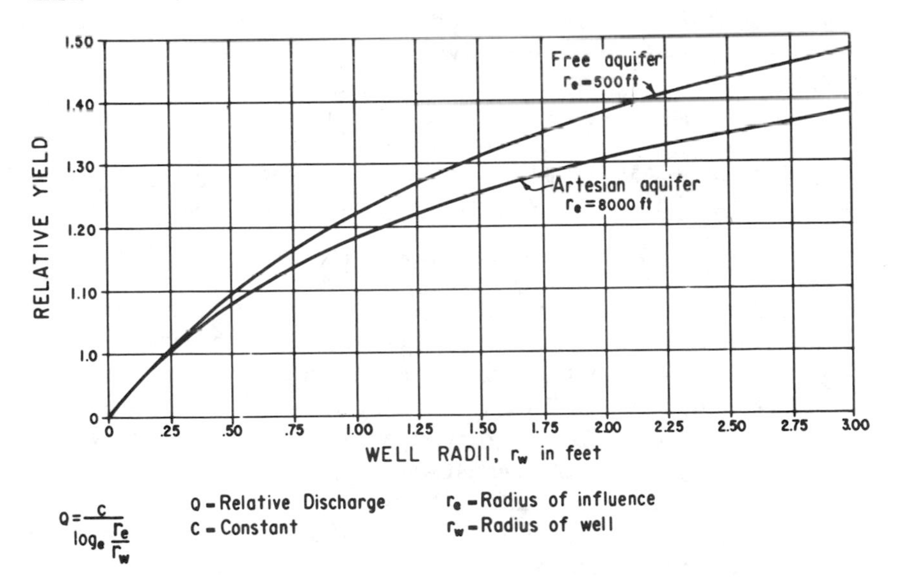

FIGURE 3–12.—Comparison of well diameter and relative yield of a well. 103–D–1421.

There appears to have been little uniformity used in the past concerning the selection of screen diameters. In an attempt to rationalize the selection of screen diameters, many items such as head losses in pipes of various diameters, well and aquifer relationships, probable effects of screen length, slot size, and patterns have been evaluated for many wells of various discharges, diameters, and efficiencies. The results, which are imprecise and strictly empirical, are given in table 11–9 in section 11–4(b). They are presented not as a rigid requirement but as a suggested tentative standard for efficient well design considering initial construction and operation and maintenance costs.

3–5. Well Penetration and Yield.—The total depth of a well may be relatively unimportant as far as yield is concerned [6]. As noted previously, the important factor is the percentage of saturated thickness of the aquifer penetrated by and open to the well; that is, the percentage of open hole. In an artesian aquifer the specific capacity of a well will vary with the percentage of open hole and is maximum when the entire thickness of the aquifer is penetrated by a screen or open hole (see fig. 2–6). A somewhat similar relationship exists in a free aquifer (see figs. 3–13 and 3–14), but because of the thinning of the aquifer in the direction of the well, the increase is limited somewhat depending on the amount of drawdown experienced and the desirability of limiting it to 60 to 65 percent of the aquifer thickness. But in either case, if sufficient additional aquifer thickness is available, the least expensive way to increase specific capacity and yield is usually to increase either the depth of the well and the percentage of open hole or both, if possible [1].

3–6. Entrance Velocity.—A generally acceptable principle of well design holds that the average entrance velocity, based on the percentage of open area of the screen and the desired yield $\frac{Q}{A}$, should be 0.1 foot per second or less. The hydraulic theory behind this criteria holds that at such low velocities flow is entirely laminar, thus turbulence will not contribute to well loss. However, the average entrance velocity concept may be misleading. Soliman [9] and Li [12] analyzed flow to a well and they showed that the entrance velocity in the upper 10 percent of a screen was about 70 times that of the lower 10 percent in an ideal aquifer. In every screened well, part of the entrance area is blocked by the screen and aquifer material or gravel pack. Depending upon the slot size and size and gradation of the grains in the aquifer, as much as 78 percent of the open area may be lost, although it is generally accepted that 50 percent is a more practical estimate. Furthermore, individual zones in an aquifer may

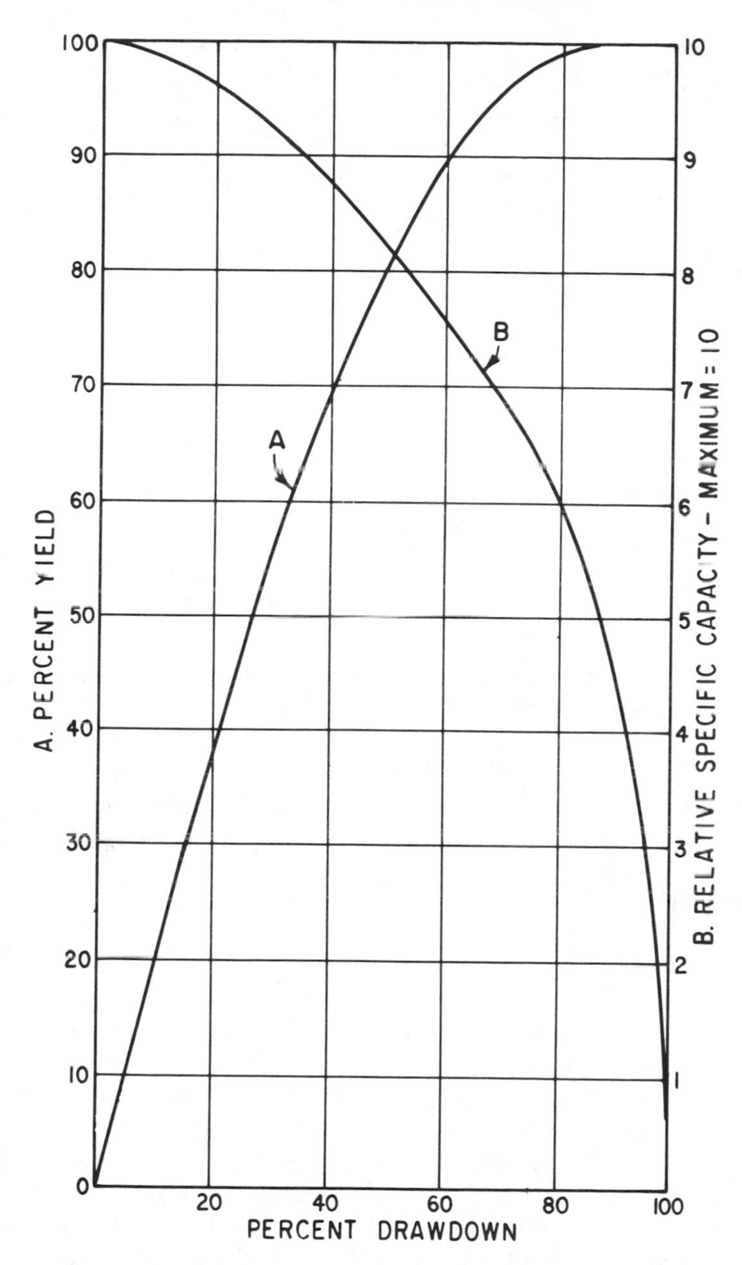

FIGURE 3-13.—Relationship between yield and drawdown in a free aquifer with 100-percent penetration. 103-D-1422.

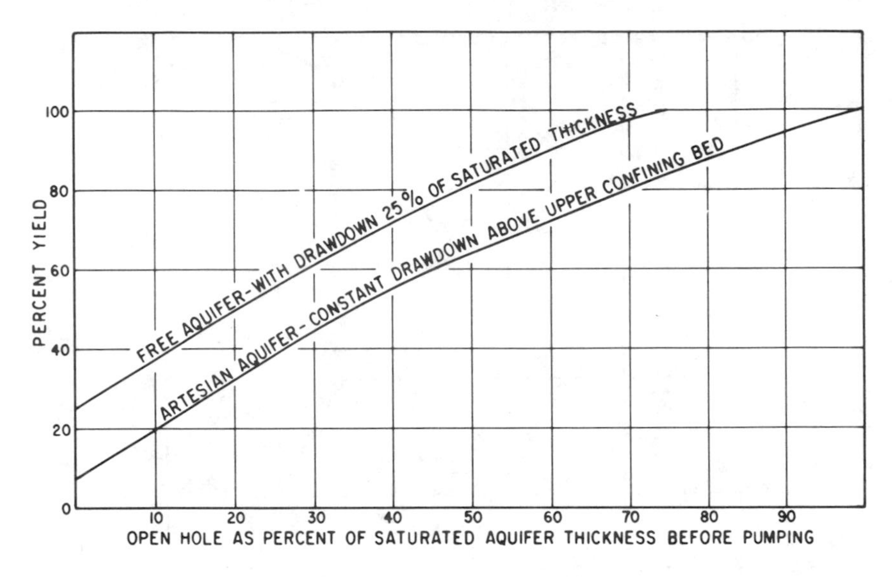

FIGURE 3–14.—Relationship between yield and percent of open hole in free and artesian aquifers. 103–D–1423.

have different permeabilities, and the volume of water delivered to the screen face, other factors being equal, is a function of the permeability of the adjacent aquifer. However, despite these many unknown and indeterminate variations from the average entrance velocity, the concept and practice has proved to be worthwhile in maintaining well efficiency and life. Where conditions and economics permit, the lowering of the entrance velocity to less than 0.1 foot per second, consistent with other criteria, would be advantageous.

3–7. Percentage of Open Area of Screen.—One of the major factors controlling head loss through a screen is the percentage of open area. Factors which control the percentage of open area include basic design, materials, fabrication processes, and strength requirements. In a screen of a given length and diameter, the head loss decreases rapidly with an increase in percentage of open area up to about 15 percent, less rapidly up to about 25 percent, and relatively slowly between about 25 and 60 percent. Beyond an open area of about 60 percent, practically no increase in efficiency is obtained. For practical purposes, a percentage of open area of about 15 percent is acceptable and easily obtained with many commercial screens, although not with perforated casing. If it is not possible to obtain at least 15 percent of open area, other criteria may have to be modified to maintain the maximum 0.1 foot per second entrance velocity. This may be obtained by increasing the diameter or length of the screen or perforated casing used and, hence, the total open area for any pattern of perforations.

Peterson, *et al.* [7], and Vaadia and Scott [11] experimentally determined a relationship among the length, head loss, and diameter; percentage of open area; and the type of slot of a screen. Their findings, however, have limited usefulness under field conditions.

The percentage of open area in slotted pipe ranges from about 1.0 percent for 0.020-inch slots to about 12 percent for 0.250-inch slots in slotting patterns which do not seriously weaken the pipe. Punched or slotted screens have open areas between 4 and 18 percent, depending upon pattern and size of slots. Open areas of louvered screens range from about 3 percent for 0.020-inch slots to about 33 percent for 0.200-inch slots. The open area of cage-type wire wound screens ranges from about 2 percent for 0.006-inch slots to as much as 62 percent for 0.150-inch slots.

3-8. Screen Slot Sizes and Patterns.—Uniform axial flow of water in a pipe is characterized by a stagnant zone at the wall of the pipe, and velocity and turbulence increase toward the center of the pipe. Conditions in a screen are different, however, because a screen performs as a header or collector in which each perforation operates as a radially directed jet.

As a result of this jet inflow, the stagnant layer is absent or only partially present and velocity of the axial flow is not uniform but increases from the bottom of the screen to the top. This distribution of flow has not been studied thoroughly but it can be assumed that with screens of the same diameter and equal percentage of open area, the one with the smaller and more numerous slots would have lower velocity of flow through the slot and the smaller head loss. Parallel-sided slots such as are found in many saw- or machine-perforated casings appear to be the least efficient type of orifice. In addition, they are more subject to clogging by sand grains. The thinner the edge, the more efficient the slot; thus, the V-shaped slot found in most wire-wound screens and on some perforated pipe has a small hydraulic advantage in addition to its self-cleaning properties. Furthermore, a sharp, clean, smooth slot edge not only contributes to better hydraulic efficiency, but also may reduce the rate of corrosion and encrustation.

The convergence of flow lines and continual acceleration of the water in radial flow to a well have been discussed previously (sec. 3-2). The spacing of screen slots can add measurably to the convergence. With two screens having equal open area, the one with large widely spaced slots will require more convergence of flow lines and greater acceleration of the streamflow through the aquifer to the well, with a consequent greater loss of head than would be experienced with the one having small closely spaced slots (see fig. 3-15).

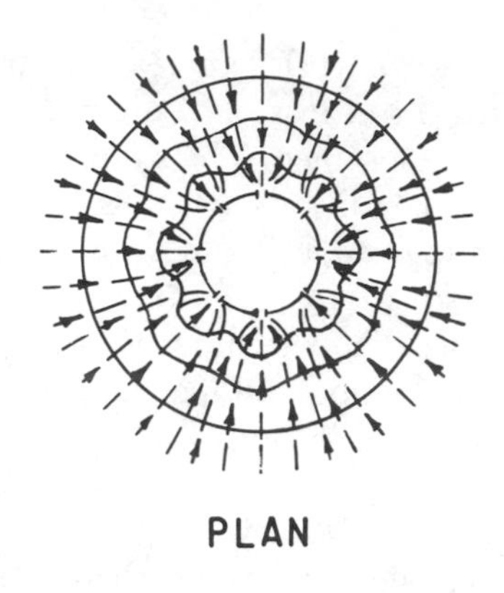

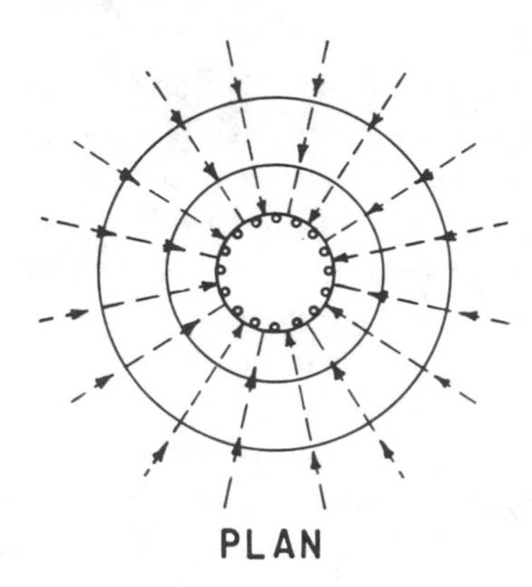

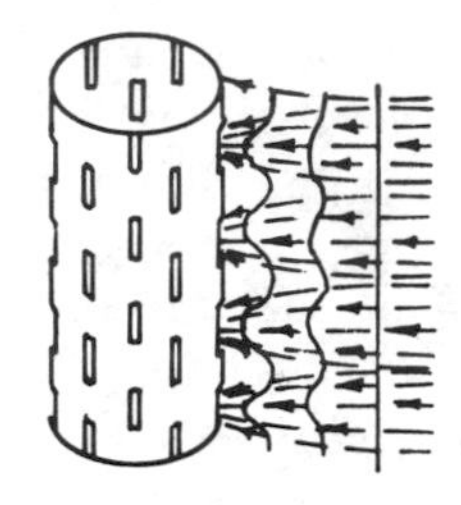

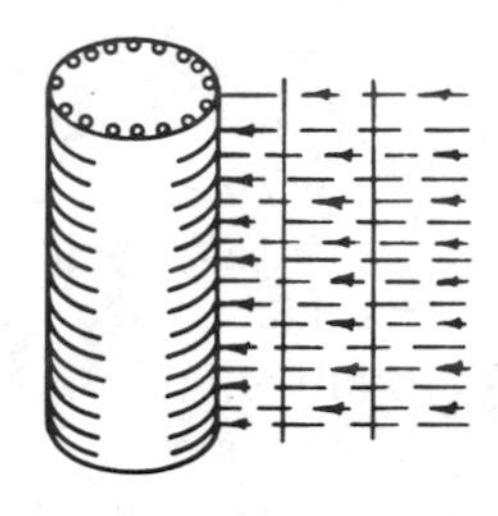

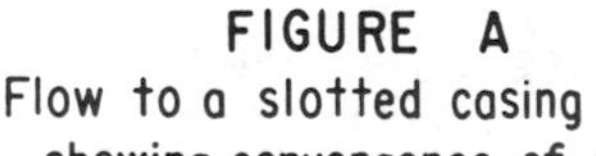

FIGURE A
Flow to a slotted casing showing convergence of flow lines and distortion of equipotential lines.

FIGURE B
Flow to a continuous slot screen showing less convergence of flow lines and distortion of equipotential lines.

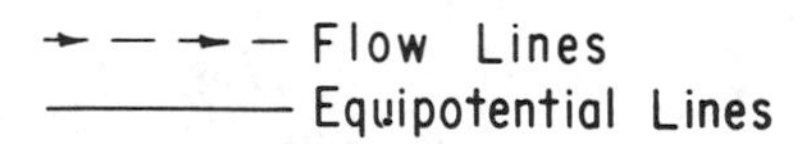

FIGURE 3–15.—Distribution of flow to well screens. 103–D–1424.

Use of very fine slots results in a small percentage of open area even in wire-wound screens. In addition, the small cross section of the slots results in high friction losses and may tend to promote encrustation. Slot sizes of 0.006 to 0.020 inch generally should be limited to use in small, low-capacity wells. A gravel pack is advisable in large-capacity wells if a slot size smaller than 0.030 inch would be required to stabilize the base material.

3–9. Gravel Packs.—The theory of gravel packs holds that surrounding the screen with a more permeable material than the aquifer increases the effective diameter of the well. Since the theoretical permeabilities of packs may be from 10 to 1,000 times as large as that of the aquifer, this is true. However, the benefit derived is questionable since tripling the effective diameter theoretically would increase the yield only about 20 percent. If the aquifer were sufficiently thick, a similar increase possibly could be obtained much more inexpensively by increasing the depth of the well.

Where gravel packing is required, the grain size of the pack should be uniform and as large as possible commensurate with stabilization of the aquifer and ease of installation. The screen slot size should permit only a small percentage of the pack material to pass, and the pack should consist of firm, well-rounded grains. All these factors together with a low entrance velocity contribute to better hydraulic efficiency (see sec. 11--11).

3–10. Well Efficiency.—Well efficiency is a function of the loss of head resulting from flow through the screen and pack and axially in the well to the pump. Thus, in a 100-percent efficient well, all drawdown results from head losses in the aquifer and would be unrelated to the presence or design of the well. A reasonably accurate method is available for estimating the efficiency of a fully penetrating artesian well with 100 percent open holes if values of r_w and T and S of the aquifer are known. In such a well, values of the measurable drawdown s_m and t can be determined from a pumping test. By inserting the values of T, S, r_w, and t in the nonequilibrium equation, the theoretical drawdown s_c can be computed. The efficiency then would be the computed drawdown, s_c, divided by the measured drawdown, s_m, times 100, or $E=\frac{s_c}{s_m}100$.

If, however, the well does not have 100 percent open hole, the effects of partial penetration and anisotropy are difficult or even impossible to determine, but may have a major influence on the computed efficiency.

The problem is even greater for a well in a free aquifer. Not only is the result influenced by the value of r_w, but drawdown negates the effectiveness of 100 percent open hole even if used. Furthermore, anisotropy may have an adverse effect regardless of the percent of open hole and the effects may be compounded if the well does not fully penetrate the aquifer.

It appears, therefore, that any attempt to accurately determine well efficiency is futile unless conditions are ideal. The most practical procedure is to apply the theoretical and empirical factors accepted as being good design practice in conjunction with adequate well development and disregard theoretical well efficiency as a significant factor.

A step test analysis and determination of the apparent efficiency by the methods of Jacob and Rorabaugh [3,8] may be useful in comparing variations in apparent efficiency of an individual well with time as an aid in recognizing deterioration and possible need of rehabilitation (see subsec. 5-14(a)).

3-11. Bibliography.—

[1] Bentall, Ray (compiler), "Methods of Determining Permeability, Transmissibility, and Drawdown," U.S. Geological Survey Water-Supply Paper 1536-I, 1963.

[2] Ferris, J. G., Knowles, D. B., Brown, R. H., and Stallman, R. W., "Theory of Aquifer Tests," U. S. Geological Survey Water-Supply Paper 1536-E, pp. 69-174, 1962.

[3] Jacob, C. E., "Drawdown Test to Determine the Effective Radius of an Artesian Well," Transactions of the ASCE, vol. 112, pp. 1047-1070, 1947.

[4] ——, "Radial Flow in a Leaky Artesian Aquifer," Transactions of the American Geophysical Union, vol. 27, No. II, pp. 198-208, April 1946.

[5] Meinzer, O. E. (editor), "The Physics of the Earth—IX, Hydrology," Dover Publications, New York, 1949.

[6] Muskat, Morris, "The Flow of Homogeneous Fluids Through Porous Media," J. W. Edwards, Ann Arbor, Mich., 1946.

[7] Petersen, J. S., Rohwer, C., and Albertson, M. L., "Effect of Well Screens on Flow into Wells," Proceedings of the ASCE, vol. 79, Separate No. 365, December, 1953.

[8] Rorabaugh, M. I., "Graphical and Theoretical Analysis of Step-Drawdown Test of Artesian Well," Proceedings of the ASCE, vol. 79, Separate No. 362, December 1953.

[9] Soliman, M. M., "Boundary Flow Considerations in the Design of Wells," Proceedings of the ASCE, Journal of the Irrigation and Drainage Division, vol. 91, No. IR1, pp. 159-177, March 1965.

[10] Theis, C. V., "Relation Between the Lowering of the Piezometer Surface on the Rate and Duration of Discharge of a Well Using Ground Water Storage," Transactions of the American Geophysical Union, vol. 16, pp. 519-524, 1935.

[11] Vaadia, Y., and Scott, V. H., "Hydraulic Properties of Perforated Well Casing, " Proceedings of the ASCE, Journal of the Irrigation and Drainage Division, vol. 84, No. IR1, Paper 1505, January 1958.

[12] Li, W. H., "Interaction Between Well and Aquifer," Proceedings of the ASCE, vol., 80, Separate No. 578, December 1954.

[13] Zangar, C. N., "Theory and Problems of Water Percolation," Bureau of Reclamation, Engineering Monograph No. 8, p. 47, 1953.

«Chapter IV

PLANNING GROUND-WATER INVESTIGATIONS AND PRESENTATION OF RESULTS

4-1. Purpose and Scope.—Planning a ground-water investigation or project requires a thorough appreciation of the purpose, the scope of the work required, the areal extent and geologic complexity of the area involved, and the limitations imposed by available financing and allotted time. Since ground-water hydrology is a dynamic and inexact science, the accuracy and reliability of acquired data usually increase with the time available for observation and interpretation, and much of the success and value of such an investigation depends on the imagination, experience, and judgment of the ground-water technical specialists involved. In addition to often being poorly definitive and requiring long time spans, ground-water investigations generally are costly because of the time factor and the need for extensive subsurface and other data that require drilling and other costly operations.

Some typical purposes of a ground-water investigation include:

- Locating a small domestic or stockwater well
- Designing a large well field to furnish irrigation, industrial, or municipal water
- Lowering the water table where drainage is required
- Locating and designing ground-water recharge facilities
- Estimating the safety and economic aspects of water loss and effect on adjacent lands of seepage from a reservoir or canal
- Estimating the average annual volume of water recoverable and the storage space available in a ground-water reservoir
- Dewatering an excavation for construction purposes
- Planning conjunctive surface and ground-water utilizations

Each purpose may present unique problems and require different concepts, data, approaches, fundings, and time considerations. The location of a single small well may require only a cursory reconnaissance of an area and an examination of a few existing wells, all of which may be accomplished in a day or two. Investigations leading to dewatering of an excavation of limited size may require one or more test wells and a pumping test, which usually can be completed in several weeks to several months. In other instances, where conditions are complex and over a large area, the work may entail many months or even years of study and investigations. Layout of sizeable well fields for any purpose may require a comprehensive ground-water inventory to determine the relationships between climate, longtime ground-water fluctuations, ground and surface water interaction, spatial variation in aquifer characteristics, recharge and discharge, and other similar factors.

Ground-water data based on short-term investigations may be more indicative than substantive. When reliable quantitative information is required, provision should be made for refinement of data by continued observation and data collection.

4-2. Accessibility of Study Areas.—In ground-water investigations involving field work, consideration should be given to accessibility to the area by personnel and equipment. Such aspects of the area as location, extent, topography, transportation facilities, land ownership patterns, cultural development, and climate should be determined prior to planning of field investigations.

4-3. Previous Pertinent Investigations.—In estimating funding, time, and manpower requirements, a review of previous reports on an area is essential. The U.S. Geological Survey reports on geology and hydrology often provide valuable information. Other pertinent data may be found in the records of the National Weather Service and in U.S. Department of Agriculture reports. A search of engineering and geological bibliographies may provide additional references. Many State engineers, State geological surveys, water resource centers, and similar agencies have records of wells and other subsurface investigations which may include location, logs, yields, and methods of construction. Section 4-12 presents a checklist to aid in the selection of pertinent data. The references obtained should be abstracted, analyzed, and summarized. A determination can then be made of the additional data required, methods of acquisition, and the time, manpower, and funds necessary to accomplish the work.

4-4. Planning Field Programs.—Sufficient data are frequently available from previous investigations to solve a problem or develop a plan without additional field work. However, on extensive and complex studies, additional field investigation and testing are usually necessary. After completion of the review, abstraction, and summarization of available information, a field reconnaissance of the area should be made. The review of previous work provides a basis for planning additional work, whereas the reconnaissance field survey provides the information needed to determine field conditions, obstacles, limits, and possible alternative methods for completing any additional work contemplated. Additional field work may be required including:

- Preparation of new or supplemental planimetric, topographic, and geologic maps of suitable scale.
- Geologic field mapping to obtain, clarify, or add information on structure, stratigraphy, and lithology.
- Inventories of wells and similar facilities.
- Initiation of ground-water level measurement and sampling programs.
- Location of test sites and the selection of existing wells or the design, preparation of specifications, and construction of new wells for pump tests together with the location, design, construction, and sampling of exploration holes, observation wells, and piezometers.
- The establishment of gaging stations on streams and springs.
- Determination of the location and measuring point elevations on wells and springs.
- The logging of all new drill holes and wells.
- The selection of lines for and type of geophysical surveys.
- The desirability and type of electrical resistivity, gamma ray, and other similar borehole logs.
- Provision for mechanical analysis of drill hole samples.
- Chemical and bacterial analysis of water samples.

When the required field work has been tentatively determined, the minimum number and type of field personnel, cooperative arrangements with other offices, necessary equipment, and the time and fund requirements can be estimated. Adjustments can then be made if desirable or necessary to conform to the requirements of the overall project. The program and plan should be kept flexible, allowing for curtailment or expansion as determined from information acquired as the investigation progresses.

4-5. Field and Office Coordination.—Field data should be compiled, analyzed, and evaluated in close coordination with field operations. Such a simultaneous program of investigation and analysis permits the recognition of deficiencies and allows revisions of the planned program to be made when desirable or necessary. In some instances field work may be seasonal because of climatic conditions, requiring that the same people perform field work and analysis in alternate seasons.

4-6. Subsurface Investigations.—Of primary interest in most ground-water investigations is the information on the stratigraphy, structure, and hydraulic characteristics of the subsurface materials and on water table and piezometric surface levels and fluctuations.

This information can be obtained from logs of wells previously drilled in the area, samples of material from wells, well pump tests, and records of levels of the water table or piezometric surface. Some of this information may be available from local well drillers, but care must be exercised in using it. The Water Resources Division of the U.S. Geological Survey may be a source of data, and State engineer's offices or other State water regulatory agencies or State colleges may also have reports and records on the area under investigation.

4–7. Climatic Data.—In major ground-water investigations, records of precipitation, temperatures, wind movement, evaporation, and humidity may be essential or useful supplemental data. The source of such records in the United States is the National Weather Service. In ground-water studies, climatic data are used principally for estimating the seasonal variations and amounts of precipitation which may be available for ground-water recharge. This must be determined for any complete estimate of ground-water availability. However, in many studies of limited extent such detail is not necessary or justifiable. If it becomes necessary to make the determination, the detailed methods can be found in any complete text on hydrology [1,3,5,6,7,8,9,10,11,12].[1]

4–8. Streamflow and Runoff.—Streamflow data may be essential in solving the ground-water equation since seepage to or from streams is a major element of discharge or recharge of ground water. The best records are those obtained from continuously recording gages but some information can be obtained from staff gages and rating curves if the gages have been read frequently. If the study is sufficiently critical, the installation of continuous recorders may be justified. The Water Resources Division of the U.S. Geological Survey and State and local water resource agencies are sources of streamflow data [1,2,4,6,7,8,9,11,12].

4–9. Soils and Vegetative Cover.—Soil maps and reports are not usually as readily available as topographic and geologic maps and vary more widely with respect to the quantity and quality of the information they contain. Soil maps and reports supply information on soil characteristics and surface gradients which influence runoff and infiltration.

Vegetative cover maps serve a multiple purpose. They may show areas of phreatophytes where the ground water is close to the surface and may indicate the density and type of vegetation which intercepts

[1] Numbers in brackets refer to items in the bibliography, section 4–16.

precipitation, retards runoff, and transpires moisture. Both soil and vegetative cover maps can usually be obtained from the U.S. Department of Agriculture, State colleges, or other Federal and State agencies interested in forestry, grazing, and agriculture. Such information is rarely needed in detail for most studies and, where maps are not available, field observations and notes are adequate for interpretative purposes.

4-10. Geophysical Data.—Geophysical surveys combined with test drilling may provide valuable information on subsurface conditions including approximate depth to water and bedrock. Also, geophysical well logs may be helpful. Prior to undertaking geophysical investigations, an experienced geophysicist should be consulted regarding the probable value of geophysics and the best procedures to use in solving a particular problem. Federal and State agencies and oil and mining companies are the best sources of geophysical data.

4-11. Water Analyses.—The chemical and bacterial qualities of water may be items of necessary information in ground-water investigations. For water intended for human consumption, the bacterial and chemical qualities of the water must be known to determine its suitability and also to furnish a guide for the type and intensity of treatment required to make it potable. The chemical quality must also be known for industrial and irrigation water supplies, since the presence of selected chemical constituents may not only make water unfit for consumption by either humans or livestock, but unsuitable for industrial or irrigation use.

Chemical analyses are also helpful in preparing well and pump designs and specifications for permanent facilities where corrosive or encrusting waters are known to be or suspected of being present. In addition, chemical analyses can often be used to determine the source of the water or its contaminants. State or local health agencies may have records of bacterial and chemical analyses of ground water within their area of responsibility. Reference [13] contains an excellent discussion on the interpretation of the chemical characteristics of natural water.

4-12. Basic Data Checklist.—A complete list of required basic data is impossible to compile because of the variety of ground-water problems encountered. However, as a guide to normally available data in published and unpublished reports and records and as an aid in planning the data to be obtained by field investigations and tests, the following list may be of assistance:

A. Maps, Cross Sections, and Fence Diagrams

(1) Planimetric
(2) Topographic
(3) Geologic
  (a) structure
  (b) stratigraphy
  (c) lithology
(4) Hydrologic
  (a) location of wells, observation holes, and springs
  (b) ground-water table and piezometric contours
  (c) depth to water
  (d) quality of water
  (e) recharge, discharge, and contributing areas
(5) Vegetative cover
(6) Soils
(7) Aerial photographs

B. Data on Wells, Observation Holes, and Springs

(1) Location, depth, diameter, types of well, and logs
(2) Static and pumping water level, hydrographs, yield, specific capacity, quality of water
(3) Present and projected ground-water development and use
(4) Corrosion, incrustation, well interference, and similar operation and maintenance problems
(5) Location, type, geologic setting, and hydrographs of springs
(6) Observation well networks
(7) Water sampling sites

C. Aquifer Data

(1) Type, such as unconfined, artesian, or perched
(2) Thickness, depths, and formational designation
(3) Boundaries
(4) Transmissivity, storativity, and permeability
(5) Specific retention
(6) Discharge and recharge
(7) Ground and surface water relationships
(8) Aquifer models

D. Climatic Data

(1) Precipitation
(2) Temperature
(3) Evapotranspiration
(4) Wind velocities, directions, and intensities

E. Surface Water

(1) Use
(2) Quality
(3) Runoff distribution, reservoir capacities, inflow and outflow data
(4) Return flows, section gain or loss
(5) Recording stations

F. Local Drilling Facilities and Practices

(1) Size and types of drilling rigs locally available
(2) Logging services locally available
(3) Locally used materials, well designs, and drilling practices
(4) State or local rules and regulations

4–13. Preparation of Basic Illustrations.—Analysis and evaluation of subsurface data for a ground-water study are readily performed using maps, cross sections, fence diagrams, and other similar illustrations. The size, scale, and symbols used for such illustrations during the investigation stage are largely a matter of convenience and ease of use. Many drawings are maintained in an incomplete stage with new data being added as they become available until the work is practically completed. However, consistent with Reclamation practice, the size, scale, and symbols used in final illustrations intended for inclusion in final reports should conform to the specifications of Appendix A, Drafting Standards, and the Engineering Geology Manual, Technical Instructions No. 1. Whenever feasible, the scales of such illustrations showing related or interconnected information should be uniform to permit ready comparison and interpretation through overlays and other similar means.

The number and types of illustrations may vary depending upon the scope and intensity of the work and the complexity of the area. The information presented in section 4–14 summarizes the maps more commonly used in ground-water studies and interpretations.

4–14. Types of Ground-Water Study Maps, Hydrographs, and Sections.—(a) *Topographic Maps.*—Although topographic maps may not be necessary for all ground-water studies, appreciation and understanding of topography are useful if not essential. For some reconnaissance studies, either a good planimetric map or aerial photographs may be used in the field study instead of a topographic map. However, for more detailed studies, good topographic maps are a necessity. Such maps supply information on surface gradients and drainage patterns and are used as the basis for construction of cross sections and maps showing geology, depth to water, surface and water table gradients, contributing and recharge areas, and related features and phenomena. Depending upon the type of terrain and the detail required, scales of satisfactory topographic maps range from 1/2 inch to the mile (1:126,700) to 4 inches to the mile (1:15,800). At times, maps with a scale of 1 inch to 400 feet (1:4,800) may be desirable for the detailed study of local phenomena within larger areas of interest. Desirable contour intervals range from 1 foot in areas of low relief or for large-scale detailed maps to 25 to 50 feet for rugged areas or small scale maps.

The U.S. Geological Survey (USGS) is the primary source of topographic maps, but other Federal agencies, including the Department of Agriculture and the Army Corps of Engineers as well as various State agencies, are also sources of suitable maps. If a satisfactory map is not available, one may need to be prepared.

(b) *Aerial Photographs.*—In many areas aerial photographs must serve as a substitute for topographic maps. Such photographs are available either as contact prints or enlargements at scales ranging from 1:20,000 to 1:4,000. Where the photographs have been taken with sufficient overlap, they may be used with a stereoscope to obtain a three-dimensional view of the terrain. Also, mosaics compiled from numerous individual pictures covering large areas are frequently available.

The Department of Agriculture and the USGS are major sources of aerial photographs and mosaics. These agencies, as well as Reclamation, will usually have access to other sources of photographs. In addition to conventional black and white and color photography, side-looking radar, infrared photography, thermal scanner imagery, and other remote sensing techniques are becoming operational and may provide more widely available data in the future.

(c) *Geologic Maps and Sections.*—Geologic maps and sections (fig. 4–1 is an example), especially when accompanied by adequate reports, are useful in most ground-water investigations and are essential where complex stratigraphy and structures are involved. Analyses of reports and maps give information on recharge areas, possible aquifers, water level conditions, structural and stratigraphic control of water movement, and related factors. The USGS and State geological agencies are primary sources of such materials. Universities and colleges, geological societies, oil and mining companies, and other similar organizations also have data which may be obtainable. In areas for which there are no geologic reports or maps, a reconnaissance geologic investigation may be necessary as a minimum alternative.

(d) *Water Table Contour Maps.*—A water table contour map is the most commonly constructed and most useful map for studies of unconfined ground water. It is a topographic map of the water table, and the contour lines are usually lines of equal elevation (see fig. 4–2).

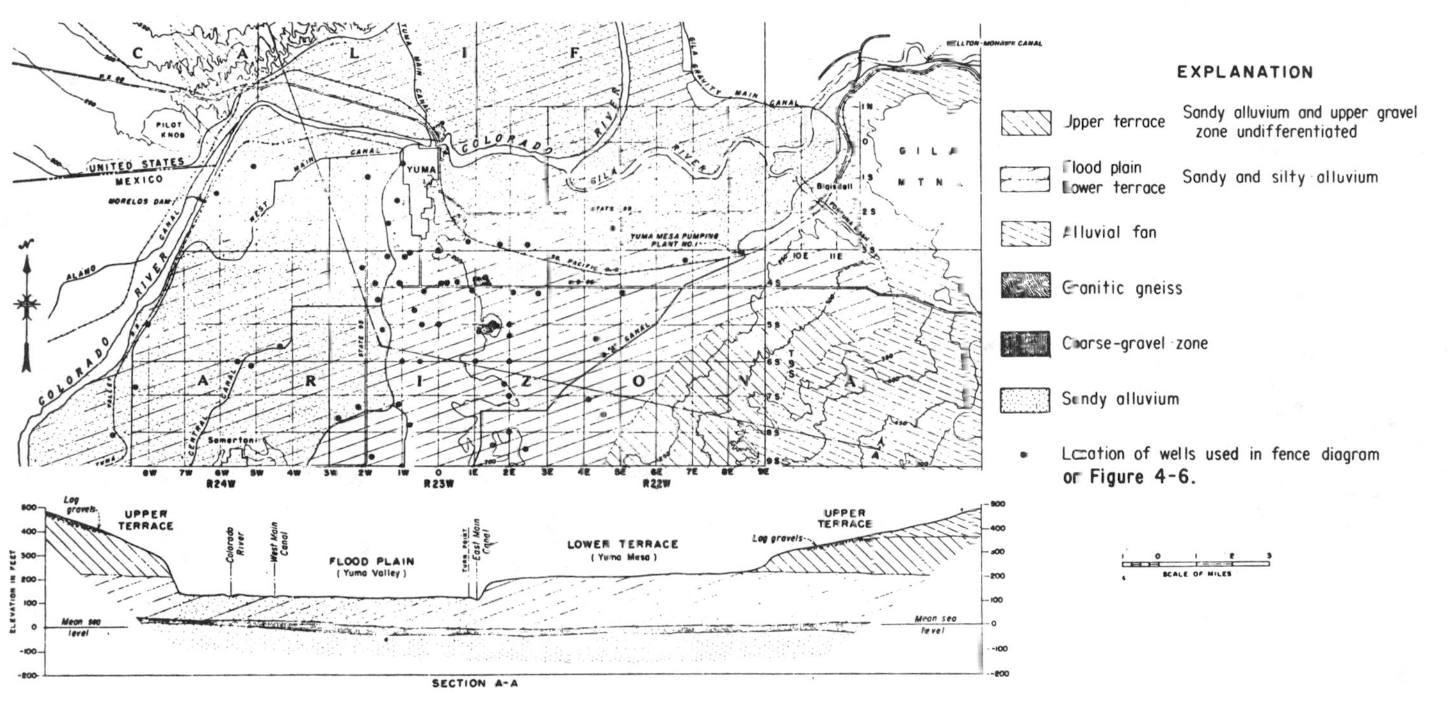

FIGURE 4-1.—Generalized geologic map and cross section of a portion of the Yuma, Ariz., area. 103-D-700.

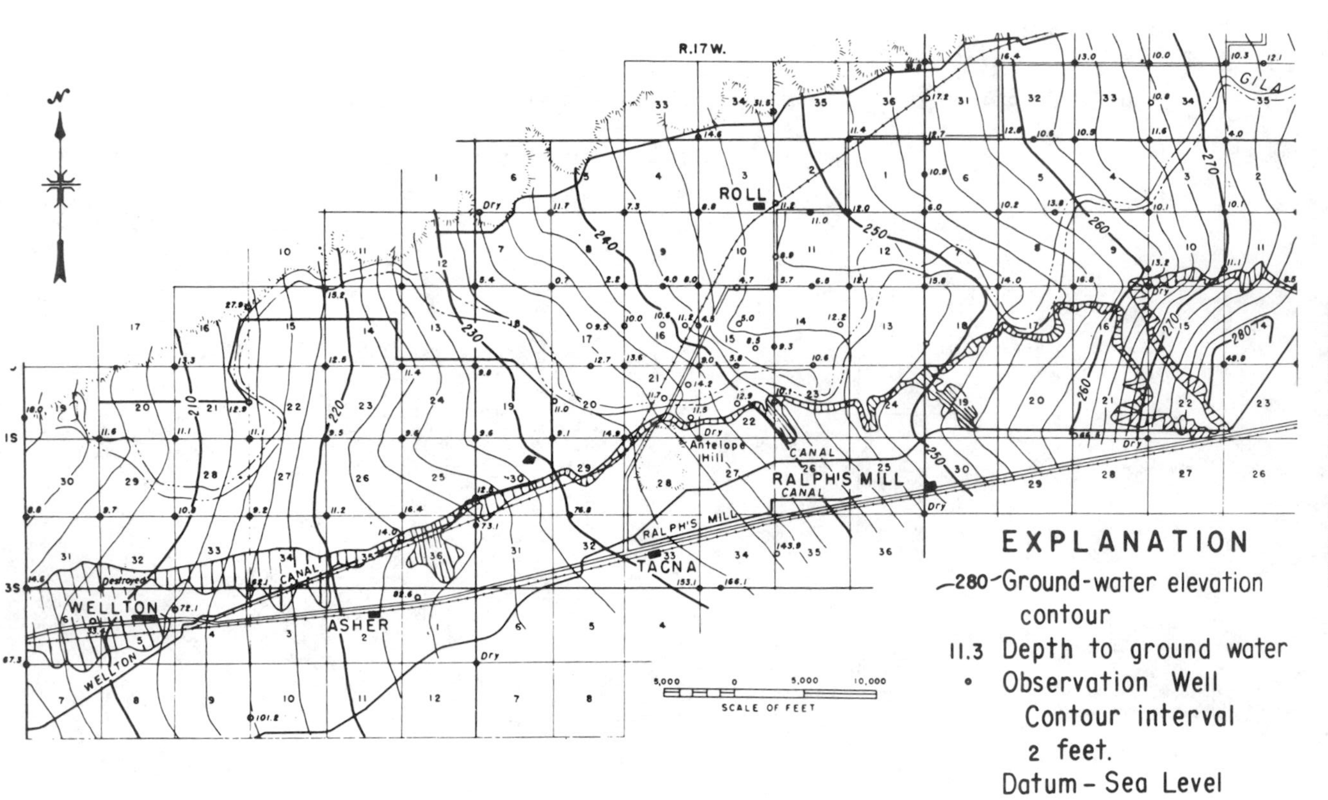

FIGURE 4-2.—Water table contour map. 103-D-1425.

The map is constructed using water level elevations in observation wells, stream and lake surfaces, and spring discharge points for controls.

(e) *Piezometric Surface Maps.*—A piezometric surface map is similar to a water table contour map, except that it is based on the piezometric potential developed in piezometers or tightly sealed wells which penetrate a single confined aquifer (see fig. 4–3).

(f) *Depth-to-Water Table Maps.*—Depth-to-water table maps are of particular interest when considering drainage and dewatering problems. Such maps show the depth to water from the ground surface (see fig. 4–4). They are most easily prepared by overlaying a water table contour map on a surface topographic map. The points at which the contours intersect are a whole number of feet apart in elevation and are the control points for drawing a contour map of depth to water. They can also be prepared by calculating the depth to water from the ground surface and placing this depth figure on a map at the location of the observation well. Contours are then drawn connecting these points.

Care should be exercised in the preparation, use, and evaluation of ground-water level and depth maps. Initially, it should be remembered that only a limited number of spaced control points (observation wells, etc.) can normally be used and that ground-water conditions between the points may deviate widely from the expected. Furthermore, unless the control point facilities are constructed to reflect a specific condition, a composite condition such as a combined water table and piezometric level may be reflected. This could yield erroneous and misleading data.

(g) *Profiles or Cross Sections.*—Vertical geologic and hydrogeologic profiles drawn through lines of wells or drill holes depict information on subsurface conditions by spatially relating surface features and subsurface conditions (see fig. 4–5). At each location the geologic log of the hole is plotted vertically to show the top and bottom of each stratum that can be identified, and adjacent holes are compared to show continuity of strata. Free water table or piezometric surface levels can also be plotted at each well location for one reading or for a series of readings taken over a period of time. This will show the relative location of the free water table or piezometric surface and its fluctuation during the period of the readings. Cross sections should be referenced to a map for convenience in location. The horizontal scale of the section should conform to that of the map but the vertical scale generally will need to be larger than the horizontal scale to make the drawing understandable.

(h) *Thickness-of-Aquifer Map.*—The thickness-of-aquifer map is also known as an isopach map. It shows by contours the thickness of

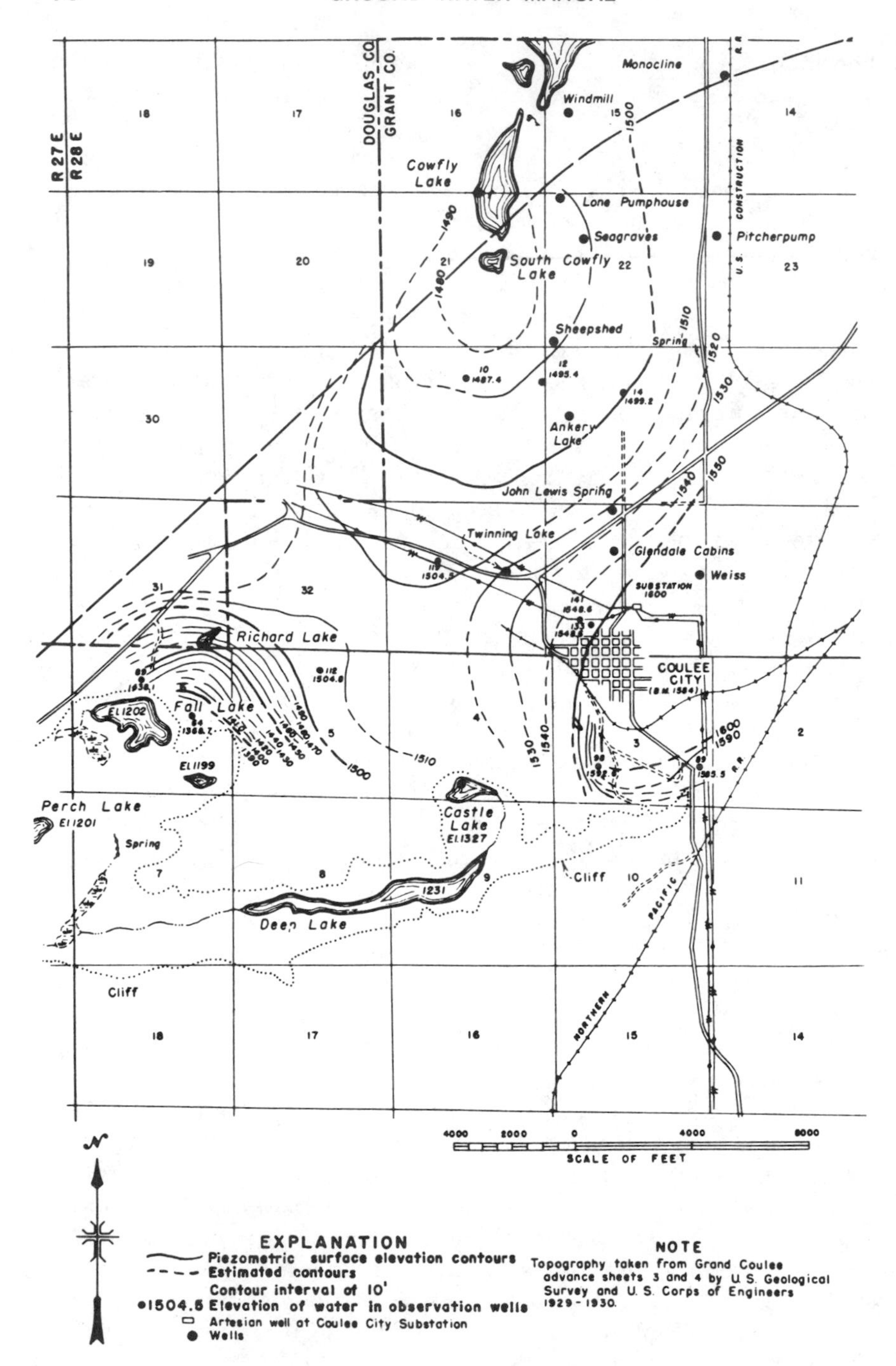

FIGURE 4-3.—Piezometric surface contour map. 103-D-1426.

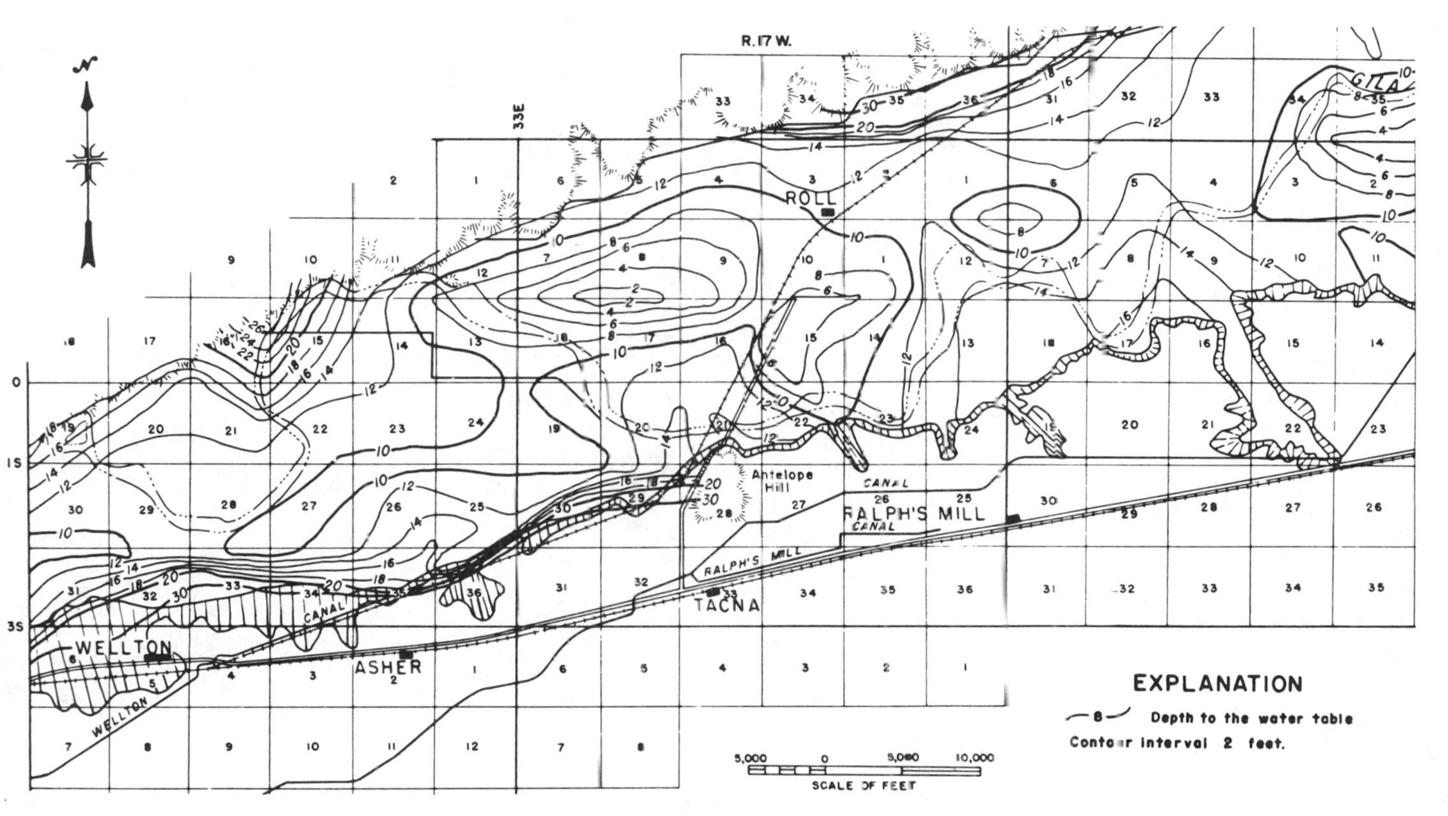

FIGURE 4-4.—Depth-to-water table contour map 103-D-1427.

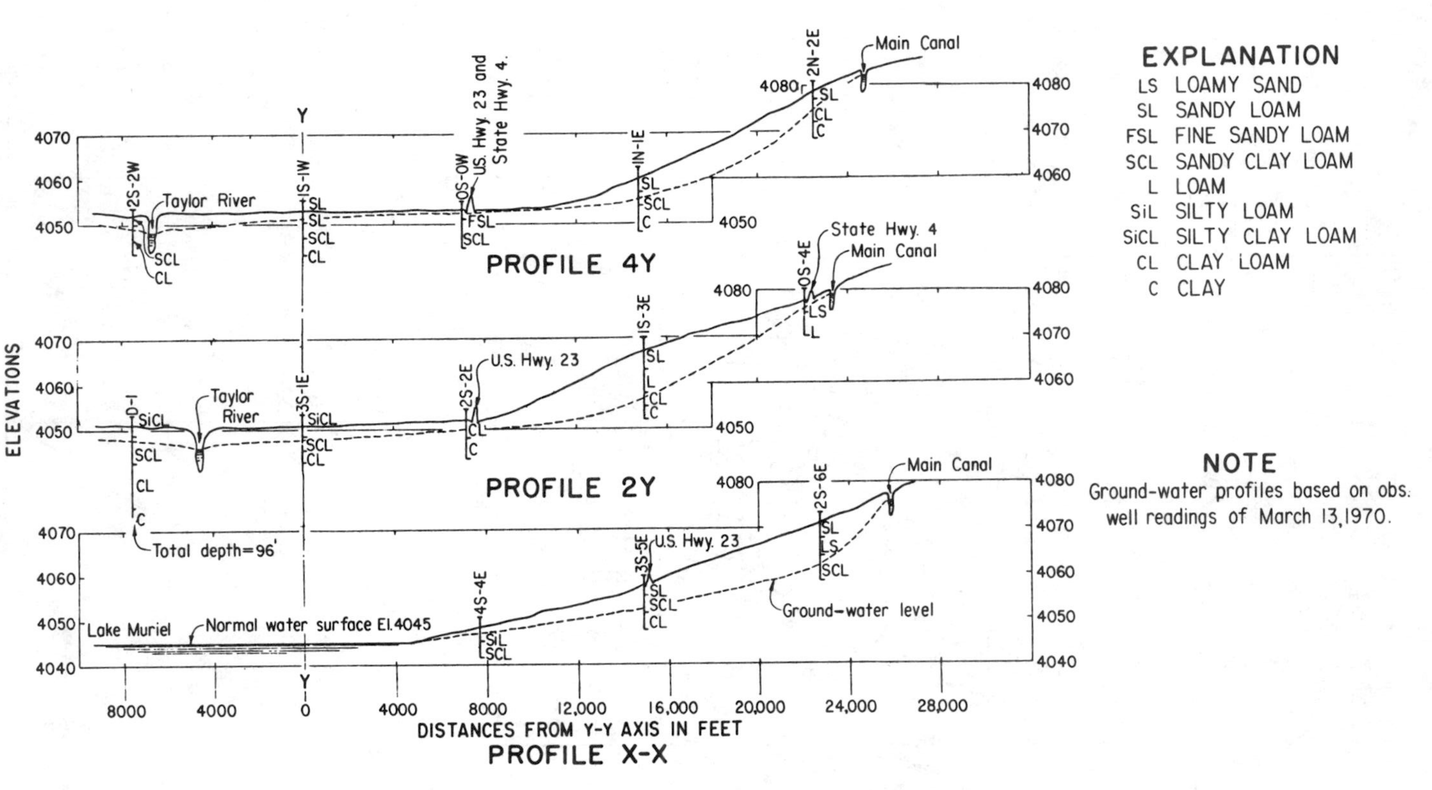

FIGURE 4-5.—Hydrogeologic profiles. 103-D-1428.

saturated materials of a free aquifer or the thickness of an artesian aquifer between the upper and lower confining beds. A similar map may be drawn to show the thickness of a confining bed. Construction of maps of this type, of course, depends upon the availability of the logs of holes and wells that fully penetrate the beds of interest.

(i) *Structure Contour Maps.*—Structure contour maps are drawn to show the upper surface of a particular stratum or formation. These maps are primarily useful in conjunction with stratigraphy in interpreting structural features, such as faults and folds, which may control ground-water movement beneath an area.

(j) *Fence Diagrams.*—Fence diagrams are three-dimensional cross sections that are helpful in presenting an areal picture of geologic and ground-water conditions. As with the sections, they are based on the logs of the holes, measurements of ground-water levels, and topography (see fig. 4–6).

(k) *Hydrographs.*—Hydrographs of individual observation wells and piezometers are essential in depicting ground-water fluctuations, trends, and other time-related factors. Hydrographs are plotted on cross-section paper with water elevations as the ordinate and time as the abscissa (see fig. 4–7).

4–15. Some Guides in Ground-Water Map Interpretation.—The basic principle of ground-water flow holds that water moves from a higher level or potential toward the lower. The contours on ground-water elevation contour maps are those of equal potential and the direction of movement is at right angles to the contours. This is true whether the contours are of a free water surface or of a piezometric surface. In a free aquifer the contours often tend to parallel the land surface contours. In many instances, however, there is little apparent relationship between surface and subsurface flow.

Ground-water mounds can result from downward seepage of surface water or upward leakage from deeper artesian aquifers in areas of local recharge. In an ideal aquifer, gradients from the center of a recharge mound will decrease radially and at a declining rate. An impermeable boundary or change in transmissivity will affect this pattern and may provide clues in determining such changes.

Analysis of conditions revealed by ground-water contours is made in accordance with Darcy's law, $Q=KiA$, which is discussed in section 2–1. Accordingly, the spacing of contours (the gradient) is dependent on the flow rate and on the aquifer thickness and permeability. If continuity of the flow rate is assumed, the spacing depends only on aquifer thickness and permeability. Thus, areal changes in contour spacing may be indicative of changes in aquifer conditions. However, in view of the heterogeneity of most aquifers, changes in

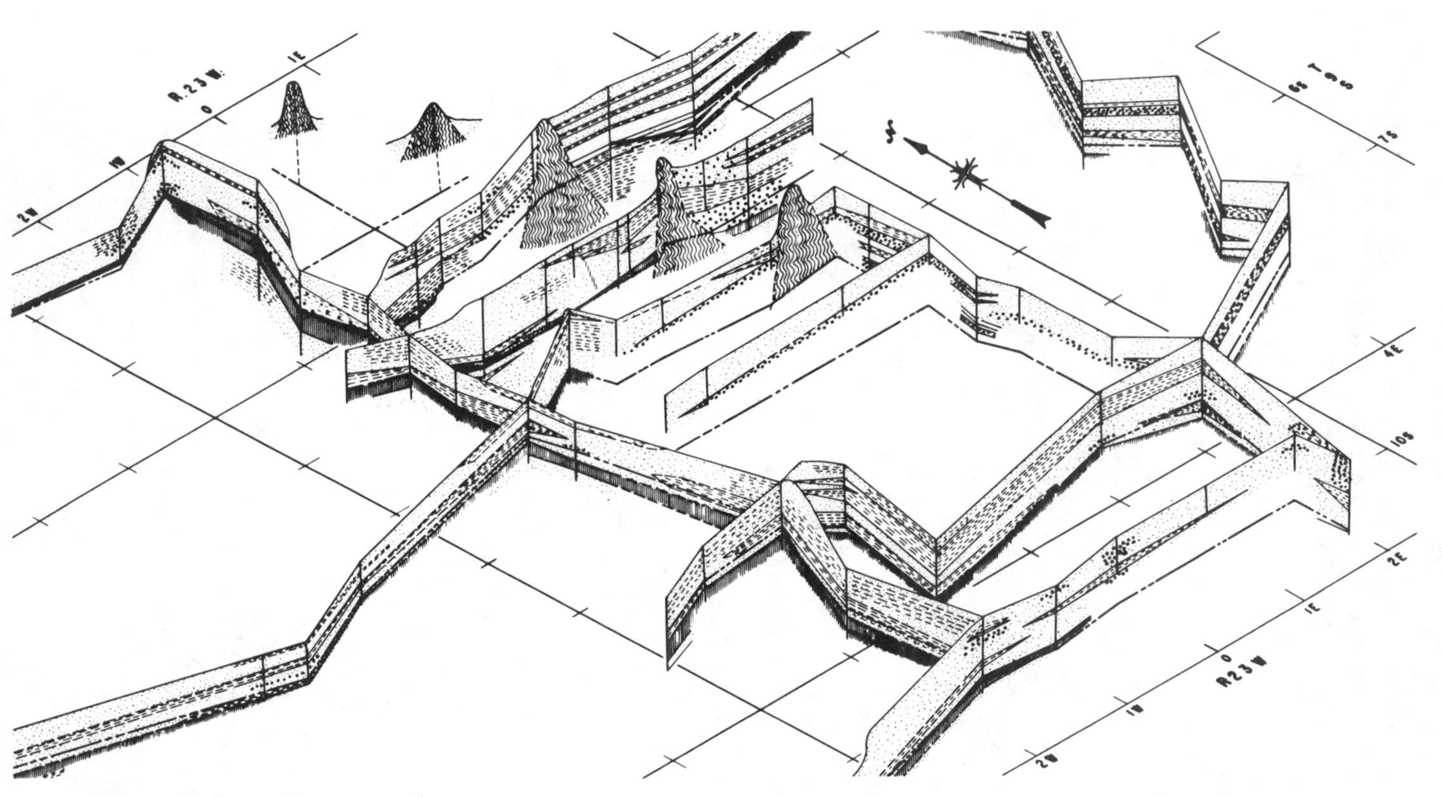

FIGURE 4-6.—Isometric fence diagram showing subsurface stratigraphy of a Yuma, Ariz., area. (Sheet 1 of 2.) 103-D-1429-1.

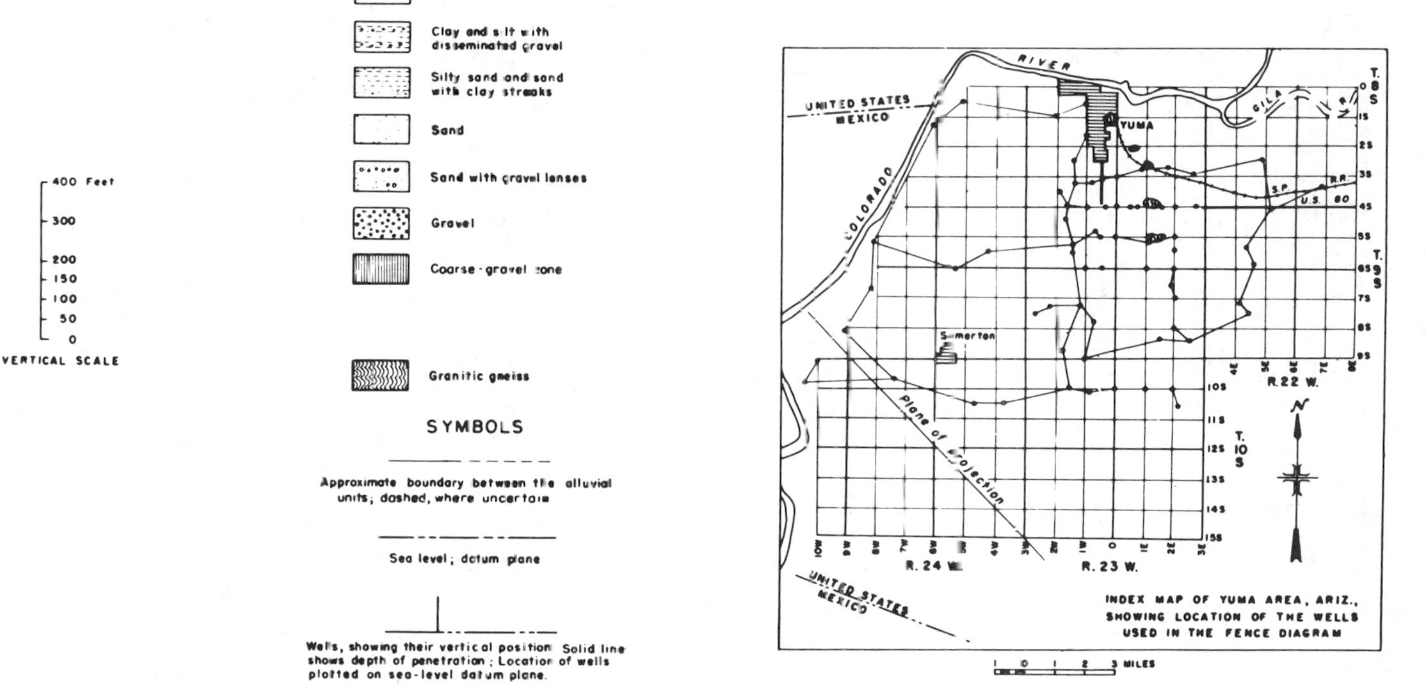

FIGURE 4-6.—Isometric fence diagram showing subsurface stratigraphy of a Yuma, Ariz., area. (Sheet 2 of 2.) 103-D-1429-2.

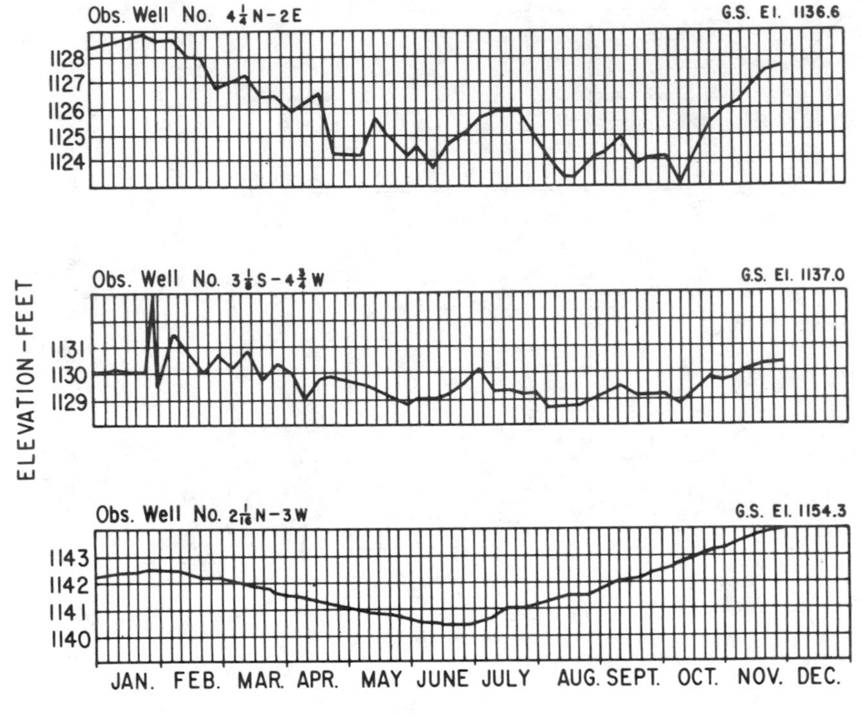

FIGURE 4-7.—Hydrographs of observation wells. 103-D-1430.

gradients must be carefully interpreted with consideration of all possible combinations of factors.

Pumping from a relatively small area of an extensive aquifer may cause little change in static water level over the unpumped area, while the water level in the pumped portion continues to lower rapidly. This is the result of the pumpage exceeding the ability of the aquifer to transmit water to the pumped area, a condition that can be recognized by contours of the water levels within the aquifer.

An overlay of two ground-water contour maps made from measurements taken at different times permits an estimate of the change in ground-water storage which has occurred in the interval between the two series of measurements if the storativity (sec. 2-2) is known. Similarly, the same volume of change multiplied by the porosity gives an estimate of the change in gross storage. The latter is useful only in the event of a rising water table which saturates a volume that previously was relatively dry. The volume of water required to saturate the material can be estimated in this manner. To obtain the volume of water released from storage when the water table lowers, it is necessary to apply the storativity factor, since the entire pore space will not be evacuated.

If the permeability and cross-sectional area (or transmissivity and width) of the aquifer are known and the gradient is available from a contour map, an estimate may be obtained of the rate of flow by applying Darcy's law.

Since aquifers act both as reservoirs and conduits, periodic estimates of the change in storage during the year may permit an estimate of the annual recharge. Similar estimates for a number of years may give an estimate of the average annual recharge.

The accuracy of the foregoing estimates depends upon the uniformity of the aquifer and the overall applicability of the aquifer characteristics as determined from pumping or other tests. While the theory is simple, the heterogeneity of most aquifers necessitates caution and requires considerable judgment in the application of resultant data.

4-16. Bibliography.—

[1] "Hydrology Handbook," ASCE Manual of Engineering Practice No. 28, American Society of Civil Engineers, New York, 1952.

[2] Butler, S. S., "Engineering Hydrology," Prentice-Hall, Englewood Cliffs, N. J., 1957.

[3] Criddle, W. D., "Methods of Computing Consumptive Use of Water," Proceedings of the ASCE, Journal of Irrigation and Drainage Division, vol. 84, No. IR1, Paper 1507, January 1958.

[4] Davis, S. N., and DeWiest, R. J. M., "Hydrogeology," John Wiley & Sons, New York, 1966.

[5] Hamon, W. R., "Estimating Potential Evapotranspiration," Proceedings of the ASCE, Journal of the Hydraulics Division, vol. 87, No. HY3, pp. 107–120, May 1961.

[6] Skeat, W. O. (editor), "Manual of British Water Engineering Practice," fourth edition, vol. II, "Engineering Practice," W. Heffer and Sons, Cambridge, 1969.

[7] Kazmann, R. G., "Modern Hydrology," Harper and Row, New York, 1965.

[8] Lindsley, R. K., Jr., Kohler, M. A., and Paulhus, J. L. H., "Applied Hydrology," McGraw-Hill, New York, 1949.

[9] ———, "Hydrology for Engineers," McGraw-Hill, New York, 1958.

[10] Lowry, R. L., Jr., and Johnson, A. F., "Consumptive Use of Water for Agriculture," Transactions of the ASCE, vol. 107, Paper 2158, pp. 1243–1266, 1942.

[11] Rouse, H. (editor), "Engineering Hydraulics," Proceedings of the Hydraulics Conference, University of Iowa, Iowa City, 1949.

[12] Wisler, C. O., and Brater, E. R., "Hydrology," second edition, John Wiley & Sons, Inc., New York, 1959.

[13] Hem, J. D., "Study and Interpretation of the Chemical Characteristics of Natural Water," U.S. Geological Survey Water-Supply Paper 1473, 1959.

«Chapter V

ANALYSIS OF DISCHARGING WELL AND OTHER TEST DATA

5–1. Background Data.—Quantitative data on hydraulic characteristics of aquifers including transmissivity, storativity, and boundary conditions are essential to the understanding and solution of aquifer problems and the proper evaluation and utilization of ground-water resources. Field tests provide the most reliable method of obtaining these data. Such tests involve the removal from or the addition of water to a well and subsequent observation of the reaction of the aquifer to the change. The normal change in the water level is the creation of a cone-like zone of depression or buildup surrounding the well. This cone is unique in shape and lateral extent and is dependent primarily on the time since the start of testing, volume or rate of water withdrawn or added, and the hydraulic characteristics of the aquifer.

Analyses of results of systematic observations of water level changes and of other test data yield values of aquifer characteristics. The extent and reliability of these analyses are dependent on features of the test including duration of test, number of observation wells, and method of analysis. Two general types of analyses are available for determination of aquifer characteristics: (1) steady state or equilibrium methods which yield values of transmissivity and related permeability, and (2) transient or nonequilibrium methods which also yield storativity and boundary conditions. The principal difference between the two methods is that the transient method permits analysis of ground-water conditions which change with time and involve storage, whereas the steady state method does not.

Test analyses also require an understanding and appreciation of the hydrologic and geologic setting of the aquifer. Conditions that should be known include: location, character, and distance of nearby bodies of surface water; depth, thickness, and stratigraphic conditions of the aquifer; and construction details of the test well and of observation wells, if used. However, despite knowledge of all apparent conditions that tend to influence a test, deviation of aquifer conditions from the ideal on which analyses are based and imperfections of the testing procedure generally rule out precise results.

5–2. Steady State Equations.—The Theim-Forchheimer or equilibrium equations [1] [1] are based on the following assumptions:

- Aquifer is homogeneous, isotropic, and of uniform thickness
- The discharging well penetrates and receives water through the entire thickness of the aquifer
- Coefficient of transmissivity or permeability (hydraulic conductivity) is constant at all times and at all locations
- Discharging has continued for a sufficient duration for the hydraulic system to reach a steady state
- Flow to the well is horizontal, radial, and laminar, and originates from a circular open water source with a fixed radius and elevation which surrounds the well
- Rate of discharge from the well is constant

The equilibrium equations (see fig. 5–1), which yield values of permeability and transmissivity, were used for years as the only equations available for the analysis of discharging well tests. The general test procedure is to simultaneously pump from a test well at a constant, known rate and to periodically measure the drawdown in two or more nearby observation wells. The test is normally continued until the plot (for each observation well) of time since start of pumping plotted on log scale, versus drawdown plotted on arithmetic scale, falls on a straight line. Another check on adequacy of duration of the test consists of a similar straight-line plot of distance versus drawdown at the same period of time in three or more observation wells. Often, because of the absence of one or more of the ideal aquifer conditions on which the analyses are based, straight-line plots do not develop within a reasonable time limit of pumping. Nevertheless, the test results may be adequate for the purposes intended.

Rearranged for determination of permeability, K, the equilibrium equations are:

For a confined (artesian) aquifer,

$$K=\frac{Q \log_e (r_2/r_1)}{2\pi M(s_1-s_2)} \tag{1}$$

and

$$T=\frac{Q \log_e (r_2/r_1)}{2\pi(s_1-s_2)}. \tag{2}$$

For an unconfined (free) aquifer,

$$K=\frac{Q \log_e (r_2/r_1)}{\pi(h_2{}^2-h_1{}^2)} \tag{3}$$

[1] Numbers in brackets refer to items in the bibliography, section 5–20.

and

$$T=\frac{QM \log_e (r_2/r_1)}{\pi(h_2^2-h_1^2)}. \tag{4}$$

where:

$\log_e$=natural log=(common log×2.303),
K=permeability or hydraulic conductivity, L/t,
Q=discharge of the test well, L^3/t,
M=saturated thickness of the aquifer, L,
$T=KM$=transmissivity of the aquifer, L^2/t,
L=unit length,
t=unit time,
$r_1, r_2, \ldots r_n$=horizontal distances from centerline of the test well to centerline of observation wells 1, 2, . . . n, L,
$h_1, h_2, \ldots h_n$=saturated thicknesses or piezometric heads of the aquifer at distances $r_1, r_2, \ldots r_n$ from the test well, L, and
$s_1, s_2, \ldots s_n$=drawdown in observation wells at distances $r_1, r_2, \ldots r_n$ from the test well, L.

Based on the previous equations, values of K and T can be computed using values of drawdown, s, at the same time in two or more observation wells located at different distances from the test well. For example, pumping test data on three observation wells from table 5–1 are shown plotted on figure 5–2. Table 5–1 presents only a partial record of the test data.

The three companion curves on figure 5–2 represent plots of time versus drawdown for the three observation wells shown in table 5–1 where:

Q=2.7 ft³/s or 161.8 ft³/min and
M=50 ft.

At t=960 minutes in observation wells 1, 2, and 3, respectively:

s_1=1.89 ft at r_1=100 ft,
s_2=1.36 ft at r_2=200 ft, and
s_3=0.80 ft at r_3=400 ft,

and

$h_1=M-s_1$=48.11 ft,
$h_2=M-s_2$=48.64 ft, and
$h_3=M-s_3$=49.20 ft.

By using equation (3), and assuming an unconfined aquifer,

$$K=\frac{161.8 \log_e \frac{200}{100}}{\pi[(48.64)^2-(48.11)^2]}=0.70 \text{ ft/min, and}$$

$$T=KM=35 \text{ ft}^2\text{/min}.$$

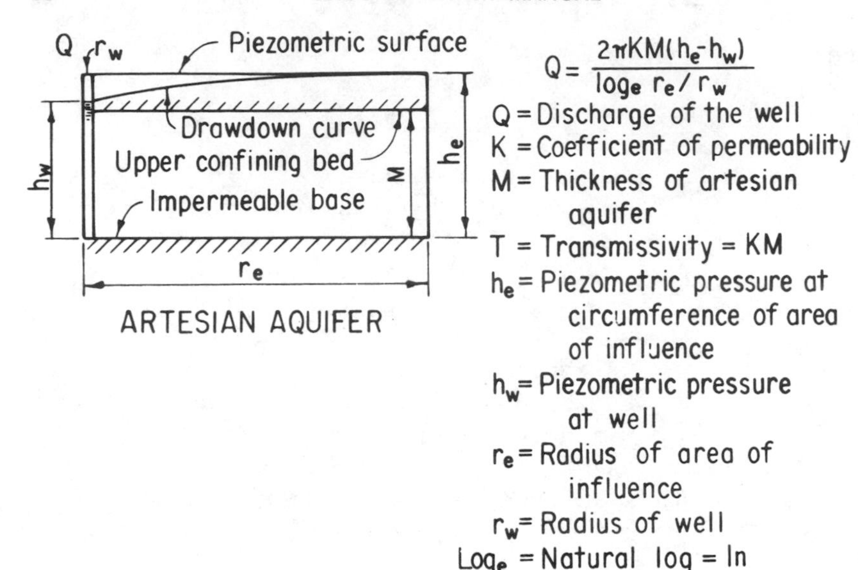

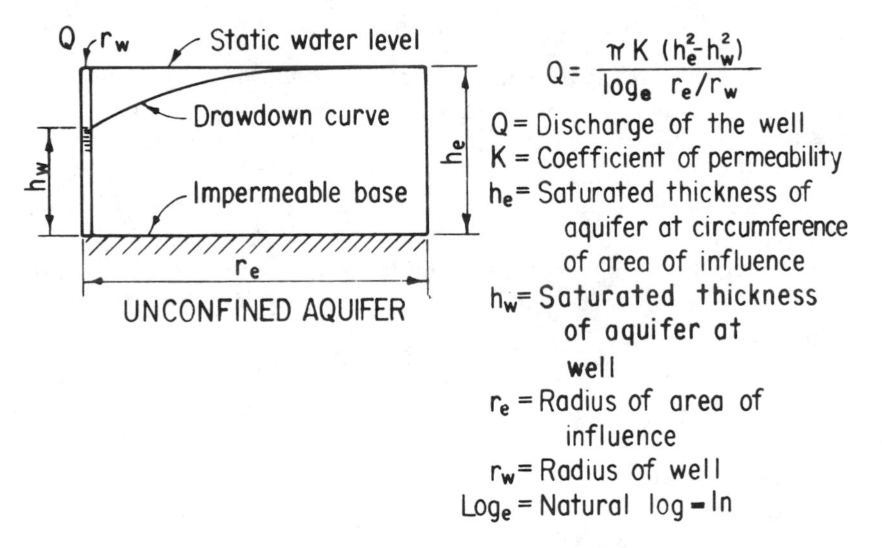

FIGURE 5-1.—Application of the equilibrium equations. 103-D-1431.

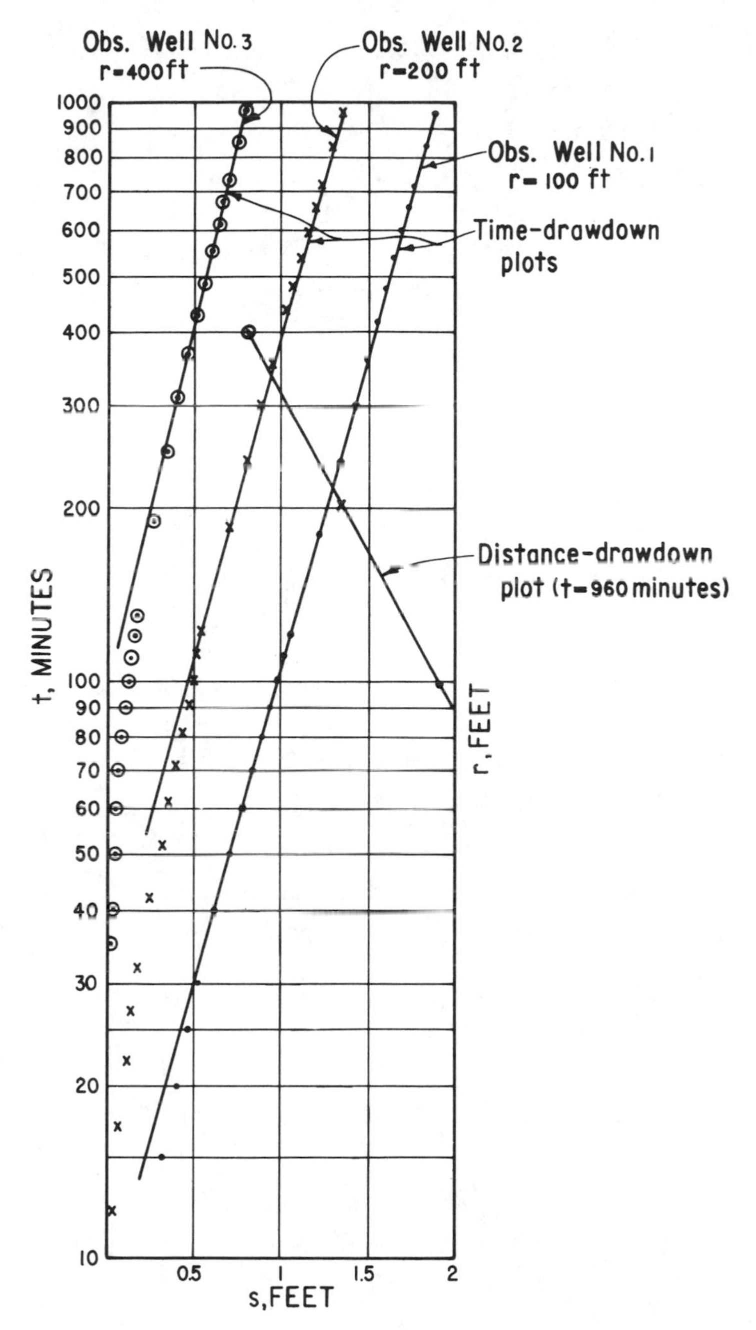

FIGURE 5-2.—Plots of water levels in three observation wells during a pumping test. 103-D-1432.

TABLE 5-1.—*Tabulated discharging well test data for solution of equilibrium and nonequilibrium equations*

Project: Sioux Flats
Feature: Wide Gap damsite
Location: NW ¼ sec. 4, T. 59 N., R. 13 E.

Discharge measured by: 1-ft Parshall flume
Drawdown measured by: Electrical sounder
Reference point: North side of casing collar

PUMP TEST NO. 1, TEST WELL

Date	Time	Depth to water, ft	Drawdown s, ft	Gage reading, ft	Discharge, ft³/s	Remarks
5-16	0840	60.99				
5-17	0830	61.01				
5-18	0845	61.00				
5-19	0820	60.98				
	0840	[1] 60.99	0.0			Pump started.
	0900	72.30	11.3	0.79	2.70	
	1000	72.60	11.6	.79	2.70	
	1100	72.80	11.8	.79	2.70	Pump off 5-21 at 0730.
	1155	72.80	11.8	.80	2.75	
	1255	72.80	11.8	.80	2.75	Avg Q=2.7 ft³/s.
	1355	73.00	12.0	.80	2.75	
	1455	73.20	12.2	.80	2.75	
	1555	73.20	12.2	.80	2.75	M=50 ft.
	1655	73.20	12.2	.80	2.75	
	1800	73.20	12.2	.80	2.75	
	1856	73.30	12.3	.80	2.75	
	1948	73.40	12.4	.80	2.75	
	2057	73.40	12.4	.80	2.75	
	2203	73.40	12.4	.80	2.75	
	2300	73.60	12.6	.80	2.75	
	2358	73.50	12.5	.80	2.75	
5-20	0104	73.60	12.6	.80	2.75	
	0204	73.60	12.6	.80	2.75	
	0259	73.60	12.6	.80	2.75	
	0400	73.80	12.8	.80	2.75	
	0501	74.00	13.0	.80	2.75	
	0602	73.90	12.9	.80	2.75	
	0702	73.90	12.9	.80	2.75	
	0759	73.80	12.8	.80	2.75	
	0855	73.80	12.8	.80	2.75	
	0955	73.80	12.8	.80	2.75	
	1055	73.80	12.8	.80	2.75	

[1] Static water level.

TABLE 5-1.—*Tabulated discharging well test data for solution of equilibrium and nonequilibrium equations*—Continued

PUMP TEST NO. 1, OBSERVATION WELL NO. 1, $r=100$ FT

Date	Time	Depth to water, ft	Drawdown s, ft	t, min	r^2/t, ft²/min	Remarks
5-16	0845	60.43				
5-17	0825	60.45				
5-18	0840	60.43				
5-19	0815	60.42				
	0841	60.42	0.00			Pump started at 0840.
	0845	60.50	.08	5	2,000	
	0850	60.64	.22	10	1,000	Pump off 5-21 at 0730.
	0855	60.74	.32	15	670	
	0900	60.83	.41	20	500	
	0905	60.90	.48	25	400	
	0910	60.96	.54	30	333	$M=50$ ft.
	0920	61.06	.64	40	250	
	0930	61.14	.72	50	200	
	0940	61.20	.78	60	170	
	0950	61.27	.85	70	140	
	1000	61.32	.90	80	125	
	1010	61.36	.94	90	110	
	1020	61.40	.98	100	100	
	1030	61.44	1.02	110	91	
	1040	61.47	1.05	120	83	
	1140	61.62	1.20	180	56	
	1240	61.73	1.31	240	42	
	1340	61.83	1.41	300	33	
	1440	61.90	1.48	360	28	
	1540	61.96	1.54	420	24	
	1640	62.01	1.59	480	21	
	1740	62.05	1.63	540	19	
	1840	62.09	1.67	600	17	
	1940	62.14	1.72	660	15	
	2040	62.17	1.75	720	14	
	2240	62.26	1.84	840	12	
5-20	0040	62.31	1.89	960	10	
	1845	62.59	2.17	2,045	4.9	

TABLE 5-1.—*Tabulated discharging well test data for solution of equilibrium and nonequilibrium equations*—Continued

Project: Sioux Flats
Feature: Wide Gap damsite
Location: NW ¼ sec. 4, T. 29 N., R. 13 E.

Drawdown measured by: Electric sounder
Reference point: East side of casing collar

PUMP TEST NO. 1, OBSERVATION WELL NO. 2, r=200 FT

Date	Time	Depth to water, ft	Draw-down s, ft	t, min	r^2/t, ft²/min	Remarks
5-16	0835	58.41				
5-17	0820	58.39				
5-18	0820	58.40				
5-19	0810	58.41				
	0838	[1] 58.41				Pump started at 0840.
	0847	58.41	0.00	7	5,720	
	0852	58.44	.03	12	3,332	Pump off 5-21 at 0730.
	0857	58.48	.07	17	2,352	
	0902	58.52	.11	22	1,820	
	0907	58.56	.15	27	1,480	M=50 ft.
	0912	58.59	.18	32	1,252	
	0922	58.66	.25	42	952	
	0932	58.72	.31	52	768	
	0942	58.77	.36	62	644	
	0952	58.81	.40	72	556	
	1002	58.85	.44	82	488	
	1012	58.89	.48	92	436	
	1022	58.92	.51	102	392	
	1032	58.95	.54	112	357	
	1042	58.98	.57	122	328	
	1142	59.12	.71	182	220	
	1242	59.22	.81	242	165	
	1342	59.30	.89	302	132	
	1442	59.38	.97	362	110	
	1542	59.44	1.03	422	94	
	1642	59.49	1.08	482	83	
	1742	59.53	1.12	542	72	
	1842	59.57	1.16	602	66	
	1942	59.61	1.20	662	60	
	2042	59.64	1.23	722	55	
	2242	59.73	1.32	842	44	
5-20	0042	59.77	1.36	962	40	
	1845	60.01	1.65	2,045	20	

[1] Static water level

TABLE 5-1.—*Tabulated discharging well test data for solution of equilibrium and nonequilibrium equations*—Continued

Project: Sioux Flats
Feature: Wide Gap damsite
Location: NW ¼ sec. 4, T. 59 N., R. 13 E.

Drawdown measured by: "Popper"
Reference point: East side of casing collar

PUMP TEST NO. 1, OBSERVATION WELL NO. 3, r=400 FT

Date	Time	Depth to water, ft	Drawdown s, ft	t, min	r^2/t, ft²/min	Remarks
5-16	0850	58. 47				
5-17	0830	58. 48				
5-18	0835	58. 48				
5-19	0820	58. 47				
	0838	[1] 58. 47				Pump on at 0840.
	0855	58. 47	0. 00	15	10, 720	Pump off 5-21 at 0730.
	0900	58. 47	. 00	20	8, 000	
	0905	58. 47	. 00	25	6, 400	
	0910	58. 47	. 00	30	5, 280	M=50 ft.
	0915	58. 48	. 01	35	4, 640	
	0920	58. 49	. 02	40	4, 000	
	0930	58. 50	. 03	50	3, 200	
	0940	58. 52	. 05	60	2, 720	
	0950	58. 54	. 07	70	2, 240	
	1000	58. 55	. 08	80	2, 080	
	1010	58. 57	. 10	90	1, 760	
	1020	58. 59	. 12	100	1, 600	
	1030	58. 61	. 14	110	1, 456	
	1040	58. 62	. 15	120	1, 328	
	1050	58. 64	. 17	130	1, 232	
	1150	58. 73	. 26	190	848	
	1250	58. 81	. 34	250	640	
	1350	58. 87	. 40	310	512	
	1450	58. 93	. 46	370	432	
	1550	58. 98	. 51	430	368	
	1650	59. 02	. 55	490	320	
	1750	59. 06	. 59	550	288	
	1850	59. 10	. 63	610	256	
	1950	59. 12	. 65	670	240	
	2050	59. 15	. 68	730	224	
	2250	59. 22	. 75	850	192	
5-20	0050	59. 27	. 80	970	160	
	1845	59. 54	1. 07	2, 045	78	

[1] Static water level.

Two precautions are recommended when using the equilibrium equations: (1) in a free aquifer, drawdowns exceeding 10 percent of the aquifer thickness should not be used in the calculations, and (2) measurements should not be used at points where the slope of the cone of depression exceeds 15°.

Basic assumptions of the equilibrium equations provide that the test well is fully penetrating, has 100 percent open hole (or screen), and that flow to the well is horizontal. In many tests, these conditions are not met; however, as discussed in section 3–2, the distribution of flow to a well approaches that of the assumed horizontal condition at a distance from the well equal to approximately 1.5 times the thickness of the aquifer. Accordingly, to minimize effects of convergent flow on test results, the nearest observation well should be located at least 1.5 times the aquifer thickness from the test well unless a large aquifer thickness provides for distances that are unreasonably large. In this event, pairs of piezometers located at reasonable distances from the well may be substituted. A piezometer should be set in both the lower and upper 15 percent of the aquifer and the drawdowns in these piezometers should be averaged for computational purposes. Where this is not feasible, observation wells with screened or open hole zones duplicating those of the test well may be substituted. These procedures are empirical, but they serve to minimize errors caused by vertical flow convergence resulting from partially penetrating wells. Mathematical methods of correcting for flow convergence have been developed, but their usefulness may be questionable when applied to field conditions [13,15,26,27,41,45,55,65].

5–3. Transient Equations.—Transient equations permit analysis of aquifer conditions that vary with time and involve storage. The assumptions on which the equations are based include:

- Aquifer is confined, horizontal, homogeneous, isotropic, of uniform thickness, and of infinite areal extent
- Pumping well is of infinitesimal diameter and fully penetrates the aquifer
- Flow to the well is radial, horizontal, and laminar
- All water comes from storage in the aquifer within the area of influence and is released from storage instantaneously with decline in pressure
- Transmissivity and storativity of the aquifer are constant in time and space

The transient equations are directly applicable to confined conditions and are suitable for use, with limitations, in unconfined aquifers. These limitations are related to the percentage of drawdown in observation wells as related to the total aquifer thickness. If the drawdown exceeds 25 percent of the aquifer thickness, the transient equations should not be used. However, if the percentage is less than 10, little error is introduced. For values between 10 and 25 percent, the following correction factor derived by C. E. Jacob [1] should be applied:

$$s' = s - \frac{s^2}{2M} \tag{5}$$

where:

s = measured drawdown in an observation well,
M = saturated thickness of the aquifer prior to pumping, and
s' = corrected drawdown.

Figure 5–3 uses data from table 5–2 to illustrate this method of correction and its application to semilog plots of drawdown versus time and drawdown versus distance. Table 5–2 presents only a partial record of the test data.

Deviations of most aquifers from assumptions of uniform thickness, horizontality, and homogeneity, are usually minor or are averaged; therefore, such deviations do not seriously affect the accuracy of aquifer test results. Also, most aquifers are anisotropic, but if the test well and observation wells are fully penetrating, the effect on drawdowns is minor. However, if the aquifer is strongly anisotropic and the test well is not fully penetrating, drawdowns in observation wells may be misleading because only that portion of the aquifer actually penetrated by the well contributes flow to the well. In such cases, use of total screen or open hole length as a substitute for total aquifer thickness will give reasonably reliable results. But the storativity computed from such data is usually too low.

The assumption that aquifers are infinitely extensive is never realized because all have boundaries. Nevertheless, most aquifers may be considered extensive when related to the time period during which pumping tests are normally run. However, if boundaries are close to the test well, straightforward application of transient equations may yield unreliable results. In such cases, methods as described in section 5–11 should be used.

The fact that a well is not fully penetrating may be partially compensated for by design and location of observation wells as described in section 3–2. The diameter of the well, unless it is large and the discharge small, is seldom an adverse factor in interpretation of aquifer tests.

The assumption that all water pumped from a well comes from storage within the aquifer is seldom realized because few, if any, aquifers are totally isolated. In addition to storage within the aquifer,

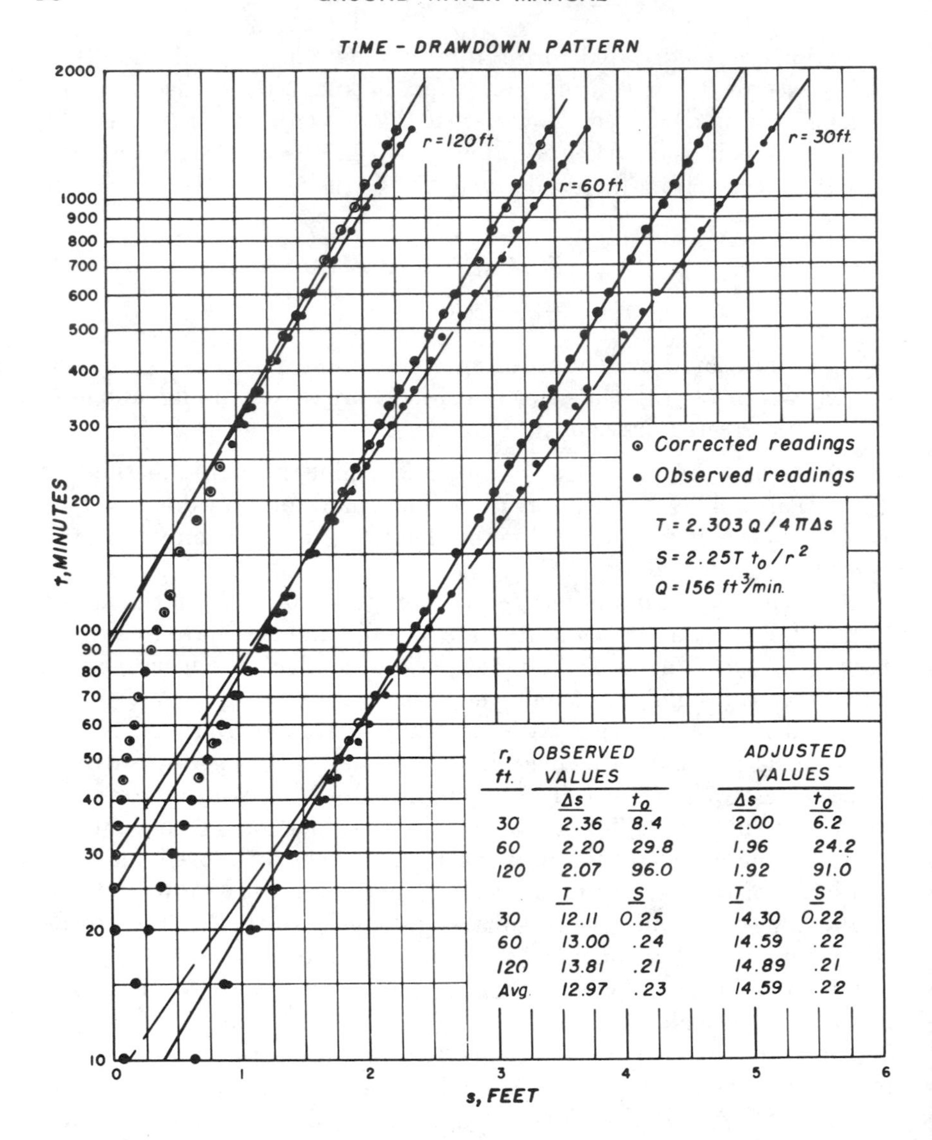

r, ft.	OBSERVED VALUES Δs	OBSERVED VALUES t_0	ADJUSTED VALUES Δs	ADJUSTED VALUES t_0
30	2.36	8.4	2.00	6.2
60	2.20	29.8	1.96	24.2
120	2.07	96.0	1.92	91.0
	T	S	T	S
30	12.11	0.25	14.30	0.22
60	13.00	.24	14.59	.22
120	13.81	.21	14.89	.21
Avg.	12.97	.23	14.59	.22

FIGURE 5-3.—Effects of correcting drawdown readings in a free aquifer. (Sheet 1 of 2.) 103-D-789.

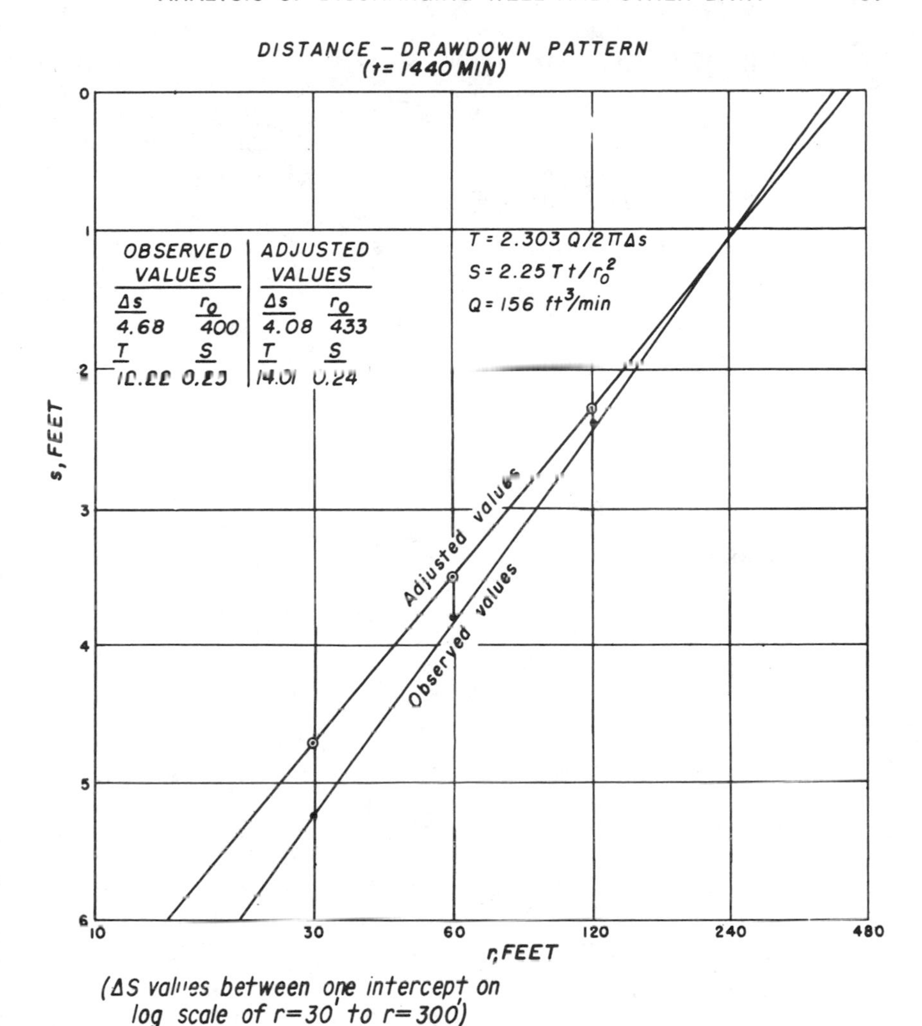

FIGURE 5-3.—Effects of correcting drawdown readings in a free aquifer. (Sheet 2 of 2.) 103–D–789.

TABLE 5-2.—*Tabulated discharging well test data for obtaining drawdown corrections*

Project: Mesa
Feature: Drainage investigation
Location: 30 feet north of pumped well

Drawdown measured by: "Popper"
Reference point: North side of casing collar

PUMP TEST NO. 1, OBSERVATION WELL NO. 1, $r=30$ FT

Date	Time	Elapsed time, min	Depth to water, ft	Drawdown s, ft	s^2, ft ^{2}	$s^2/2M$, ft	s', ft	Remarks
12-6	0800	0	15. 45	0	0	0	0	Pump on.
	0802	2	15. 49	0. 04	0. 0016	0	0. 04	
	0804	4	15. 64	. 19	. 0361	0	. 19	
	0806	6	15. 80	. 35	. 1225	0	. 35	
	0808	8	15. 96	. 51	. 2601	0. 01	. 50	
	0810	10	16. 08	. 63	. 3969	0. 01	. 62	Pump off
	0815	15	16. 35	. 90	. 8100	. 02	. 88	12-7 at
	0820	20	16. 56	1. 11	1. 232	. 02	1. 09	0800.
	0825	25	16. 73	1. 28	1. 638	. 03	1. 25	
	0830	30	16. 88	1. 43	2. 045	. 04	1. 39	
	0835	35	17. 01	1. 56	2. 434	. 05	1. 51	$Q=156$ ft^3/
	0840	40	17. 12	1. 67	2. 789	. 05	1. 62	min.
	0845	45	17. 22	1. 77	2. 133	. 06	1. 71	Avg $M=26$
	0850	50	17. 31	1. 86	3. 460	. 07	1. 79	ft.
	0855	55	17. 39	1. 94	3. 764	. 07	1. 87	$2M=52$ ft.
	0900	60	17. 47	2. 02	4. 080	. 08	1. 94	$s'=s-s^2/2M$.
	0910	70	17. 61	2. 16	4. 666	. 09	2. 07	
	0920	80	17. 73	2. 28	5. 198	. 10	2. 18	
	0930	90	17. 85	2. 40	5. 760	. 11	2. 29	
	0940	100	17. 95	2. 50	6. 250	. 12	2. 38	
	0950	110	18. 04	2. 59	6. 708	. 13	2. 46	
	1000	120	18. 12	2. 67	7. 129	. 14	2. 53	
	1030	150	18. 33	2. 88	8. 294	. 16	2. 72	
	1100	180	18. 51	3. 06	9. 364	. 18	2. 88	
	1130	210	18. 66	3. 21	10. 304	. 20	3. 01	
	1200	240	18. 80	3. 35	11. 223	. 22	3. 13	

TABLE 5-2.—*Tabulated discharging well test data for obtaining drawdown corrections*—Continued

Project: Mesa
Feature: Drainage investigation
Location: 60 feet north of pumped well

Drawdown measured by: "Popper"
Reference point: North side of casing collar

PUMP TEST NO. 1, OBSERVATION WELL NO. 2, $r=60$ FT

Date	Time	Elapsed time, min	Depth to water, ft	Drawdown s, ft	s^2, ft^2	$s^2/2M$, ft	s', ft	Remarks
12-6	0800	0	18.10	0	0	0	0	Pump on.
	0802	2	0	0	0	0	0	
	0804	4	0	0	0	0	0	
	0806	6	18.12	0.02	0	0	0.02	
	0808	8	18.14	.04	0	0	.04	
	0810	10	18.18	.08	0.01	0	.08	Pump off 12-7 at 0800.
	0815	15	18.27	.17	.03	0	.17	
	0820	20	18.27	.27	.07	0	.27	
	0825	25	18.47	.38	.14	0	.38	
	0830	30	18.57	.47	.22	0	.47	
	0835	35	18.65	.55	.30	0.01	.54	$Q=156$ ft^3/min.
	0840	40	18.73	.63	.40	.01	.62	
	0845	45	18.78	.68	.46	.01	.67	Avg $M=26$ ft.
	0850	50	18.86	.76	.58	.01	.75	
	0855	55	18.94	.84	.71	.01	.83	$2M=52$ ft.
	0900	60	19.00	.90	.81	.02	.88	$s'=s-s^2/2M$.
	0910	70	19.10	1.00	1.00	.02	.98	
	0920	80	19.22	1.11	1.23	.02	1.09	
	0930	90	19.30	1.20	1.44	.03	1.17	
	0940	100	19.38	1.28	1.64	.03	1.25	
	0950	110	19.46	1.36	1.85	.04	1.32	
	1000	120	19.52	1.42	2.02	.04	1.38	
	1030	150	19.81	1.61	2.59	.05	1.56	
	1100	180	19.87	1.77	3.13	.06	1.71	
	1130	210	20.01	1.91	3.65	.07	1.84	
	1200	240	20.12	2.02	4.08	.08	1.94	

TABLE 5–2.—*Tabulated discharging well test data for obtaining drawdown corrections*—Continued

Project: Mesa
Feature: Drainage investigation
Location: 120 feet north of pumped well

Drawdown measured by: "Popper"
Reference point: North side of casing collar

PUMP TEST NO. 1, OBSERVATION WELL NO. 3, r=120 FT

Date	Time	Elapsed time, min	Depth to water, ft	Drawdown s, ft	s^2, ft^2	$s^2/2M$, ft	s', ft	Remarks
12–6	0800	0	17.95	0	0	0	0	Pump on.
	0802	2	17.95	0	0	0	0	
	0804	4	17.95	0	0	0	0	
	0806	6	17.95	0	0	0	0	
	0808	8	17.95	0	0	0	0	
	0810	10	17.95	0	0	0	0	Pump off 12–7 at 0800.
	0815	15	17.95	0	0	0	0	
	0820	20	17.96	0.01	0	0	0.01	
	0825	25	17.97	.02	0	0	.02	
	0830	30	17.99	.04	0	0	.04	
	0835	35	18.00	.05	0	0	.05	Q=156 ft^3/min.
	0840	40	18.02	.07	0	0	.07	
	0845	45	18.05	.10	0.01	0	.10	Avg M=26 ft.
	0850	50	18.07	.12	0.01	0	.12	
	0855	55	18.10	.15	.02	0	.15	$2M$=52 ft.
	0900	60	18.13	.18	.03	0	.18	$s'=s-s^2/2M$.
	0910	70	18.16	.21	.04	0	.21	
	0920	80	18.22	.27	.07	0	.27	
	0930	90	18.27	.32	.10	0	.32	
	0940	100	18.32	.37	.14	0	.37	
	0950	110	18.38	.43	.18	0	.43	
	1000	120	18.42	.47	.22	0	.47	
	1030	150	18.55	.60	.36	0.01	.59	
	1100	180	18.65	.70	.49	.01	.69	
	1130	210	18.76	.81	.66	.01	.80	
	1200	240	18.85	.90	.81	.02	.88	

water may originate as leakage from overlying or underlying aquifers or from recharge, precipitation, irrigation, or hydraulically connected bodies of surface water. Methods of treating leaky aquifers and recharge as boundary conditions are given in sections 5–8 and 5–11, respectively.

Also, the assumption that water is discharged from storage instantaneously with a decline in head is seldom realized, especially in a free aquifer. Normally there is a lag caused by slow drainage and, as a consequence, the apparent storativity increases with time and approaches a constant value.

For most tests that are run long enough for drawdowns to reach apparent stability, the storage value determined at that time is sufficiently accurate and reliable for most applications. If drainage is unusually slow, Boulton's Analysis (sec. 5–10) [5,6] should be applied. Although transmissivity and storativity are not constant everywhere within most aquifers, pumping tests tend to average out these values.

The most widely used transient equation [38,72] is that by Theis:

$$s=\frac{Q}{4\pi T}\int_{\frac{r^2S}{4Tt}}^{\infty}\frac{e^{-u}du}{u} \tag{6}$$

where:

s=drawdown in an observation well located at a given radius from the test well at a specific time, L,

Q=uniform discharge from the well, $\frac{L^3}{t}$,

T=transmissivity of the aquifer, $\frac{L^2}{t}$,

r=radius of the observation well, L,

S=storativity of the aquifer (nondimensional),

t=time since start of pumping, t, and

$$u=\frac{r^2S}{4Tt} \text{ (nondimensional).} \tag{7}$$

5–4. Type Curve Solutions of the Transient Equation.—Since T appears twice in equation (6), a mathematical solution for each individual problem becomes tedious. Theis provided a graphical method of solution that gives satisfactory results when applied to tests in aquifers which conform approximately to ideal conditions.

The integral expression in the Theis equation is given by the series:

$$\int_u^{\infty}\frac{e^{-u}}{u}du=W(u)=-0.5772-\log_e u+u-\frac{u^2}{2\cdot 2!}+\frac{u^3}{3\cdot 3!}-\frac{u^4}{4\cdot 4!}\cdot\cdot\cdot \tag{8}$$

where $W(u)$ is the well function or exponential integral of u.

Equation (6) may be rewritten:

$$s = \frac{QW(u)}{4\pi T} \tag{9}$$

Values of $W(u)$ versus u for u values of 10^{-15} to 9.9 are given in table 5–3 where the value of u is expressed as some number (N) between 1 and 9.9, multipled by 10 with a range of appropriate exponents. For example, if u has a value of 0.0027, this is 2.7×10^{-3}. The value of $W(u)$ is found by reading across the table opposite $N=2.7$ and under the column headed $N \times 10^{-3}$. The value of $W(u)$ is 5.3400. References [43] and [56] have more complete tabulations of the $W(u)$ function.

Equations (9) and (7), respectively:

$$s = \frac{Q}{4\pi T} W(u)$$

and

$$u = \frac{r^2 S}{4Tt}$$

may be expressed in common logarithmic form:

$$\log s = \left[\log \frac{Q}{4\pi T}\right] + \log W(u) \tag{10}$$

$$\log \frac{r^2}{t} = \left[\log \frac{4T}{S}\right] + \log u \tag{11}$$

The bracketed values in equations (10) and (11) are constants for any given test. Logarithmic plots of test data for s versus $\frac{r^2}{t}$ will be similar to a logarithmic plot of $W(u)$ versus u, which is referred to as a type curve. If a test data curve is superimposed on the type curve while keeping the coordinate axes of the sheets parallel and the test data curve shifted to the point of best fit on the type curve, the displacements of common points of the test data curve and the type curve will be equal to the constants in brackets as shown in figure 5–4. The displacement values are used in equations (7) and (9), respectively, to solve for S and T:

$$S = \frac{4Tut}{r^2}$$

$$T = \frac{Q}{4\pi s} W(u)$$

A type curve of u versus $W(u)$ has been prepared on 3- by 5-cycle logarithmic paper (fig. 5–5). **A full-scale drawing of figure 5–5 has been placed in a pocket at the back of this manual.** Field data plotted on 3- by 5-cycle logarithmic paper can be used as an overlay on this drawing to determine the best fit on the type curve. Table 5–1 (sec. 5–2) shows a portion of the recorded data from a pumping test and figures 5–6, 5–7, and 5–8 show plots of the data.

There are two procedures for solving the nonequilibrium equation using the type curve of figure 5–5 [49,61,66]:

(a) *Time-Drawdown Solution.*—The time-drawdown data curve can be prepared by plotting s in an observation well against the reciprocal of t or against $\frac{r^2}{t}$ on log-log paper and fitting this curve to the type curve as shown on figure 5–6. An easier procedure which avoids the need for calculating reciprocals is to plot s against t (fig. 5–7) in a similar manner and fit this data plot against a type curve which has been reversed end-for-end. This end-for-end switch of the type curve is the equivalent of plotting the reciprocal of $\frac{1}{t}$ or t on the data curve. The data curve plot of each observation well is matched to the type curve while keeping the axes of the two sheets parallel. The curves may match only in a given segment because results from the test may depart from the ideal conditions on which the type curve is based. Such factors as free aquifer conditions, boundaries, and leakage result in departure from the type curve. Any common index point, usually where u and $W(u)$ are equal to 1 to simplify computations, on the overlapping sheets is marked and the values of u, $W(u)$, t, and s are recorded. Transmissivity and storativity are then computed using these values in the following equations:

$$T=\frac{Q}{4\pi s}W(u) \tag{12}$$

$$S=\frac{4Tu}{\frac{r^2}{t}} \text{ or } \frac{4Ttu}{r^2} \tag{13}$$

(b) *Distance-Drawdown Solution.*—The data curve is prepared by plotting the drawdowns measured at the same time in three or more observation wells against the distances squared divided by time, s versus $\frac{r^2}{t}$. The three points enclosed in squares on figure 5–6 illustrate this procedure. The three points should be fitted to the type curve, the

TABLE 5-3.—*Value of W (u) for values of u between 10^{-15} and 9.9.* 103-D-1455.

N \ u	$N\times10^{-15}$	$N\times10^{-14}$	$N\times10^{-13}$	$N\times10^{-12}$	$N\times10^{-11}$	$N\times10^{-10}$	$N\times10^{-9}$	$N\times10^{-8}$	$N\times10^{-7}$	$N\times10^{-6}$	$N\times10^{-5}$	$N\times10^{-4}$	$N\times10^{-3}$	$N\times10^{-2}$	$N\times10^{-1}$	N
1.0	33.9616	31.6590	29.3564	27.0538	24.7512	22.4486	20.1460	17.8435	15.5409	13.2383	10.9357	8.6332	6.3315	4.0379	1.8229	0.2194
1.1	33.8662	31.5637	29.2611	26.9585	24.6559	22.3533	20.0507	17.7482	15.4456	13.1430	10.8404	8.5379	6.2363	3.9436	1.7371	.1860
1.2	33.7792	31.4767	29.1741	26.8715	24.5689	22.2663	19.9637	17.6611	15.3586	13.0560	10.7534	8.4509	6.1494	3.8576	1.6595	.1584
1.3	33.6992	31.3966	29.0940	26.7914	24.4889	22.1863	19.8837	17.5811	15.2785	12.9759	10.6734	8.3709	6.0695	3.7785	1.5889	.1355
1.4	33.6251	31.3225	29.0199	26.7173	24.4147	22.1122	19.8096	17.5070	15.2044	12.9018	10.5993	8.2968	5.9955	3.7054	1.5241	.1162
1.5	33.5561	31.2535	28.9509	26.6483	24.3458	22.0432	19.7406	17.4380	15.1354	12.8328	10.5303	8.2278	5.9266	3.6374	1.4645	.1000
1.6	33.4916	31.1890	28.8864	26.5838	24.2812	21.9786	19.6760	17.3735	15.0709	12.7683	10.4657	8.1634	5.8621	3.5739	1.4092	.08631
1.7	33.4309	31.1283	28.8258	26.5232	24.2206	21.9180	19.6154	17.3128	15.0103	12.7077	10.4051	8.1027	5.8016	3.5143	1.3578	.07465
1.8	33.3738	31.0712	28.7686	26.4660	24.1634	21.8608	19.5583	17.2557	14.9531	12.6505	10.3479	8.0455	5.7446	3.4581	1.3089	.06471
1.9	33.3197	31.0171	28.7145	26.4119	24.1094	21.8068	19.5042	17.2016	14.8990	12.5964	10.2939	7.9915	5.6906	3.4050	1.2649	.05620
2.0	33.2684	30.9658	28.6632	26.3607	24.0581	21.7555	19.4529	17.1503	14.8477	12.5451	10.2426	7.9402	5.6394	3.3547	1.2227	.04890
2.1	33.2196	30.9170	28.6145	26.3119	24.0093	21.7067	19.4041	17.1015	14.7989	12.4964	10.1938	7.8914	5.5907	3.3069	1.1829	.04261
2.2	33.1731	30.8705	28.5679	26.2653	23.9628	21.6602	19.3576	17.0550	14.7524	12.4498	10.1473	7.8449	5.5443	3.2614	1.1454	.03719
2.3	33.1286	30.8261	28.5235	26.2209	23.9183	21.6157	19.3131	17.0106	14.7080	12.4054	10.1028	7.8004	5.4999	3.2179	1.1099	.03250
2.4	33.0861	30.7835	28.4809	26.1783	23.8758	21.5732	19.2706	16.9680	14.6654	12.3628	10.0603	7.7579	5.4575	3.1763	1.0762	.02844
2.5	33.0453	30.7427	28.4401	26.1375	23.8349	21.5323	19.2298	16.9272	14.6246	12.3220	10.0194	7.7172	5.4167	3.1365	1.0443	.02491
2.6	33.0060	30.7035	28.4009	26.0983	23.7957	21.4931	19.1905	16.8880	14.5854	12.2828	9.9802	7.6779	5.3776	3.0983	1.0139	.02185
2.7	32.9683	30.6657	28.3631	26.0606	23.7580	21.4554	19.1528	16.8502	14.5476	12.2450	9.9425	7.6401	5.3400	3.0615	.9849	.01918
2.8	32.9319	30.6294	28.3268	26.0242	23.7216	21.4190	19.1164	16.8138	14.5113	12.2087	9.9061	7.6038	5.3037	3.0261	.9573	.01686
2.9	32.8968	30.5943	28.2917	25.9891	23.6865	21.3839	19.0813	16.7788	14.4762	12.1736	9.8710	7.5687	5.2687	2.9920	.9309	.01482
3.0	32.8629	30.5604	28.2578	25.9552	23.6526	21.3500	19.0474	16.7449	14.4423	12.1397	9.8371	7.5348	5.2349	2.9591	.9057	.01305
3.1	32.8302	30.5276	28.2250	25.9224	23.6198	21.3172	19.0146	16.7121	14.4095	12.1069	9.8043	7.5020	5.2022	2.9273	.8815	.01149
3.2	32.7984	30.4958	28.1932	25.8907	23.5880	21.2855	18.9829	16.6803	14.3777	12.0751	9.7726	7.4703	5.1706	2.8965	.8583	.01013
3.3	32.7676	30.4651	28.1625	25.8599	23.5573	21.2547	18.9521	16.6495	14.3470	12.0444	9.7418	7.4395	5.1399	2.8668	.8361	.008939
3.4	32.7378	30.4352	28.1326	25.8300	23.5274	21.2249	18.9223	16.6197	14.3171	12.0145	9.7120	7.4097	5.1102	2.8379	.8147	.007891
3.5	32.7088	30.4062	28.1036	25.8010	23.4985	21.1959	18.8933	16.5907	14.2881	11.9855	9.6830	7.3807	5.0813	2.8099	.7942	.006970
3.6	32.6806	30.3780	28.0755	25.7729	23.4703	21.1677	18.8651	16.5625	14.2599	11.9574	9.6548	7.3526	5.0532	2.7827	.7745	.006160
3.7	32.6532	30.3506	28.0481	25.7455	23.4429	21.1403	18.8377	16.5351	14.2325	11.9300	9.6274	7.3252	5.0259	2.7563	.7554	.005448
3.8	32.6266	30.3240	28.0214	25.7188	23.4162	21.1136	18.8110	16.5085	14.2059	11.9033	9.6007	7.2985	4.9993	2.7306	.7371	.004820
3.9	32.6006	30.2980	27.9954	25.6928	23.3902	21.0877	18.7851	16.4825	14.1799	11.8773	9.5748	7.2725	4.9735	2.7056	.7194	.004267
4.0	32.5753	30.2727	27.9701	25.6675	23.3649	21.0623	18.7598	16.4572	14.1546	11.8520	9.5495	7.2472	4.9482	2.6813	.7024	.003779
4.1	32.5506	30.2480	27.9454	25.6428	23.3402	21.0376	18.7351	16.4325	14.1299	11.8273	9.5248	7.2225	4.9236	2.6576	.6859	.003349
4.2	32.5265	30.2239	27.9213	25.6187	23.3161	21.0136	18.7110	16.4084	14.1058	11.8032	9.5007	7.1985	4.8997	2.6344	.6700	.002969
4.3	32.5029	30.2004	27.8978	25.5952	23.2926	20.9900	18.6874	16.3884	14.0823	11.7797	9.4771	7.1749	4.8762	2.6119	.6546	.002633
4.4	32.4800	30.1774	27.8748	25.5722	23.2696	20.9670	18.6644	16.3619	14.0593	11.7567	9.4541	7.1520	4.8533	2.5899	.6397	.002336
4.5	32.4575	30.1549	27.8523	25.5497	23.2471	20.9446	18.6420	16.3394	14.0368	11.7342	9.4317	7.1295	4.8310	2.5684	.6253	.002073
4.6	32.4355	30.1329	27.8303	25.5277	23.2252	20.9226	18.6200	16.3174	14.0148	11.7122	9.4097	7.1075	4.8091	2.5474	.6114	.001841
4.7	32.4140	30.1114	27.8088	25.5062	23.2037	20.9011	18.5985	16.2959	13.9933	11.6907	9.3882	7.0860	4.7877	2.5268	.5979	.001635
4.8	32.3929	30.0904	27.7878	25.4852	23.1826	20.8800	18.5774	16.2748	13.9723	11.6697	9.3671	7.0650	4.7667	2.5068	.5848	.001453
4.9	32.3723	30.0697	27.7672	25.4646	23.1620	20.8594	18.5568	16.2542	13.9516	11.6491	9.3465	7.0444	4.7462	2.4871	.5721	.001291

5.0	32.3521	30.0495	27.7470	25.4444	23.1418	20.8392	18.5366	16.2340	13.9314	11.6289	9.32[illegible]3	7.0242	4.7261	2.4679	.5598	.001148
5.1	32.3323	30.0297	27.7271	25.4246	23.1220	20.8194	18.5168	16.2142	13.9116	11.6091	9.30[illegible]5	7.0044	4.7064	2.4491	.5478	.001021
5.2	32.3129	30.0103	27.7077	25.4051	23.1026	20.8000	18.4974	16.1948	13.8922	11.5896	9.28[illegible]1	6.9850	4.6871	2.4306	.5362	.0009086
5.3	32.2939	29.9913	27.6887	25.3861	23.0835	20.7809	18.4783	16.1758	13.8732	11.5706	9.26[illegible]1	6.9659	4.6681	2.4126	.5250	.0008086
5.4	32.2752	29.9726	27.6700	25.3674	23.0648	20.7622	18.4596	16.1571	13.8545	11.5519	9.24[illegible]4	6.9473	4.6495	2.3948	.5140	.0007198
5.5	32.2568	29.9542	27.6516	25.3491	23.0465	20.7439	18.4413	16.1387	13.8361	11.5336	9.23[illegible]0	6.9289	4.6313	2.3775	.5034	.0006409
5.6	32.2388	29.9362	27.6336	25.3310	23.0285	20.7259	18.4233	16.1207	13.8181	11.5155	9.21[illegible]0	6.9109	4.6134	2.3604	.4930	.0005708
5.7	32.2211	29.9185	27.6159	25.3133	23.0108	20.7082	18.4056	16.1030	13.8004	11.4978	9.19[illegible]3	6.8932	4.5958	2.3437	.4830	.0005085
5.8	32.2037	29.9011	27.5985	25.2959	22.9934	20.6908	18.3882	16.0856	13.7830	11.4804	9.17[illegible]9	6.8758	4.5785	2.3273	.4732	.0004532
5.9	32.1866	29.8840	27.5814	25.2789	22.9763	20.6737	18.3711	16.0685	13.7659	11.4633	9.16[illegible]8	6.8588	4.5615	2.3111	.4637	.0004039
6.0	32.1698	29.8672	27.5646	25.2620	22.9595	20.6569	18.3543	16.0517	13.7491	11.4465	9.144[illegible]	6.8420	4.5448	2.2953	.4544	.0003601
6.1	32.1533	29.8507	27.5481	25.2455	22.9429	20.6403	18.3378	16.0352	13.7326	11.4300	9.12[illegible]5	6.8254	4.5283	2.2797	.4454	.0003211
6.2	32.1370	29.8344	27.5318	25.2293	22.9267	20.6241	18.3215	16.0189	13.7163	11.4138	9.11[illegible]2	6.8092	4.5122	2.2645	.4366	.0002864
6.3	32.1210	29.8184	27.5158	25.2133	22.9107	20.6081	18.3055	16.0029	13.7003	11.3978	9.09[illegible]2	6.7932	4.4963	2.2494	.4280	.0002555
6.4	32.1053	29.8027	27.5001	25.1975	22.8949	20.5923	18.2898	15.9872	13.6846	11.3820	9.079[illegible]	6.7775	4.4806	2.2346	.4197	.0002279
6.5	32.0898	29.7872	27.4846	25.1820	22.8794	20.5768	18.2742	15.9717	13.6691	11.3665	9.064[illegible]	6.7620	4.4652	2.2201	.4115	.0002034
6.6	32.0745	29.7719	27.4693	25.1667	22.8641	20.5616	18.2590	15.9564	13.6538	11.3512	9.048[illegible]	6.7467	4.4501	2.2058	.4036	.0001816
6.7	32.0595	29.7569	27.4543	25.1517	22.8491	20.5465	18.2439	15.9414	13.6388	11.3362	9.033[illegible]	6.7317	4.4351	2.1917	.3959	.0001621
6.8	32.0446	29.7421	27.4395	25.1369	22.8343	20.5317	18.2291	15.9265	13.6240	11.3214	9.018[illegible]	6.7169	4.4204	2.1779	.3883	.0001448
6.9	32.0300	29.7275	27.4249	25.1223	22.8197	20.5171	18.2145	15.9119	13.6094	11.3608	9.004[illegible]	6.7023	4.4059	2.1643	.3810	.0001293
7.0	32.0156	29.7131	27.4105	25.1079	22.8053	20.5027	18.2001	15.8976	13.5950	11.2924	8.989[illegible]	6.6879	4.3916	2.1508	.3738	.0001155
7.1	32.0015	29.6989	27.3963	25.0937	22.7911	20.4885	18.1860	15.8834	13.5808	11.2782	8.975[illegible]	6.6737	4.3775	2.1376	.3668	.0001032
7.2	31.9875	29.6849	27.3823	25.0797	22.7771	20.4746	18.1720	15.8694	13.5668	11.2642	8.961[illegible]	6.6598	4.3636	2.1246	.3599	.00009219
7.3	31.9737	29.6711	27.3685	25.0659	22.7633	20.4608	18.1582	15.8556	13.5530	11.2504	8.947[illegible]	6.6460	4.3500	2.1118	.3532	.00008239
7.4	31.9601	29.6575	27.3549	25.0523	22.7497	20.4472	18.1446	15.8420	13.5394	11.2368	8.934[illegible]	6.6324	4.3364	2.0991	.3467	.00007364
7.5	31.9467	29.6441	27.3415	25.0389	22.7363	20.4337	18.1311	15.8286	13.5260	11.2234	8.920[illegible]	6.6190	4.3231	2.0867	.3403	.00006583
7.6	31.9334	29.6308	27.3282	25.0257	22.7231	20.4205	18.1179	15.8153	13.5127	11.2102	8.907[illegible]	6.6057	4.3100	2.0744	.3341	.00005886
7.7	31.9203	29.6178	27.3152	25.0126	22.7100	20.4074	18.1048	15.8022	13.4997	11.1971	8.894[illegible]	6.5927	4.2970	2.0623	.3280	.00005263
7.8	31.9074	29.6048	27.3023	24.9997	22.6971	20.3945	18.0919	15.7893	13.4868	11.1342	8.8817	6.5798	4.2842	2.0503	.3221	.00004707
7.9	31.8947	29.5921	27.2895	24.9869	22.6844	20.3818	18.0792	15.7766	13.4740	11.1714	8.8689	6.5671	4.2716	2.0386	.3163	.00004210
8.0	31.8821	29.5795	27.2769	24.9744	22.6718	20.3692	18.0666	15.7640	13.4614	11.1589	8.8563	6.5545	4.2591	2.0269	.3106	.00003767
8.1	31.8697	29.5671	27.2645	24.9619	22.6594	20.3568	18.0542	15.7516	13.4490	11.1464	8.8439	6.5421	4.2468	2.0155	.3050	.00003370
8.2	31.8574	29.5548	27.2523	24.9497	22.6471	20.3445	18.0419	15.7393	13.4367	11.1342	8.8317	6.5298	4.2346	2.0042	.2996	.00003015
8.3	31.8453	29.5427	27.2401	24.9375	22.6350	20.3324	18.0298	15.7272	13.4246	11.1220	8.8195	6.5177	4.2226	1.9930	.2943	.00002699
8.4	31.8333	29.5307	27.2282	24.9256	22.6230	20.3204	18.0178	15.7152	13.4126	11.1101	8.8076	6.5057	4.2107	1.9820	.2891	.00002415
8.5	31.8215	29.5189	27.2163	24.9137	22.6112	20.3086	18.0060	15.7034	13.4008	11.0982	8.7957	6.4939	4.1990	1.9711	.2840	.00002162
8.6	31.8098	29.5072	27.2046	24.9020	22.5995	20.2969	17.9943	15.6917	13.3891	11.0865	8.7840	6.4822	4.1874	1.9604	.2790	.00001936
8.7	31.7982	29.4957	27.1931	24.8905	22.5879	20.2853	17.9827	15.6801	13.3776	11.0750	8.7725	6.4707	4.1759	1.9498	.2742	.00001733
8.8	31.7868	29.4842	27.1816	24.8790	22.5765	20.2739	17.9713	15.6687	13.3661	11.0635	8.7610	6.4592	4.1646	1.9393	.2694	.00001552
8.9	31.7755	29.4729	27.1703	24.8678	22.5652	20.2626	17.9600	15.6574	13.3548	11.0523	8.7497	6.4480	4.1534	1.9290	.2647	.00001390
9.0	31.7643	29.4618	27.1592	24.8566	22.5540	20.2514	17.9488	15.6462	13.3437	11.0411	8.7386	6.4368	4.1423	1.9187	.2602	.00001245
9.1	31.7533	29.4507	27.1481	24.8455	22.5429	20.2404	17.9378	15.6352	13.3326	11.0300	8.7275	6.4258	4.1313	1.9087	.2557	.00001115
9.2	31.7424	29.4398	27.1372	24.8346	22.5320	20.2294	17.9268	15.6243	13.3217	11.0191	8.7166	6.4148	4.1205	1.8987	.2513	.000009988
9.3	31.7315	29.4290	27.1264	24.8238	22.5212	20.2186	17.9160	15.6135	13.3109	11.0083	8.7058	6.4040	4.1098	1.8888	.2470	.000008948
9.4	31.7208	29.4183	27.1157	24.8131	22.5105	20.2079	17.9053	15.6028	13.3002	10.9976	8.6951	6.3934	4.0992	1.8791	.2429	.000008018
9.5	31.7103	29.4077	27.1051	24.8025	22.4999	20.1973	17.8948	15.5922	13.2896	10.9870	8.6845	6.3828	4.0887	1.8695	.2387	.000007185
9.6	31.6998	29.3972	27.0946	24.7920	22.4895	20.1869	17.8843	15.5817	13.2791	10.9765	8.6740	6.3723	4.0784	1.8599	.2347	.000006439
9.7	31.6894	29.3868	27.0843	24.7817	22.4791	20.1765	17.8739	15.5713	13.2688	10.9662	8.6637	6.3620	4.0681	1.8505	.2308	.000005771
9.8	31.6792	29.3766	27.0740	24.7714	22.4688	20.1663	17.8637	15.5611	13.2585	10.9559	8.6534	6.3517	4.0579	1.8412	.2269	.000005173
9.9	31.6690	29.3664	27.0639	24.7613	22.4587	20.1561	17.8535	15.5509	13.2483	10.9458	8.6433	6.3415	4.0479	1.8320	.2231	.000004637

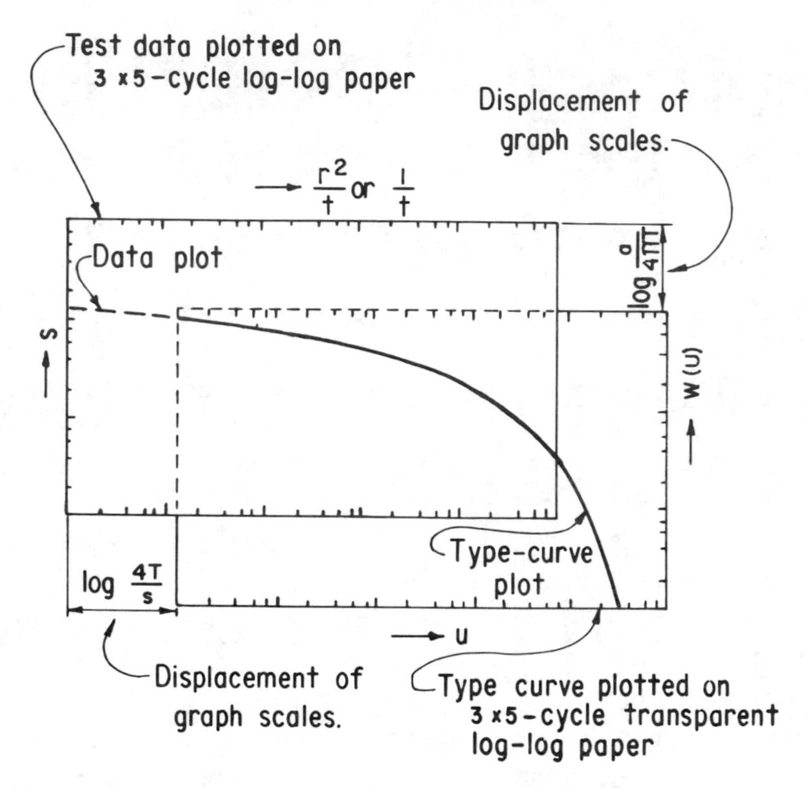

FIGURE 5-4.—Superimposition of the type curve on test data for graphic solution of the nonequilibrium equation. 103-D-1433.

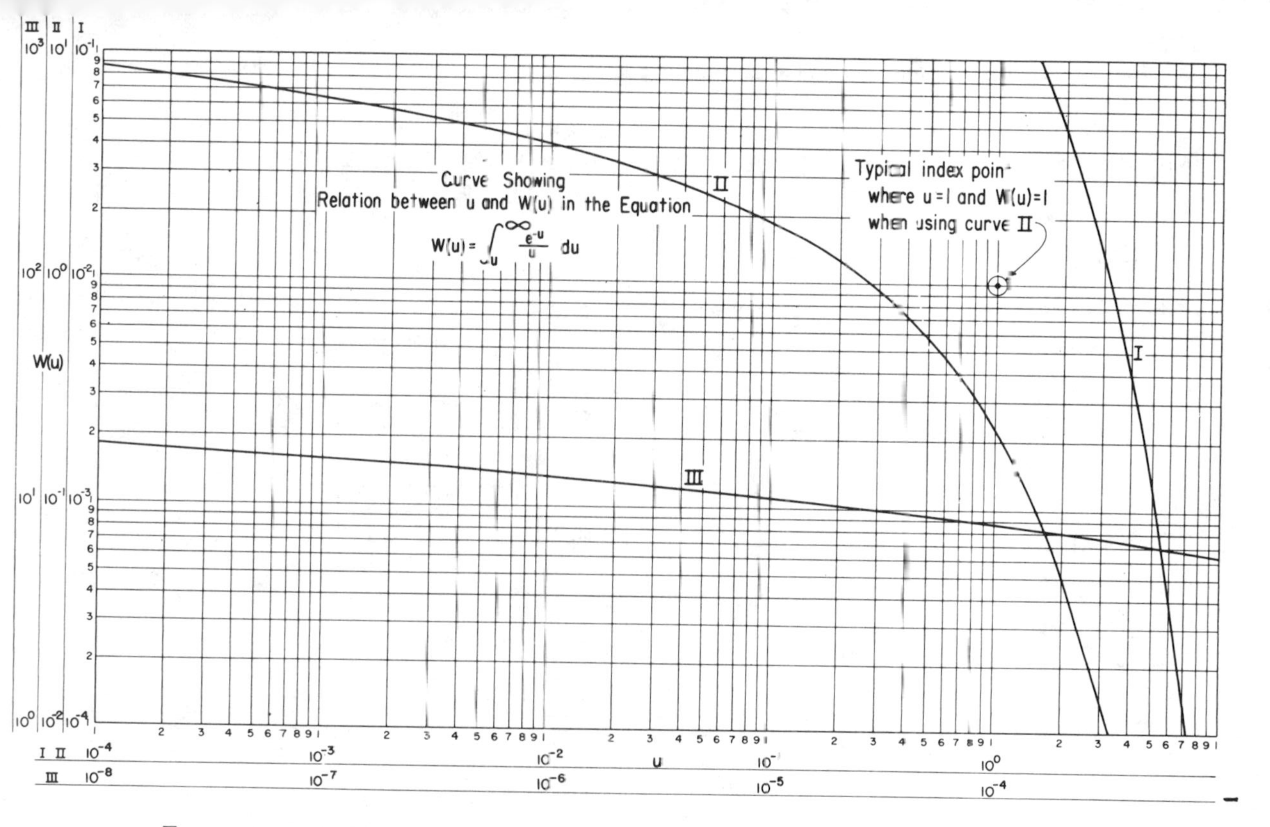

FIGURE 5-5.—Type curve resulting from the plotting of u versus $W(u)$. 103-D-1034.

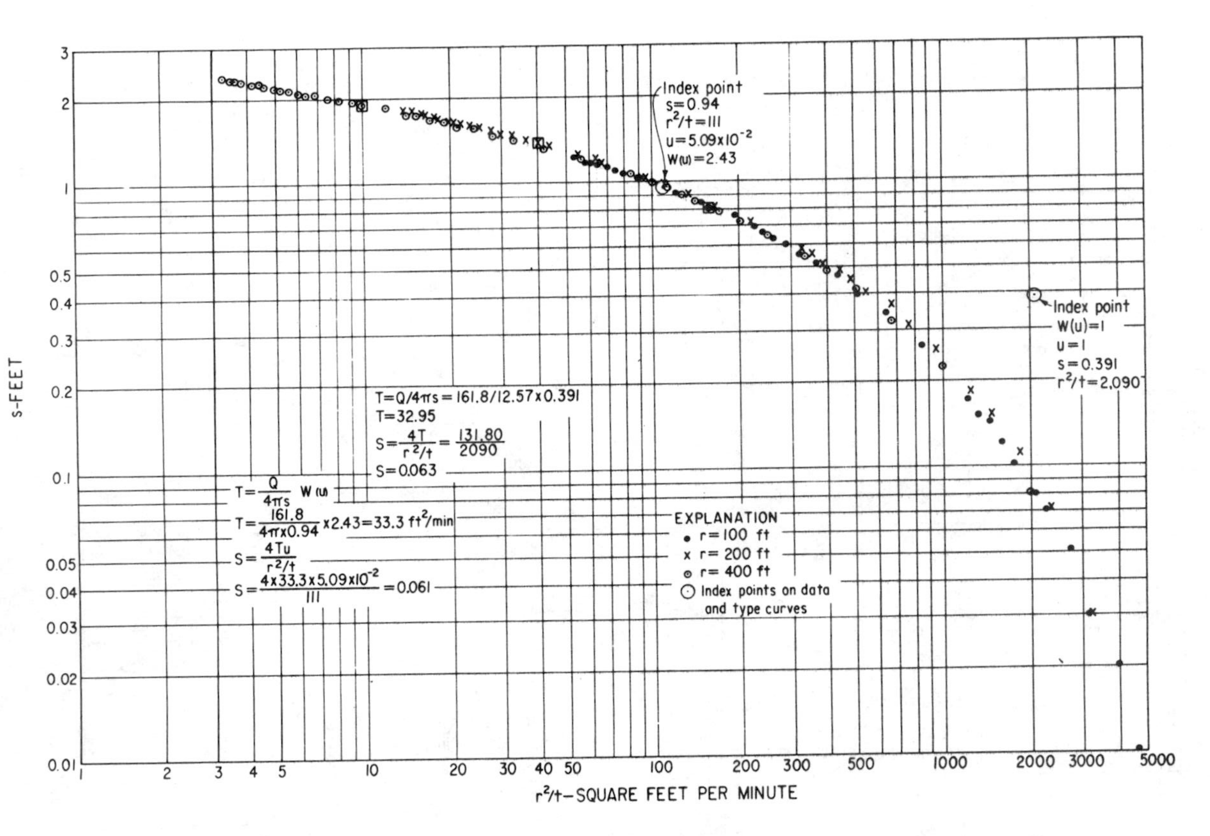

FIGURE 5-6.—Type curve solution of the nonequilibrium equation using s versus r^2/t. 103-D-792.

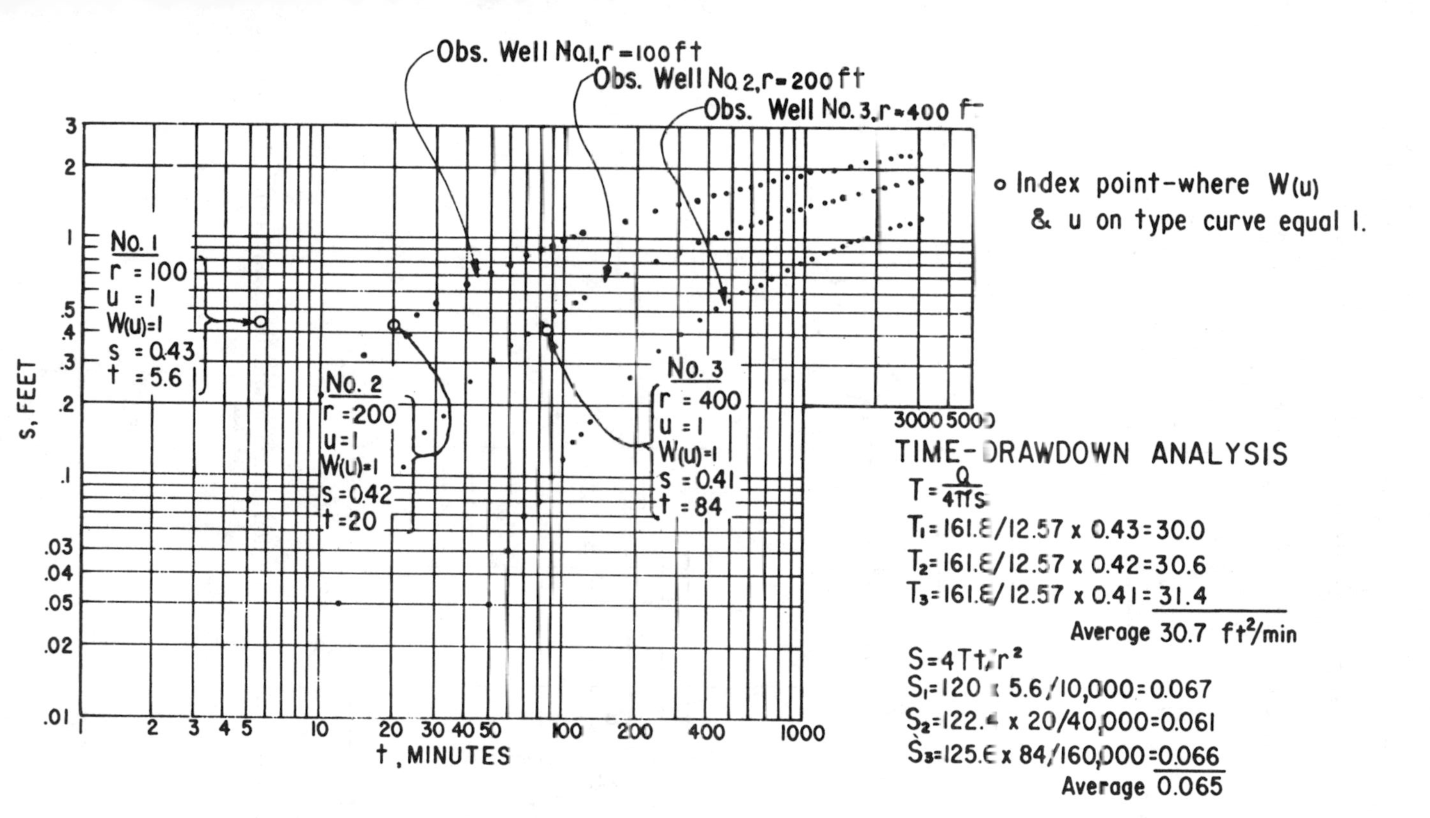

FIGURE 5-7.—Type curve solution of the nonequilibrium equation using s versus t. 103-D-1435.

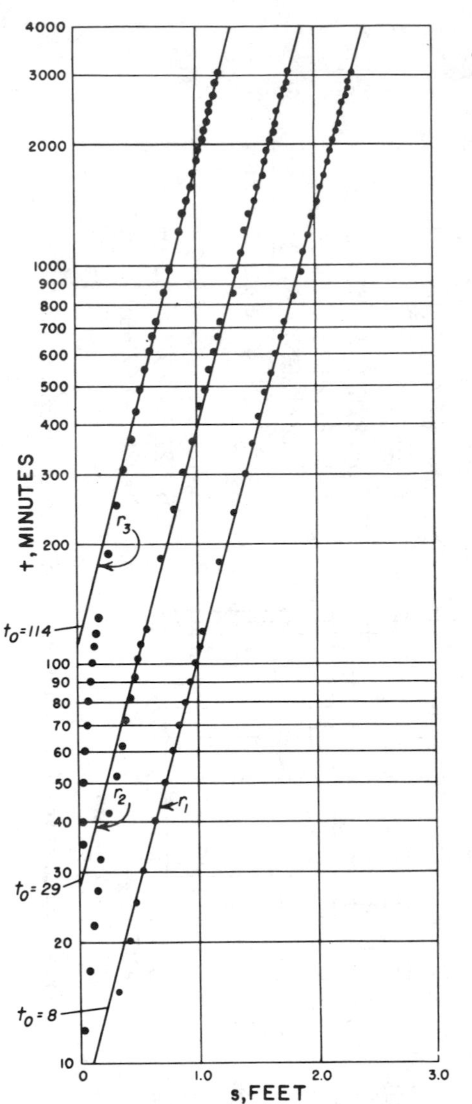

TIME - DRAWDOWN ANALYSIS

JACOB'S APPROXIMATION

$Q = 1210$ gal/min $= 161.8$ ft³/min

	Δs	t_0
$r_1 = 100'$	$1.90 - 1.0 = 0.90$	8
$r_2 = 200'$	$1.38 - 0.49 = 0.89$	29
$r_3 = 400'$	$1.68 - 0.80 = 0.88$	114

$T = 2.303Q/4\pi\Delta s$

$S = 2.25 T t_0 / r^2$

$$\frac{2.303Q}{4\pi} = \frac{372.6}{12.5664} = 29.65$$

$$T_1 = \frac{29.65}{0.90} = 33.0 \text{ ft}^2/\text{min}$$

$$T_2 = \frac{29.65}{0.89} = 33.3 \text{ ft}^2/\text{min}$$

$$T_3 = \frac{29.65}{0.88} = 33.7 \text{ ft}^2/\text{min}$$

AVG. $T = 33.3$ ft²/min

$S_1 = 2.25 \times 33.0 \times 8/10{,}000 = 0.060$

$S_2 = 2.25 \times 33.3 \times 29/40{,}000 = 0.055$

$S_3 = 2.25 \times 33.7 \times 114/160{,}000 = 0.054$

Average = 0.056

FIGURE 5–8.—Straight-line solution of the nonequilibrium equation. (Sheet 1 of 2.) 103–D–793.

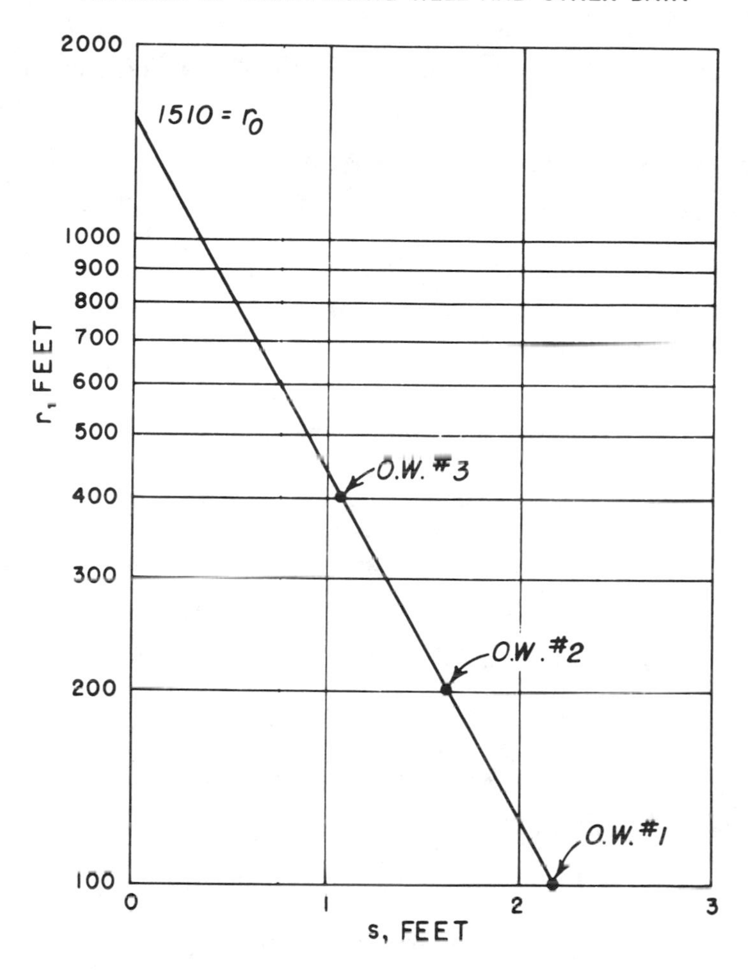

DISTANCE-DRAWDOWN ANALYSIS

JACOB'S APPROXIMATION

$Q = 1210$ gal/min $= 161.8$ ft^3/min

$t = 2045$ minutes

$\Delta s = 2.17 - 0.33 = 1.84$ ft

$r_0 = 1510$ ft

$T = 2.303Q / 2\pi\,\Delta s$

$T = 372.6 \;/\; 2\pi \times 1.84 = 32.2$ ft^2/min

$S = 2.25Tt / r_0^2 = 0.065$

FIGURE 5-8.—Straight-line solution of the nonequilibrium equation. (Sheet 2 of 2.) 103-D-793.

values of s, $\frac{r^2}{t}$, $W(u)$, and u obtained, and the values of T and S computed using equations (12) and (13).

A more tedious but reliable analysis involves plotting s versus $\frac{r^2}{t}$ for each drawdown measurement in each observation well as shown in figure 5–6. This method enables better determination of departures from the type curve which may indicate aquifer conditions that are less than the ideal [1, 7, 8, 12, 36, 65, 74, 75, 76, 80].

5–5. Jacob's Approximation Solutions for the Nonequilibrium Equation.—In addition to solution by the type curve methods previously described, a method developed by Cooper and Jacob [11] permits an approximate solution to the Theis nonequilibrium equation using a straight line graphical approach, which is simple and may offer advantages over the type curve methods.

When the value of u in equation (8) is less than 0.01, that is, when $\frac{r^2}{t}$ becomes very small, the terms following the first two terms of the series equation for $W(u)$ may be neglected and equation (9) may then be approximated by:

$$s=\frac{Q}{4\pi T}\left[\log_e\left(\frac{1}{u}\right)-0.5772\right] \tag{14}$$

which can be reduced to:

$$s=\frac{Q}{4\pi T}\left(\log_e\frac{2.25Tt}{r^2S}\right) \tag{15}$$

by solving for the value of u necessary to make s equal to zero.

Converting to common logarithms, equation (15) may be rewritten:

$$s=\frac{2.303Q}{4\pi T}\left(\log\frac{2.25Tt}{r^2S}\right) \tag{16}$$

Inasmuch as a plot of the drawdown s versus the logarithm of distance r or time t is a straight line, two simple semilog graphical methods can be used to solve for transmissivity and storativity.

(a) *Distance-Drawdown Solution.*—In this method, drawdowns taken at the same time in each of three or more observation wells are plotted against the distance on a log scale of each observation well from the test well as shown in figure 5–8. The straight-line portion of this curve is projected to cover at least one log cycle and the zero drawdown axis. Transmissivity is calculated by:

$$T=\frac{2.303Q}{2\pi\Delta s} \tag{17}$$

where:

Δs is the difference in drawdown over one log cycle.

Storativity is determined by projecting the straight line to the zero drawdown interception which defines the distance r_o, and substituting the value of r_o into:

$$S = \frac{2.25Tt}{r_o^2} \tag{18}$$

Figure 5–8 shows an example of a distance versus drawdown solution.

(b) *Time-Drawdown Solution.*—In this method the drawdown in each observation well since pumping began is plotted against time on the log scale. The straight-line portion of the plot is projected to intercept one or more log cycles and the zero drawdown axis.

The values of transmissivity and storativity are computed using the following equations:

$$T = \frac{2.303Q}{4\pi\Delta s} \tag{19}$$

$$S = \frac{2.25Tt_o}{r^2} \tag{20}$$

where:

Δs = drawdown over one log cycle,
t_o = time at zero drawdown intercept, and
r = distance from the test well to the observation well.

Figure 5–8 shows an example of a time-drawdown analysis in which three observation wells are used.

In using Jacob's straight-line methods, the plots of s versus log plot of r, and s versus log plot of t will not fall on a straight line until pumping has continued for a sufficient time for u to become less than 0.01. The time required to reach this condition increases with distance from the test well.

Because of this time requirement and where boundaries, delayed drainage, or leaky aquifer conditions affect the shape of the drawdown curve early in the pumping period, the straight line segment may not occur.

Interception of a boundary by the ever-widening cone of depression results in a change of the slope of the drawdown curve. For an impermeable (negative) boundary, the slope theoretically increases by a factor of two and for a recharge (positive) boundary, it reduces by a factor of two. Two dissimilar boundaries at equal distances from the observation well tend to cancel each other. The change in slope resulting from a linear recharge boundary such as a stream may approach but never attain a condition of zero additional drawdown with time; whereas, the change resulting from a leaky aquifer or delayed drainage may attain a stable drawdown condition. Recognition and analysis of complex boundary, leaky aquifer, and delayed drainage conditions are often difficult and sometimes impossible.

In both the distance-drawdown and time-drawdown analyses, if Δs is large, minor deviations in extrapolating a straight line to the zero drawdown may result in large errors in the values of r_o or t_o, which result in a large error in the value of S. Under such circumstances, an alternative solution by the type curve method may be advisable to assure a more reliable value of storativity. Another alternative is the application of equations developed by Lohman [49].

5–6. Use of Data from Test Wells.—Data from a test well being pumped or allowed to discharge by artesian flow can be analyzed by either the type curve or straight-line time versus drawdown analyses. However, since the effective diameter of such a well cannot be determined without observation wells, the equations for the solution of S are not applicable. Also, because of head losses inherent in flow to a discharging well, the drawdown in the well may greatly exceed the drawdown in the aquifer immediately adjacent to the well and result in erroneous test data. In such instances, use of the straight-line solution is preferable because a log-log data plot may be so flat it would be difficult to fit to the type curve.

5–7. Recovery Analyses.—A well can be pumped or allowed to flow at a constant rate for a known period, shut off, and allowed to recover to gather data for analysis [1,2,7,8,15,73]. The residual drawdown at any instant and at any point on the cone of depression will be the same as if the well had continued to discharge, but as if a recharge well of equal flow has been introduced at the same point at the instant the discharge stopped [15]. The residual drawdown at any time during the recovery period is the difference between the observed water level and the static water level. Hence, the residual drawdown, s', at any instant is:

$$s'=\frac{Q}{4\pi T}\left[\int_u^\infty \frac{e^{-u}}{u}\,du-\int_{u'}^\infty \frac{e^{-u'}}{u'}\,du'\right] \tag{21}$$

where:

$$u=\frac{r^2S}{4Tt} \text{ and } u'=\frac{r^2S}{4Tt'},$$

t is the time since pumping started, and t' is the time since pumping stopped. Q, T, S, and r have been previously defined in equation (6).

The value:

$$\frac{r^2S}{4Tt'}$$

decreases as t' increases. Therefore, the recovery equation can be written in the form:

$$T = \frac{2.303Q}{4\pi s'} \log\left(\frac{t}{t'}\right). \tag{22}$$

To solve equation (22) graphically using semilog paper, data from table 5-4 are plotted on figure 5-9 where $\frac{t}{t'}$ is plotted on the log scale against residual drawdown s' on the arithmetic scale. When t' becomes large, the plot of the observed data should fall on a straight line. The slope of the line gives the value of the quantity,

$$\frac{\log \frac{t}{t'}}{s'}$$

in equation (22). The value of $\frac{t}{t'}$ is usually chosen over one log cycle so that $\log t/t'$ is unity and then equation (22) becomes:

$$T = \frac{2.303Q}{4\pi \Delta s'} \tag{23}$$

where $\Delta s'$ is the change in residual drawdown over one log cycle t/t'. The storativity cannot be determined because the effective radius of the test well cannot be determined. The recovery method may give a slightly high value of transmissivity in free aquifers, but it is reasonably accurate when used with data from an artesian aquifer. In areas where boundary conditions are known or suspected, the recovery method should be used with caution because of the difficulty in separating the influence of boundaries.

The recovery method may be used to analyze the recovery of observation wells to determine T and S. Values of $\frac{t}{t'}$ are plotted on semilog paper against residual drawdowns. (See fig. 5-10 and table 5-4) The value of T is determined as described for recovery of a test well. Storativity can be estimated from recovery data in an observation well by using the equation:

$$S = \frac{2.25\ Tt'/r^2}{\log^{-1}[(s_p - s')/\Delta(s_p - s')]} \tag{24}$$

where:

s_p = pumping period drawdown projected to time t',
s' = residual drawdown at t', and
$(s_p - s')$ = recovery at t'.

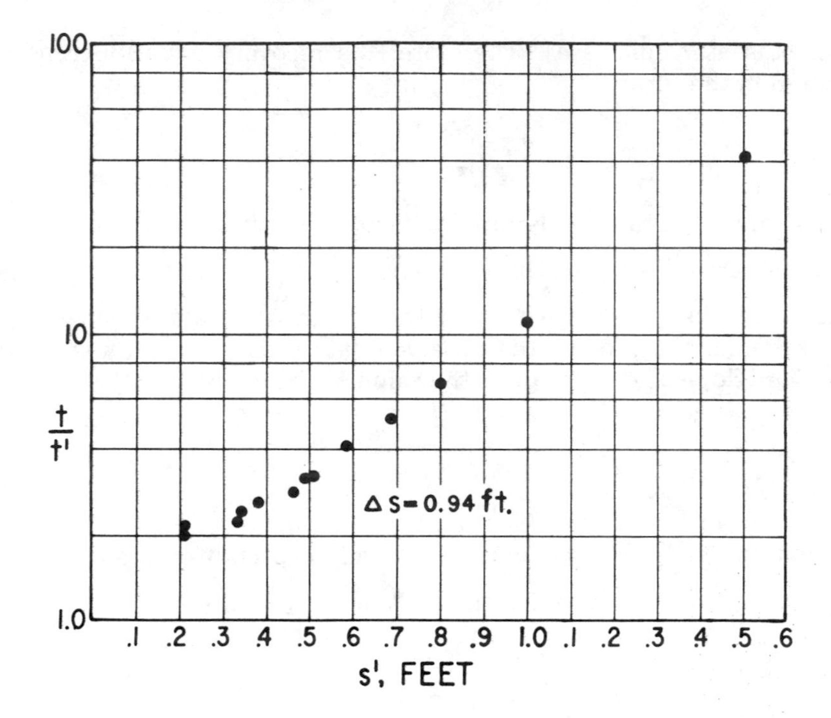

$$T = \frac{2.303Q}{4\pi \Delta s'} = \frac{(2.303)(162.9)}{(4\pi)(0.94)} = 31.7 \text{ ft}^2/\text{min}$$

FIGURE 5–9.—Recovery solution for transmissivity in a discharging well. 103–D–1436.

TABLE 5–4.—*Tabulated data from a discharging and recovering well test.*

Project: Las Vegas	Discharge measured by:	Parshall flume
Feature: Pichaco Dam	Drawdown measured by:	M scope
	Reference point:	North side of casing collar

Date	Time	Pump test No. 1			Test well		Remarks
		Time, min	Depth to water, ft	Drawdown, ft	Gage reading, ft	Discharge, ft³/s	
5–16	0840		60. 99				
5–17	0820		61. 01				
5–18	0845		61. 00				
5–19	0820		60. 98				
	0840	0	60. 99				Pump started.
	0843	3	71. 2	10. 2			Q=162.9 ft³/min.
	0848	8	71. 2	10. 6			
	0853	13	71. 3	10. 8			
	0900	20	72. 3	11. 3	0. 79	2. 70	
	1000	80	72. 6	11. 6	. 79	2. 70	
	1100	140	72. 8	11. 8	. 79	2. 70	
	1155	195	72. 8	11. 8	. 80	2. 75	
	1255	255	72. 8	11. 8	. 80	2. 75	
	1355	315	73. 0	12. 0	. 80	2. 75	
	1455	375	73. 2	12. 2	. 80	2. 75	
	1555	435	73. 2	12. 2	. 80	2. 75	
	1655	495	73. 2	12. 2	. 80	2. 75	
	1800	560	73. 2	12. 2	. 80	2. 75	
	1856	616	73. 3	12. 3	. 80	2. 75	
	1958	668	73. 4	12. 4	. 80	2. 75	
	2057	737	73. 4	12. 5	. 80	2. 75	
	2200	800	73. 5	12. 5	. 80	2. 75	Pump off.

TABLE 5-4.—*Tabulated data from a discharging and recovering well test*—Continued

RECOVERY OF TEST WELL

Date	Time	t, min	t', min	t/t'	Depth to water, ft	Residual draw-down, ft	Remarks
5-19----	2200	800	0	0. 0	73. 5	12. 5	Pump off.
	2203	803	3	268. 0	41. 0±	+20. 0	
	2208	808	8	101. 0	56. 0±	+5. 0	
	2213	813	13	62. 5	60. 5	+0. 5	
	2220	820	20	41. 0	62. 49	1. 5	
	2320	880	80	11. 0	61. 99	1. 0	
5-20----	0020	940	140	6. 7	61. 79	0. 80	
	0115	995	195	5. 1	61. 68	. 69	
	0215	1, 055	255	4. 1	61. 58	. 59	
	0315	1, 115	315	3. 5	61. 50	. 51	
	0415	1, 175	375	3. 1	61. 48	. 49	
	0515	1, 235	435	2. 8	61. 45	. 46	
	0615	1, 295	495	2. 6	61. 37	. 38	
	0720	1, 360	560	2. 4	61. 33	. 34	
	0816	1, 416	616	2. 3	61. 32	. 33	
	0908	1, 418	668	2. 2	61. 32	. 33	
	1017	1, 527	727	2. 1	61. 21	. 22	
	1120	1, 600	800	2. 0	61. 21	. 22	

TABLE 5–4.—*Tabulated data from a discharging and recovering well test*—Continued

PUMP TEST NO. 1, OBSERVATION WELL NO. 1, r=100 FT

Date	Time	Time, min	Depth to water, ft	Drawdown, ft	Remarks
5–16	0845		61. 20		
5–17	0825		61. 21		
5–18	0840		61. 21		
5–19	0815		61. 20		
	0835		61. 20		
	0840		61. 20		Pump started.
	0845	5	61. 28	0. 08	
	0850	10	61. 42	. 22	
	0855	15	61. 55	. 33	
	0900	20	61. 61	. 41	
	0905	25	61. 70	. 50	
	0910	30	61. 75	. 55	
	0920	40	61. 86	. 66	
	0930	50	61. 93	. 73	
	0940	60	62. 00	. 80	
	0950	70	62. 06	. 86	
	1000	80	62. 12	. 92	
	1010	90	62. 16	. 96	
	1020	100	62. 20	1. 00	
	1030	110	62. 24	1. 04	
	1040	120	62. 27	1. 07	
	1140	180	62. 44	1. 24	
	1240	240	62. 55	1. 35	
	1340	300	62. 65	1. 45	
	1440	360	62. 72	1. 52	
	1540	420	62. 79	1. 59	
	1640	480	62. 85	1. 65	
	1740	540	62. 91	1. 71	
	1840	600	62. 93	1. 73	
	1940	660	62. 97	1. 77	
	2046	720	63. 01	1. 81	
	2200	800	63. 06	1. 86	Pump off.

TABLE 5-4.—*Tabulated data from a discharging and recovering well test*—Continued

RECOVERY OF OBSERVATION WELL

Date	Time	t, min	t', min	t/t'	Depth to water, ft	Residual drawdown, ft
5-19	2200	800	0	0. 0	63. 06	1. 86
	2205	805	5	161. 0	62. 98	1. 78
	2210	810	10	81. 0	62. 84	1. 64
	2215	815	15	54. 3	62. 73	1. 53
	2220	820	20	41. 0	62. 65	1. 45
	2225	825	25	33. 3	62. 57	1. 37
	2230	830	30	27. 7	62. 52	1. 32
	2240	840	40	21. 0	62. 52	1. 22
	2250	850	50	17. 0	62. 35	1. 15
	2300	860	60	14. 3	62. 29	1. 09
	2310	870	70	12. 4	62. 23	1. 03
	2320	880	80	11. 0	62. 17	0. 97
	2330	890	90	9. 88	62. 14	. 94
	2340	900	100	9. 00	62. 10	. 90
	2350	910	110	8. 27	62. 07	. 87
5-20	2400	920	120	7. 67	62. 05	. 85
	0100	980	180	5. 44	61. 90	. 70
	0200	1, 040	240	4. 33	61. 81	. 61
	0300	1, 100	300	3. 67	61. 74	. 54
	0400	1, 160	360	3. 22	61. 69	. 49
	0500	1, 220	420	2. 90	61. 66	. 46
	0600	1, 280	480	2. 67	61. 60	. 40
	0700	1, 340	540	2. 48	61. 56	. 36
	0800	1, 400	600	2. 33	61. 56	. 36
	0900	1, 460	660	2. 21	61. 54	. 34
	1000	1, 520	720	2. 11	61. 51	. 31
	1120	1, 600	800	2. 00	61. 49	. 29

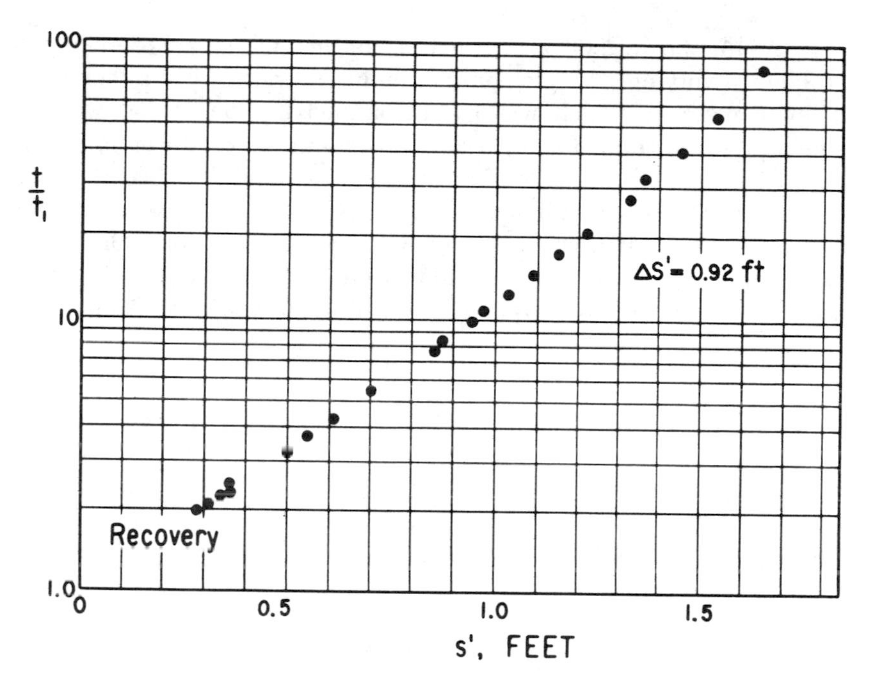

$$T = \frac{2.303\,Q}{4\pi \Delta s'} = \frac{(2.303)(162.9)}{4\pi\,(0.92)} = 32.1\ \text{ft}^2/\text{min}$$

$$S = \frac{2.25\,(T t'/r^2)}{\log^{-1}[(s_p - s')/\Delta(s_p - s')]} = \frac{(2.25)(32.1)\left(\frac{600}{10{,}000}\right)}{\log^{-1}(1.75/0.99)} = 0.07$$

Note: Semilog plots necessary to determine $(s_p - s') = 1.75$ and $\Delta(s_p - s') = 0.99$ not shown.

FIGURE 5-10.—Recovery solution for transmissivity and storativity in an observation well. 103-D-1437.

5-8. Leaky Aquifer Solutions.—Under sufficient head, even apparently impermeable geologic materials will transmit water, and confining layers enclosing artesian aquifers are no exception. Where two or more aquifers are separated by a confining layer, pumping from one aquifer may disturb the mutual hydraulic balance and result in an increase or decrease in leakage between the aquifers. Such leakage is a boundary condition. Theoretically, the area of influence of a discharging well expands until leakage into the aquifer induced by the well equals the well discharge. At this point the area of influence stabilizes and the drawdown becomes constant with time. Conversely, if the discharge from a well in an aquifer that is losing water by leakage balances the amount of leakage, the area of influence will stabilize.

In 1946, Jacob [39] published a mathematical solution to the problem involving a single confined aquifer overlain by a leaky confining bed above which was an unconfined aquifer. Later work by Glover, Moody, and Tapp [17,85] of the Bureau of Reclamation developed simplified methods of analysis using a family of type curves. These curves, shown on figure 5–11, should be superimposed on a log-log plot of drawdown versus time from three observation wells (fig. 5–12) taken from a tabulation of field measurements and data (table 5–5). Table 5–5 presents only a partial record of the test data. This fitting of curves is similar to the procedure described earlier for using the Theis-type curve. The values of s and t are applied to the equations:

$$u=\frac{s}{\frac{Q}{2\pi KM}} \tag{25}$$

When $u=1$, equation (25) can be written:

$$s=\frac{Q}{2\pi KM}$$

Also, since $T=KM$, equation (25) can be written:

$$T=\frac{Qu}{2\pi s}$$

Similarly,

$$\eta=t\left(\frac{K'}{SM'}\right) \tag{26}$$

When $\eta=1$, equation (26) can be written:

$$t=\frac{SM'}{K'}$$

The terms of the above equations are as defined on figure 5–13.

Each of the family of curves has an x value as noted on figure 5–11. When the closest fit is found between the data curve and the type curve, and the values of s and t have been obtained, the x value of the type curve to which the fit is made is noted. If the fit falls between curves, the x value is interpolated. The value x is related to r, K', T, and M' by:

$$x=r\sqrt{\frac{K'}{TM'}} \tag{27}$$

or

$$\frac{K'}{M'}=T\left(\frac{x}{r}\right)^2.$$

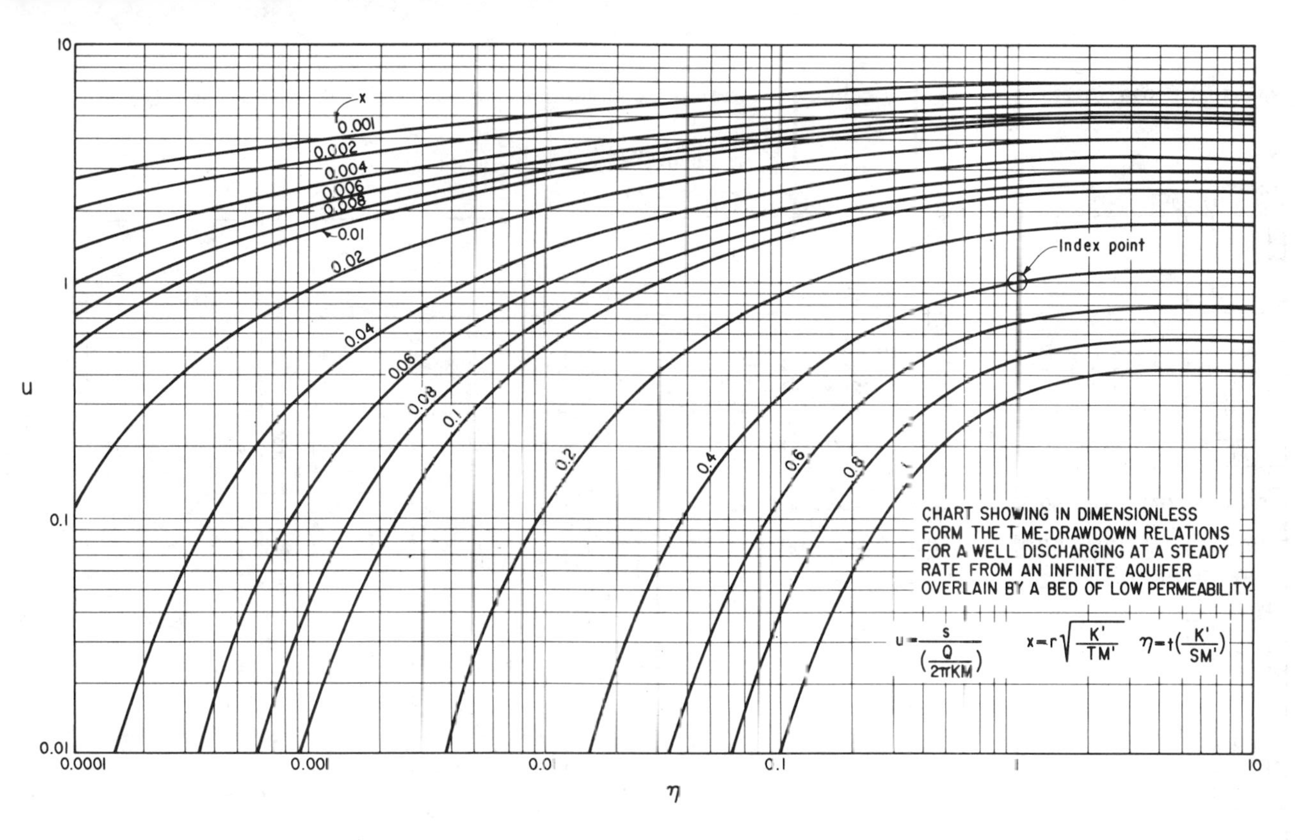

FIGURE 5-11.—Leaky aquifer type curves. 103-D-1438.

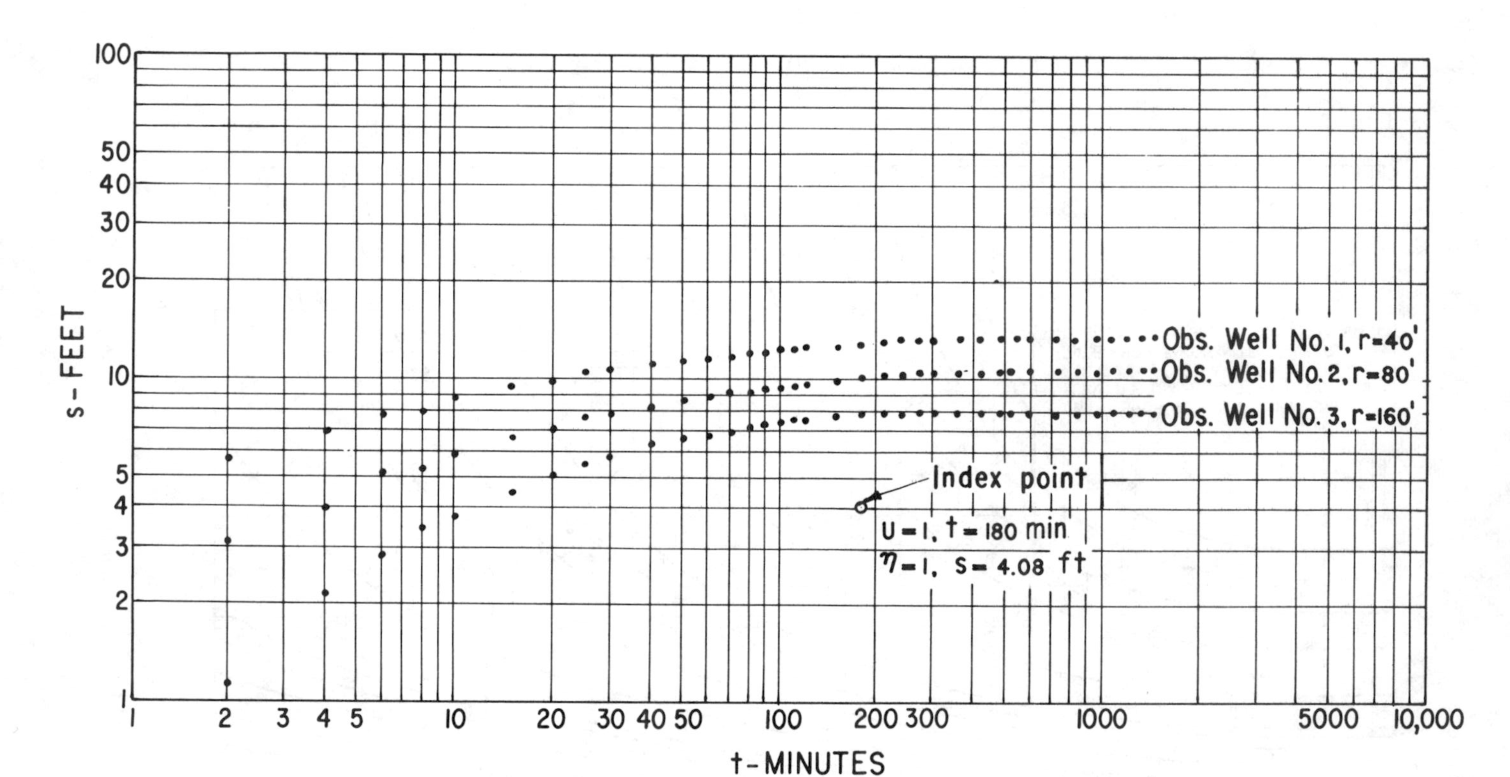

FIGURE 5-12.—Leaky aquifer type curve solution for transmissivity, storativity, and permeability of the leaky bed. 103-D-800.

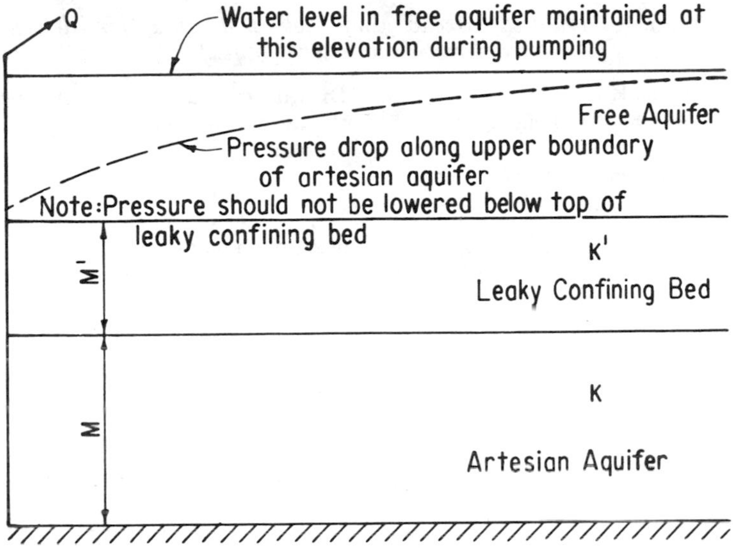

Q

Water level in free aquifer maintained at this elevation during pumping

Free Aquifer

Pressure drop along upper boundary of artesian aquifer

Note: Pressure should not be lowered below top of leaky confining bed

M'

K'

Leaky Confining Bed

M

K

Artesian Aquifer

K = Permeability of artesian aquifer .... L/t

M = Thickness of artesian aquifer L

T = KM Transmissivity of artesian aquifer .... L^2/t

K' = Vertical permeability of leaky confining bed .... L/t

M' = Thickness of leaky confining bed L

$\frac{K'}{M'}$ = Leakance 1/t

Q = Discharge from well .... L^3/t

S = Storativity of artesian aquifer, dimensionless

FIGURE 5-13.—Leaky aquifer definitions. 103-D-1439.

TABLE 5–5.—*Tabulated discharging well test data for determining characteristics of a leaky aquifer*

Project: Texas Hill
Feature: Salt Flat drainage program

Location: 200 feet north SW. corner sec. 10, T. 20 N., R. 5 W.

Drawdown measured by: "Popper"
Distance from pumped well: 40 feet north

PUMP TEST NO. 1, OBSERVATION WELL NO. 1

Date	Time	Elapsed, time, min	Depth to water, ft	Draw-down s, ft	Remarks
8–24	1050		26. 59	0	
8–25	1055		26. 58	0	
8–26	1045		26. 59	0	Static water level.
	1100	0	26. 59	0	Pump on.
	1102	2	32. 24	5. 65	
	1104	4	33. 55	6. 96	
	1106	6	34. 31	7. 72	
	1108	8	34. 59	8. 00	
	1110	10	35. 30	8. 71	Pump off 8–28 at 1600.
	1115	15	36. 06	9. 47	
	1120	20	36. 58	9. 99	
	1125	25	36. 94	10. 35	
	1130	30	37. 29	10. 70	
	1140	40	37. 73	11. 14	Avg discharge:
	1150	50	38. 05	11. 46	=4,488 gal/min.
	1200	60	38. 21	11. 62	=600 ft³/min.
	1210	70	38. 45	11. 86	
	1220	80	38. 61	12. 02	
	1230	90	38. 85	12. 26	
	1240	100	38. 92	12. 33	
	1250	110	38. 96	12. 37	
	1300	120	39. 00	12. 41	
	1330	150	39. 28	12. 69	
	1400	180	39. 44	12. 85	
	1430	210	39. 68	13. 09	
	1500	240	39. 72	13. 13	
	1530	270	39. 84	13. 25	
	1600	300	39. 92	13. 33	
	1700	360	39. 96	13. 37	
	1800	420	40. 00	13. 41	

TABLE 5-5.—*Tabulated discharging well test data for determining characteristics of a leaky aquifer*—Continued

Project: Texas Hill
Feature: Salt Flat drainage program
Location: 240 feet north SW. corner sec. 10, T. 20 N., R. 5 W.

Drawdown measured by: "Popper"
Distance from pumped well: 80 feet north

PUMP TEST NO. 1, OBSERVATION WELL NO. 2

Date	Time	Elapsed time, min	Depth to water, ft	Draw down s, ft	Remarks
8-24	1045		26. 54	0	
8-25	1050		26. 54	0	
8-26	1040		26. 54	0	Static water level.
	1100	0	26. 54	0	Pump on.
	1102	2	29. 64	3. 10	
	1104	4	30. 56	4. 02	
	1106	6	31. 59	5. 05	
	1108	8	31. 83	5. 29	
	1110	10	32. 51	5. 97	Pump off 8-28 at 1600.
	1115	15	33. 26	6. 72	
	1120	20	33. 70	7. 16	
	1125	25	34. 14	7. 60	
	1130	30	34. 50	7. 96	
	1140	40	34. 90	8. 36	Avg discharge:
	1150	50	35. 17	8. 63	=4,488 gal/min.
	1200	60	35. 45	8. 91	=600 ft³/min.
	1210	70	35. 73	9. 19	
	1220	80	35. 85	9. 31	
	1230	90	36. 01	9. 47	
	1240	100	36. 09	9. 55	
	1250	110	36. 17	9. 63	
	1300	120	36. 29	9. 75	
	1330	150	36. 49	9. 95	
	1400	180	36. 61	10. 07	
	1430	210	36. 73	10. 19	
	1500	240	36. 81	10. 27	
	1530	270	36. 89	10. 35	
	1600	300	36. 93	10. 39	
	1700	360	36. 93	10. 34	
	1800	420	36. 96	10. 42	

TABLE 5–5.—*Tabulated discharging well test data for determining characteristics of a leaky aquifer*—Continued

Project: Texas Hill
Feature: Salt Flat drainage program
Location: 320 feet north SW. corner sec. 10, T. 20 N., R. 5 W.

Drawdown measured by: "Popper"
Distance from pumped well: 160 feet north

PUMP TEST NO. 1, OBSERVATION WELL NO. 3

Date	Time	Elapsed, time, min	Depth to water, ft	Draw-down *s*, ft	Remarks
8–24	1040		26. 60	0	
8–25	1045		26. 61	0	
8–26	1035		26. 60	0	
	1100	0	26. 60	0	Pump on.
	1102	2	27. 71	1. 11	
	1104	4	28. 75	2. 15	
	1106	6	29. 46	2. 86	
	1108	8	30. 06	3. 46	
	1110	10	30. 38	3. 78	Pump off 8–28 at 1600.
	1115	15	31. 18	4. 58	
	1120	20	31. 69	5. 09	
	1125	25	32. 09	5. 49	
	1130	30	32. 45	5. 85	
	1140	40	32. 97	6. 37	Avg discharge:
	1150	50	32. 24	6. 64	=4,488 gal/min.
	1200	60	33. 40	6. 80	=600 ft^3/min.
	1210	70	33. 56	6. 96	
	1220	80	33. 76	7. 16	
	1230	90	33. 96	7. 36	
	1240	100	34. 04	7. 44	
	1250	110	34. 12	7. 52	
	1300	120	34. 16	7. 56	
	1330	150	34. 24	7. 64	
	1400	180	34. 48	7. 88	
	1430	210	34. 52	7. 92	
	1500	240	34. 56	7. 96	
	1530	270	34. 56	7. 96	
	1600	300	34. 56	7. 96	
	1700	360	34. 55	7. 95	
	1800	420	34. 56	7. 96	

Since T, x, and r are known, the value of $\frac{K'}{M'}$ can be calculated from equation (27), and then S is calculated by solving equation (26) when $\eta=1$. If M' is approximately known, K' can be calculated.

Using data from table 5-5 and figures 5-11, 5-12, and 5-13, some sample calculations for material in this section could be:

Q=pump discharge=600 ft³/min

x_1=x value for type curve which fits the plotted s-t relation for observation well No. 1=0.04

x_2=x value for well No. 2=0.08

x_3=x value for well No. 3=0.16

M'=20 ft

At the index point shown on figure 5-12:

$$u=1,\ \eta=1,\ t=180 \text{ min, and } s=4.08 \text{ ft}$$

$$T=\frac{Qu}{2\pi s}=\frac{600(1)}{2\pi(4.08)}=23.4 \text{ ft}^2/\text{min}$$

$$x_1=r_1\sqrt{\frac{K'}{TM'}} \text{ or } \frac{K'}{M'}=\frac{Tx_1^2}{r_1^2}$$

$$\frac{K'}{M'}=\frac{(23.4)(0.04)^2}{(40)^2}=2.34\times10^{-5} \text{ min}^{-1}$$

$$S=\frac{tK'}{M'}=180\ (2.34\times10^{-5})=0.004.$$

If the aquifer is overlain and underlain by other aquifers from which it is separated by confining layers, the assumed M' value may be erroneous and it may be impossible to separate the contribution from each aquifer. Then, the leakage factor $\frac{K'}{M'}$ computed is a combined value of both layers.

In free aquifers, delayed drainage may cause the plot of s versus t on log-log paper to appear similar to data plotted from a leaky aquifer. In extreme cases of delayed drainage, several weeks of pumping may be required to differentiate between the two conditions. Such a lengthy testing period generally cannot be justified from an economic standpoint so analysis must be based on judgment of short term test results, known and inferred aquifer conditions, and knowledge of well construction details [16,17,21,22,23,24,25,34,39,80].

Additionally, the presence of the lateral boundaries may further influence the drawdown curves so that a reliable interpretation of aquifer conditions and calculation of aquifer characteristics are virtually impossible.

5–9. Constant Drawdown Solutions.—Aquifer test procedures discussed thus far have relied on the use of constant discharge and variable drawdown of a well. Under some conditions, such as with a flowing artesian well, it is simpler to test the well by permitting it to discharge at a variable rate but with constant drawdown. This can be done by shutting off discharge until the pressure inside the well stabilizes and then permitting resumption of discharge. During this discharge period, the rate of discharge is recorded periodically.

Jacob and Lohman [42] derived equations for tests of this type, including a straight-line approximation solution to determine T and S:

$$T=\frac{2.303}{4\pi\Delta\dfrac{\left(\dfrac{s_w}{Q}\right)}{\Delta\log\left(\dfrac{t}{r_w^2}\right)}} \tag{28}$$

where:

T=transmissivity, $\frac{L^2}{t}$

s_w=constant drawdown, L,
(the difference in feet of water between the static head and the top of the casing or center of the discharge valve)

Q=weighted average discharge during a timed interval, $\frac{L^3}{t}$,

t=elapsed time since start of test, t,.

r_w=radius of the well, L,

$\Delta\left(\frac{s_w}{Q}\right)$=change in the ratio $\frac{s_w}{Q}$ over a time period, $\frac{1}{L^2}$, and

$\Delta\ log\left(\frac{t}{r_w^2}\right)$=change in base 10 logarithm over a time period.

Table 5–6 shows the tabulated data from a constant drawdown test and the straight-line method of solution. In this table, column ① shows the times at which measurements were made, column ② shows the average rate of discharge during the time interval shown in column ③, column ④ shows the total discharge during the time interval, column ⑤ shows the elapsed time since start of discharge, column ⑥ is the constant drawdown divided by rate of discharge in column ②, and column ⑦ is time shown in column ⑤ divided by the square of the well radius.

TABLE 5-6.—*Constant drawdown test data.* From [86]. 103-D-1441.

Field data for flow test on Artesia Heights well near Grand Junction, Colo., Sept. 22, 1948
(Valve opened at 10:29 a.m. s_w = 92.33 ft.; r_w = 0.276 ft. Data from Lohman (1965, tables 6 and 7, well 28))

①	②	③	④	⑤	⑥	⑦
Time of observation	Rate of flow, gal/min	Flow interval, min	Total flow during interval, gal	Time since flow started, min	$\frac{s_w}{Q}$ ft-min/gal	$\frac{t}{r_w^2}$ min/ft²
10:30	7.28	1	7.28	1	12.7	13.1
10:31	6.94	1	6.94	2	13.3	26.3
10:32	6.88	1	6.88	3	13.4	39.4
10:33	6.28	1	6.28	4	14.7	52.6
10:34	6.22	1	6.22	5	14.8	65.7
10:35	6.22	1	6.22	6	15.1	78.8
10:37	5.95	2	11.90	8	15.5	105
10:40	5.85	3	17.55	11	15.8	145
10:45	5.66	5	28.30	16	16.3	210
10:50	5.50	5	27.50	21	16.8	276
10:55	5.34	5	26.70	26	17.3	342
11:00	5.34	5	26.70	31	17.3	407
11:10½	5.22	10.5	54.81	41.5	17.7	345
11:20	5.14	9.5	48.83	51	18.0	670
11:30	5.11	10	51.10	61	18.1	802
11:45	5.05	15	75.75	76	18.3	999
12:00 (noon)	5.00	15	75.00	91	18.5	1,190
12:12	4.92	12	59.04	103	18.8	1,354
12:22	4.88	10	53.68	113	18.9	1,485
Total[1] ----------		114	596.98	----------		

[1] 596.98 gal/114 min = 5.23 gal/min, weighted average discharge.

In the straight-line method of solution, the values for $\frac{s_w}{Q}$ are plotted on an arithmetic scale against values of $\frac{t}{r_w^2}$ for the corresponding time on a log scale. A straight line is then fitted through the points as shown on figure 5–14.

The change in value of $\frac{s_w}{Q}$ is determined over one log cycle of $\frac{t}{r_w^2}$ and is set equal to $\Delta\frac{s_w}{Q}$. In the example on figure 5–14, $\Delta\frac{s_w}{Q}$ =18.40-15.38=3.02 feet-minute per gallon. Measurements must be converted to consistent units and inserted into equation (28) to give T in square feet per minute as follows:

$$T=\frac{2.303}{\dfrac{4\pi\Delta\left(\dfrac{s_w}{Q}\right)}{\Delta\log\left(\dfrac{t}{r_w^2}\right)}}=\frac{2.303}{\dfrac{4\pi(22.59)}{1}}=0.008\ \text{ft}^2/\text{min.}$$

If r_w is known, the straight line on the plot can be extended to intercept $\frac{s_w}{Q}=0$, and S can be computed from the equation:

$$S=2.25T\left(\frac{t}{r_w^2}\right)$$

Transmissivity, T, can be roughly checked by recovery measurements made after the discharge has been stopped. Equations (22) and (23) and the weighted average discharge per period of discharge are used with the recovery data.

5–10. Delayed Drainage Solutions.—The early response of an unconfined aquifer to a discharging well is especially dependent on the degree of isotropy present. Tests of short duration in such aquifers may be unreliable because of the delayed drainage effect produced by anisotropy. The usual plots of log t versus log s may show a steep initial slope, then a flat segment followed by another slope. The general shape of the curve is that of an elongated letter S (see table 5–7 and fig. 5–15). The early part of the curve may be influenced by several or all of the following factors:

- Changing storativity caused by delayed drainage
- Expansion of water below the water table resulting from reduction in pressure
- Vertical flow components

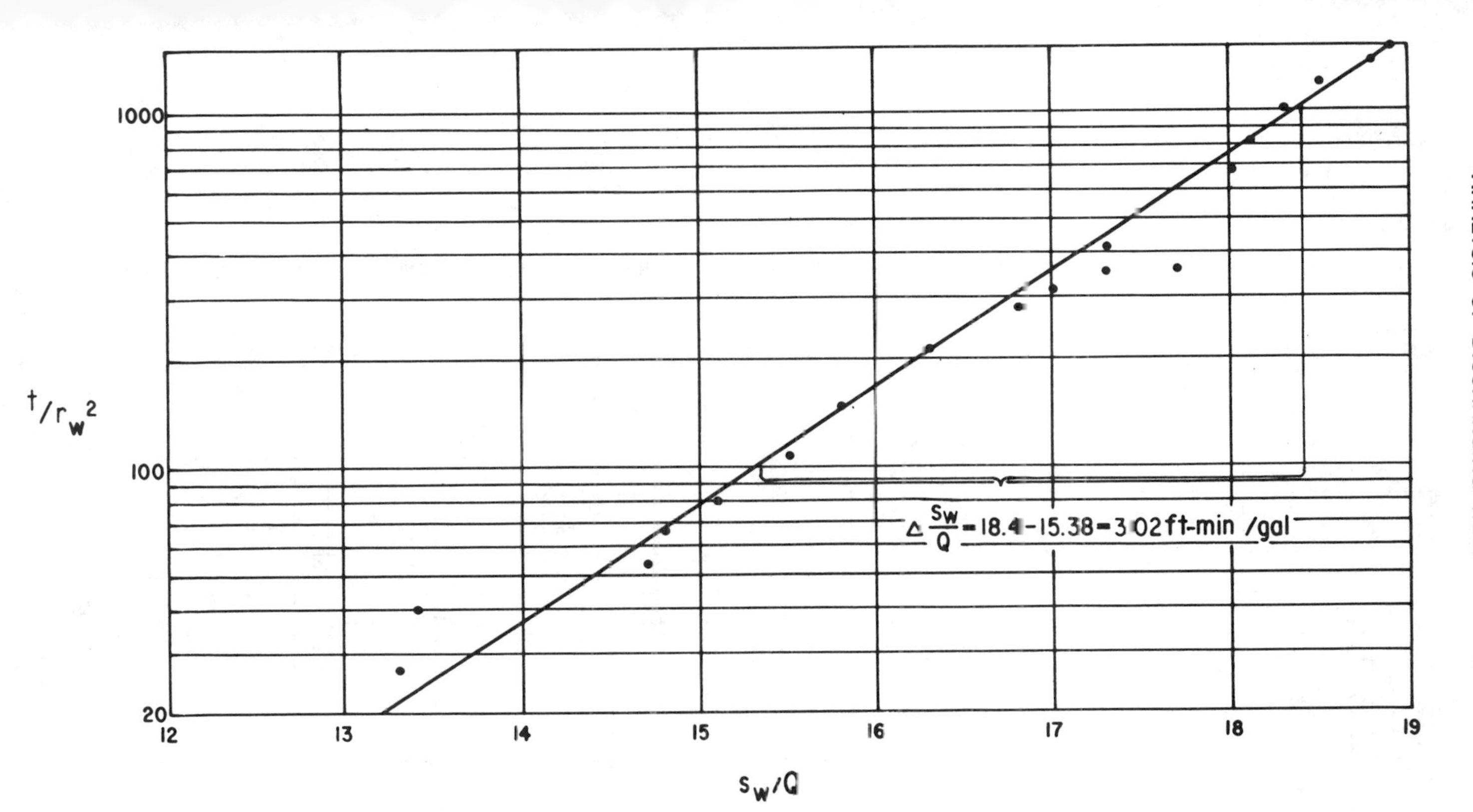

FIGURE 5-14.—Constant drawdown solution for transmissivity and storativity. 103-D-1442.

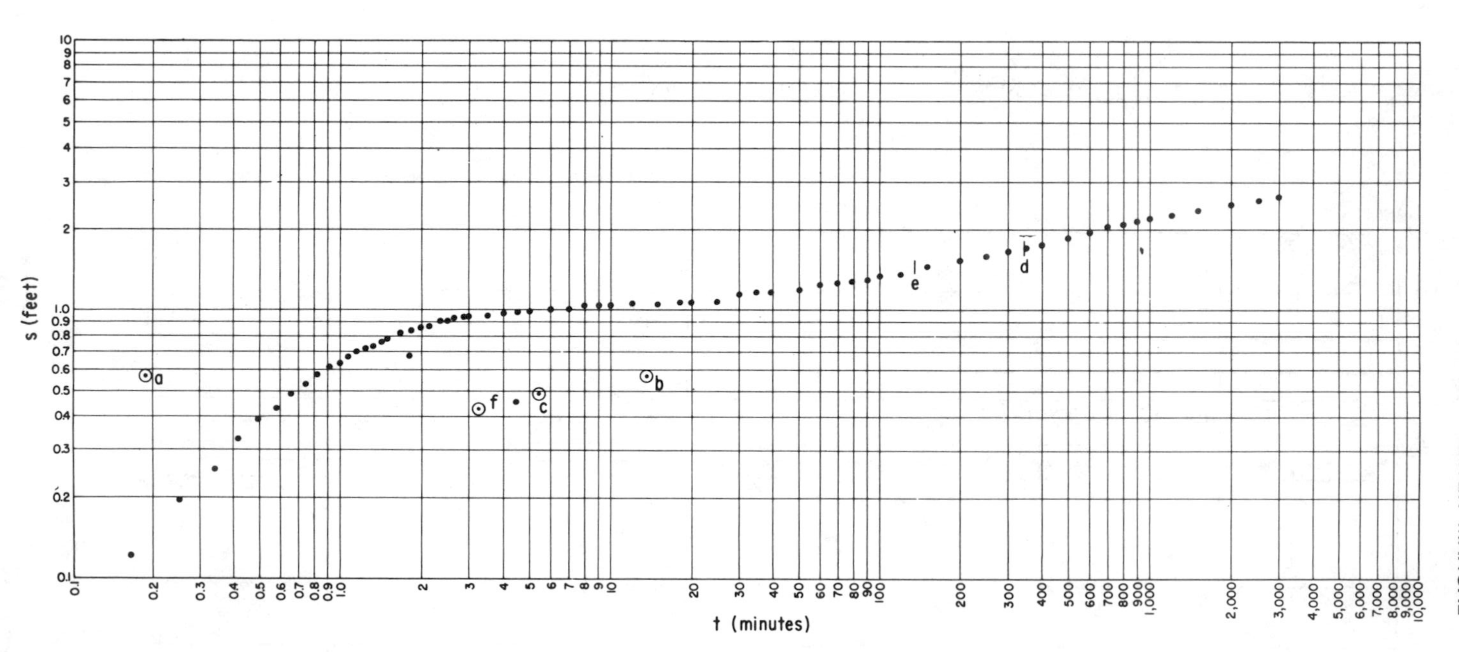

FIGURE 5-15.—Delayed drainage type curve solution for transmissivity and storativity. From table 5-7. 103-D-1444.

TABLE 5–7.—*Adjusted field data for delayed yield analysis on Fairborn, Ohio, well* (from Lohman) [49]

Time since pumping began *t*, min	Corrected drawdowns *s*, ft	Time since pumping *t*, min	Corrected drawdowns *s*, ft	Time since pumping *t*, min	Corrected drawdowns *s*, ft
0. 165	0. 12	2. 65	0. 92	80	1. 28
. 25	. 195	2. 80	. 03	90	1. 29
. 34	. 255	3. 00	. 94	100	1. 31
. 42	. 33	3. 50	. 95	120	1. 36
. 50	. 39	4. 00	. 97	150	1. 45
. 58	. 43	4. 50	. 975	200	1. 52
. 66	. 49	5. 00	. 98	250	1. 59
. 75	. 53	6. 00	. 99	300	1. 65
. 83	. 57	7. 00	1. 00	350	1. 70
. 92	. 61	8. 00	1. 01	400	1. 75
1. 00	. 64	9. 00	1. 015	500	1. 85
1. 08	. 67	10. 00	1. 02	600	1. 95
1. 16	. 70	12. 00	1. 03	700	2. 01
1. 24	. 72	15. 00	1. 04	800	2. 09
1. 33	. 74	18. 00	1. 05	900	3. 15
1. 42	. 76	20. 00	1. 06	1, 000	2. 20
1. 50	. 78	25. 00	1. 08	1, 200	2. 27
1. 68	. 82	30. 00	1. 13	1, 500	2. 35
1. 85	. 84	35. 00	1. 15	2, 000	2. 49
2. 00	. 86	40. 00	1. 17	2, 500	2. 59
2. 15	. 87	50. 00	1. 19	3, 000	2. 66
2. 35	. 90	60. 00	1. 22		
2. 50	. 91	70. 00	1. 25		

- Thinning of the saturated zone as drawdown increases
- Observation well lag
- Aquifer heterogeneity

Where these factors are present, a log-log plot of s versus t from an observation well appears similar to plots of test data from a leaky aquifer except that the drawdown seldom stabilizes with time, as may be the case with the leaky aquifer.

The early part of a data curve, influenced by delayed drainage, may permit a good fit to the Theis type curve and give a reasonably reliable value for T, but S may be so small as to be in the artesian range. As pumping continues, the area of influence and the drawdown decrease in growth at a logarithmic rate, and the delayed drainage tends to catch up with these portions of the response, resulting in the relatively flat portion of the data curve. Eventually, the various factors reach a balance and the final slope develops from which reliable values of

T and S may be computed using the Theis nonequilibrium solution. This solution, however, may require pumping for an excessively long period and involves a tedious trial-and-error process.

Boulton [5,6] developed equations for treating delayed drainage influence; Prickett [61] and Stallman [70] improved on Boulton's work to make it more amenable to practical application. The abbreviated form of Boulton's equations are:

$$s=\frac{Q}{4\pi T}W\left(u_{av}\frac{r}{M}\right) \tag{29}$$

$$u_a=\frac{r^2S}{4Tt} \tag{30}$$

$$u_y=\frac{r^2S_y}{4Tt} \tag{31}$$

$$d=\frac{\left(\frac{r}{M}\right)^2\frac{1}{u_a}}{4t}\text{ type A curve} \tag{32}$$

$$d=\frac{\left(\frac{r}{M}\right)^2\frac{1}{u_y}}{4t}\text{ type B curve} \tag{33}$$

where in consistent units:

s=drawdown at time t and at a distance r, L,

Q=rate of discharge, $\frac{L^3}{t}$,

T=transmissivity, $\frac{L^2}{t}$,

$W\left(u_{av}\frac{r}{M}\right)$=well function of u when η tends to infinity,

t=time since start of pumping, t,

r=distance of observation well from test well, L,

S=early time coefficient of storage (nondimensional),

S_y=true specific yield or coefficient of storage (nondimensional),

d=reciprocal of the delay index, t,

$\eta=\frac{S+S_y}{S}$ (nondimensional), and

M=aquifer thickness, L.

On figure 5–16, the type curve consists of two families of curves. Type A curves are shown to the left of the $\frac{r}{M}$ values and type Y curves are shown to the right of the $\frac{r}{M}$ values.

The type A curves are applicable to early time pumping and type Y curves to later time pumping when response of the aquifer is in accord with the Theis nonequilibrium assumptions. Values for $\frac{1}{u_a}$ are shown at the top of the type curves and $\frac{1}{u_y}$ values are at the bottom.

The method of using the type curves for finding values of T, S, S_y, η, and d, and the delay index, that is, the reciprocal of d, will now be discussed.

The field data s and t (table 5–7) are plotted on log-log paper (fig. 5–15) to the same scale as the type curves. This time drawdown curve is analogous to two of the families of free aquifer-type curves shown on figure 5–16. The field data curve is superimposed on the type curves, while keeping the axes of both sheets parallel, and moved horizontally and vertically until the best possible fit is found for type A curves. If necessary, interpolate a position and $\frac{r}{M}$ value between the type curves. The $\frac{r}{M}$ value of the match-type curve is noted. In this match position, a point on the intersection of the major axes of the type curve is selected and the corresponding point on the data curve is marked. The point selected may be anywhere on the type curve provided it overlies part of the data curve. The coordinates of the common match points are s, $\frac{1}{u_a}$, $W\left(u_{av}\frac{r}{M}\right)$, and t. These values are substituted in equation (29) to determine T. The value for S is then calculated using the determined value of T and the values of $\frac{1}{u_a}$ and t from the match points using equation (30).

The curves are then moved horizontally with respect to each other and as much as possible of the late field data curve is matched to the type Y curve. The type Y curve should have the same $\frac{r}{M}$ value as was used when matching to the type A curve. A similar match of the field data curve to the type curve is made. The coordinates of the match point s, $\frac{1}{u_y}$, $W\left(u_{ay}\frac{r}{M}\right)$, and t are substituted in equations (29) and (31) to determine T and S_y. The value of T is calculated using equation (29) and coordinates $W\left(u_{ay}\frac{r}{M}\right)$, and s. This T value should be similar to that obtained in matching the early time data. The

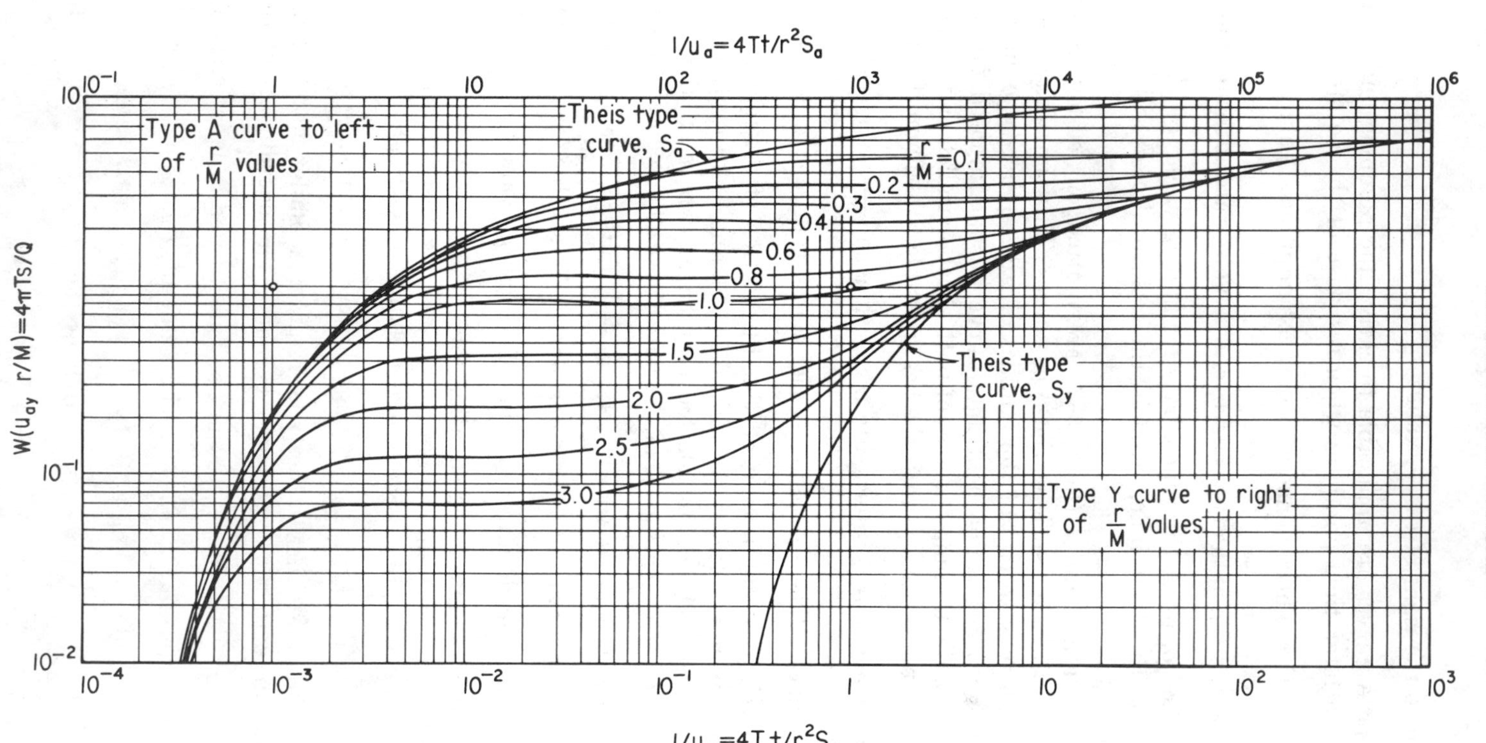

FIGURE 5–16.—Delayed drainage type curves. 103–D–1443.

value of S_y is determined using equation (31), the calculated value of T, and coordinates $\frac{1}{u_y}$ and t from the later time match.

The reciprocal of the delay index, d, is calculated by substituting the $\frac{r}{M}$ value with the later time, match point coordinates $\frac{1}{u_a}$, and t in equation (32).

$$\text{If } \eta = \frac{S+S_y}{S} > 6.5,$$

values of the aquifer characteristics determined are probably of acceptable reliability whether Jacob's corrections for drawdown and partial penetration have or have not been made. If $\eta < 6.5$ and corrections have not been made, they should be computed, a new data curve prepared, and the aquifer characteristics recalculated. If the recalculation results in $\eta > 6.5$, the new values are probably sufficiently reliable for most applications. However, if η still remains less than 6.5, Boulton's equation (equation 6 in reference [61]) should be used.

As an example of a delayed drainage solution, figure 5–15 was matched to the A-type curve on figure 5–16. Where

$$\frac{1}{u_a}=1 \text{ and } W\left(u_{ay}\frac{r}{M}\right)=1$$

on the type curve, $s=0.56$ foot and $t=0.18$ minute (point a on figure 5–15). In addition, the test data showed $Q=144.4$ ft^3/min and $r=73$ ft. Inserting the appropriate values in equation (29):

$$0.56=\frac{144.4\times 1}{12.57\times T} \text{ and } T=20.5 \text{ ft}^2/\text{min}.$$

Referring to equation (30)

$$u_a=\frac{r^2S}{4Tt}$$

when:

$u_a=1$, $r=73$ ft, $t=0.18$ min, and $T=20.5$ ft^2/min, then:

$$S=\frac{4\times 20.5\times 0.18}{73^2\times 1}=0.003$$

Moving the data curve to the right on the type curve to the best late-time match where $s=0.56$ foot (point b on figure 5–15) gives a value of 13.8 min for:

$$W\left(u_{av}\frac{r}{M}\right)=1 \text{ and } \frac{1}{u_y}=1$$

on the type curve. Inserting the appropriate values in equation (29) does not change the value of T, but using equation (31):

$$S_y=\frac{4\times 20.5\times 13.8}{73^2\times 1}=0.21$$

5-11. Determination of Aquifer Boundaries.—The equilibrium equation is based on the concept of an aquifer of infinite areal extent, which obviously does not exist. Finite boundaries (sec. 2-9) in one or more directions complicate application of the equation. Suitably located image wells serve to simulate hydraulically the flow regime caused by such boundaries and may permit the hydraulic system to be analyzed as being in an aquifer of infinite areal extent.

Although most boundaries are not abrupt nor do they follow a straight line, it is usually possible to treat them as abrupt changes along a straight line. The more distant the boundary from the discharging well, the less the magnitude of influence will be on the drawdown and the longer the time will be before drawdown is influenced. Figures 5-17 and 5-18 show boundary relationships to image wells.

Figure 5-18 illustrates a recharge boundary where an aquifer is bounded on one side by a fully penetrating stream. As the cone of depression expands and eventually encounters the boundary, both the rate of expansion of the cone and the rate of drawdown in the discharging well are slowed. With continuation of pumping, the cone of depression expands along the stream until the well receives a major contribution from the stream.

To analyze the influence of a recharge boundary, a recharge image well is hypothetically located on the opposite side of the boundary from the real discharging well and at an equal distance from the boundary. The image well recharges water to the aquifer at the same rate as the real well discharges. Consequently, the water level buildup by the image well cancels the drawdown at the boundary resulting from the real well. This satisfies the limits of the problem.

Under actual field conditions, the stream channel is seldom fully penetrating and hydraulic continuity between the stream and aquifer may be restricted by partial plugging of the streambed by fine materials. However, such deviations from ideal conditions merely tend to cause a shifting of the boundary farther from the discharging well.

Similarly, in the case of an impermeable boundary, figure 5-17 shows a fault which has brought impermeable material in contact with an aquifer. In such a situation, there is no flow across the

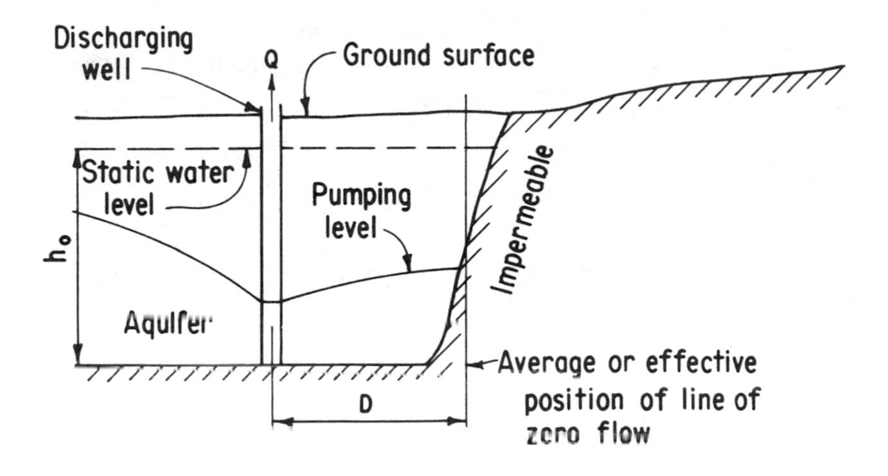

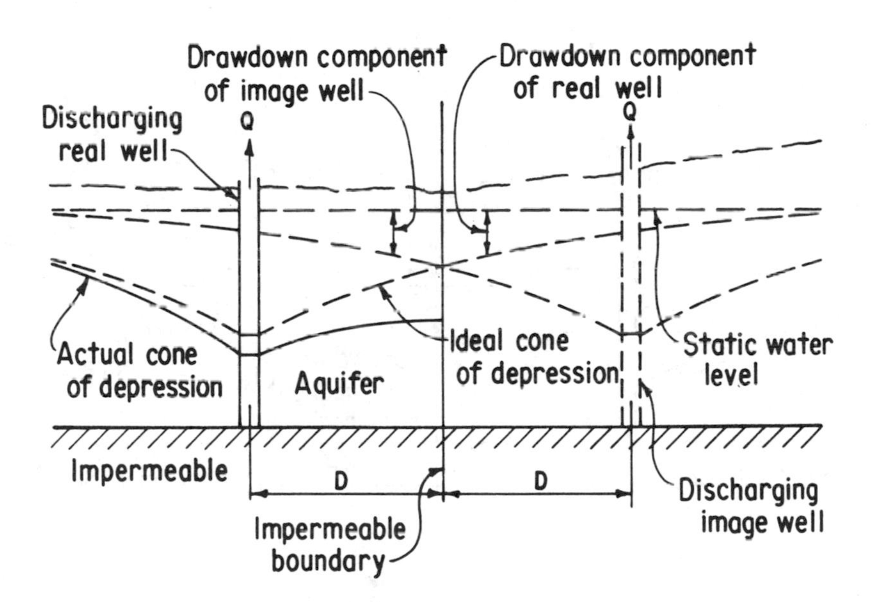

(After U.S. Geological Survey)

FIGURE 5-17.—Relationship of an impermeable boundary and an image well 103-D-1446.

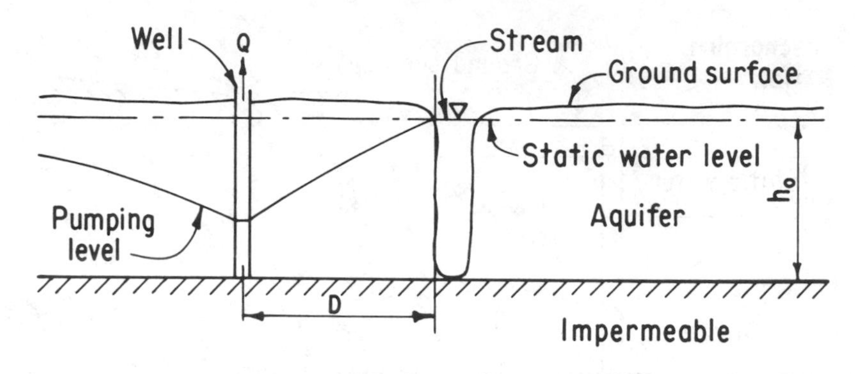

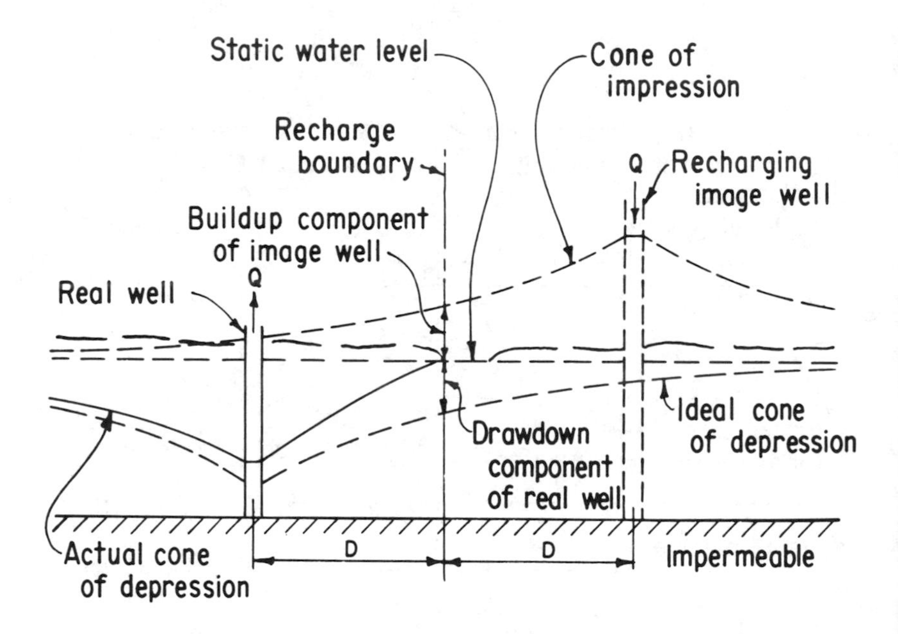

(after U.S. Geological Survey)

FIGURE 5–18.—Relationship of a recharge boundary and an image well. 103–D–1445.

boundary. A discharging image well is hypothetically located an equal distance on the opposite side of the boundary from the real discharging well. The image well is assumed to discharge at the same rate as the real well. Theoretically, a ground-water divide forms along the line of the boundary as the real and image well cones of depression intercept. This satisfies the limits of the problem.

To solve problems involving image wells, the principle of superposition as described in section 2–20 is used. This method provides that when two or more cones of depression interfere or overlap, the resultant drawdown can be determined by adding algebraically the individual drawdowns involved at the point of interest. Figures 5–17 and 5–18 show the theoretical, individual drawdowns of the real and image wells and the resultant or actual drawdowns.

In many instances, sufficient geologic and hydrologic information is available to permit anticipation of the direction and distance to boundaries from a well. Where such information is not available, the influence of the well must be analyzed. The resultant cones of depression have typical shapes, depending on the type of boundary, which may aid in analyzing the boundaries. Analysis by either the Theis type curve solution (sec. 5–4) or Jacob's approximation (sec. 5–5) will provide recognition criteria and estimated distance and direction of boundaries.

The approximation method is advantageous for most analyses because changes in slope of a straight line are usually easier to recognize than in a log-log curve. Some data curves may present problems in differentiating between positive boundary, leaky aquifer, or delayed drainage conditions [14,15,80].

(a) *Boundary Location by Type Curve Analyses.*—Figure 5–19 shows the relationship between a type curve and actual data from observation wells for a recharge boundary; figure 5–20 shows similar relationships for an impermeable boundary. These figures were plotted from tabulated data that are partially reproduced in tables 5–8 and 5–9. Figure 5–19 shows s versus t plotted on log-log paper, whereas figure 5–20 shows s versus $1/t$ plotted on log-log paper to show the two different methods plotted. Either method can be used.

To determine the distance to a boundary from a discharging well, the data curve is fitted to the type curve as shown on figures 5–19 and 5–20. Then, successive steps to read the data are: (1) difference in drawdown between the type curve and the data curve, s_i, is noted for any time, t_i; (2) this difference is spotted on the s axis and the time, t_R, noted where the s_i value intersects the trace of the type curve. The values of t_i and t_R are the times of equal drawdown resulting from the image well and real well, respectively. On figure 5–19, for observation

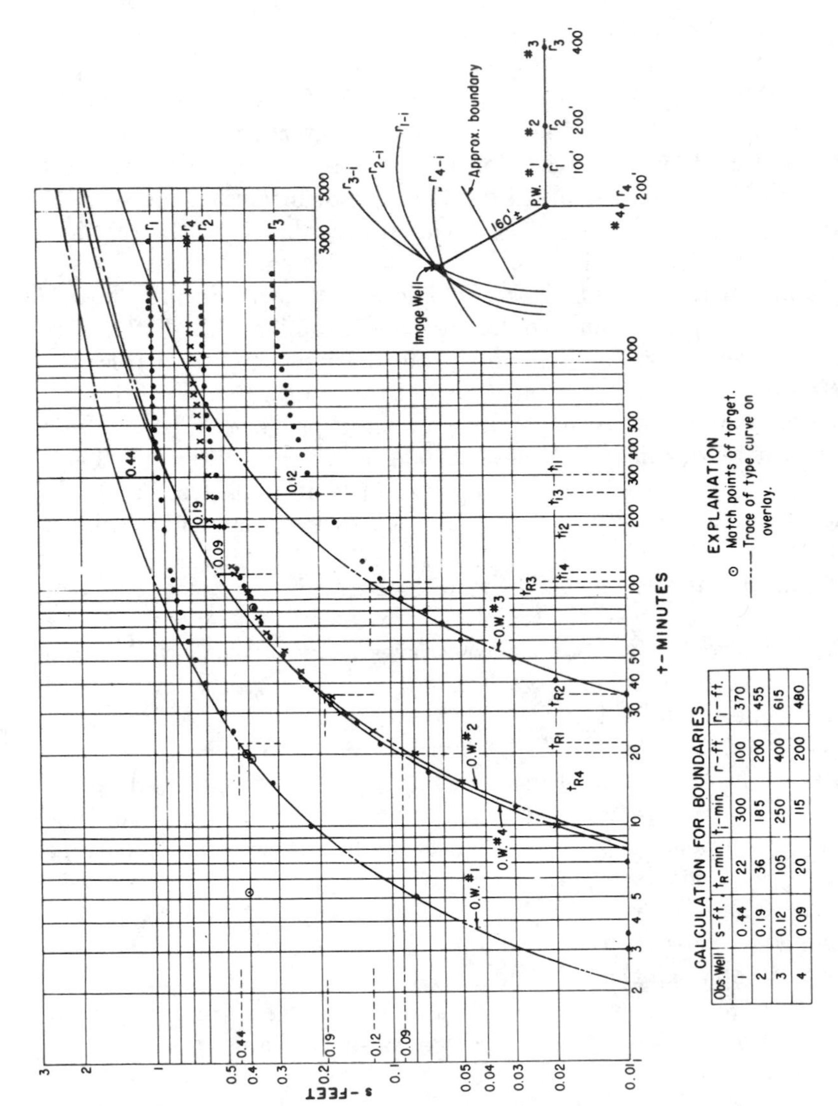

CALCULATION FOR BOUNDARIES

Obs. Well	s-ft.	t_R-min.	t_i-min.	r-ft.	r_i-ft.
1	0.44	22	300	100	370
2	0.19	36	185	200	455
3	0.12	105	250	400	615
4	0.09	20	115	200	480

FIGURE 5-19.—Recharge boundary location by type curve analysis. 103-D-796.

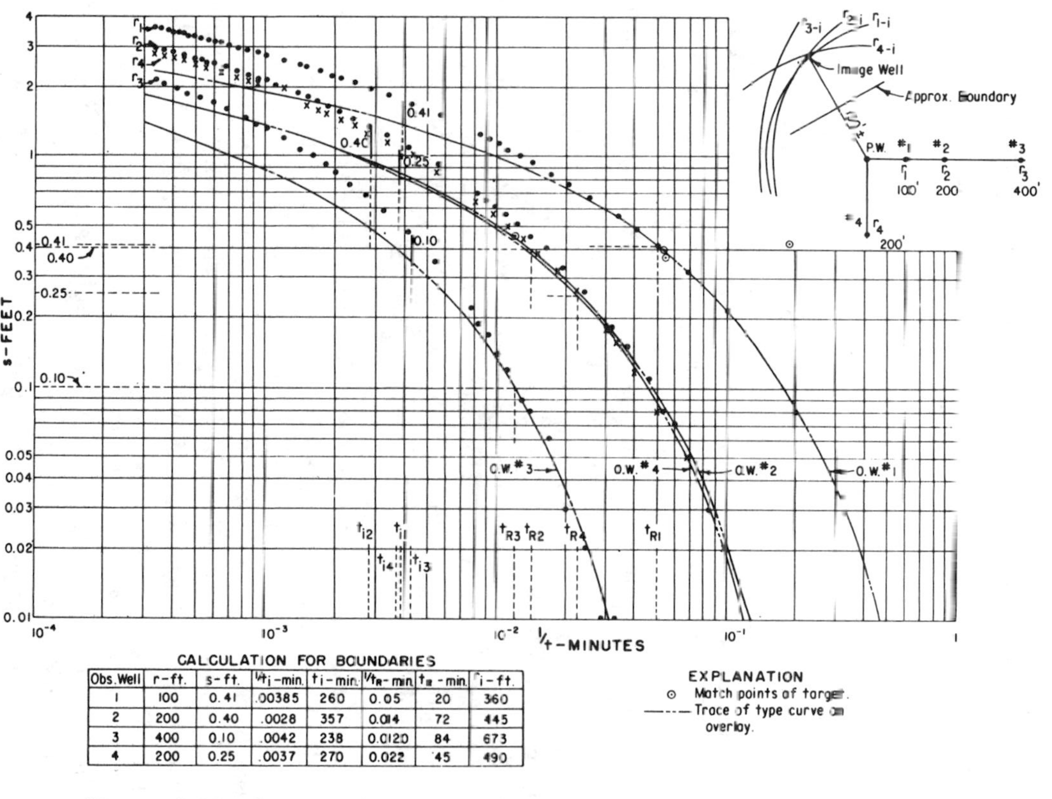

Obs. Well	r-ft.	s-ft.	$1/t_i$-min.	t_i-min.	$1/t_R$-min.	t_R-min.	r_i-ft.
1	100	0.41	.00385	260	0.05	20	360
2	200	0.40	.0028	357	0.014	72	445
3	400	0.10	.0042	238	0.0120	84	673
4	200	0.25	.0037	270	0.022	45	490

FIGURE 5–20.—Impermeable boundary location by type curve analysis. 103–D–797.

well No. 1 at a radius of 100 feet from the test well, $s_i=0.44$ foot when $t_i=300$ minutes. Transferring $s_i=0.44$ foot to the s axis of the graph and reading across to the type curve, $t_R=22$ minutes. The time values are related by:

$$\frac{r_i^2}{t_i}=\frac{r_R^2}{t_R} \tag{34}$$

The distance r_R from the real well to the closest observation well being known, then:

$$r_i=\sqrt{\frac{(100)^2\times 300}{22}}=370 \text{ feet.}$$

This process is followed for each observation well and a circle is drawn on a scaled plan of the test site layout using a radius equal to r_i distance computed for each observation well. The intersection of the circles marks the approximate location of the image well. If the observation wells lie in a straight line, especially if normal to the boundary, the circles may be nearly tangent to each other and the location of the image well may be weakly determined. This problem can be minimized by offsetting at least one observation well, as well No. 4 was offset in the plan views of figures 5–19 and 5–20.

To locate the boundary, a line is drawn between the real well and the image well. The boundary is located at the midpoint and is normal to the line. Values of T and S may be calculated using the same procedures explained previously. However, the values of t and s must be taken from the early portion of the data curve that coincides with the type curve before the deviation caused by boundary conditions is evident.

Either a leaky roof aquifer or a number of boundaries near the test area will occasionally cause the data curve to depart from the type curve in a manner that makes analysis difficult. In such instances, analysis by the leaky roof method should be tried if the curves appear similar. If this is not satisfactory, analysis by either the straight-line approximation or delayed storage method may permit a reliable interpretation.

(b) *Boundary Location by Straight-Line Approximation.*—Equation (16) in section 5–5 on the straight-line approximation method of solution shows that the slope of the line of the semilog graph is dependent only upon the rate of pumping and the transmissivity of the aquifer. In a pumping test, the discharge is held constant and the transmissivity is assumed to be constant. The plot of drawdown in an observation well versus time on a log scale initially follows a curve which changes gradually to a straight line as pumping continues.

TABLE 5-8.—*Tabulated discharging well test data for determining a recharge boundary*

Project: Tongue Valley

Feature: Drainage tests, Area 1

Location: 1,280 ft E. of SW. corner sec. 10, T. 25 N., R. 16 W.

Discharge measured by: 7- by 10-inch orifice

Drawdown measured by: Electrical sounder

Reference point: North side of casing collar

PUMP TEST NO. 1, PUMPING WELL

Date	Time	Depth to water, ft	Drawdown *s*, ft	Elapsed time, min	Manometer reading, in	Discharge, gal/min	Remarks
2-14	1130	38. 42					
2-15	1120	38. 47					
2-16	1135	38. 53					
2-17	1125	38. 59					
2-18	0710	[1] 38. 64					
	0805	38. 64	0. 00	0			Pump started.
	0807	42. 36	3. 72	2	34. 5	1, 230	
	0809	42. 67	4. 03	4	34. 0	1, 220	
	0811	42. 85	4. 21	6	34. 0	1, 220	
	0813	42. 97	4. 33	8	34. 0	1, 220	
	0815	43. 09	4. 45	10	33. 7	1, 214	Pump off 2-20 at 1115.
	0820	43. 27	4. 63	15	33. 7	1, 214	
	0825	43. 39	4. 75	20	33. 4	1, 208	
	0830	43. 49	4. 85	25	33. 5	1, 210	Avg discharge 1,210 gal/min.
	0835	43. 55	4. 91	30	33. 5	1, 210	
	0845	43. 67	5. 03	40	33. 4	1, 208	
	0855	43. 75	5. 11	50	33. 4	1, 208	
	0905	43. 80	5. 16	60	33. 6	1, 212	
	0915	43. 85	5. 21	70	33. 5	1, 210	
	0925	43. 89	5. 25	80	33. 4	1, 208	
	0935	43. 90	5. 26	90	33. 5	1, 210	
	0945	43. 94	5. 30	100	33. 5	1, 210	
	1005	43. 97	5. 33	120	33. 5	1, 210	
	1105	44. 03	5. 39	180	33. 5	1, 210	
	1205	44. 06	5. 42	240	33. 6	1, 212	
	1305	44. 09	5. 45	300	33. 7	1, 214	
	1405	44. 12	5. 48	360	33. 8	1, 216	Adjusted discharge value.
	1505	44. 15	5. 51	420	33. 5	1, 210	
	1605	44. 14	5. 50	480	33. 5	1, 210	
	1705	44. 14	5. 50	540	33. 5	1, 210	

[1] Static water level.

TABLE 5–8.—*Tabulated discharging well test data for determining a recharge boundary*—Continued

Project: Tongue Valley
Feature: Drainage tests, Area 1
Location: 100 ft N. of pumping well sec. 10, T. 25 N., R. 16 W.

Drawdown measured by: "Popper"
Reference point: West side of casing collar

PUMP TEST NO. 1, OBSERVATION HOLE NO. 1, r=100 FT

Date	Time	Depth to water, ft	Drawdown s, ft	Correction [1]	Corrected drawdown, ft	Elapsed time, min	Remarks
2–14---	1125	38. 39					
2–15---	1115	38. 44					
2–16---	1130	38. 50					
2–17---	1120	38. 56					
2–18---	0705	[2] 38. 61					
	0805	38. 61	0. 00			0	Pump started.
	0810	38. 69	. 08			5	
	0820	38. 83	. 22			10	
	0825	38. 93	. 32			15	
	0830	39. 02	. 41			20	
	0835	39. 08	. 47			25	Pump off 2–20 at 1115.
	0845	39. 14	. 53			30	
	0855	39. 22	. 61			40	
	0905	39. 28	. 67			50	
	0915	39. 32	. 71			60	
	0925	39. 36	. 75			70	
	0935	39. 39	. 78			80	
	0945	39. 41	. 80			90	
	0955	39. 43	. 82			100	
	1005	39. 44	. 83			110	
	1105	39. 46	. 85			120	
	1205	39. 50	. 89	0. 01	0. 90	180	
	1305	39. 54	. 93	. 01	. 93	240	
	1405	39. 55	. 94	. 01	. 95	300	
	1505	39. 55	. 94	. 02	. 96	360	
	1605	39. 56	. 95	. 02	. 97	420	
	1705	39. 59	. 96	. 02	. 98	480	
	1805	39. 59	. 96	. 02	. 98	540	
	1905	39. 57	. 96	. 03	. 99	600	
	2005	39. 57	. 96	. 03	. 99	660	

[1] Static water level rising.
[2] Static water level.

TABLE 5-9.—*Tabulated discharging well test data for determining an impermeable boundary*

Project: Dry Lake

Feature: Playa Reservoir

Location: SW ¼ sec. 10, T. 4 N., R. 21 E.

Discharge measured by: 7- by 10-inch orifice

Drawdown measured by: Electrical sounder

Reference point: North side of casing collar

PUMP TEST NO. 1, PUMPING WELL

Date	Time	Depth to water, ft	Drawdown s, ft	Manometer reading, in	Discharge, gal/min	Elapsed time, min	Remarks
7-15	0810	30. 90					
7-16	0815	30. 91					
7-17	0810	30. 90					
7-18	0840	[1] 30. 90					
	0900					0	Pump started.
	0902	34. 20	3. 30	34. 0	1, 220	2	Pump off 7-21 at 0900.
	0904	34. 90	4. 00	34. 5	1, 230	4	
	0906	35. 00	4. 10	34. 0	1, 220	6	
	0908	35. 20	4. 30	34. 0	1, 220	8	Avg $Q=$ 1,220 gal/min. 161.8 ft³/min.
	0910	35. 30	4. 40	33. 7	1, 214	10	
	0915	35. 40	4. 50	33. 7	1, 214	15	
	0920	35. 60	4. 70	33. 4	1, 208	20	
	0925	35. 70	4. 80	33. 5	1, 210	25	
	0930	35. 80	4. 90	33. 5	1, 210	30	
	0940	35. 90	5. 00	33. 5	1, 210	40	
	0950	36. 00	5. 10	33. 5	1, 210	50	
	1000	36. 10	5. 20	33. 5	1, 210	60	
	1010	36. 20	5. 30	33. 5	1, 210	70	
	1020	36. 30	5. 40	33. 5	1, 210	80	
	1030	36. 40	5. 50	33. 5	1, 210	90	
	1040	36. 50	5. 60	33. 5	1, 210	100	
	1100	36. 60	5. 70	33. 4	1, 208	120	
	1200	36. 90	6. 00	33. 4	1, 208	180	
	1300	37. 10	6. 20	33. 5	1, 210	240	
	1400	37. 30	6. 40	33. 6	1, 212	300	
	1500	37. 40	6. 50	33. 6	1, 212	360	
	1600	37. 50	6. 60	33. 5	1, 210	420	
	1700	37. 60	6. 70	33. 5	1, 210	480	
	1800	37. 70	6. 80	33. 5	1, 210	540	
	1900	37. 80	6. 90	33. 5	1, 210	600	
	2000	37. 90	7. 00	33. 5	1, 210	660	
	2100	38. 00	7. 10	33. 4	1, 208	720	
	2200	38. 01	7. 20	33. 5	1, 210	780	

[1] Static water level.

TABLE 5-9.—*Tabulated discharging well test data for determining an impermeable boundary*—Continued

Project: Dry Lake
Feature: Playa Reservoir
Location: SW ¼ sec. 10, T. 4 N., R. 21 E.

Drawdown measured by: "Popper"
Reference point: North side of casing collar
$r = 100$ ft north of pumped well

PUMP TEST NO. 1, OBSERVATION WELL NO. 1

Date	Time	Depth to water, ft	Draw-down s, ft	Elapsed time, min	Remarks
7-15	0815	30.87			
7-16	0820	30.87			
7-17	0815	30.88	0.00		
7-18	0830	[1] 30.88			
	0900			0	Pump started.
	0905	30.96	.08	5	
	0910	31.10	.22	10	
	0915	31.20	.32	15	
	0920	31.29	.41	20	Pump off 7-21 at 0900.
	0925	31.37	.49	25	
	0930	31.44	.56	30	
	0940	31.55	.67	40	
	0950	31.65	.77	50	
	1000	31.73	.85	60	
	1010	31.83	.95	70	
	1020	31.89	1.01	80	
	1030	31.96	1.08	90	
	1040	32.02	1.14	100	
	1050	32.08	1.20	110	
	1100	32.13	1.25	120	
	1200	32.39	1.51	180	
	1300	32.58	1.70	240	
	1400	32.75	1.87	300	
	1500	32.87	1.99	360	
	1600	32.98	2.10	420	
	1700	33.08	2.20	480	
	1800	33.16	2.28	540	
	1900	33.24	2.36	600	
	2000	33.34	2.46	660	
	2100	33.38	2.50	720	
	2300	33.51	2.63	840	
7-19	0100	33.65	2.77	960	

[1] Static water level.

When the area of influence of a discharging well reaches an impermeable boundary, the rate of drawdown will be doubled. This results from the addition of the equal influence caused by the hypothetical image well to the influence of the real well. Influence from a second impermeable boundary will triple the drawdown. A plot of drawdown versus time on a log scale shows straight sections of lines (or legs) each having a slope that is an approximate multiple of the initial slope. The transition from each leg to another is not sharp, but follows a curve.

If the influence from two or more boundaries located at approximately equal distances reaches an observation well almost simultaneously, the drawdown data will plot on a path of increasing curvature and the straight-line portion of the plot may be three or more times steeper than the initial slope. Approximate locations of image wells may be obtained by drawing tangents with double or triple (or one-half or one-third) slope values to the initial slope of the data.

To determine the location of an image well by using the straight-line approximation method, the various legs of the curve are extended through the plotted points that fall on a straight line. Figure 5–21 shows an impermeable boundary determination. The time, t_i, is determined for a certain drawdown difference between the second leg and the extension of the first leg. This point should not be chosen where the plotted points are on a curve. In the example on figure 5–21, the location is chosen where $s=1$ foot. This drawdown of 1 foot is then located on the straight line of the first leg and the time, t_R, noted. If a third leg appears, the same procedure is followed in determining the amount of drawdown between the second and third legs, but referring the time and amount of drawdown to the fifth leg. Since the distance from the discharging well to each observation well is known, the distance to the image well causing the same amount of drawdown (or recharge for a recharge boundary) can be determined by the relationship shown in the previous equation (34):

$$\frac{r_i^2}{t_i}=\frac{r^2_R}{t_R}$$

In figure 5–19 for the observation well at a radius r_i of 100 feet and a drawdown $s=1$ foot, the values of t are: $t_i=1{,}380$ minutes and $t_R=94$ minutes. From this, the radius of the image well from the observation well was calculated as 385 feet, and the boundaries can now be located.

(c) *Multiple Boundaries.*—Where two or more boundaries are present, image wells should be added as shown on figure 5–22 to

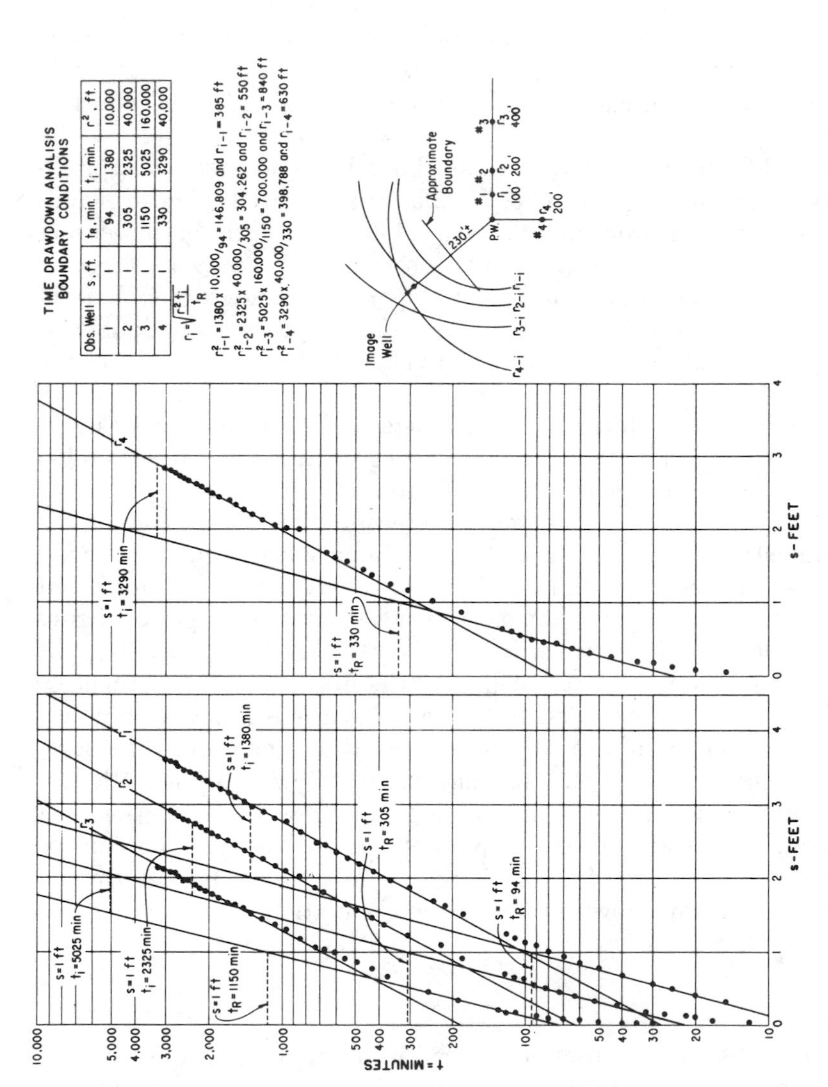

FIGURE 5–21.—Impermeable boundary determination by straight-line analysis. 103–D–798.

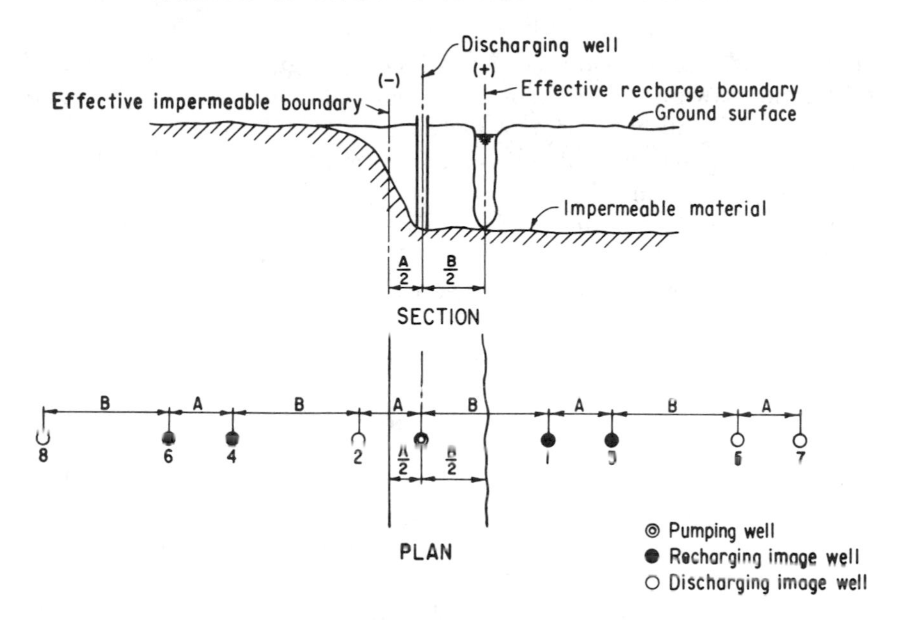

FIGURE 5–22.—Use of multiple image wells for boundary determination. 103–D–1447.

maintain the condition of no flow across an impermeable boundary and no drawdown along a recharge boundary. Figure 5–22 also illustrates a discharging well between impermeable and recharge boundaries. This figure shows that when the first two image wells have been located, a repetitious pattern of image well spacings and discharge-recharge characteristics are present which permit the location and type of additional image wells. Theoretically, image wells extend to infinity in either direction, but in practice the number is limited to that which results in an acceptable effect at the boundaries. The analysis then consists of analyzing the system of real and image wells as though they were in an ideal infinite aquifer. References [2,14,74,76,80] discuss various complex boundary conditions and their treatment.

5–12. Interference and Well Spacing.—When a well discharges, a cone of depression is formed with its axis and lowest point at the well. As discharging continues, the circumference of the cone extends farther and farther from the well and the cone continues to deepen. Theoretically, the cone continues to expand to infinity; however, under actual conditions this is not the case. Recharge may balance discharge thereby stabilizing the cone both horizontally and vertically, and the rate and amount of drawdown at some distance becomes too small to measure. This distance may be anywhere from several hundred feet to several miles, depending on time and rate of discharge and aquifer characteristics. The distance to the point of negligible drawdown (radius of influence) can be calculated by assuming a very small value for s and using the Theis nonequilibrium equation (6) in section 5–3. The values of Q, t, S, and T must be known to use the equation.

Wells in a well field designed for water supply should be spaced as far apart as possible so their areas of influence will have a minimum of interference with each other. Because drawdown interference is additive, the capacity of wells with intersecting areas of influence will be reduced or drawdowns increased.

The solid lines on figure 5–23 show drawdown in wells A and B, assuming that only one well is discharging at a time and that the aquifer is infinite. When both wells are discharging at the same time, drawdown curves between the wells would be additive and their combined drawdown is shown by the dotted lines. The drawdown curve with both wells discharging can be estimated for any point, see line segment 0–0 on figure 5–23. Drawdown caused by well A is shown by line segment 1–2, well B drawdown is segment 1–3, and the drawdown caused by both wells is shown by line segment 1–4, or segment 1–3 plus segment 1–2. This is the principle of superposition or mutual interference and applies in well hydraulics whether the wells are real or images. For drainage wells, it may be desirable to lay out the well spacing to intentionally cause interference, thereby increasing the drainage effect at the midpoints between wells. In any case, the curve of the cone of depression should be known either by calculation or field measurement so that proper spacing of wells can be established [1,2,15,28,46,52,53,62,80].

5–13. Barometric Pressure and Other Influences on Water Levels.— Water levels in wells in many artesian aquifers respond to changes in atmospheric pressure. An increase in atmospheric pressure causes the water level to decline and a decrease causes the water level to rise. The barometric efficiency of an aquifer may be expressed:

$$BE=\frac{s_w}{s_b}\times 100 \tag{35}$$

where:

BE = the barometric efficiency (nondimensional),
s_w = the water level change, L, and
s_b = the barometric pressure change, L.

Barometric efficiency can be estimated by plotting the water level changes as ordinates and barometric pressure changes as abscissas on rectangular coordinate paper as shown on figure 5–24. The slope of a straight line fitted through the plotted points is the barometric

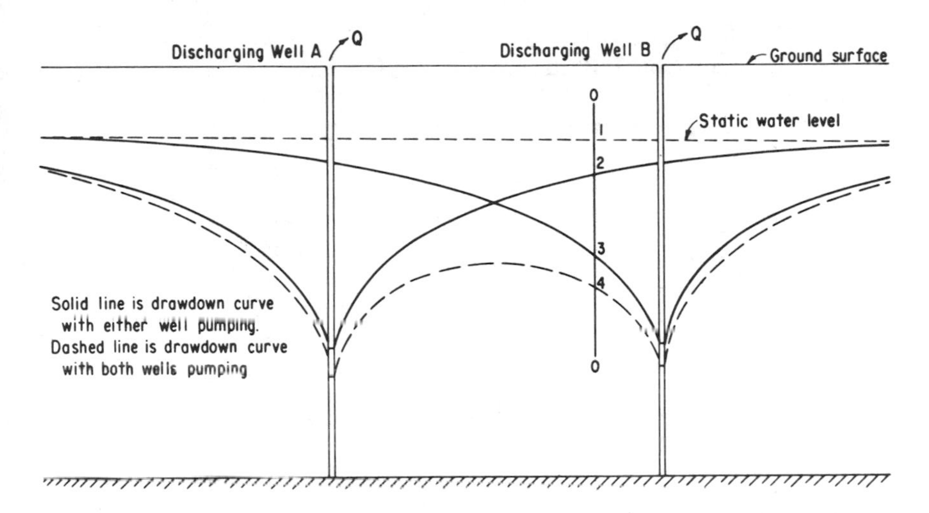

FIGURE 5-23.—Interference between discharging wells 103-D-1448.

efficiency which may be as high as 80 percent. In conducting discharging well tests in artesian aquifers when drawdowns are expected to be small, records of water levels and barometric pressures should be kept for several days prior to the start of the test to determine the influence of pressure changes on water levels. Barometric readings are continued during the test and measured drawdowns corrected accordingly by applying the barometric efficiency (table 5-10). For example, a well which showed a 50-percent barometric efficiency would have a water rise of 0.05 foot for each decrease of 0.10 foot in barometric pressure measured in head of water and conversely. Such values should be added to or subtracted from the measured drawdowns to eliminate the influence of atmospheric pressure changes.

Similar fluctuations in water levels may result from ocean and earth tidal fluctuations, earthquakes, and passing trains [15,73].

5-14. Well Performance Tests.—The tests described thus far are intended primarily to determine aquifer characteristics, including transmissivity, storativity, and boundary conditions. Similar tests are also conducted to determine well characteristics. The two principal well characteristics of well performance, yield and drawdown, are measures of the capacity of the well to produce water; both should be taken into account when capacity is considered. More specifically, the tests are conducted for the following reasons: (1) to determine general adequacy of development prior to completion; (2) upon completion, to determine general capacity, establish a baseline for later tests, and to determine correct pump capacity and setting; and (3) to determine deterioration of the well following a period of use.

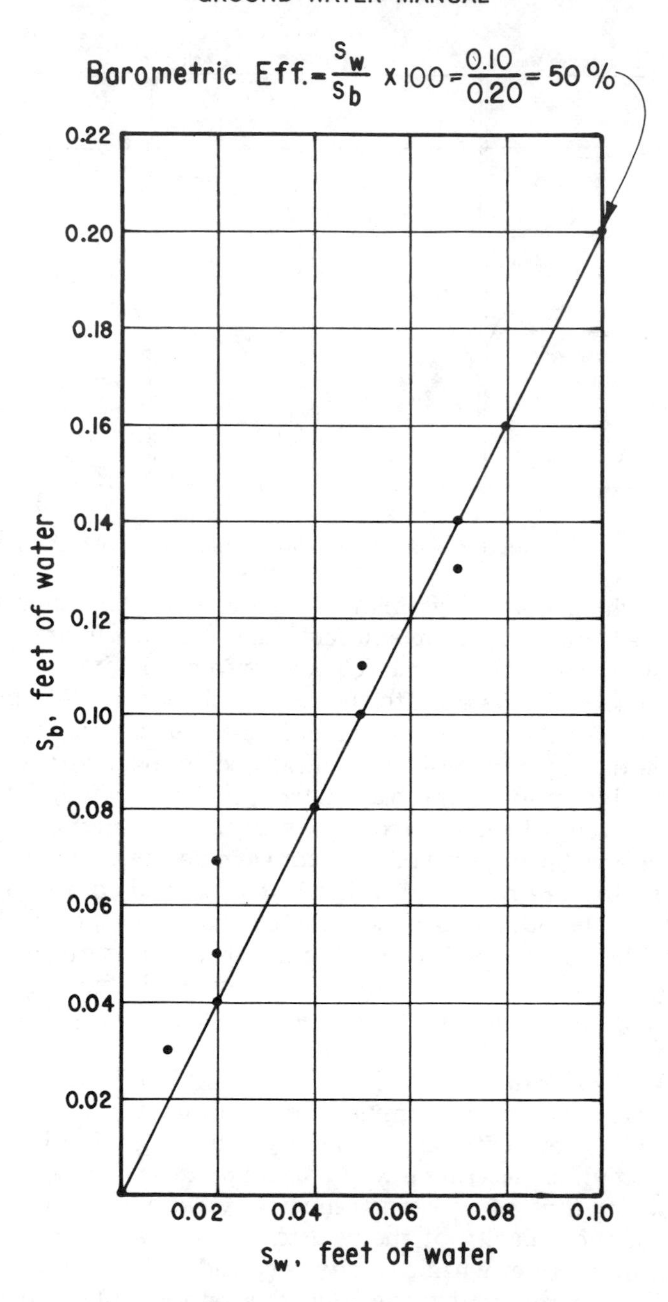

FIGURE 5–24.—Relationship of water levels and barometric pressure. 103–D–1449.

TABLE 5-10.—*Water level and barometric pressure data*

Time	Inches of mercury [1]	Feet of water	Change in barometric pressure, ft	Summation of change in barometric pressure, ft	Depth to water level, ft	Water level elevation, ft	Change in water level, ft	Summation of change in water level, ft
0800	29. 67	33. 65	0. 0	0. 00	37. 45	5263. 30	0. 00	0. 00
0900	29. 65	33. 62	−0. 03	−0. 03	37. 46	+0. 31	+0. 01	+0. 01
1000	29. 64	33. 61	−0. 01	−0. 04	37. 47	+0. 32	+0. 01	+0. 02
1100	29. 63	33. 60	−0. 01	−0. 05	37. 47	+0. 32	+0. 00	+0. 04
1200	29. 61	33. 57	−0. 03	−0. 08	37. 49	+0. 34	+0. 02	+0. 04
1300	29. 58	33. 54	−0. 03	−0. 11	37. 50	+0. 35	+0. 01	+0. 05
1400	29. 56	33. 52	−0. 02	−0. 13	37. 52	+0. 37	+0. 02	+0. 07
1500	29. 50	33. 45	−0. 07	−0. 20	37. 55	+0. 40	+0. 03	+0. 10
1600	29. 53	33. 49	+0. 04	−0. 16	37. 53	−0. 38	+0. 02	+0. 08
1700	29. 55	33. 51	+0. 02	−0. 14	37. 52	+0. 37	−0. 01	+0. 07
1800	29. 56	33. 52	+0. 01	−0. 13	37. 52	+0. 37	−0. 00	+0. 07
1900	29. 58	33. 54	+0. 03	−0. 10	37. 50	+0. 35	−0. 02	+0. 05
2000	29. 60	33. 57	+0. 03	−0. 07	37. 47	+0. 32	−0. 03	+0. 02

[1] 1 inch of mercury = 1.134 feet of water.

(a) *Step Tests.*—Step tests are used principally to determine the comparative specific capacity; that is, yield versus drawdown, of a well at different rates of discharge. Usually the tests are started at a low step, such as 25 percent of design capacity, and increased in three or four incremental steps, such as 50, 75, 100, and 125 percent of capacity. Ideally, a recovery period should follow each step, but the costs of such a procedure often cannot be justified. More commonly, the entire test is run without pause. The first step is continued until the water level reaches approximate stabilization. This usually is reached within a period ranging from 1 to 4 hours. Subsequent steps are run for the same length of time as the first.

The specific capacity of wells declines with increasing discharge and length of pumping time. The decline caused by increasing discharge is usually small in wells in artesian aquifers, but may be large in wells in free aquifers. If a step test of a new well shows increasing specific capacity with increasing discharge, it probably indicates that the well is continuing to develop and that the original development was inadequate.

In the determination of most favorable capacity, if an exact design capacity is not necessary, a plot of drawdown versus yield for each step may show a point at which yield is optimum. Also, such plots, when combined with constant yield test data, will yield data on correct pump settings.

The yield of most wells declines with use because of general deterioration of the well through buildup of encrustation on screens, plugging of aquifers and gravel packs, and other similar factors. If a step test was conducted upon completion of the well, the running of a duplicate test may yield data on the extent of the deterioration and clues as to the nature.

Analyses of data from step tests of discharging wells have been used by Jacob [40] and Rorabaugh [63] to determine the efficiency of wells. Mogg [51] contends, however, that determination of well efficiency should properly be based on the theoretical specific capacity of a well which is a function of transmissivity of the aquifer. This requires test data from a fully penetrating test well and an observation well.

(b) *Constant Yield Tests.*—For wells of low capacity and intermittent operation, such as domestic or stock wells, a bailing or pumping test of several hours' duration may be adequate to determine the yield and correct pump capacity and setting.

Wells of larger capacity which must operate for prolonged periods should be tested at a rate approximating the intended capacity for a long enough period to simulate actual production pumping conditions. Normally, unless aquifer conditions are simple and uniform or are well known, such a test should be continued at a uniform rate for a minimum of 72 hours.

5-15. Streamflow Depletion by a Discharging Well.—Where an aquifer is hydraulically connected to a stream or lake, discharge from a nearby well will result in depletion of the surface waterbody. It is often important to know, because of possible water rights conflicts, the timing and extent of such depletion. A method of estimating surface water depletion resulting from operation of a nearby well has been developed by Glover and Balmer [20] in an analysis made by the images method. Figure 5-25 is a graph from which the portion of the well discharge contributed by a stream can be estimated. In this graph, q is the discharge of the well drawn from the stream, Q is the total well discharge, X is the distance from the well to the stream, or more accurately, the computed distance to the recharge boundary, T is the transmissivity, t is the time since discharge began, and S is the specific yield. If values of X, Q, T, and S are known, the percentage of water pumped from the well that originates from the stream can be estimated for any time. Similarly, by assuming a small value such as 0.01 (it cannot be zero) on the abscissa, an estimate may be made of the time at which practically all water pumped would come from the stream. In an unconfined aquifer, the drawdown in the well should not be less than 50 percent of the original saturated thickness of the aquifer [16,18,19,20,30,31,54,64,77,78,79,80].

5-16. Estimates of Future Pumping Levels and Well Performance.— In well design, an estimate of minimum pumping levels during the prospective well life may be necessary to provide for an adequately deep pump setting and head and power requirements. Initially, hydrographs of water levels in nearby wells are examined to determine long-term ground-water trends and seasonal fluctuations. Potential additional ground-water development is evaluated and, if possible, the effects of such development on future ground-water levels estimated. Also, in the case of free aquifers, the effect of decrease in aquifer thickness on transmissivity is estimated from the relationship $T=KM$. Using the estimated or determined value of S, the value of T (corrected as necessary), and estimated pumping schedules, the maximum levels are estimated which might be encountered in a well in which the drawdown in the well exceeds the theoretical drawdown by 20 to 30 percent. This 20- to 30-percent factor compensates for losses incurred in the well. After completion and testing of a well, step test analyses are used to refine the original estimate of minimum water levels. The same procedure can be used on existing or rehabilitated wells.

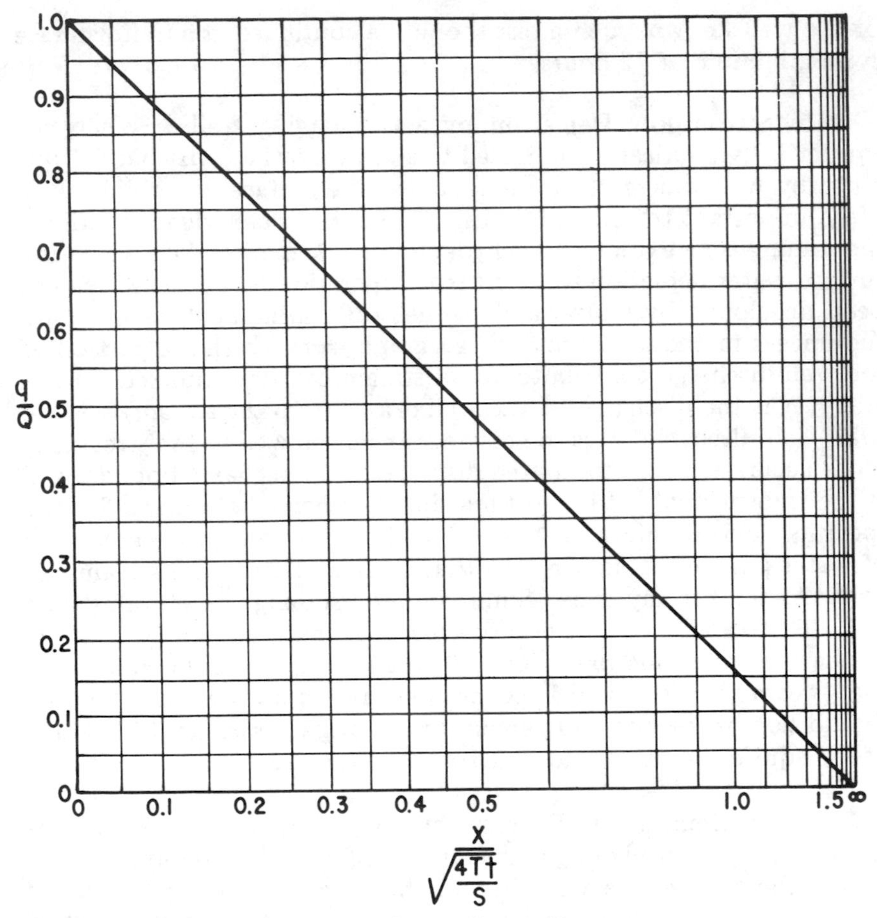

q – Flow that comes from the stream

Q – Discharge of the well

x – Distance from the well to the stream or the computed distance to a recharge boundary

T – Transmissivity

t – Time from beginning of pumping

S – Specific yield

FIGURE 5–25.—Analysis of streamflow depletion by a well. 103–D–1451.

Common problems for analysis are to predict long-term performance of a well or group of wells, or the reaction of an aquifer to various distribution of wells and rates of pumping. The computed values of T and S may be substituted in the various equations based on the non-equilibrium equation to arrive at solutions to such problems. For example, it may be necessary to determine the drawdown in a well pumping at a given rate for a given period of time. S, T, and r (radius of the well) would be known. This permits solving for u. Using this value of u, we can find $W(u)$ from figure 5–5 or table 5–3 in section 5–4. Substituting the value for $W(u)$ in equation (9), we can then solve for s. The value obtained for s will be the theoretical drawdown just outside the well casing. The drawdown inside the well will be greater, depending on the efficiency of the well. The drawdown at any distance from the well can be found by setting a value for r. When more than one well is involved, the drawdown at any point is equal to the sum of the drawdowns for each well at that point.

Other uses of equations (7) and (9) would be to determine:

(1) length of time a well can be pumped before drawdown reaches a given value,

(2) rate at which a well can be pumped for a given period before the top of the screen is exposed or the pump breaks suction, and

(3) distance from a well at which drawdown reaches a selected value in a given time. This is especially useful in locating drainage wells or determining interference between wells.

Determining the feasible yield from a well or well field requires an economic analysis involving the amount of water needed, specific capacity of the wells, the cost of power, and the cost of wells and pumps.

5–17. Estimating Transmissivity from Specific Capacity.—For some investigations, especially those of a reconnaissance nature, those covering large areas, or investigations that are limited by funds, estimates of approximate transmissivity values may be acceptable. The drilling of test and observation wells and the running of tests may be unnecessary or unjustified if there are existing wells in the area. Limited data on the yields and drawdown of such wells may be used to calculate approximate transmissivity values from specific capacity data.

In the previously discussed methods of determining transmissivity by discharging well tests, two essential terms in the analyses are the well discharge, Q, and drawdown, s, or more specifically $\frac{Q}{s}$. In a closely controlled test, Q and s are related to values of time, distance, and storativity in an observation well to obtain a transmissivity value that is as accurate and reliable as possible. However, within limits, these three factors may be ignored and the simple $\frac{Q}{s}$ relationship used to determine approximate transmissivity values. Figure 2–4 in section 2–3 shows the general relationship between specific capacity, in gallons per minute of discharge versus feet of drawdown, and transmissivity. If the specific capacity values are based on pumping periods of several hours or more, the transmissivity values tend to be conservatively low because the inefficiency of the well probably overshadows the effects of drawdown increasing with time.

Figure 2–4 may also be used to determine the value of specific capacity when the transmissivity is known. In all instances, however, it should be understood that the values from figure 2–4 are approximations only. Walton [80] gives a more thorough analysis of the specific capacity method of determining transmissivity.

5–18. Flow Nets.—Forchheimer [84] evolved a graphical solution for complex ground-water flow problems. His methods are used primarily for analysis of flow under and around foundations, through dams and similar structures at which relatively uniform conditions prevail. He considered that such problems could be analyzed by flow nets in a two-dimensional flow system through a unit cross section. In 1937, Arthur Casagrande [9] published an English discussion of the principle and application of flow nets [15,59,80].

Solutions are based on the law of continuity, which states that the volume of water flowing into a saturated element of soil is equal to the volumes flowing out of it, and on Darcy's law, where boundaries are fixed and flow is steady. The two-dimensional flow can be represented by orthogonal families of curves. One set of curves represents flow lines and the other equipotential lines. The two sets of curves intersect each other at right angles. Since the number of lines is infinite, there is no unique way of constructing a flow net.

The equipotential lines, which are drawn one unit apart, represent lines of an equal water or piezometric level. The flow lines are drawn to represent equal flow rates through the zone bounded by the lines.

Boundaries on all flow nets are either equipotential or flow lines. Flow lines start at right angles to recharge boundaries and impermeable boundaries are represented by flow lines. Theoretically, there can be no flow across a flow line.

The two families of lines are drawn to form a net of orthogonal squares which are not usually squares, but possess the property that the ratio of the sum of the opposite sides of each square bounded by two flow lines and two equipotential lines is unity and the lines intersect at right angles. If η_q equals the number of flow tubes, η_e the number of equipotential drops, and h the head difference between the inlet and outlet of the net, the flow Q is represented by:

$$Q=Kh\frac{\eta_q}{\eta_e} \tag{36}$$

where: $\frac{\eta_q}{\eta_e}$ is known as the shape factor and K is the permeability.

Forchheimer's solution can be applied to qualitative analysis and sometimes to quantitative determination of ground-water flow distribution where conditions are uniform, flow is steady, and sufficient data are available to permit drawing a reasonably accurate flow net. However, the basic assumptions are seldom met, so the Forchheimer method must be used with care.

The ground-water gradient, i, can be estimated by measuring the distance, L, between equipotential lines h_1 and h_2. By using the gradient equation,

$$i=\frac{h_1-h_2}{L},$$

and the area A of a plane normal to the direction of flow which can be estimated from the width and thickness of the flow zone, the flow rate can be calculated using Darcy's law: $Q=KiA$.

Figure 5–26 is a water-table elevation contour map on which flow lines have been sketched. A ground-water mound caused by deep percolation is shown near the upper, right corner. Water flows out radially from the mound. The distance between adjacent flow lines, W, increases with distance from the mound and with flattening gradient. Flow lines and equipotential lines form an orthogonal network of cells, the size of which increase as the gradient decreases. Theoretically, the flow through any one cell equals the flow through any other cell.

Since there can be no flow across a flow line and if the aquifer is uniformly thick,

$$Q_1=Q_2=K_1W_1i_1=K_2W_2i_2 \text{ or}$$

$$\frac{K_1}{K_2}=\frac{W_2i_2}{W_1i_1}. \tag{37}$$

5–19. Drainage Wells.—Drainage wells are occasionally used as an alternative to or supplement for deep, open, or buried pipe drains for subsurface drainage. Such wells differ little from conventional water supply wells except that the provision of water is incidental to the lowering of the water table. However, conditions must be

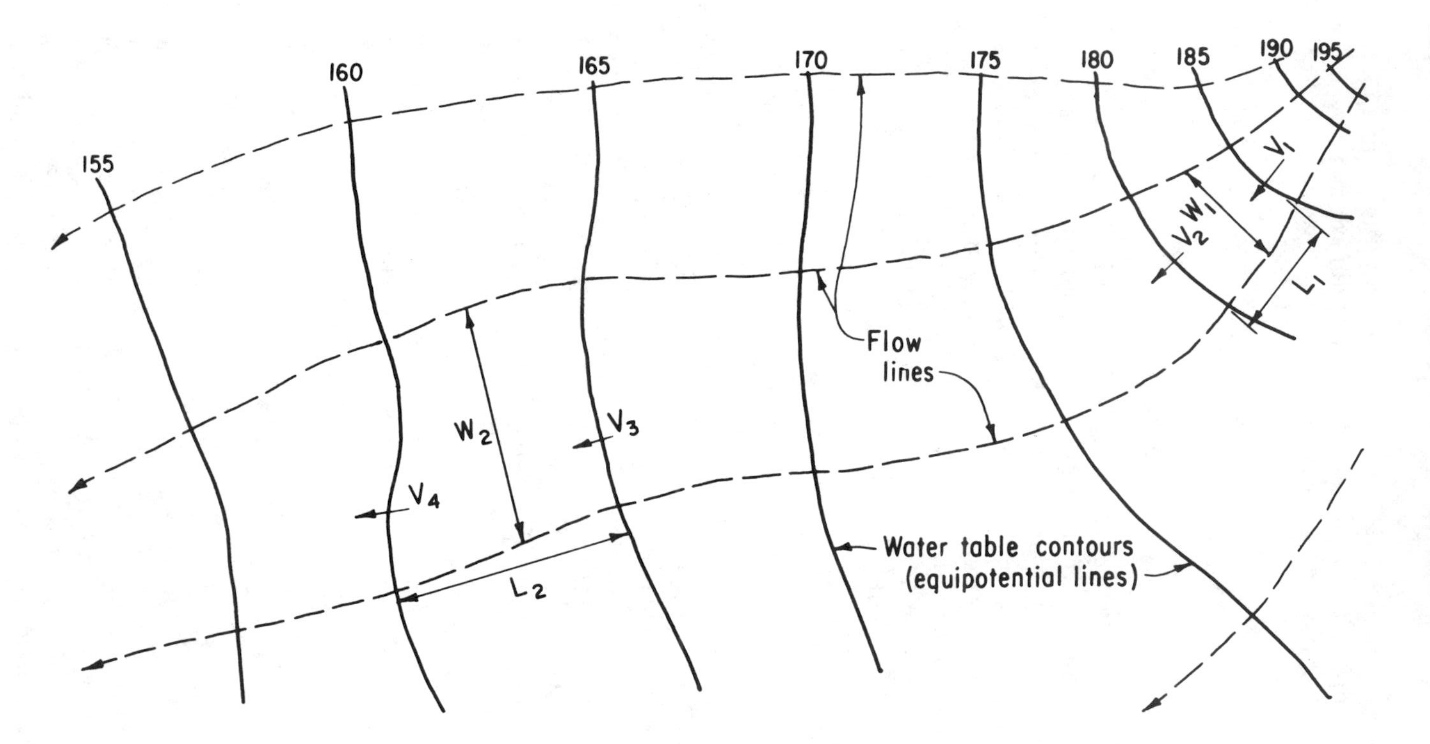

FIGURE 5–26.—Flow net analysis. 103–D–1452

especially favorable to justify consideration of drainage wells as an alternative to conventional drains. Such conditions include:

- Presence of a productive aquifer
- Presence of adequate vertical permeability in all materials between the root zone and the aquifer
- Availability of pumping energy (unless artesian flow is adequate to provide the necessary drainage)

In multiwell drainage, the wells are usually located and spaced to provide mutual interference to maximize the drawdown at the critical midpoint areas between wells.

Investigations necessary to determine the feasibility of using drainage wells generally are extensive and costly. Primary basic data needed include:

- Aquifer transmissivity and storativity
- Areal extent, thickness, and homogeneity of the aquifer
- Mode of occurrence of ground water—whether unconfined, confined, or leaky
- Estimated recharge caused by precipitation, irrigation, and other sources
- Boundary conditions
- Quality of water data

Where conditions are favorable, a method developed by Hantush [28] may be useful in determining the feasibility of using drainage wells, general features, layout of facilities, and operating criteria. Although the equation is steady state, averaging transient conditions during long-term pumping permits its use for initial estimates of required well discharge and spacing.[2]

Hantush's equations are (see fig. 5–27):

$$f_o=\frac{K}{W}\left(\frac{h_e}{r_w}\right)^2\left[1-\left(\frac{h_w}{h_e}\right)^2\right] \tag{38}$$

$$\log\left(\frac{r_e}{r_w}\right)=0.464\log f_0-0.157 \tag{39}$$

$$r_e=\frac{M}{d} \tag{40}$$

$$Q=\pi W r_e^2 \tag{41}$$

where, using consistent units:

f_o = a pure number whose value depends on the aquifer characteristics and geometry of the well,

---

[2] These equations were developed for a single well with an impervious boundary at radius r_e. They do not allow for well interference and are therefore conservative.

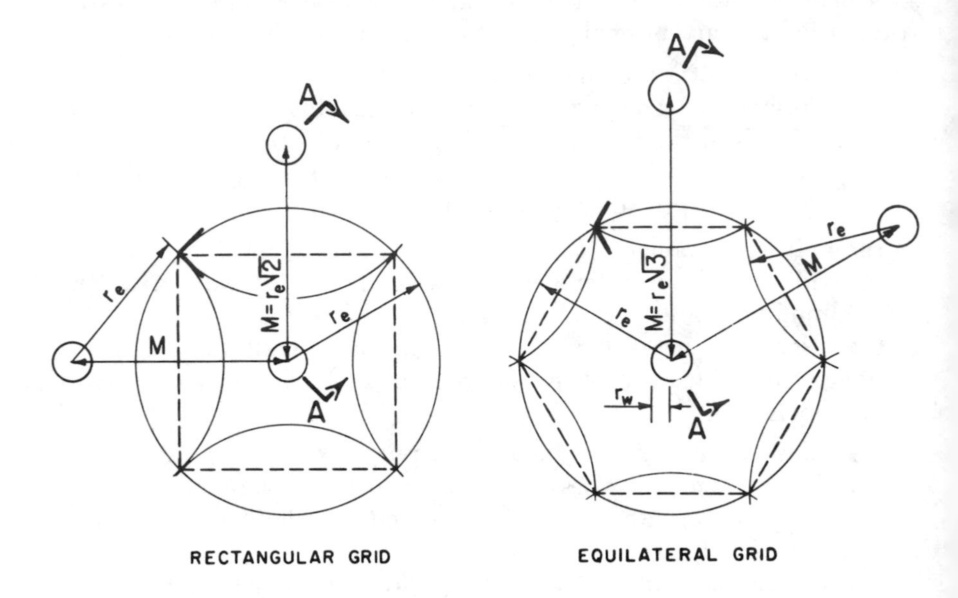

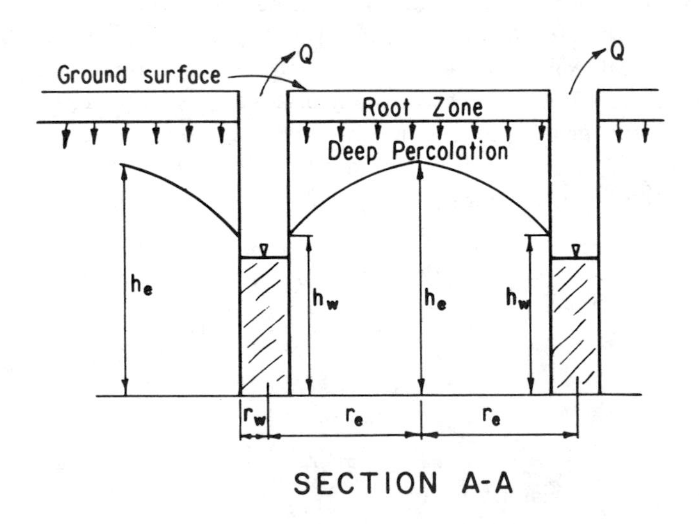

FIGURE 5-27.—Analysis of drainage well requirements. 103-D-1453.

W=rate of uniform deep percolation, $\frac{L}{t}$,

K=hydraulic conductivity or permeability of the aquifer, $\frac{L}{t}$,

h_e=maximum permissible thickness at the center of each cell with an area drained by a network of wells, L, (the maximum thickness of the saturated aquifer between the dewatered root zone and the base of the aquifer),

r_w=effective radius of a well, L,

h_w=saturated thickness at a pumping well which completely penetrates the aquifer, L,

r_e=radius of a circle circumscribing the diversion area of each of a large number of equally spaced wells, L,

M=spacing of the wells, L,

d=dimensionless constant which depends on the grid on which wells are located, (for a rectangular grid, it is $\sqrt{2}=1.414$ and for an equilateral grid it is $\sqrt{3}=1.732$), and

Q=steady discharge of an individual well in the grid, $\frac{L^3}{t}$.

The Hantush analysis gives a preliminary estimate of the drainage requirements. Further refinement may be necessary, depending on the reliability of the data and complexity of the hydrologic system.

The total capacity of wells in a drainage well field must exceed the estimated average recharge to the aquifer to maintain an acceptable depth to the water table. Capacity up to 1.5 to 2 times greater than the recharge may be necessary in some cases to: (1) establish an initial gradient to the wells; (2) permit rapid lowering of the water table prior to the irrigation season where pumping is not continued year around; (3) permit rapid lowering of the water table in local areas where unforeseen rises in water levels are caused by above normal precipitation, surface flooding, well failure, or other similar factors; and (4) permit adequate control within the drained area.

5-20. Bibliography.—

[1] Bentall, R. (compiler), "Methods of Determining Permeability, Transmissibility, and Drawdown," U.S. Geological Survey Water-Supply Paper 1536–I, 1963.

[2] ———, "Short Cuts and Special Problems in Aquifer Test," U.S. Geological Survey Water-Supply Paper 1545–C, 1963.

[3] Boulton, N. S., "The Flow Pattern Near a Gravity Well in a Uniform Water-Bearing Medium," Journal of the Institution of Civil Engineers (London), vol. 36, No. 10, pp. 534–550, December 1951.

[4] ———, "The Drawdown of the Water Table under Non-Steady Conditions Near a Pumped Well in an Unconfined Formation," Institution of Civil Engineers Proceedings (London), vol. 3, part III, No. 2, Paper No. 5979, pp. 564–579, August 1954.

[5] ———, "Unsteady Radial Flow to a Pumped Well Allowing for Delayed Yield for Storage," International Association of Scientific Hydrology, Assemblee Génèrale de Roma, Tome II, Publication No. 37, Rome, pp. 473–477, 1955.

[6] ———, "Analysis of Data from Non-Equilibrium Pumping Tests Allowing for Delayed Yield from Storage," Institution of Civil Engineers Proceedings (London), vol. 26, Paper No. 6693, pp. 469–482, November 1963. Also, "Discussion of . . ." in vol. 28, pp. 603–610, August 1964.

[7] Brown, R. H., "Selected Procedures for Analyzing Aquifer Test Data," Journal of the American Water Works Association, vol. 45, No. 8, pp. 844–866, 1953.

[8] Bruin, J., and Hudson, H. E., Jr., "Selected Methods for Pumping Test Analysis," Illinois State Water Survey Report of Investigation No. 25, Urbana, 1955.

[9] Casagrande, A., "Seepage Through Dams," Journal of the New England Water Works Association, vol. 51, pp. 131–172, 1937.

[10] Chow, V. T., "On the Determination of Transmissibility and Storage Coefficient from Pumping Test Data," Transactions of the American Geophysical Union, vol. 33, No. 3, pp. 397–404, June 1952.

[11] Cooper, H. H., Jr., and Jacob, C. E., "A Generalized Graphical Method for Evaluating Formation Constants and Summarizing Well-Field History," Transactions of the American Geophysical Union, vol. 27, No. IV, pp. 526–534, August 1946.

[12] Davis, S. N., and DeWiest, R. J. M., "Hydrogeology," John Wiley & Sons, New York, 1966.

[13] DeWiest, R. J. M., "Geohydrology," John Wiley & Sons, New York, 1965.

[14] Ferris, J. G., "Ground Water Hydraulics as a Geophysical Aid," Michigan Department of Construction, Technical Report No. 1, Lansing, 1948.

[15] Ferris, J. G., Knowles, D. B., Brown, R. H., and Stallman, R. W., "Theory of Aquifer Tests," U.S. Geological Survey Water-Supply Paper 1536–E, 1962.

[16] Glover, R. E., "Ground-Water Movement," Bureau of Reclamation, Engineering Monograph No. 31, 1964.

[17] Glover, R. E., Moody, W. T., and Tapp, W. N., "Till Permeabilities as Estimated from the Pump-Test Data Obtained

During the Irrigation Well Investigations." Memorandum to Drainage and Ground-water Engineer, Bureau of Reclamation, June 11, 1954.

[18] Glover, R. E., "Ground Water Surface Water Relationship," Water Research Conference, Colorado State University, Fort Collins, 1960.

[19] ———, "The Pumping Well," Colorado State University Agricultural Experiment Station, Technical Bulletin No. 100, Fort Collins, 1968.

[20] ———, and Balmer, G. G., "River Depletion Resulting from Pumping a Well near a River," Transactions of the American Geophysical Union, vol. 35, No. 3, pp. 468–470, June 1954.

[21] Hantush, M. S., "Analysis of Data from Pumping Tests in Leaky Aquifers," Transactions of the American Geophysical Union, vol. 37, No. 6, pp. 702–714, December 1956.

[22] ———, "Modification of the Theory of Leaky Aquifers," Journal of Geophysical Research, vol. 65, No. 11, pp. 3713–3725, November 1960.

[23] ———, "Tables of the Functions W(UB)," New Mexico Institution of Mining and Technology, Professional Paper No. 103, Socorro, 1961.

[24] ———, "Tables of the Functions W(UB)," New Mexico Institution of Mining and Technology, Professional Paper No. 104, Socorro, 1961.

[25] ———, "Nonsteady Flow to Flowing Wells in Leaky Aquifers," Journal of Geophysical Research, vol. 64, No. 8, pp. 1043–1052, August 1959.

[26] ———, "Drawdown Around a Partially Penetrating Well," Proceedings of the ASCE, Journal of the Hydraulics Division, vol. 87, No. HY4, pp. 83–98, July 1961.

[27] ———, "Aquifer Tests on Partially Penetrating Wells," Proceedings of the ASCE, Journal of the Hydraulics Division, vol. 87, No. HY5, pp. 171–195, September 1961.

[28] ———, "Supplement to Peterson's Design of Replenishment Wells," Proceedings of the ASCE, Journal of Irrigation and Drainage Division, vol. 90, No. IR1, part 1, pp. 67–76, March 1964.

[29] ———, "Drainage Wells in Leaky Water-Table Aquifers," Proceedings of the ASCE, Journal of the Hydraulics Division, vol. 88, No. HY2, pp. 123–137, March 1962.

[30] ———, "Depletion of Storage, Leakage, and River Flow by Gravity Wells in Sloping Sand," Journal of Geophysical Research, vol. 69, No. 12, pp. 2551–2560, June 15, 1964.

[31] Hantush, M. S., "Wells near Streams with Semipervious Beds," Journal of Geophysical Research, vol. 70, No. 12, pp. 2829–2838, June 15, 1965.

[32] ———, "Wells in Homogeneous Anisotropic Aquifers,' Water' Resources Research, vol. 2, No. 2, pp. 273–279, 1966.

[33] ———, "Analysis of Data from Pumping Tests in Anisotropic Aquifers," Journal of Geophysical Research, vol. 71, No. 2, pp. 421–426, January 15, 1966.

[34] ———, and Jacob, C. E., "Non Steady Radial Flow in an Infinite Leaky Aquifer," Transactions of the American Geophysical Union, vol. 36, No. 1, pp. 93–100, February 1955.

[35] ———, "Steady Three-Dimensional Flow to a Well in a Two-Layered Aquifer," Transactions of the American Geophysical Union, vol. 36, No. 2, pp. 286–292, April 1955.

[36] Heath, R. C., and Trainer, F. W., "Introduction to Ground-Water Hydrology," John Wiley & Sons, New York, 1968.

[37] Skeat, W. O. (editor), "Manual of British Water Engineering Practice," W. Heffner & Sons, Cambridge, 1969.

[38] Jacob, C. E., "On the Flow of Water in an Elastic Artesian Aquifer," Transactions of the American Geophysical Union, Reports and Papers, vol. 21, pp. 574–586, 1940.

[39] ———, "Radial Flow in a Leaky Artesian Aquifer," Transactions of the American Geophysical Union, vol. 27, No. II, pp. 198–208, April 1946.

[40] ———, "Drawdown Test to Determine the Effective Radius of an Artesian Well," Transactions of the ASCE, vol. 112, pp. 1047–1070, 1947.

[41] ———, "Adjustment for Partial Penetration of a Pumping Well," U.S. Geological Survey Open File Report, 1945.

[42] ———, and Lohman, S. W., "Nonsteady Flow to a Well of Constant Drawdown in an Extensive Aquifer," Transactions of the American Geophysical Union, vol. 33, No. 4, pp. 559–569, August 1952.

[43] Kazmann, R. G., "Inverse Tables of the Exponential Integral," U.S. Geological Survey, Water Resources Branch, Division of Ground Water, December 1941.

[44] ———, "Modern Hydrology," Harper & Row, New York, 1965.

[45] Kruseman, G. P., and DeReder, N. A., "Analysis and Evaluation of Pumping Test Data," International Inst. for Land Reclamation and Improvement Book II, Wagenenger, Netherlands, 1970.

[46] Lang, S. M., "Methods for Determining the Proper Spacing of Wells in Artesian Aquifers," U.S. Geological Survey Water-Supply Paper 1545–B, 1961.

[47] Lennox, D. H., "Reader's Comment on Step Down Tests," Drillers Journal, UOP Johnson Division, St. Paul, Minn., January/February 1969.

[48] ———, "Analysis and Application of the Step-Drawdown Test," Proceedings of the ASCE, Journal of the Hydraulics Division, vol. 92, No. HY6, November 1966.

[49] Lohman, S. W., "Ground Water Hydraulics," U.S. Geological Survey Professional Paper 708, 1972.

[50] Meinzer, O. E., "Outline of Methods for Estimating Ground-Water Supplies," U.S. Geological Survey Water-Supply Paper 638-C, 1932.

[51] Mogg, J. L., "Step Drawdown Test Needs Critical Review," Drillers Journal, UOP Johnson Division, pp. 3–8, 11, July/August 1968.

[52] Moody, W. T., "Determination of Minimum Drawdown Within an Array of Wells," Memorandum to T. Ahrens, Bureau of Reclamation, March 11, 1955.

[53] ——— "Determination of the Drawdown at the Center of an Array of Wells," Memorandum to T. P. Ahrens, Bureau of Reclamation, February 16, 1955.

[54] Moulder, E. A., and Jenkins, C. T., "Analog Digital Models of Stream Aquifer Systems," U.S. Geological Survey Water Resources Division, Open File, March 1969.

[55] Muskat, Morris, "The Flow of Homogeneous Fluids through Porous Media," J. W. Edwards, Ann Arbor, Mich., 1946.

[56] "Tables of Sine, Cosine, and Exponential Integrals," vols. 1 & 2, National Bureau of Standards, 1940.

[57] Norris, S. E., and Fidler, R. E., "Use of Type Curves Developed from Electric Analog Studies of Unconfined Flow to Determine the Vertical Permeability of an Aquifer at Piketown, Ohio," Ground Water, vol. 4, No. 3, pp. 43–48, July 1966.

[58] Papadopulos, I. S., "Nonsteady Flow to Multiaquifer Wells," Journal of Geophysical Research, vol. 71, No. 20, pp. 4791–4797, October 15, 1966.

[59] Peattie, K. R., "A Conducting Paper Technique for the Construction of Flow Nets," Civil Engineering and Public Works Review (GB), vol. 51, No. 595, pp. 62–64, January 1956.

[60] Prickett, T. A., and Longquist, C. G., "Comparison Between Analog and Digital Simulation Techniques for Aquifer Evaluation," Illinois State Water Survey Reprint Series No. 114, Urbana, December 1968.

[61] Prickett, T. A., "Type-Curve Solution to Aquifer Tests Under Water-Table Conditions," Ground Water, vol. 3, No. 3, pp. 5–14, July 1965.

[62] Remson, I., McNeary, S. S., and Randolph, J. R., "Water Levels Near a Well Discharging from an Unconfined Aquifer," U.S. Geological Survey Water-Supply Paper 1536–B, 1961.

[63] Rorabaugh, M. I., "Graphical and Theoretical Analysis of Step-Drawdown Tests of Artesian Wells," Proceedings of the ASCE, vol. 79, Separate No. 362, December 1953.

[64] ———, "Stream Bed Percolation in Development of Water Supplies," International Association of Scientific Hydrology Publication No. 33, Brussels, 1951.

[65] Rouse, H. (editor), "Engineering Hydraulics," Proceedings of the Hydraulics Conference, University of Iowa, Iowa City, 1949.

[66] Schicht, R. J., "Selected Methods of Aquifer Test Analysis," Water Resources Bulletin, vol. 8, No. 1, pp. 175–187, February 1972.

[67] Sheahan, N. T., "Type-Curve Solution of Step-Drawdown Test," Ground Water, vol. 9, No. 1, pp. 25–29, January/February 1971.

[68] Stallman, R. W., "The Significance of Vertical Flow Components in the Vicinity of Pumping Wells in Unconfined Aquifers," U.S. Geological Survey Professional Paper 424–B, 1961.

[69] ———, "Effects of Water Table Conditions on Water Level Changes Near Pumping Wells," Water Resources Research, vol. 1, No. 2, pp. 295–312, 1965.

[70] ———, "Electric Analog of Three-Dimensional Flow to Wells and Its Application to Unconfined Aquifers," U.S. Geological Survey Water-Supply Paper 1536–H, 1963.

[71] Stone, R. F., "Reader's Comment on Step Down Tests," Drillers Journal, UOP Johnson Division, pp. 10–12, January/February 1969.

[72] Theis, C. V., "The Relation Between the Lowering of the Piezometric Surface and the Rate and Duration of Discharge of a Well Using Ground-Water Storage," Transactions of the American Geophysical Union, vol. 16, part II, pp. 519–524, August 1935.

[73] Todd, D. K., "Ground Water Hydrology," John Wiley & Sons, New York, 1959.

[74] "Methods and Techniques of Ground Water Investigations and Developments," UNESCO, Water Resources Series, No. 33, New York, 1967.

[75] "Ground Water and Wells," UOP Johnson Division, St. Paul, Minn., 1966.

[76] Walton, W. C., "Selected Analytical Methods for Well and Aquifer Evaluation," Illinois State Water Survey, Department of Regulation and Education Bulletin No. 49, Urbana, 1962.

[77] ———, "Estimating the Infiltration Rate of a Stream Bed by Aquifer Test," International Association of Scientific Hydrology Publication No. 63, Berkeley, Calif., 1963.

[78] ———, "Effect of Induced Stream Bed Infiltration of Water Levels on Wells During Aquifer Tests," Water Resources Research Center, University of Minnesota, Bulletin No. 2, 1966.

[79] ———, "Recharge from Induced Stream Bed Infiltration Under Varying Water Level and Stream Stage Conditions," Water Resources Research Center, University of Minnesota, Bulletin No. 6, 1967.

[80] ———, "Ground Water Resource Evaluation," McGraw-Hill, New York, 1970.

[81] Weeks, E. P., "Field Methods of Determining Vertical Permeability and Aquifer Anisotropy," U.S. Geological Survey Professional Paper 501–D, 1964.

[82] Wenzel, L. K., "Methods for Determining Permeability of Water-Bearing Materials," U.S. Geological Survey Water-Supply Paper 887, 1942.

[83] ———, and Greenlee, A. L., "A Method of Determining Transmissibility- and Storage-Coefficients by Tests of Multiple-Well Systems," Transactions of the American Geophysical Union, vol. 24, part II, pp. 547–560, January 1944.

[84] Forchheimer, Philipp, "Hydraulik," third edition, B. G. Teubner Verlagsgesellschaft, Berlin, 1930.

[85] Glover, R. E., "Studies of Ground Water Movement," Bureau of Reclamation Technical Memorandum 657, March 1960.

[86] Lohman, S. W., "Geology and Artesian Water Supply of the Grand Junction Area, Colorado," U.S. Geological Survey Professional Paper 451, 1965.

«Chapter VI

ESTIMATES OF AQUIFER YIELD, HYDROLOGIC BUDGETS, AND INVENTORIES

6–1. Introduction.—Proper management of a resource such as ground water requires knowledge of the magnitude, distribution, and depletion and replenishment, if any, of the resource. Without such an assessment the effects of past development and predictions of the influences of future development cannot be adequately determined. Budgets and inventories provide the means of assessment of ground-water resources and involve such factors as storage, recharge, and discharge. Because of the interrelationship of surface and ground water, comprehensive, quantitative budgets and inventories must consider both modes of water occurrence.

An adequate estimate of the availability of ground water in storage beneath an area requires determination of the ground-water basin boundaries, both vertical and horizontal, and of aquifer dimensions and characteristics. Where more than a nominal area is involved, the aquifer characteristics and related factors are seldom sufficiently uniform to be applied to the entire area. The areal extent of the variations must be estimated and delineated by a study and analysis of the subsurface geologic and hydrologic conditions, well capacities, and similar factors. Since ground water is a dynamic resource with constantly changing water levels caused by natural or artificial influences, interpretation is facilitated by careful analysis of water level fluctuations as related to such influences. Such interpretation may need to be supplemented by pumping tests located on the basis of the initial studies to clarify localized variations in aquifer characteristics and boundaries.

Such data are essential to aquifer analysis, including electric analog or digital computer analysis of aquifer response to development. For assessments of longtime aquifer yield and performance, evaluations are usually based on an average annual basis and maximum high and low water conditions. The basic results of a ground-water inventory are the determination of the total water in storage and the annual change. Further studies involve the response of the system to various schemes of development, possibilities of induced additional recharge due to development, artificial recharge, desirability and probable life of a water mining operation, design of wells, conjunctive use of surface and ground water, and possibilities of subsidence.

The techniques of a ground-water evaluation are relatively subjective and the degree of accuracy and the reliability of initial results are often questionable. Many evaluation studies involve a continued reassessment and refinement of the estimates as more data on actual response of the aquifer to development become available.

As previously mentioned, a ground-water study involves consideration not only of ground water but of surface water. The boundaries of a ground-water reservoir may or may not coincide with those of an overlying surface water basin. If they do, the study may be simplified. In many investigations, the setting of arbitrary boundaries may be required.

Most ground-water inventory methods were devised for application in semiarid and arid zones, generally in response to obvious over-development. There is no wholly standardized investigation procedure because of variations in conditions and needs. Methods will vary depending upon areal development; complexity of the geology; climate; availability of existing data; time, funds, equipment, and manpower available to obtain data; and similar factors. The following sections summarize many of the factors involved and methods of procedure.

6–2. Hydrologic Budgets [12,13,21].[1]—The hydrologic budget is a quantitative evaluation of the total water gained or lost from a basin or part of a basin during a specific period of time. It considers all water, whether surface or ground water, entering, leaving, or stored within the area of study. The hydrologic budget is summarized in the following equation:

$$\Delta S = P - E \pm R \pm U \tag{1}$$

where:

ΔS = changes in storage in channels and reservoirs, in ground-water storage, and in soil moisture,
P = precipitation on the area of study (+),
E = evapotranspiration from the area (−),
R = difference between stream outflow (−) and inflow (+), and
U = ground-water outflow (−) and inflow (+).

The components of ΔS are: ΔS_s, changes in surface storage which may be available in the form of reservoir or lake capacity curves; ΔS_c, changes in stream channel storage which are of minor importance in a longtime budget (usually ignored in the budget analysis); ΔS_m, changes in soil moisture which are also of minor importance, hence ignored; and ΔS_g, changes in ground-water storage which can be

[1] Numbers in brackets refer to items in the bibliography, section 6–7.

estimated from contour maps of the water table or piezometric surface and the storativity of the aquifer at the beginning, during, and end of the study period or periods [1,8,13,21,32,33].

Precipitation consists of all the rain and snow falling on the area. Records are usually available from weather stations in or adjacent to the area of study. Methods of analysis of the various factors are discussed in several references [4,6,8,9,34].

The estimating of long-term evapotranspiration has been studied and discussed by many authors and numerous methods have been used, all of which are approximate. In humid areas, potential evaporation is sometimes used as a preliminary estimate, but in arid lands this loss may exceed all contributions. Evaporation pan records, when corrected, give a value of evaporation from open water, but evapotranspiration from soils is more complex.

Reference should be made to standard discussions on the subject and possible methods of estimation considered in view of available records and controlling factors [1,3,5,6,7,8,9,10,11,19,24,34].

Streamflow R consists of surface runoff of precipitation within the area R_s, surface inflow to the area R_t, and ground-water seepage to streams R_g. The value of R_t can be estimated from stream gaging records. A number of methods for estimating R_g are discussed in the literature, particularly for individual basins [1,4,6,8,9,10,11,19,34].

Ground-water flow components are, U_t, the underflow from and U_o, the underflow to adjacent basins. Flow can be estimated by determining the width of the flow path from knowledge of the aquifer dimensions, the gradient from water level contour maps, and the transmissivity from results of pumping tests or other sources. These factors can be applied to the solution of $Q=KiA$ to determine total underflow.

6–3. Ground-Water Inventories.—The ground-water components of the hydrologic cycle used in estimating a ground-water budget are summarized in the equation:

$$G-D=\Delta S_g \tag{2}$$

where:

G=recharge to the aquifer,
D=discharge from the aquifer, and
ΔS_g=change in storage in the aquifer.

The components of G may include: deep percolation from precipitation; seepage from surface water bodies; ground-water underflow from adjacent areas; artificial recharge including deep percolation from irrigation, sewage disposal facilities, and recharge wells; and leakage through confining beds.

Components of D may include evapotranspiration; seepage to surface water bodies; ground-water underflow to adjacent areas; discharge of springs; artificial discharge including drainage systems, wells, and infiltration galleries; and discharge through confining beds [12,13,21,27,33].

Adequate records to permit an accurate and reliable appraisal of all the factors involved are seldom available.

Changes in ground-water storage are reflected by fluctuations in the ground-water levels. Since most assessments are based on long-term averages with the beginning and end of the study period at about the same season of the year, changes in soil moisture can usually be ignored.

Estimates on the portion of rainfall which may enter an aquifer are generally made by an analysis of the fluctuations of ground-water level as the result of a specific isolated storm or on the basis of a long-time correlation between water level hydrographs and precipitation records. The deep percolation from an individual storm is influenced by the intensity and duration of the precipitation and the deficiency of soil moisure at the beginning of the storm. The long-period correlation will therefore usually give a more nearly effective average. In using such analyses, it must be recognized that during the recharge there is usually continued discharged from the aquifer, and the net change in the water table represents the difference between the recharge and discharge. Where artificial recharge, specifically for recharge of aquifers, is practiced, adequate records of rate and volume of inflow usually are available. Recharge from irrigation may be estimated from the difference between the consumptive use and water deliveries less the surface waste.

Where surface and ground-water basins do not coincide, underflow may be a major factor in an inventory. Estimates of such flow should be based on dimensions of the aquifer, gradient, and transmissivity as described earlier.

Rough estimates of ground-water withdrawals for irrigation during a given period can be made by several methods including:

- Survey of well owners to determine rates and duration of pumping of all sizable wells in the area.
- Survey of landowners and agricultural agencies to determine total acreages of common crops and normal water application for such crops.
- Survey of utility companies to determine installed ratings of all sizable pumps and the power usage of these pumps.

These data can be combined with other data such as areawide depths to ground water and pump efficiencies to arrive at reasonably reliable rough estimates of withdrawals. Similar information for wells serving municipal and industrial uses may be available from owners, utility companies, and local or State regulatory agencies.

Evapotranspiration can be estimated by use of Blaney-Criddle [5], Lowry-Johnson [35], or similar equations if data on crop types and acreages and vegetative cover maps are available.

Estimates of seepage to or from streams can be made based on the difference between surface inflow and surface outflow plus evapotranspiration. Based on ground-water contour maps, reaches of the stream where seepage predominates can be segregated and inflow-outflow measurements of each section made. Similar estimates of ground-water seepage for lakes or reservoirs are sometimes possible on the basis of inflow-outflow data corrected for evaporation plus or minus the change in storage.

Surface discharge of springs often can be measured. They are seldom uniform, however, so periodic measurements taken when the water table elevations are measured are recommended to obtain a value of average discharge or to correlate the discharge with ground-water elevation.

As discussed previously, ground-water inventory studies are often subjective and influenced by local conditions, availability of data, time and funds, and climatic variations.

6-4. Perennial Yield Estimates of Aquifers.—An estimate is often desired of the probable perennial yield of an aquifer. A number of methods that have been derived for such estimates in arid areas are summarized by Todd [25].

6-5. Model Analyses of Aquifer Performance.—(a) *Mathematical Models* [15,17,30,32].—Complex environments, lack of data, and the need for long and tedious calculations have often precluded quantitative description and analyses of ground water conditions over large areas. However, the development of the digital and other similar computers and models now makes such analyses feasible, particularly when geohydrologic factors can be idealized and simplified. The mathematical model is based on the hydraulic properties of the aquifer, image well theory, and basic ground-water equations.

Records of past pumpage and water levels may be used to determine whether the mathematical model satisfactorily approximates the geohydrologic limits and characteristics of the aquifer. This can be accomplished by computing the historical changes in water level at several points using the hydraulic properties of the simplified boundaries and checking the results against historical records.

The time-consuming computations associated with mathematical models can usually be reduced to simple standardized procedures which can be handled by digital computers. Details of computer programming are beyond the scope of this manual but are described for some systems in the references in the bibliography, section 6–7.

(b) *Electric Analog Models* [14,15,16,17,20,22,23,31,32].—Darcy's law, which can be expressed as

$$V=K\frac{dh}{dL}$$

where V is flow velocity, K is the hydraulic conductivity, and $\frac{dh}{dL}$ is the hydraulic gradient, is analogous to Ohm's law governing the flow of electricity through an electric conductor,

$$I=R^{-1}\frac{dE}{dS}.$$

In this law, I=the electric current (amperes) per unit cross-sectional area, R is the resistance (ohms), and $\frac{dE}{dS}$ is the change in electromotive force (voltage gradient) over the distances. This analogy makes it possible to simulate the laminar flow of a viscous liquid through a porous medium by the flow of an electric current through a conductor. The voltage E then is analogous to the hydraulic head h; the voltage gradient $\frac{dE}{dS}$ to the hydraulic gradient $\frac{dh}{dL}$, the reciprocal of the resistivity R to the hydraulic conductivity K, and the electrical current I to the flow velocity V. In addition, there is a similarity between the flow lines of an electric current and the flow lines of a liquid and between lines of equal voltage and equipotential lines of the flow of a liquid.

The above discussion is applicable only to steady state flow analyses. Where storage is involved, capacitors must be included in the circuits to simulate storativity.

(c) *Other Analog Models*.—Other analog models, such as sand, elastic membranes, and thermal, are also used. A model analog is not a universal instrument for solving all problems. Each model is normally designed for and adapted to a particular problem or problems. Model technique is practically a science in itself and is outside the scope of this manual. Cited references give details and applications of specific model types [6,16,18,23,26].

6–6. Ground-Water Reports.—A ground-water report may range from less than a page containing a statement of the problem and a conclusion or recommendation, or both, to a voluminous work of many pages containing a text, numerous maps, charts, graphs, and tables. The importance and complexity of the task, and the time and funds expended, generally determine the length and content of the report.

The author of the report should exercise judgment in determining the type of information that is necessary and the amount of detail required. The main body of the report may contain some or all of the following data in greater or lesser detail as required to clearly state the problems, conclusions, and recommendations. An outline of a typical report could be as follows:

A. Problem or purpose of study.
B. Location and size of the area of interest.
C. Cultural features of the area:
    1. Public utilities.—Electric power availability, location of existing lines, number of phases, and power rate schedule.
    2. Natural gas facilities.—Location, capacity, and rate schedule.
    3. Water supplies.—Domestic, municipal, industrial, irrigation, and stock. Sources, capacities, quality of raw and treated water, reliability of sources, and rate schedules.
    4. Sewage disposal.—Location and capacity of treatment plant, type of treatment, method of disposal of effluent, quality of effluent, and method of disposal of residue.
    5. Transportation.—Highways, roads, railroads, and shipping points.
    6. Settlement.—History if pertinent, location and size of towns, land ownership, and present and contemplated use.
    7. Cover and crops.—Vegetative cover, natural types and densities, crops, and crop acreages.
    8. Irrigation.—Extent, practices, and trends.
D. Climatic summary:
    1. Amount, rates, distribution, seasonal occurrence, and type of precipitation.
    2. Temperature extremes, monthly means, length of growing season.
    3. Wind directions, velocity, and seasonal occurrence.
    4. Humidity.
    5. Evapotranspiration.

E. Surface hydrology:

1. Natural surface drainage, channel characteristics, runoff volumes and characteristics, flood potential, location of gaging stations, and the losing and gaining reaches of channels.
2. Surface water bodies including natural lakes, swamps, reservoirs, etc., with their location, size, capacity, and fluctuations in water levels.
3. Present and proposed canals and drains: location, size, length, capacity, lining, losing and gaining reaches, and physical condition.
4. Quality of surface water: chemical, bacteriological, seasonal fluctuations, and trends in quality.

F. Geology and Geomorphology:

1. Summary of the physiography.
2. Elevations and relief.
3. Surface gradients.
4. Summary of the regional geology.
5. Stratigraphy and lithology.
6. Geologic structure.
7. Summary of the more important hydrogeologic provinces.
8. Unstable formations from standpoint of well drilling, construction, and design.
9. Areas of possible subsidence.
10. Earthquake danger or potential.

G. Ground-water hydrology:

1. Location, depth, thickness, lithology, areal extent, and type of aquifer or aquifers present.
2. Water table and piezometric surface gradients, direction of flow, recharge and discharge areas, areas of artesian pressure, contributing areas.
3. Seasonal and annual fluctuations in ground-water levels, extremes, and longtime trends.
4. Present ground-water development, including: number of wells, locations, depths, screen diameters, settings, lengths, and types; casing diameter, type, and weights; yields; drawdowns; pumping lifts; and annual pumpage.
5. Well history, average life, experience with incrustation, corrosion, sand pumping, collapse, and surface caving.
6. Transmissivity and storativity of aquifers.
7. Quality of ground water, chemical, bacterial, and trends; corrosivity of water and soils.

8. Suitability of aquifers for proposed development or use.

H. Proposed ground-water development program:
   1. Number, location, and spacing of proposed production wells.
   2. Probable capacities, pump lifts, and horsepower requirements of proposed wells.
   3. Proposed well design.
   4. Recharge possibilities; location, type, and design of facilities required; sources, volume, and quality of recharge water; and probable O&M problems.

I. Factors and facilities for ground-water development:
   1. Number of drilling contractors in area; and number, type, and capacities of rigs.
   2. State and local laws and regulations governing ground-water rights, drilling permits, design and construction of wells, and licensing of drillers; name and location of State or local offices administering such rules and regulations.
   3. Water well supply dealers, pipe dealers, chemical supply houses, well logging and geophysical survey companies, laboratories capable of making mechanical analyses of samples, chemical and bacterial water analyses and soil tests, and sources of materials such as gravel for packs.

J. Maps:
   1. The location of the problem area is usually shown as an insert at a small scale on a larger scale map showing the general study area within a State. A larger scale map of the project or study area may show location of construction features relating to cultural features in the study area.
   2. Maps of study and adjacent area (usually of uniform scale):
      a. Planimetric.—Shows county and land office subdivision lines, existing location of wells, towns, highways, railroads, public utilities, etc. May be used as a base for other maps.
      b. Topographic.—At same scale as planimetric map. May sometimes be used as base for geologic map.
      c. Geologic.—Usually shows surficial geology with symbols indicating structural features,

cross-section lines, geophysical surveys, etc. (See fig. 4–1.)

d. Ground-water and piezometric surface maps showing water surface elevation contours at minimum and maximum periods; the location and elevation of measuring points with water elevation notations at each point of measurement; and contributing, recharge, and discharge areas and flowing well areas, if present. Several maps may be used to show changes with time or season. (See fig. 4–2.)

e. Isobathic or depth of water.—Similar to d. above but showing depth to water by contours. (See fig. 4–4.)

f. Isopachic of aquifer or aquifers.—Similar to e. but showing thickness of aquifers by contours.

g. Surface water map showing natural surface drainage, surface water bodies, existing and proposed dams, canals and drains, and stream gaging stations.

h. Land ownership.—Farm unit boundaries.

i. Land use and vegetative cover.

j. Quality of water.—Chemical and bacteriological.

k. Aquifer characteristics.—Variations in transmissivity and storativity values by contours or by areas containing values within a given range.

l. Isohyetal or Theisen polygones for precipitation showing location of weather stations.

m. Well field and service area, ground-water facilities, and plans.

K. Cross sections, fence diagrams, and hydrographs:

1. Geologic.—Includes control exploration hole designations.
2. Hydrologic.—Could be several cross sections showing seasonal variations including measuring point well locations. (See fig. 4–5.)

L. Graphs, charts, and tables:

1. Temperature range and growing season.
2. Average annual monthly precipitation.
3. Annual precipitation, minimum, mean, and extreme.
4. Cumulative precipitation.
5. Stream and lake hydrographs—base flows.

6. Quality of water both areal and seasonal.
7. Projections of water use, population, power, etc.
8. Ground and surface water use.
9. Mechanical analyses of aquifer samples.
10. Chemical and bacterial analyses of water samples.
11. Pump test measurements and analyses.
12. Well logs.—Drillers, geologist, resistivity, etc.
13. Geophysical surveys.
14. Evapotranspiration.—Records or estimates.

M. Drawings.
1. Well and infiltration gallery designs.
2. Test site layouts.
3. Special equipment designs.

6-7. Bibliography.—

[1] "Hydrology Handbook," ASCE Manual of Engineering Practice No. 28, American Society of Civil Engineers, New York, 1952.

[2] Bennett, G. D., and Patten, E. P., Jr., "Methods of Flow Measurement in Well Bores," U.S. Geological Survey Water-Supply Paper 1544-C, 1962.

[3] Behnke, J. J., and Maxey, G. B., "An Empirical Method of Estimating Monthly Potential Evapotranspiration in Nevada," Journal of Hydrology, vol. VIII, No. 4, North Holland Publishing Co., Amsterdam, pp. 418–430. August 1969.

[4] Butler, S. S., "Engineering Hydrology," Prentice-Hall, Englewood Cliffs, N.J., 1957.

[5] Criddle, W. D., "Methods of Computing Consumptive Use of Water," Proceedings of the ASCE, Journal of Irrigation and Drainage Division, vol. 84, No. IR1, Paper 1507, January 1958.

[6] DeWiest, R. J. M., "Geohydrology," John Wiley & Sons, New York, 1966.

[7] Hamon, W. R., "Estimating Potential Evapotranspiration," Proceedings of the ASCE, Journal of the Hydraulics Division, vol. 87, No. HY3, pp. 107–120, May 1961.

[8] Skeat, W. O., (editor), "Manual of British Water Engineering Practice," fourth edition, vol. II, "Engineering Practice," W. Heffer and Sons, Cambridge, 1969.

[9] Kazmann, R. G., "Modern Hydrology," Harper and Row, New York, 1965.

[10] Lindsley, R. K., Jr., Kohler, M. A., and Paulhus, J. L. H., "Applied Hydrology," McGraw-Hill, New York, 1949.

[11] Lindsley, R. K., Jr., Kohler, M. A., and Paulhus, J. L. H., "Hydrology for Engineers," McGraw-Hill, New York, 1958.

[12] Meinzer, O. E., "Outline of Methods for Estimating Ground-Water Supplies," U.S. Geological Survey Water-Supply Paper 638–C, 1932.

[13] ———, and Stearns, N. D., "A Study of Ground Water in the Pomperaug Basin, Connecticut," U.S. Geological Survey Water-Supply Paper 597–B, 1929.

[14] Monter, C. L., Ribbens, R. W., and Phillipps, H. B., "Electric Analog Studies of Ground Water Conditions on Portales Valley, Portales Project, New Mexico," Office of Chief Engineer, Bureau of Reclamation, 1967.

[15] Moulder, E. A., and Jenkins, C. T., "Analog Digitial Models of Stream Aquifer Systems," U.S. Geological Survey Water Resources Division, Open File Report, March 1969.

[16] Peattie, K. R., "A Conducting Paper Technique for the Construction of Flow Nets," Civil Engineering and Public Works Review (GB), vol. 51. No. 595, pp. 62–64, January 1956.

[17] Prickett, T. A., and Longquist, C. G., "Comparison Between Analog and Digital Simulation Techniques for Aquifer Evaluation," Illinois State Water Survey Reprint Series No. 114, Urbana, December 1968.

[18] Rorabaugh, M. I., "Stream Bed Percolation in Development of Water Supplies," International Association of Hydrology, Publication No. 33, Brussels, 1951.

[19] Rouse, H. (editor), "Engineering Hydraulics," Proceedings of the Hydraulics Conference, University of Iowa, Iowa City, 1949.

[20] Schicht, R. J., "Selected Methods of Aquifer Test Analysis," Water Resources Bulletin, vol. 8, No. 1, pp. 175–187, February 1972.

[21] ———, and Walton, W. C., "Hydrologic Budgets for Three Small Water Sheds in Illinois," Illinois State Water Survey Report of Investigation No. 46, Urbana, 1961.

[22] Skibitzke, H. E., "Electric Computers as an Aid to the Analysis of Hydrologic Problems," General Assembly, International Union of Geodessy and Geophysics, Helsinki, 1960.

[23] Stallman, R. W., "Electric Analog of Three-Dimensional Flow to Wells—Its Application to Unconfined Aquifers," U.S. Geological Survey Water-Supply Paper 1536–H, 1963.

[24] Thornthwaite, C. W., "An Approach to a Rational Classification of Climate," Geographical Review, vol. 38, pp. 55–74, 1948.

[25] Todd, D. K., "Ground Water Hydrology," John Wiley & Sons, New York, 1959.

[26] Todd, D. K., "Flow in Porous Media Studied by Hele-Shaw Channel," Civil Engineering, vol. 25, No. 2, p. 51, February 1955.

[27] Turner, S. F., and Halpenny, L. C., "Ground-Water Inventory in the Upper Gila River Valley, New Mexico and Arizona," Transactions of the American Geophysical Union, vol. 22, part III, pp. 738–744, 1941.

[28] "Development of the Ground Water Resources with Special Reference to Deltaic Areas," UNESCO, Water Resources Series No. 24, New York, 1963.

[29] "Methods and Techniques of Ground Water Investigations and Developments," UNESCO, Water Resources Series No. 33, New York, 1967.

[30] Walton, W. C., and Neil, J. C., "Analyzing Ground Water Problems with Mathematical Models and Digital Computers," Illinois State Water Survey Reprint Series 1961–B.

[31] ———, and Prickett, T. A., "Hydrogeologic Electric Analog Computer," Proceedings of the ASCE., Journal of the Hydraulics Division, vol. 89, No. HY6, pp. 67–91, November 1963.

[32] ———, "Ground Water Resources Evaluation," McGraw-Hill, New York, 1970.

[33] Williams, C. C., and Lohman, S. W., "Methods Used in Estimating Groundwater Supply in the Wichita, Kansas, Well-Field Area," Transactions of the American Geophysical Union, vol. 28, No. 1, pp. 120–131, February 1947.

[34] Wisler, C. O., and Brater, E. F., "Hydrology," second edition, John Wiley & Sons, New York. 1959.

[35] Lowry, R. L., Jr., and Johnson, A. F., "Consumptive Use of Water for Agriculture," Transactions of the ASCE, vol. 107, Paper 2158, pp. 1243–1302, 1942.

«Chapter VII

INITIAL OPERATIONS AND COLLECTION OF CORRELATIVE DATA

7–1. Initial Operations.—Before initiating field work, the field force should be familiar with the requirements and procedures outlined in chapters IV, V, and VI, and should review the basic data available for the area involved. A preliminary reconnaissance survey should then be made of the area, at which time particular attention should be given to those subareas for which there is a lack of data in published reports or the reports indicate the existence of problems such as ground-water overdevelopment, waterlogging or poor drainage, and saline or alkaline soils. In addition, observations should be made of geomorphological features which might influence the occurrence of ground water: surface elevations and gradients; soil and rock textures; stream pattern, gradients, and bed characteristics; springs, seeps, and marshy areas; vegetation types and densities; distribution, density, and type of water wells; land-use patterns, size of farm units, and land ownership; and present water use and the relationships of these features to the general geology. During the survey, information on the location and capabilities of laboratory facilities, drilling contractors, well service companies, and similar organizations whose services might be required should be obtained and, if possible, initial contacts made. The survey should permit a delineation of those subareas requiring additional geologic, topographic or other mapping, and the outline of an initial program for the work to be done.

On large projects, much of the required information on political, social, and economic factors, such as utilities, land use, and ownership, will have previously been prepared within the Bureau or by other agencies. Similarly, these offices may have data on climate and surface water hydrology. When the data are not readily available and the ground-water hydrologist is unfamiliar with surface water hydrology, the type of information that may be required should be carefully determined and knowledgeable advice sought on obtaining these data.

7–2. Records of Wells, Springs, Seeps, and Marshes.—One of the early activities usually undertaken in a ground-water investigation is an inventory of existing ground-water facilities and collection of well logs. In many States, the State Engineer's office, the Water Resources Board, or similar organizations will usually have files of well records giving location, depth, formation, logs, casing and screen used, static water level, pumping water level, yield, date drilled, the driller, and similar data. Copies of the records should be obtained and the data for each well entered on a form similar to the form shown on figure 7–1. Each well should be numbered and a map showing its location should be prepared.

The records in the State offices are often incomplete or questionable, so each well should be checked in the field. If not included in the State records, the ground surface elevation at the well should be determined either by leveling or, if a suitable topographical map is available, by observation and interpolation. In the case of especially useful wells, the owner should be contacted for additional data and permission to measure the depth of water in the well and, if required, to make a pumping test. Also, the driller (if available) can often furnish data.

Wells may often be found in the field which are not on the State records, and such wells should be inventoried. Also, data on new wells drilled during the study should be obtained from the driller and owner.

For location and mapping purposes, wells are usually tied to the state plane coordinate system or the township-range system used by the U.S. Geological Survey (USGS) shown on figure 7–2.

One of the first steps normally taken during a ground-water investigation is the measurement of water levels in wells. Because ground-water levels are dynamic, measurements should be made in all wells in as short a period as possible. Figure 7–3 is a typical form used in making such a survey. On completion of the first complete ground water elevation measuring program, a ground-water contour map should be prepared, preferably using a topographic map as a basis. A study of the map will permit recognition of those points where control is poor or lacking and will serve as a guide in locating observation wells that must be drilled. In addition, the map can be used to indicate possible ground and surface water interrelationships. It also indicates locations for observation wells near streams, canals, lakes, reservoirs, and other bodies of water, as well as locations for staff gages [3].[1] Staff gage locations are also plotted on the map and a form similar to the one on figure 7–4 can be completed.

The presence of springs, seeps, and marshes is usually the result of the water table intersecting the ground surface or of leakage from an artesian aquifer. Accordingly, the location, discharge, and water level elevation of such features may be significant. During the initial well survey, all springs, seeps, and marshes should be visited, given a distinctive symbol and number, plotted on the map, and recorded on a form similar to the one on figure 7–4. The approximate water level elevation should be determined, a gage or weir installed,

---

[1] Numbers in brackets refer to items in the bibliography, section 7-10.

WELL RECORD

Region____ Date________19____ Record by______________

Project or Unit____________________ Well No. or Name______________

Source of data______________________________

1. Location: State______________ County______________
   Map______________________________
   _____¼_____¼ sec. _____T _____N/S R_______E/W
2. Owner:__________________ Address______________
   Tenant__________________ Address______________
   Driller__________________ Address______________
3. Topography____________________
4. Elevation_______ft. above/below _______
5. Type: Dug, drilled, driven, bored, jetted ___19___
6. Depth: Rept._______ft. Meas._______ft.
7. Casing: Diam.____in., to ____ in., type ____
   Depth_____ft., Finish ______________
8. Chief aquifer______________ From ______ft. to ______ft.
   Others______________________________
9. Water level________ft rept./meas. ________19____ above/below _____
   ______________ which is ______ft. above/below surface
10. Pump: Type____________ Capacity________gal/min _____
    Power: Kind______________ Horsepower__________
11. Yield: Flow__ gal/min, Pump___gal/min, Meas., Rept. est.____
    Drawdown____ft. after______hours pumping______gal/min
12. Use: Dom., Stock, PS., R.R., Ind., Irr., Obs. ______________
    Adequacy, permanence ______________________
13. Quality ______________________ Temp __________°F.
    Taste, odor, color______________ Sample yes/no _______
    Unfit for______________________________
14. Remarks: (Log, Analyses, etc.)______________________
    ____
    ____

FIGURE 7–1.—Typical well record form. 103–D–1440.

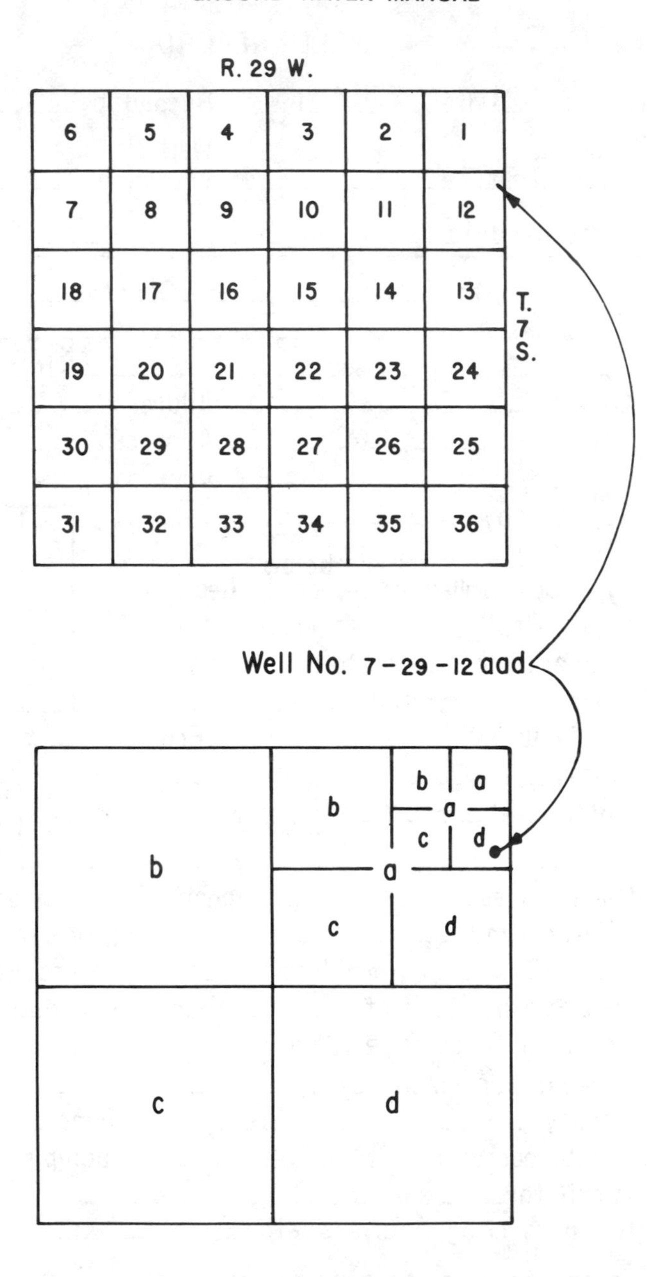

FIGURE 7-2.—USGS township-range numbering system. 103-D-1456.

WATER LEVEL MEASUREMENTS (Field)

Measured by

Location of project

DATE	HOUR	WELL	TAPE READING AT—		DEPTH TO WATER	REMARKS
			MEAS. POINT	WATER LV.		

FIGURE 7-3.—Typical water level record form. 103-D-1457.

STAFF GAGE or WEIR RECORD (field)

Date: _______________

Region: _________________ Recorded by: __________

Project on Unit: __________ Staff Gage or Weir No. __

____

1. Location: State __________ County __________

   Map ____________________________________

____ 1/4 _____ 1/4 sec ______ T ______ N/S R _____ E/W ____

2. Owner of property: __________________________
   Address: ____________________________________
   Tenant of property: _________________________
   Address: ____________________________________
   Installed by: _______________________________
3. Gage name and type: ___________________________
   (Staff gage, cipoletti weir, etc.)
4. Feature measured, name and type: _______________
   (Stream, lake, marsh, spring, etc.)
5. Elevation, 0 index on gage or weir crest: __________ ft.
6. Measured water surface: ______________________ ft.
7. Elevation measured water surface: _______________ ft.
8. Discharge if determined: ____________________ ft^3/s
9. Date of measurement: ______________________ 19__.
10. Measured by: ___________________________________
11. Remarks: _______________________________________

____

____

FIGURE 7-4.—Staff gage or weir record form. 103-D-1458.

and an effort made to determine the geologic and hydrologic conditions giving rise to the feature.

7-3. Initiation and Frequency of Ground Water Level Measurements.—Water level measurements may be made of the water table or the piezometric surface under numerous conditions. Records may be required of continuous measurements such as those supplied by a continuous water stage recorder or of periodic measurements with a time interval extending from less than 1 minute (for a pump test) to 6 months. The frequency of measurement should be adjusted to the circumstances. In some instances, only a few measurements are possible or expedient to make, but in other instances frequent measurements over a long period of time may be required. The possibility of error in interpretation decreases as the frequency of measurement and length of record increase. In ground-water inventory and drainage investigations, water level observations may continue for many years. Initially, measurements are made often until the annual regimen is established. The frequency may then be reduced to about four a year with the exception of a few carefully selected observation wells. These may be read 6 to 24 times a year or equipped with continuous recorders.

Where water storage structures or new irrigated areas are contemplated, it is advisable to install a number of observation wells at least 2 years prior to construction and to take measurements monthly or bimonthly in order to determine preexisting ground-water conditions. The program should be continued after construction and full operation to permit comparison of preexisting and post-facility conditions. Such data may be invaluable in the event of claims or suits for damage.

7-4. Water Level Measuring Devices.—Measurements may be made with a number of different devices and procedures (fig. 7-5). Probably the most common device for measuring static water levels is the chalked steel tape which has a weight attached on the lower end. The weight keeps the tape taut and aids in lowering it into the well. The tape is chalked with carpenter's chalk, ordinary blackboard chalk, or dry soil which changes shade upon becoming wet. The line of the color change denotes the length of tape immersed in water. Subtracting this length from the reading at the measuring point gives the depth to water. Cascading water in a well may mask the mark of the true water level on the tape; however, this usually occurs only in a well that is being pumped. When this condition is encountered, another method of measuring may have to be used. In small-diameter wells, the volume of the weight may cause the water level to rise in the pipe, and the measurement may be somewhat inaccurate.

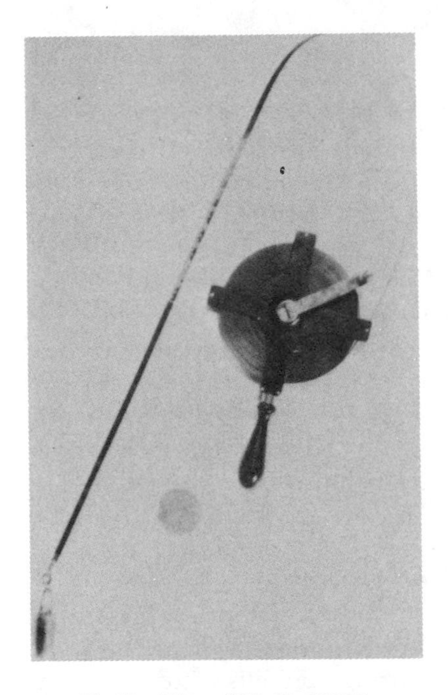

Chalked line. PX–D–25997.

Popper. PX–D–25996.

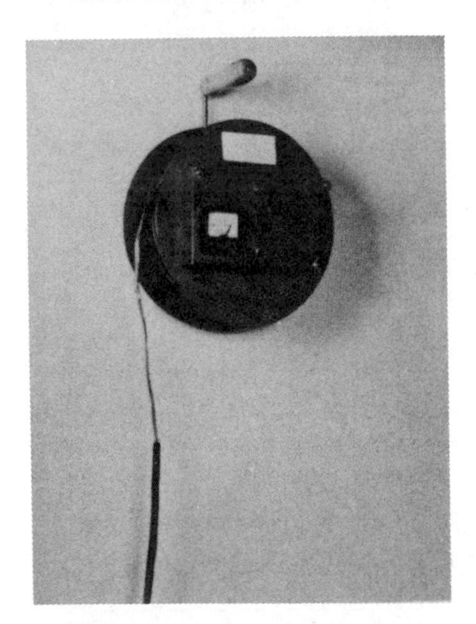

Electric sounder. PX–D–25177.

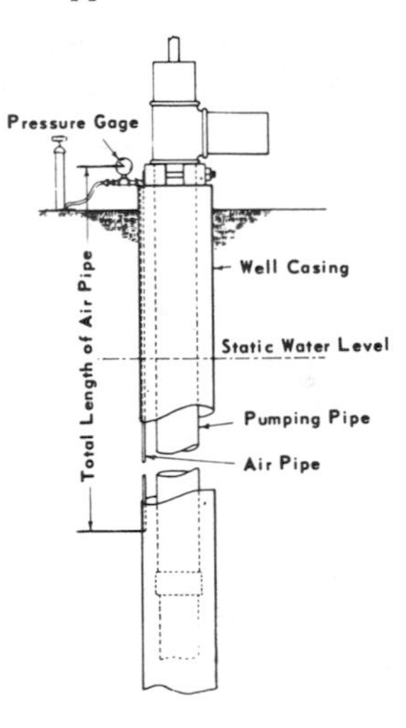

Air line.PX–D–25995.

FIGURE 7–5.—Devices for measuring depth to water in well.

Electric sounders may also be used to measure the depth to water in wells. There are a number of commercial models available, none of which is entirely reliable. Many sounders use brass or other metal indicators clamped around a conductor wire at 5-foot intervals to indicate the depth to water when the meter indicates contact. The spacing of these indicators should be checked periodically with a surveyor's tape to assure accurate and reliable readings.

Some electric sounders use a single-wire line and probe, and rely on grounding to the casing to complete the circuit; others use a two-wire line and double contacts on the electrode. Most sounders are powered with flashlight batteries and the closing of the circuit by immersion in water is registered on a milliammeter. Experience has shown the two-wire circuits with a battery are by far the most satisfactory. Electric sounders are generally more suitable than other devices for measuring the depth to water in wells that are being pumped because they generally do not require removal from the well for each reading. However, when there is oil on the water, water cascading into the well, or a turbulent water surface in the well, measuring with an electric sounder may be difficult. Oil not only insulates the contacts of the probe, but if there is a considerable thickness of oil, it will give an erroneous reading. In some instances, it may be necessary to insert a small pipe in the well between the column pipe and the casing from the ground surface to about 2 feet above the top of the pump bowls. This pipe should be plugged at the bottom with a cork or similar seal which is blown out after the pipe is set. Measurements with the electric sounder can then be made in the smaller pipe where the disturbances are eliminated or dampened, the true water level is measured, and the insulating oil is absent. Figure 7–6 illustrates a convenient arrangement for direct measurement of drawdown during pumping tests. A marker on the sounder wire is referred to a value on the tape and the same marker is used as a reference to determine drawdown by changes on the tape when contact with the water is made. A new marker is used each time the water level drops a 5-foot increment.

A simple and reliable method for measuring the depth to water in observation holes between 1½ and 6 inches in diameter is a steel tape with a popper (see fig. 7–5). The popper is a metal cylinder 1 to 1½ inches in diameter and 2 to 3 inches long with a concave undersurface and is fastened to the end of a steel tape. The popper is raised a few inches and then dropped to hit the water surface, where it makes a distinct "pop." By adjusting the length of tape, the point at which the popper just hits the surface is rapidly determined. Poppers generally are not satisfactory for measuring pumping wells because of the operating noise and lack of clearance.

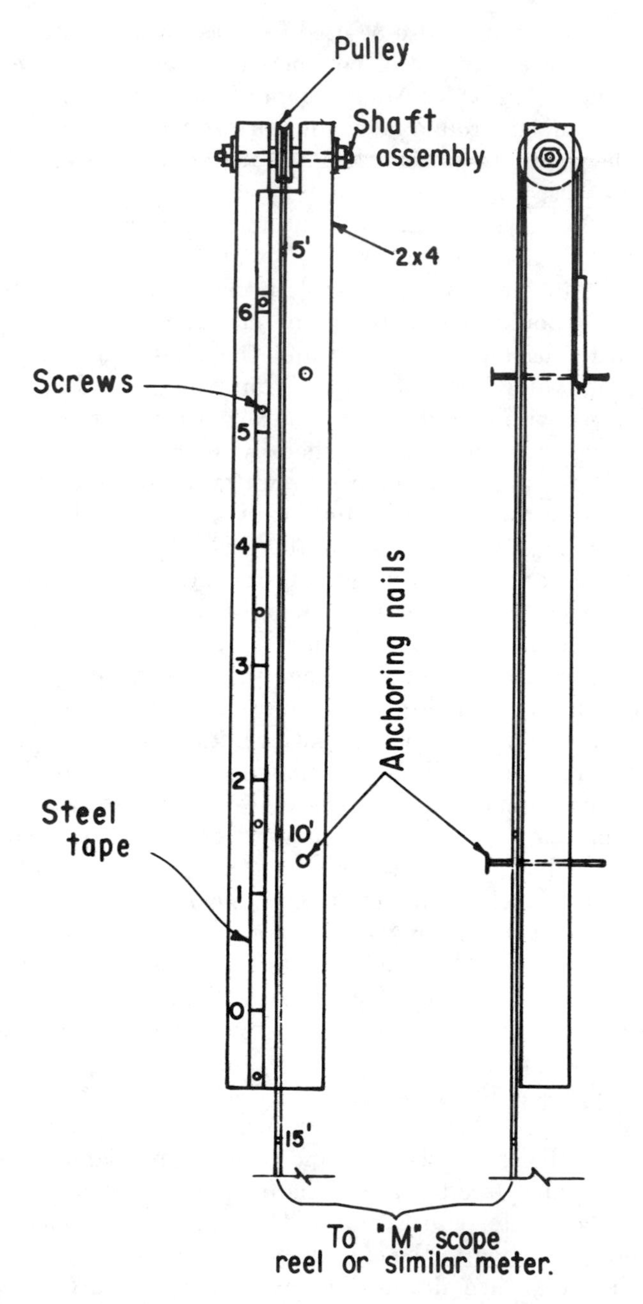

FIGURE 7-6.—Direct drawdown measuring board. 103-D-1459.

Permanent pump installations should always be equipped with an access hole for probe insertion or an air line and gage, or preferably both, to measure drawdown during pumping. An air line is accurate only to about 0.5 foot unless calibrated against a tape for various drawdowns, but is sufficiently accurate for checking well performance.

Artesian wells with piezometric heads above the ground surface are conveniently measured by capping the well with a cap that has been drilled, tapped, and fitted with a plug which is removed for the insertion of a Bordon gage or mercury manometer stem. The static level is determined from the gage or manometer reading after the pressure has stabilized. Figure 7–7 is a drawing of a mercury gage, designed by S. W. Lohman of the Geological Survey, that is particularly suited to field use, especially when running a recovery test after constant head tests of artesian aquifers (see sec. 5–9). For continuous records, a recording pressure gage may be used.

7–5. Records of Water Level Measurements.—Accurate permanent records should be kept of all water level measurements and should include: (1) identification of the well by number and location; (2) location and elevation of reference point; (3) elevation of ground surface; (4) date of measurement; (5) measured depth to water or to the bottom of the hole, if dry; (6) computed elevation of the water table or piezometric surface; (7) for piezometers, the aquifer or other zone represented by the reading; and (8) a note whether the well was being pumped when measured, was pumped recently, or whether a nearby well was pumping during the measurement.

7–6. Exploration Holes and Observation Well and Piezometer Installation.—Areas may be encountered containing wells for which logs are not available, where well construction features preclude measurement of water levels, or where the wells have not been drilled. For investigations of any size or importance and even for many individual well installations in such areas, exploratory drilling is often necessary. Such drilling should be tailored to the needs of the investigation and gaps in available data. In many instances, holes drilled for stratigraphic or other data can be converted for use as observation wells or piezometers.

- Stratigraphic holes are drilled primarily for the purpose of determining the nature, depth, and thickness of the geologic formations.
- Pilot holes are usually drilled to obtain data on which to base the design of wells.
- Observation wells are usually constructed for the purpose of measuring water levels where subsurface conditions are relatively simple.

① 1- $\frac{1}{4}$" Stainless steel stop cock.
② 4' Length of $\frac{5}{8}$" i.d. rubber hose.
③ 1- 2" Dia. ink bottle.
④ 1- 3 Holed No. 8 rubber stop.
⑤ 1- $\frac{3}{4}$" Hose coupling.
⑥ 48" Length of 2 mm i.d. glass tubing.
⑦ 45" Length of stainless steel strip with graduations which give readings in feet of water.
⑧ 1- 4" Length of $\frac{1}{4}$" o.d. stainless steel tubing with fittings.
⑨ 1- $\frac{5}{16}$" Stainless steel stop cock.
⑩ 1- 4" Length of $\frac{5}{16}$" o.d. stainless steel or plastic tubing with fitting.

Assorted lumber (marine plywood)
Assorted $\frac{1}{8}$" bolts with nuts.

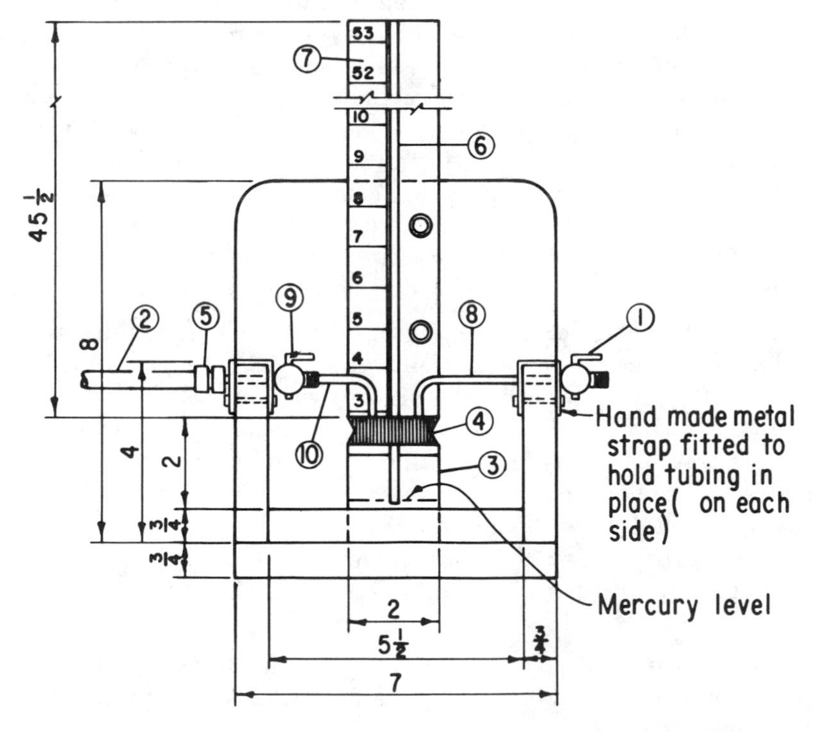

(After S.W. Lohman)

FIGURE 7-7.—Mercury manometer for measuring artesian heads. 103-D-1460.

- Piezometers are a special type of observation well so finished as to permit the measurement of the water level in a particular stratum or zone.

After exploratory or similar holes are completed, a permanent record should be made of each. This record should include an as-built drawing of the facility showing the elevation of the point from which measurements of the depth to water in the hole will be made; the elevation of the average ground surface in the vicinity of the well; the depth of hole, the length, size, and type of casing; location of seals and packers; and the location of the screen or perforations.

The record should also show subsurface geologic conditions, water level data, the location of the hole with respect to landlines or whatever land subdivision system is used in the area, and the identification number of the hole (see sec. 7–2).

7–7. Drilling of Exploratory Holes, Observation Wells, and Piezometers. There are many methods and combinations of methods of drilling exploratory holes and wells. Classified according to method of installation, the most common holes are dug, drilled, bored, driven, or jetted. Briefly, these are described as follows:

(a) *Dug holes* are usually restricted to shallow depths where information is not needed for more than a few feet below the water table. This type hole is rarely used in the United States.

(b) *Drilled holes* may be put down by any of the well drilling methods in common use, but the type of rig and tools used and the diameter of hole drilled will depend upon the materials to be drilled and the data to be obtained. In general, a 4-inch hole is about the smallest that is satisfactory in unconsolidated materials and a 3-inch hole in consolidated rock. The hole is cased if the material will not stand for the period of time required for use. Drilled holes can be put down to great depths (thousands of feet) and through any material. The quality of samples obtained depends largely upon the type of drill rig used.

(c) *Bored or augered holes* may be drilled manually or by machine-driven augers. This type of hole can only be used in unconsolidated fine- to medium-grained material. The depth limitation for hand-augered holes is about 40 feet, whereas machine-driven augers may penetrate to several hundred feet. When holes are bored in unstable material below the water table, caving may prevent further progress. Samples obtained by augering may range from nonexistent in saturated coarse-grained materials to disturbed but representative samples of fine-grained materials.

(d) *Driven holes* are put down by driving a pipe, usually equipped with a well point, into the material. Neither samples nor a log can be obtained, and this method is suitable only for measuring water levels. Installation is restricted to shallow depths in fine- and medium-grained unconsolidated materials.

(e) *Jetted holes* are similar to driven holes except that the pipe is put down by hydraulic jetting and often can be installed to greater depths. Badly mixed washed samples and a rough log may be obtained when holes are jetted.

Where conditions are uniform, it may be satisfactory to install observation holes on a grid with holes spaced at uniform intervals. Where conditions are not uniform, wells should be located to conform to the local variations in conditions.

The magnitude and type of the study will also affect the spacing and location of holes. In a reconnaissance study to obtain general information on an area, a wide spacing is satisfactory; for a detailed study, the spacing must be reduced to provide the necessary detail.

For ground-water inventory or development studies, the holes should be deep enough to penetrate at least 10 feet below the lowest water table of record or to the top of an artesian aquifer. If information on thickness of an aquifer is required, one or more holes should be drilled through the aquifer. An indication of the required hole depth can usually be obtained from an inventory of existing well records. Separate wells or piezometers may be required when two or more aquifers are involved.

For protection against damage, holes completed for observation or piezometric measurements should be located, if possible, in a fence row or adjacent to a permanent structure.

The practice of installing observation wells on a step-by-step or stage basis is recommended both from a technical and an economic standpoint.

Casing installed in observation wells should be sized to the purpose of the facility and means of obtaining data. Generally, if water levels are to be measured by a wetted tape or electric probe, a ¾- to 1¼-inch-diameter steel or plastic pipe is suitable. However, if a standard water stage recorder is to be used or water samples taken from the facility, a minimum 4-inch casing may be required. Suitable perforations should be made opposite the saturated zone to assure reliable readings.

Piezometer installations (fig. 7–8), rather than simple observation wells, are essential to a clear understanding of ground-water conditions where subsurface conditions are complex. The presence of a confined zone or several zones each with a different water level requires use of piezometers to confine and separate each level. Observation of pumping test influence may especially require the use of piezometers, even in apparently homogeneous aquifers. Installation of piezometers,

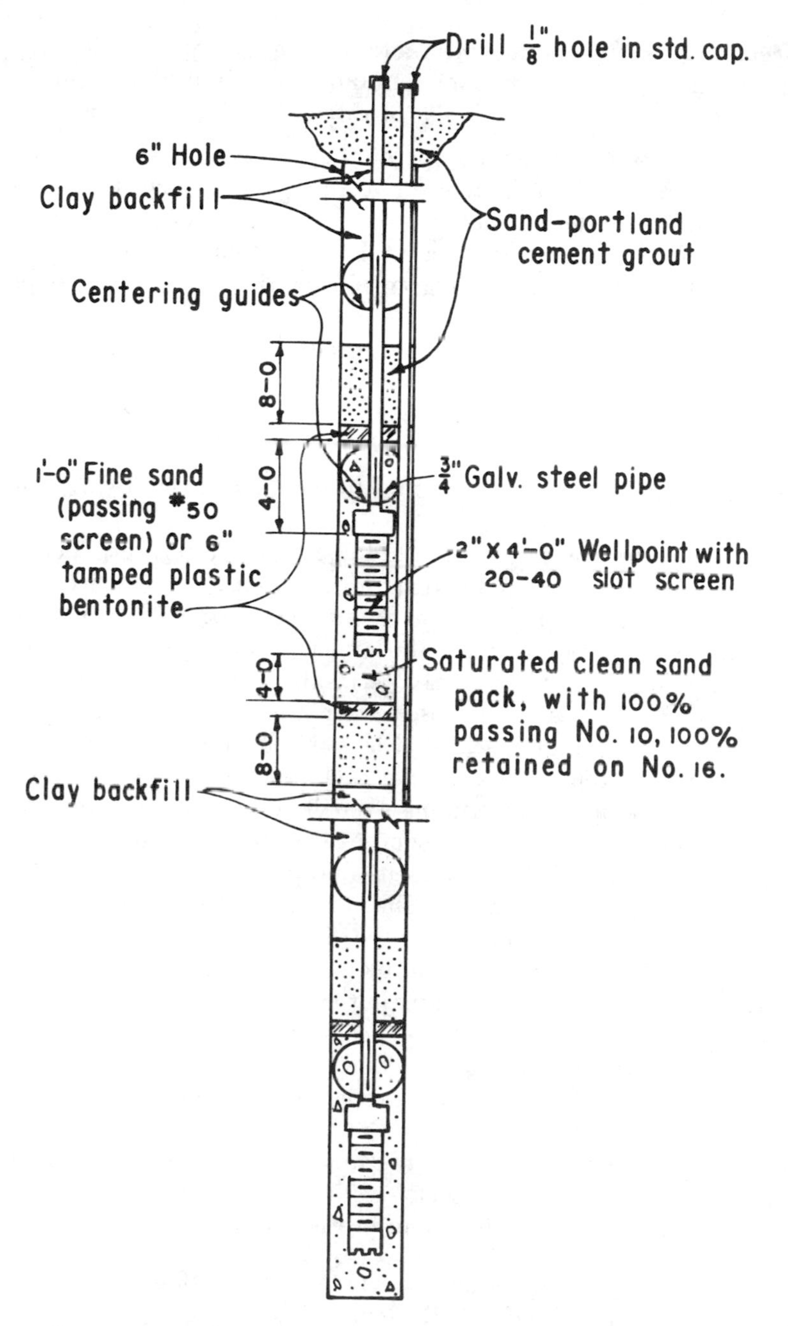

FIGURE 7-8.—Typical dual piezometer installation. 103-D-1461.

especially in slowly permeable materials, may require strict design considerations to minimize time lag and other similar problems.

Each piezometer should consist of three essential components:

(1) A watertight standpipe of the smallest possible diameter, consistent with the method of reading, attached to the tip and extending to the surface.

(2) A tip consisting of a well screen, porous tube, or other similar feature, and in fine-grained materials, a surrounding zone of filter sand.

(3) A seal consisting of cement grout, bentonite slurry, or other similar slowly permeable material placed between the standpipe and the hole to isolate the zone.

Where several piezometers are required at a given location, it may be possible, as a cost-saving feature, to install them in a single hole, as shown on figure 7–8.

In addition to the described standpipe-type piezometer, there are several commercially available instruments that are operated by hydraulic or pneumatic pressure, or by an electric signal. Such instruments may be especially valuable for unusual subsurface or monitoring conditions, such as in very slowly permeable materials.

The casing or pipes in an observation well or piezometer usually extend above the ground surface at least a foot unless pit installation is necessary. The top of the casing or each pipe should be fitted with a screwcap or locking cap containing a small hole to permit adjustment of air pressure in the pipe in response to water level fluctuations or barometric changes. Where artesian flow conditions are present, a tight-fitting cap which has been drilled and tapped for a Bourdon gage or mercury manometer should be used. If climatic conditions require protection against freezing, a suitable shelter equipped with heating facilities or replacement of the water in the upper portion of the piezometer by a non-freezing fluid may be necessary.

Facilities should be protected against standing surface water and leakage alongside the casing by proper grading and placement of grout or clay seals at the surface.

When an observation well or piezometer must be located in the open where damage by livestock or farm machinery may occur, it should be adequately identified and protected.

7–8. Formation Sampling and Logging of Exploration and Observation Holes.—The Bureau of Reclamation's *Earth Manual* [2] describes methods and equipment for drilling and sampling which are applicable to ground-water investigations. Undisturbed samples generally are not required for ground-water investigations. However, representative samples which preserve grain size and gradation relationships of granular materials are often required, especially for design criteria. So far as possible, the drilling method and equipment should be capable of yielding the necessary samples. Sampling applicable to well drilling is also described in chapter XVI.

Each hole should be carefully logged with regard to depth and material as the samples are obtained. For field logging of unconsolidated materials, the Unified Soil Classification symbols and nomenclature described in the *Earth Manual* [2] should be used. About a quart of representative samples should be saved of each of the primarily sandy or coarser materials. Samples need not be taken of clayey or predominantly silty materials unless unusual conditions are found or data needed, but such materials should be described and accurately located in the logs. If gravels larger than 1 inch in diameter are obtained in the samples, they may be removed and the size range and approximate percentage of the sample they represent should be noted.

(a) *Undisturbed Samples of Unconsolidated Material.*—Undisturbed samples taken by drive sampling or coring should be described as homogeneous, layered, stratified, etc. When layers consist of different materials such as clay and fine sand, the nature, thickness, and color of the layers should be recorded and the coarse fraction separated, if possible, and mechanically analyzed.

Samples of coarse, granular materials of a more homogeneous nature should be described on the basis of visual examination according to the Unified Soil Classification given in the *Earth Manual* [2] and mechanically analyzed.

(b) *Disturbed Samples of Unconsolidated Material.*—Disturbed samples, such as those obtained with a cable tool, rotary, or reverse circulation rig, usually represent a mixture of the materials in the interval sampled. The sample should be examined carefully for larger cohesive fragments which may indicate the nature of the material in place and any material adhering to the bit, auger, or bail should be scraped off and included with the sample unless it is obvious that the material was scraped off the hole wall while being withdrawn. Samples other than those obtained with a direct-circulation rotary rig should not be washed prior to being sent to the laboratory. Samples taken from the ditch when using a direct-circulation rotary rig and clay-based drilling fluid should be placed in a 5-gallon container filled with water, stirred vigorously, and permitted to settle for at least 20 minutes. The muddy water then should be decanted and the material from the bottom of the container taken for a sample. The total volume of cuttings representing each drilled interval should be mixed and quartered until a 2-quart volume of representative material remains.

(c) *Mechanical Analyses of Samples.*—Samples should be washed on a No. 200 sieve and the percentage of minus 200 material determined. A hydrometer analysis of the minus 200 size normally is not necessary. The plus 200 sizes should be screened through a nest of ⅜ and No. 4, 8, 16, 30, 50, and 100 sieves. Forms 7–1451 and 7–1415 illustrated on figures 7–9 and 7–10, respectively, should be prepared for each sample. The washed and sieved samples less the minus 200 sieve sizes should be recombined for visual study.

(d) *Visual Examination of Samples.*—The washed samples should be examined with a binocular microscope or hand lens and adequately described, including grain size and roundness, mineralogy, and other characteristics.

(e) *Drill Core Samples of Consolidated Rock.*—Cores should be identified regarding the type of rock, color, cementation, fractures, and other similar characteristics. Sandstone and conglomerate cores, if readily friable, should be crumbled and mechanically analyzed.

In many instances, the field logs can be refined after a mechanical analysis and a visual study of the samples.

7–9. Water Samples from Bore Holes, Wells, and Surface Sources.— The type of investigation and purpose of the study determine, to a large degree, the need for water samples, sampling locations, and the frequency of collection [4]. Ground-water quality may vary from hole to hole in the same aquifer and sometimes with depth in a relatively homogeneous aquifer.

When drilling uncased holes with augers and cable tools, a representative water sample can usually be obtained from the first water encountered by bailing the hole dry, permitting the water level to rise, and then bailing a sample from the hole. However, representative samples from levels deeper in the formation or from deeper aquifers cannot be obtained unless the hole is cased.

In rotary drilled holes, samples from individual aquifers or specific depths cannot readily be obtained except by drill stem tests or other similar procedures. On completion of the hole, flushing out the drilling fluid and pumping or bailing of the hole for a sufficient time will permit obtaining a fairly representative composite water sample. In addition, an electrical log of an uncased hole may sometimes be interpreted to give some idea of the relative quality of the water in different aquifers and at different depths [1].

In consolidated rock, where casing is not usually used, a composite water sample can be taken. When water samples are required from specific depths or aquifers, a pump equipped with inflatable packers above and below the intake screen may sometimes be a practical solution to the problem (see fig. 7–11).

LABORATORY SAMPLE NO. 1

FEATURE WELL NO. 1 AREA UPPER BENCH EXC. NO. ___ DEPTH 150' TO 155'

SAMPLE PREPARATION

PREPARED BY A.X. BOBB % MOIST + NO. 4 ___ WET WT. TOTAL SAMPLE ___

DATE SEPTEMBER 16, 19 % MOIST − NO. 4 ___ DRY WT. TOTAL SAMPLE 111.9

SIEVE SIZE	5"	3"	1-1/2"	3/4"	3/8"	NO. 4	TOTAL WT. PASSING NO. 4
WT. PAN + RETAINED MAT.					422.9		
WT. PAN					311.0		
WET WT. RETAINED							___ WET
DRY WT. RETAINED					1.1	4.0	106.8 DRY
DRY WT. PASSING					110.8	106.8	
% OF TOTAL PASSING					99.0	95.4 = W%	

SIEVE AND HYDROMETER ANALYSIS

DISH NO. 1 DRY WT. OF SAMPLE (W) = 106.8 gms. FACTOR (F) = $\frac{W\%}{W} = \frac{95.4}{106.8} = 0.893$

DRY WT. SAMPLE (SIEVED) 106.8

SIEVING TIME 15 MINUTES DATE SEPTEMBER 16, 19

SIEVE NO.	WEIGHT RETAINED	WEIGHT PASSING	F X WEIGHT PASSING = % OF TOTAL PASSING	% OF TOTAL PASSING	PARTICLE DIA (mm)	REMARKS
8	3.3	103.5		92.4	2.380	
16	13.9	89.6		80.0	1.190	
30	47.0	42.6		38.0	0.590	
50	35.4	7.2		6.4	0.297	
100	4.8	2.4		2.1	0.149	
200	0.6	1.8		1.6	0.074	
PAN	1.8					
TOTAL	106.8					

TESTED AND COMPUTED BY R.E. SMITH CHECKED BY L.R. JONES DATE SEPTEMBER 16, 19

HYDROMETER ANALYSIS

HYDROMETER NO. ___ DISPERSING AGENT ___

STARTING TIME ___ DATE ___ AMOUNT ___ ml

TIME	TEMP C°	HYD READ	HYD CORR	READ CORR	F X CORRECT READ = % OF TOTAL PASSING	% OF TOTAL PASSING	PARTICLE DIA. (mm)	REMARKS
.5 MIN*							0.050	
1 MIN							0.037	
4 MIN							0.019	
19 MIN							0.009	
60 MIN							0.005	
HR. 15 MIN*							0.002	
25 HR. 45 MIN*							0.001	

TESTED AND COMPUTED BY ___ CHECKED BY ___ DATE ___

*NOT REQUIRED FOR STANDARD TEST.

FIGURE 7-9.—Typical mechanical analysis form. 103-D-1462.

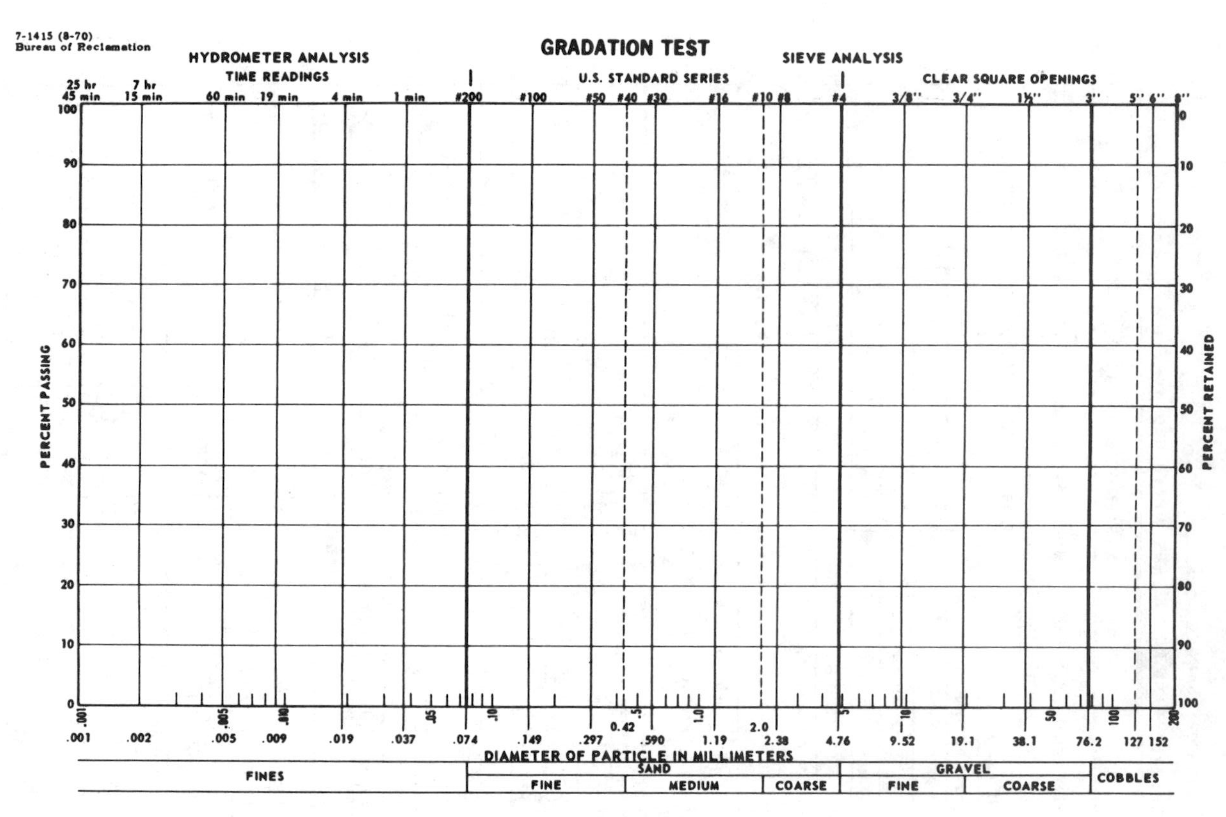

Figure 7-10.—Typical gradation test form. 103-D-1463.

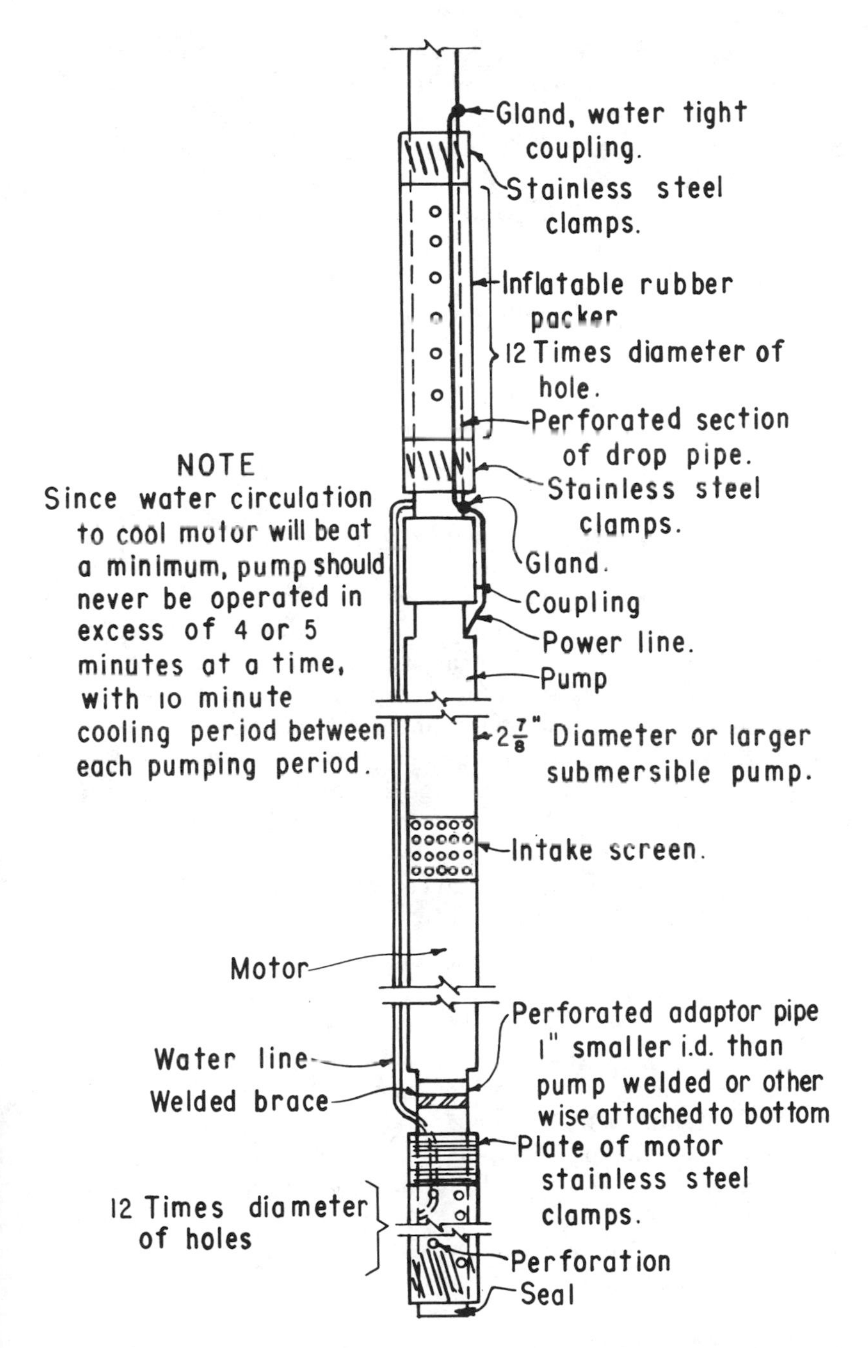

FIGURE 7–11.—Packer-equipped pump for selective sampling of water from wells. 103–D–1464.

Prior to taking a water sample, sufficient water should be bailed or pumped from the hole to assure a representative sample. An amount equal to twice that stored in the hole is normally adequate.

Samples from existing wells are usually taken at the discharge. The well should be pumped for a sufficient length of time to assure a representative sample. Temperature may be taken by inserting the thermometer in the stream as it discharges from the pipe.

One-quart water samples are usually adequate for most chemical analyses, but if pesticides and similar contaminants must be determined, several gallons may be required.

Polyethylene sample bottles are the most satisfactory for Bureau purposes. The new bottles should be thoroughly rinsed, filled with water, and allowed to stand for about a week, then emptied, rinsed once or twice with tap water, and lastly rinsed with distilled water. After drying, they should be capped and not opened until used. A similar treatment is recommended before a bottle is reused. Bottles which are discolored or contain visible deposits not readily removed by rinsing should not be reused. The polyethylene bottles should have relatively long screw caps with a positive seal lip on the bottle. They should be filled to the rim of the seal so that little or no air is contained in the bottle. Such bottles are not subject to breakage by shock or freezing, nor are they likely to lose fluid due to changes in atmospheric pressure. However, precautions against freezing should be taken since the character of the water may change due to freezing and subsequent thawing. Samples should be transported to the laboratory and analyzed as soon as possible.

Each sample bottle should be tagged or otherwise identified and the following applicable sample data recorded:

- Well or hole number and location
- Depth of well
- Source (aquifer or formation) of the water
- Method of collection and time since pumping or bailing started
- Depth or interval from which sample was taken
- Water temperature
- Date and time of collection
- Appearance at time of collection; that is, clear, milky, colorless, etc.
- Initials or name of collector
- Type of analysis required; that is, comprehensive or key constituents only
- Field analyses made, if any

Samples for bacterial analyses are usually taken in 4- to 6-ounce sterile glass bottles provided by a health agency or other laboratory

[4]. The caps should not be removed until a sample is to be taken. When taking a sample, care should be exercised not to touch the inside of the cap or bottle with the fingers, nor should water be permitted to flow over the hands or fingers and into the bottle or inside of the cap.

The samples should be kept cool or refrigerated, if possible, during transport to the laboratory. Not more than 48 hours should elapse between taking a sample and its delivery to the laboratory.

Samples for bacterial and chemical analyses are frequently taken from surface water sources. Samples should be taken in streams or channels some distance below any tributaries at points where it is apparent that thorough mixing of the water has occurred. The selection of suitable sampling points in lakes and similar water bodies is more difficult. Seasonal stratification, overturn, wind direction, protected areas, and similar factors make the taking of representative samples difficult, sometimes even impossible. In many instances, a number of samples taken at different points and from different depths may be necessary. Each sample should be identified and the following data furnished:

- Identity of water body
- Location of sampling site
- Gage height or water discharge
- Date and time of collection
- Temperature of water
- Depth from which taken
- Appearance of water sample when taken
- Type of analysis required
- Name of collector

Surface water samples should be taken with the same procedures outlined above for ground-water samples and should be treated in the same manner.

For more specialized studies, special purposes, and conditions, additional treatment of samples in the field may be required. A USGS paper outlines many of these [4].

The frequency of water sampling and the type of analyses to be made usually cannot be predetermined but should be developed on the basis of experience and needs. Samples for chemical analysis should be taken on completion of any borehole or well and a comprehensive analysis made. Figure 7–12 shows a typical chemical analysis form. The first known sample taken from an existing well should always be given a comprehensive analysis. Subsequent samples might be taken at low and high water stages, seasonally or annually. Analyses may be comprehensive or for key constituents only, depending on conditions and requirements.

REPORT OF WATER ANALYSES
Design and Construction Division
Engineering Laboratories Branch--Chemical Laboratory

Date shipped ________ By ____________________ Sheet ____ of ____
Date rec'd ________ Analyst ____________________ Date ________

Lab No.	Field identification	K x 10^6 at 25° C	pH	% Na	Parts per million, p/m										
					TDS	B	Ca	Mg	Na	K	CO_3	HCO_3	SO_4	Cl	NO_3
							Milligram equivalents per liter, me/l								
	To convert me/l to p/m, multiply by						20.0	12.2	23.0	39.1	30.0	61.0	48.0	35.5	62.0

FIGURE 7-12.—Typical chemical analyses form. 103-D-1465.

This procedure is also common for bacterial analyses. Samples are usually checked for pathogenic organisms or for indication of sewage or similar contamination. In some instances, however, the examination may require determination of the presence of sulphate-reducing or similar nonpathogenic but corrosion-fostering or other economically deleterious organisms.

7-10. Bibliography.—

[1] Pryor, W. A., "Quality of Water Estimated from Electric Resistivity Logs," Illinois State Geological Survey Circular 215, Urbana, 1956.

[2] "Earth Manual," second edition, Bureau of Reclamation, 1974.

[3] "Water Measurement Manual," second edition, Bureau of Reclamation, 1967.

[4] Rainwater, F. H., and Thatcher, L. L., "Methods for Collection and Analysis of Water Samples," U.S. Geological Survey Water-Supply Paper 1454, 1960.

«Chapter VIII

GEOPHYSICAL INVESTIGATIONS, BORE HOLE LOGS, AND SURVEYS

8-1. Geophysical Surveys.—Geophysical surveys made in conjunction with reconnaissance, surface geologic investigations, and exploratory drilling may permit a rapid and relatively low-cost evaluation of the subsurface geology and possibly the general ground-water conditions of an area. Such surveys may materially reduce drilling requirements and resultant overall costs of a ground-water investigation.

Geophysical surveys are essentially the interpretation of the variations in measured response at the surface to certain forces, either natural or artificially generated within the Earth's crust. Such variations result from differences in physical characteristics such as density, elasticity, magnetism, and electrical resistivity of the underlying materials. Measurements are usually made at predetermined distances or locations along a traverse or on a grid.

Geophysical surveys should be correlated with geologic conditions of each area to assure reliable interpretation; therefore, a survey is usually started near an outcrop or drill hole where subsurface conditions are known or can be readily interpreted. Such correlation permits projection of geologic conditions to distant areas through analysis of the geophysical measurements. The accuracy and reliability of data obtained from a geophysical survey depend in large part on the amount of subsurface control available as well as the geologic complexity of an area. The number of holes to be drilled, therefore, varies with the complexity and size of the area under study and the degree of accuracy and detail desired.

Four basic methods of geophysical surveying are available: seismic, electrical resistivity, gravimetric, and magnetic. The first two depend on the introduction of mechanical or electrical energy into the Earth's crust and measurement of the effects of subsurface materials and conditions on the transmission of this energy. The latter two methods depend on the measurement of variations in the intensity and direction of the natural forces of gravity and magnetism.

Many variations of the four basic methods have been developed and, prior to considering a survey, a qualified geophysicist should be consulted about the type of survey best adapted to obtaining the data desired. Such assistance may be available from Federal agencies, private consultants, or service firms.

(a) *Seismic Surveys.*—Seismic surveys are based on measurement of the velocity distribution of artificially generated seismic waves in the Earth's crust. The methods of application of energy range from blows with a sledge hammer to dynamite explosions in drill holes [2,7,8,9,10,18].[1]

The velocity of the seismic waves generated depends on the density and elasticity of the subsurface materials. The velocity is usually lowest for unconsolidated materials and increases with the degree of consolidation or cementation.

As seismic waves radiate from the point source, some travel through the surface layers, some are reflected from surfaces of underlying materials having different physical properties, and others are refracted as they pass through the various layers.

The seismic methods using explosives as an energy source are particularly adapted to use in large areas where deep probing is desired. However, the mechanical introduction of energy by using a hammer may be preferable because of shallow depths or for environmental reasons.

(b) *Electrical Resistivity Surveys.*—Electrical resistivity is a characteristic which makes possible the differentiation between types of earth materials. Resistivity surveys are made by sending a direct or low-frequency alternating current through the ground between two metal stakes and measuring both the current and resulting potential at other stakes or special electrodes. Electrical resistivity of the earth materials is most closely related to the moisture content of the material and its chemical characteristics. For example, dry gravel and sand have higher resistivity than saturated gravel and sand; clay and shale have very low resistivity [1,2,7,8,9,19,20,21].

For depth soundings, the electrodes or stakes are moved farther and farther apart. As a result of these increasing distances, the current penetrates progressively deeper. In this way, the resistivity of a constantly increasing volume of earth is measured and data are obtained with which to plot a resistivity curve.

Electric resistivity surveys are applicable to large or small areas and are extensively used in ground-water investigations because they are responsive to moisture conditions, equipment is readily portable, and the method is commonly more acceptable than the blasting required with seismic methods. The resistivity method, however, is not usable in the vicinity of powerplants, substations, high-tension powerlines, and similar sources of extraneous Earth currents which would adversely affect the accuracy of the field measurements.

---

[1] Numbers in brackets refer to items in the bibliography, section 8–4.

(c) *Magnetometer Surveys.*—The magnetic properties of rocks affect the Earth's magnetic field. For example, many basalts are more magnetic than sediments or acid igneous rocks. In a magnetometer survey, the strength of the vertical component of the Earth's magnetic field is measured and plotted on a map. Analysis of the results may indicate qualitatively the depth to bedrock and presence of buried dikes, sills, and similar phenomena. Additional subsurface control such as deep drill hole data can greatly improve the interpretation. The magnetometer survey is a rapid and relatively low-cost method of determining a limited amount of subsurface geologic information [7,8,9]. This method does not compare with either the seismic or resistivity for detail, and is best suited for broadly outlining a ground-water basin.

(d) *Gravimetric Surveys.*—In the gravimetric survey, the force of gravity is measured at stations along a traverse or in a grid. Gravity variations result from the contrast in density between subsurface materials of various types. This method is applicable to small or large areas. The equipment is light and portable, and field progress is relatively rapid. Field data commonly require extensive altitude corrections. Drill hole data and similar references are necessary for accurate interpretation. The gravimetric survey is a valuable tool in investigating gross features such as depth to bedrock, old erosion features on bedrock, and other features such as buried intrusive bodies [6,7,8,9,11]. The results of this method are less detailed than with seismic or resistivity methods.

8–2. Drill Hole Logs.—The term, *well log*, when unqualified, usually refers to a stratigraphic or lithologic record determined from examination and interpretation of cores, samples, or cuttings. There are numerous other types of logs, however, which in most cases consist of direct or indirect measurements of various physical properties of the formations penetrated by a borehole. These logs are obtained by lowering an instrument, connected to a surface-mounted recording device, down the hole to obtain the subsurface record. Interpretation of these records may furnish qualitative data and sometimes quantitative data on characteristics of subsurface material. Various types of logs used in exploration holes and water wells include electrical logs (both resistivity and self-potential), caliper, temperature, and radiation.

Geophysical logging and interpretation services may be available from Federal agencies, private consultants, or logging firms.

(a) *Electrical Logs.*—Electrical logs consist of two general types: the single or multiple electrode.

The single electrode [4,5,12,13] yields two curves, a spontaneous potential (SP), and a resistance curve, which are plotted on a two-pen recorder. The SP curve is a record of the variations in natural direct-current potentials which exist between subsurface materials and a static electrode at the surface. This variation is plotted against depth. The resistance curve is a record of the variations in resistance between a uniform 60-hertz alternating current impressed on the sonde[2] and a static electrode at the surface.

The logging instrument consists of a sonde with a single electrode supported by a single conductor cable leading to the two-pen recorder; an alternating-current generator; an electrode attached to the recorder and grounded at the surface to complete the circuits; and cable reels, winches, and similar accessory equipment.

The natural direct-current potentials which exist between subsurface materials vary according to the nature of the beds traversed. For example, the potential of an aquifer containing salty or brackish water is usually negative with respect to associated clay and shale, while that of a freshwater aquifer may be either positive or negative but of lesser amplitude than the salty water. This would be evident on the graphic record or log of the SP curve.

The resistivity of a material is a measure of the resistance it offers to the flow of an electric current. It is related to the nature of the material and the quantity, quality, and distribution of contained fluids. These factors vary from one material to another, so resistance measurements made between an electrode in a bore hole and one at the surface can be used to determine formation boundaries, some characteristics of the individual beds, and sometimes a qualitative evaluation of the contained water.

The single electrode log requires much less complex equipment than other types of logging methods. The curves obtained can usually be readily interpreted to show aquifer boundaries near to the correct levels, and the thickness of formations if greater than about 1 foot. The true resistivity cannot be obtained, only the relative magnitude of the resistivity of each formation. With sufficient records from a uniform area, these relative magnitudes can sometimes be interpreted qualitatively regarding the quality of water in the various aquifers.

The multiple electrode log [4,5,12,13,16] consists of three or more curves, an SP curve, and two or more resistivity curves. The SP curve is identical to that of the single electrode log. The resistivity

---

[2] Sonde—Probe or unit which is let down the hole and which contains the downhole electrode or electrodes.

curves show the variations of potential with depth of an imposed 60-hertz alternating current between electrodes spaced different distances apart on the sonde. Commonly used electrode spacings are: short-normal, 16 to 18 inches; long-normal, 64 inches; and long-lateral, 18 feet 8 inches. The radius of investigation about the hole varies with the spacing.

The logging instrument consists of a sonde with three or more electrodes spaced at various distances, supported by a multiple conductor cable leading to the recorder, an alternating-current generator, and an electrode attached to the recorder and grounded at the surface to complete the SP and resistivity circuit and cables, reels, winches, and similar accessory equipment.

Electric logs cannot be run in cased holes. Although a fluid-filled hole is more desirable, logs can be run on dry holes using special types of sondes on which the electrodes contact the hole wall. Electric logging equipment is affected by extraneous Earth currents, and at times a satisfactory log cannot be obtained in the vicinity of power stations, switchyards, and similar installations.

When electric logs are to be run in rotary drilled holes, the hole should be drilled with nonsaline water-based mud or organic drilling fluid. For good results, the hole diameter should not exceed 12 inches; 6- or 8-inch-diameter holes are best. An electric log should be run in the pilot hole of a large diameter well rather than in the reamed well hole.

(b) *Caliper Logs* [9].—Caliper logs show the variation in the diameter of uncased drill holes. Such logs are useful in interpreting electric logs in which the apparent resistivity is influenced by hole diameter variations, and in estimating the volume of cement required for grouting in a casing or the volume of gravel which may be required for a pack. They also may show the nature of the subsurface materials, since a drill hole is usually washed out to a larger diameter when poorly consolidated and noncohesive materials are penetrated by the hole. The caliper log is sometimes useful in well rehabilitation work because it will show where uncased holes have raveled or caved and where casing in cased holes has been damaged or otherwise undergone deterioration. The logs are made by running a self-actuated caliper through the hole. A recorder at the surface shows the relationship between the diameter of the hole and the hole depth.

(c) *Temperature Logs* [15,17].—Temperature logs use a sonde in which a resistance-type thermocouple is placed and calibrated to correlate resistance variations with temperature variations. Temperature logs are frequently made at the same time as electrical logs. The recorder plots a curve relating temperature to hole depth. The logs are a valuable tool in investigating inter-aquifer migration of water, adequacy of grouting, quality of water, corrosion, and other similar studies. They also contribute to the reliability of interpretation of multiple electrode electric logs.

(d) *Radiation Logs* [16].—Radiation logs are of two general types: those which measure the natural radiation of materials (gamma ray logs); and those which measure radiation transmitted through, from, or induced in the formation by a neutron-emitting source contained in the sonde (neutron log). The logs are not affected by casing in a well, so they may be used to identify formation boundaries in a cased well. Also, they can be used in a dry hole whether cased or uncased.

Nearly all rocks contain some radioactive material. Clays and shales are usually several times more radioactive than sandstones, limestones, and dolomites. The gamma ray log uses recording equipment similar to that used in an electrical log, although the sonde is different and contains a scintillometer or geiger counter. The gamma ray log is a curve relating depth to intensity of natural radiation and is especially valuable in detecting clays and other materials of high radiation. The gamma ray log is also valuable in investigating old cased wells on which there are no reliable logs to obtain information necessary for workover or rehabilitation.

Neutron logging equipment contains a neutron source in addition to a counter and is most useful in determining presence of water and saturated porosity. The neutron log is seldom used in water wells because of the costs involved and the dangers inherent in its use.

8–3. Photographic and Other Well Surveys.—Loss of well efficiency, development of sand pumping, change in quality of water, or well failure are all causes of concern and usually require well rehabilitation or replacement. A photographic survey of the well is one of the most economical and helpful tools for determining the nature of the problem and possible method of rectification.

Four general types of surveys are available commercially: vertical photographs, vertical stereo photographs, television, and motion pictures.

Prior to making any photographic survey of a well, an effort should be made to clarify and reduce the turbidity of water in the well. Many methods have been used for this problem, but none have been markedly successful. However, the following procedure is recommended:

> If the water has a pH below 7, about 2 pounds of slaked lime ($Ca(OH)_2$) per 1,000 gallons of water in the well should be added and thoroughly dispersed through the total well depth by surging before adding the coagulant as described in the following paragraph.

If the pH of the water is above 7, the alkalizing agent is not necessary and about one-half pound of alum ($Al_2(SO_4)_3$) or ferric sulfate ($Fe(SO_4)_3$) per 1,000 gallons of water in the casing and screen should be added to the well. The well should then be strongly agitated with a surge block or similar tool through the entire depth of water for at least 30 minutes for each 100 feet of water in the well.

The coagulant should be added to the well at least 3 days, and preferably a week, before inserting the camera into the hole.

If there is a layer of oil on top of the water in the well, an effort should be made to bail the oil out before adding the coagulant. This is usually not wholly successful, so the camera lens should be wetted with a strong solution of detergent as it is placed in the hole. This will prevent the lens from being coated with oil, which would considerably reduce the definition of the image.

(a) *Single Vertical Photo Surveys.*—The single vertical photographic survey has been superseded in most cases by the vertical stereo photograph or the television camera, but companies furnishing such single photographs may still be in operation. Single photographs furnish valuable information when other types of pictures are not available, but are not as informative and are more difficult to interpret. They may be used in wells 4 inches in diameter or larger.

(b) *Stereo Photo Surveys* [14].—A stereo photograph survey consists of 35-mm photographs taken simultaneously by two cameras set in the same plane with the optical axes set at a slight angle to each other. The resultant film, when examined through a stereoscopic viewer, gives a three-dimensional axial picture that is readily interpreted and which furnishes good data on corrosion indications, encrustation, casing breaks, partings, collapse, and similar features. A recent improvement permits taking pictures showing views normal to the wall of the well at selected locations. These pictures are usually taken after studying the vertical views.

Pictures are available in both black and white and color. The black and white films are negatives, and some experience is required to interpret them. They can be furnished at the well site within 45 minutes of completion of the survey. The color pictures require about a week for processing and delivery. The cameras can be used successfully in wells 6 inches or larger in diameter. Each stereo pair of photographs clearly show from 3 to 5 feet of hole below the camera and the depth at which the photographs were taken.

Stereo surveys are available from several commercial operators.

(c) *Television Surveys* [3].—Closed circuit television equipment is complex and sophisticated, but gives excellent black and white vertical views down the well and, with some equipment, horizontal views of the side of the hole. The latter feature is particularly valuable in closeup viewing of suspected corrosion or encrustation of screens or perforated zones.

Some equipment will operate in holes as small as 3 inches, but generally a 6- to 8-inch hole is required.

Television surveying services are available from several Federal agencies and commercial operators.

(d) *Motion Picture Surveys*.—Cameras are also available for taking moving pictures in wells, either in color or black and white. Lens attachments permit taking side hole pictures or vertical pictures along the well axis. The equipment usually requires a well diameter of 10 inches or more. The surveys compare in cost with television camera surveys, but are not as flexible in operation nor do they permit the detailed examination of critical areas that is possible with the television camera.

(e) *Other Well Surveys* [9].—Most of the previously discussed well logging and surveying methods and equipment are used frequently in water well investigations. In addition, there are numerous other types of well surveys which have been developed primarily in the petroleum industry for use in oil wells. These include instruments which measure the direction and degree of hole inclination or deflection and pressures at various depths, and devices for taking samples of formation through the side of the hole and for taking samples of fluids contained in the formations through the side of the hole. These types of surveys is seldom required in water wells. The alinement and plumbness as measured with a cage or dolly in the pump chamber are usually sufficiently accurate for the depths involved (see sec. 16–4). In some deep holes, the wall sampling devices may be helpful, but usually adequate information can be obtained at less cost with a multiple electrode electric log. Where the use of these more complex types of surveys is considered necessary, one of the larger oil well survey companies should be consulted to determine whether the desired information can be obtained and if so, the cost. In some instances, move-in and operation costs of some oil well survey equipment will exceed the entire cost of a water well.

8–4. Bibliography.—

[1] Buhle, M. B., and Bruilkmann, J. E., "Electrical Earth Resistivity Surveying in Illinois," Illinois State Geological Survey Circular 376, Urbana, 1964.

[2] Carpenter, G. C., and Bassarab, D. R., "Case Histories of Resistivity and Seismic Ground Water Studies," Ground Water, vol. 2, No. 1, pp. 21–25, January 1964.

[3] Callahan, J. T., Wait, R. L., and McCollum, M. J., "Television—A New Tool for Ground Water Geologists," Ground Water, vol. 1, No. 4, pp. 4–6, October 1963.

[4] Morris, T. S., "Electric Detective, Investigations of Ground Water Supplies with Electric Well Logs," Water Well Journal, vol. 11, No. 3 and 4, March and May 1957.

[5] Guyod, H. (compiler), "Logging Systems, Applications and Interpretations," Mandel Industries, Technical Bulletins, Houston, Tex.

[6] Heigold, P. C., McGinnis, L. D., and Howard, R. H., "Geologic Significance of the Gravity Field on the DeWitt-McLean County Area, Illinois," Illinois State Geological Survey Circular 369, Urbana, 1964.

[7] Heiland, C. A., "Geophysical Exploration," Prentice-Hall, Englewood Cliffs, New Jersey, 1940.

[8] Jakosky, J. J., "Exploration Geophysics," Trida Publishing Co., Los Angeles, Calif., 1950.

[9] Leroy, L. W., and Crain, H. M., "Subsurface Geologic Methods," Colorado School of Mines, Golden, 1949.

[10] Linehan, D. L., and Keith, S., "Seismic Reconnaissance for Ground Water Development," Journal of the New England Water Works Association, vol. 63, No. 1, pp. 76–92, March 1949.

[11] McGinnis, L. D., Kempton, J. R., and Heigold, P. C., "Relationship of Gravity Anomalies to a Drift-filled Bedrock Valley System in Northern Illinois," Illinois State Geological Survey Circular 354, Urbana, 1967.

[12] Pryor, W. A., "Quality of Water Estimated from Electric Resistivity Logs," Illinois State Geological Survey Circular 215, Urbana, 1956.

[13] "Log Identification Charts," Schlumberger Well Surveying Corp., Document No. 4, 1955; No. 5, 1952; and No. 8, 1958.

[14] "Photographic Diagnosis of What's Wrong Inside the Well Casing," Underground Raindrops, U.S. Electric Motor Company, vol. 3, No. 1, Los Angeles, Calif., Spring 1959.

[15] Schneider, Robert, "An Application of Thermometry to the Study of Ground Water," U.S. Geological Survey Water-Supply Paper 1544B, 1962.

[16] Patten, E. P., Jr., and Bennett, G. D., "Application of Electrical and Radioactive Well Logging to Ground Water Hydrology," U.S. Geological Survey Water-Supply Paper 1544D, 1963.

[17] Chapman, H. T., and Robinson, A. E., "A Thermal Flowmeter for Measuring Velocity of Flow in a Well," U.S. Geological Survey Water-Supply Paper 1544E, 1962.

[18] Wantland, Dart, "Seismic Investigations in Connection with the United States Geological Survey Ground Water Studies in the Gallatin River Valley, Montana," Bureau of Reclamation, Geology Report No. C–115, September 4, 1951.

[19] ———, "Geophysical Measurements of the Depth of Weathered Mantel Rock: Symposium of Surface and Subsurface Reconnaissance," ASTM Special Technical Publication No. 122, 1951.

[20] ———, "The Application of Geophysical Methods to Problems in Civil Engineering," Paper presented at the Canadian Institute of Mining and Metullurgy, Ottawa, January 21–23, 1952.

[21] ———, "Second Phase of Geophysical Investigations in Connection with the United States Geological Survey Ground Water Studies in the Gallatin River Valley, Montana," Bureau of Reclamation, Geology Report No. C–121, May 1, 1953.

[22] ———, and McDonald, H. R., "Geophysical Procedures in Ground Water Studies," Proceedings of the ASCE, Journal of the Irrigation and Drainage Division, vol. 86, No. IR3, part 1, pp. 13–26, September 1960.

Chapter IX

METHODS OF DETERMINING AQUIFER CHARACTERISTICS

9–1. Methods for Estimating Approximate Values of Aquifer Characteristics.—In practically all ground-water investigations, data are required on the aquifer characteristics, transmissivity, storativity, and boundaries. Several methods of making such tests with various degrees of accuracy are available.

When inventorying existing wells, the data collected often include yield and drawdown from which specific capacity values may be determined. Section 5–17 discusses methods of estimating the transmissivity of aquifers from such specific capacity data. The procedure is basic but must be used with judgment because well yield depends on several factors, some of which are not readily determinable. When using this method, similarly constructed wells should be grouped together and actually tested for yield and drawdown, if possible.

When wells are equipped with meters or weirs, discharge can be measured easily. Even when not metered, the discharge of wells yielding less than 100 gallons a minute can be readily measured with sufficient accuracy by using a calibrated bucket or drum and a stopwatch. Most wells are not metered, and those having larger discharges are the most significant. Several convenient methods of measuring approximate well discharge with a minimum of equipment are described in the Bureau of Reclamation *Water Measurement Manual* [1].[1]

Static water levels in wells in the vicinity of the test should be measured after wells have been shut down for some time, preferably 12 hours or more. If this condition cannot be realized, the status of such wells should be recorded.

Other methods of estimating approximate values of permeability, transmissivity, and sometimes storativity include bail tests, slug tests, and analyses of cyclic pumping or natural ground-water fluctuations. They are described in references [2,3]. However, these methods are either of limited applicability or the results are of questionable accuracy. They are mentioned only as possible alternatives when other methods are not available.

9–2. Controlled Pumping Tests to Determine Aquifer Characteristics.—The most accurate, reliable, and commonly used method

[1] Numbers in brackets refer to items in the bibliography, section 9–12.

of determining aquifer characteristics is by controlled aquifer pumping tests. Before performing such a pumping test, personnel should be acquainted with the contents of chapter V. The number of pumping tests required is determined largely by the size of the area, the uniformity and homogeneity of the aquifer or aquifers involved, and known or suspected boundary conditions. One test is usually adequate for a small area, but in an extensive area several tests may be necessary. A reasonably sophisticated test may cost from $2,000 to $10,000, not including the cost of the well, so every effort shoud be directed toward obtaining a maximum amount of accurate and reliable data

9-3. Types of Aquifers.—The investigations discussed previously in chapters IV and V will permit determination of the type of aquifer or aquifers and the interrelationships which may be involved at a particular test site. These factors should be considered in planning pumping tests.

(a) *Unconfined Aquifers.*—Relatively thin aquifers located at shallow depths are readily tested because wells and observation wells may be drilled economically to fully penetrate the aquifer. Observation wells may be located short distances from the pumped well, thus field measurements are easily and quickly made. Long-term pumping is generally not required to obtain usable drawdown measurements. The testing for deeper thin aquifers is similar except for the increased cost of the deeper holes and pump setting.

Problems arise when the aquifer is excessively thick (100 feet or more) and the water table is deep. Existing wells may not fully penetrate the aquifer, and the cost of drilling fully penetrating observation wells and a test well may not be economically justifiable.

When a pumping well does not fully penetrate an unconfined aquifer, the distorted flow pattern to the well is accentuated. In the ideal aquifer, the observation wells should be located at a minimum distance equal to 1½ to 2 times the aquifer thickness from a partially penetrating pumping well. This will result in a flow pattern equivalent to that of a fully penetrating well. Any well with an 85 percent or more open or screened hole in the saturated thickness may be considered as fully penetrating. If the aquifer is vertically anisotropic, the nearest observation well would ideally have to be from the pumped well, a distance equal to twice the thickness of the aquifer times the square root of the ratio of the horizontal to the vertical permeability before the flow pattern would be equivalent to that of a fully penetrating well. Such relationsips are of limited value because a reliable and economical method of determining vertical and horizontal permeability is unavailable, aquifer thickness may be unknown, and prolonged pumping would be required.

(b) *Confined Aquifers.*—A confined aquifer will often be overlain by an unconfined aquifer from which it is separated by a confining layer. The unconfined aquifer should always be cased off in the pumping and observation wells. If the confined aquifer is not excessively thick, the well should be screened for the entire thickness of the aquifer. The nearest observation well should be located at least 25 feet from the pumping well and should penetrate and be screened in the upper 10 percent of the aquifer at a minimum. The water level in the pumping well should not be allowed to fall below the bottom of the upper confining bed during an aquifer test.

In a thick confined aquifer where it would be uneconomical to drill a fully penetrating well, a similar relationship holds regarding the location and depth of observation wells as described above for a free aquifer. However, the area of influence in an artesian aquifer expands much more rapidly than in a free aquifer and the distance to the nearest observation well is not as critical from the standpoint of pumping time to obtain measurable drawdown.

In a confined aquifer, partial penetration by a discharging well may be compensated for by spacing the observation wells an adequate distance from the pumping well, but the same relationship of twice the aquifer thickness times the square root of the permeability ratios previously described applies. Another method is to use two piezometers at each distance with the closest pair being at least one-half the aquifer thickness distance from the pumped well. One of the piezometers in each pair is open to the upper 10 percent of the aquifer and the other to the lower 10 percent. The average of the drawdowns in each pair of piezometers is used as the effective drawdown at each distance. If the pumping well is screened through the midsection of the aquifer, the same arrangement may be used as was described previously for an unconfined aquifer under similar conditions, or a piezometer point may be set at the same elevation as the midpoint of the screen.

(c) *Composite and Leaky Aquifers.*—Many areas are underlain by an unconfined aquifer and one or more confined aquifers. The confining layers may vary from practically impermeable to moderately permeable as compared to the aquifers. In the latter case, interchange of water between aquifers may occur depending on the pressure differences which exist among them. Under such conditions, each aquifer should be pump tested separately. The pumping well should be cased through the untested sections and should be screened through the entire thickness of the tested aquifer. Observation wells and piezometers should be set to conform with the design of the pumping well. In testing such aquifers, the semilog straight-line plots of time or distance against drawdown to determine the length of the test may not apply, and if the test is run sufficiently long, the plot may become a line of zero drawdown. The field data are analyzed as described in section 5–8.

(d) *Delayed Drainage.*—Consideration should also be given to the nature of the aquifer and the probable effect of delayed drainage on a pumping test. (See sec. 5–10.) To estimate the nature of the aquifer materials, the well logs and sample cuttings should be carefully examined. From this the minimum planned time for a test should be estimated on the basis of the following tabulation:

Minimum pumping time recommended for aquifer test

Predominant aquifer material:	*Minimum pumping time, hours*
Silt and clay	170
Fine sand	30
Medium sand and coarser materials	4

In many instances, economic and other factors will rule out tests as long as 170 hours, so less than ideal test results may have to suffice.

9–4. Selection and Location of Pumping Wells and Observation Wells.—If an existing well is to be used for a test, the well should ideally closely conform to the requirements for aquifer testing. Also, the logs, data on types of construction, and performance characteristics of other wells in the area should also be examined. Other nearby wells may be suitable as observation wells, but in most cases additional observation wells will have to be drilled.

The pumping well and observation wells should be located, if at all possible, to conform to known or suspected boundaries, including deep percolation from irrigation. The wells should be located far enough away from the boundaries to permit recognition of drawdown trends before the boundary conditions influence the drawdown readings. (See secs. 5–5 and 5–11.) If more than one boundary is present, the effects of the first one should be relatively stable before the influence of the second becomes effective. Conversely, a study may involve an estimate of induced seepage from a stream or body of surface water as a result of pumping from a nearby well. In such a study, the well may be placed relatively close to the recharge boundary, and one or more observation wells should penetrate into the bed of the surface water body.

In selecting or locating sites for observation wells, an effort should be made to meet ideal conditions. If a partially penetrating well is used, the distance to the nearest observation well, r, ideally should be:

$$r > 1.5M\left(\frac{K_r}{K_z}\right)^{1/2} \tag{1}$$

where M=thickness of the aquifer, K_r=horizontal permeability, and K_z=vertical permeability. If this is not possible, or K_z cannot be determined, the suggested alternatives presented in chapter V should be considered. When laying out observation well locations, consideration should be given to the proposed duration of the test and the probable magnitude of the transmissibility and storativity. If estimates can be made of S and T; the drawdowns at various distances and at increasing times, the time when u will be less than 0.01, and the probable length of time required for the test can be estimated (see secs. 5–3 and 5–5). The drawdown at any point in the area of influence will increase with time and the rate of discharge. Conversely, the greater the diffusivity factor, $\alpha=\frac{S}{T}$, the slower the rate of expansion of the area of influence.

While any number of observation wells may be used, the recommended minimum number is four—three on a line passing through the center of the pumped well and one on a line normal to the previously mentioned one and also passing through the pumped well. The distance from the pumped well to the nearest observation well and the spacing between observation wells involve consideration of the ideal conditions, how the test conditions conform to such conditions, desirable adjustments to compensate for departures from the ideal, and feasible locations for the wells in the field.

If a well must be drilled specifically for testing, the design should reflect whether it is to be purely a test well to be abandoned after the test or whether it would fit into the final plan as a production well. In the former case, the least costly construction commensurate with the purpose should be followed. In the second, a pilot hole may be necessary and good well design from the standpoint of efficiency, long well life, and desired yield should be followed.

9–5. Disposal of Discharge.—When planning the test, the method and place of disposal of discharge from the well should be determined. The discharge from the pumped well should be transported some distance from the well for convenience and comfort during the test. However, there are other considerations of greater importance.

If the aquifer is unconfined and the unsaturated materials overlying the aquifer are relatively permeable, the discharge should be transported by pipeline to an existing drain beyond the probable area of influence that will develop during the test. Otherwise, the deep percolation may be recirculated and the test adversely affected.

If the aquifer is unconfined, the water table lies at depths of 100 feet or more, and the overlying materials are of low permeability, an existing ditch or drain that will remove the flow rapidly from the area may be used safely. Discharge from a confined aquifer may be treated similarly.

9–6. Preparations for Pumping Test.—If possible, tests should be run when there is the least likelihood of heavy rains occurring. Infiltration and deep percolation of precipitation may adversely affect a test.

For a few days before starting the test, water levels in the pumping well and observation wells should be measured at about the same time each day to determine whether there is a measurable trend in groundwater levels. If such a trend is apparent, a curve of the change in depth versus time should be prepared and used to correct the water levels read during the test.

In areas of severe winter climate where the frostline may extend to depths of several feet, pumping tests should be avoided during the winter where the water table is less than about 12 feet from the surface. Under some circumstances, the frozen soil acts as a confining bed, combined with leaky aquifer and delayed storage characteristics that may make the results of the test unreliable.

If the aquifer is confined, barometric changes may affect water levels in wells (see sec. 5–13). An increase in barometric pressure may cause a decrease in the water levels and a decrease in pressure may cause an increase in water levels. If water levels and barometric pressures are measured several times daily for at least 4 days prior to running a test and both measurements are expressed in feet of water (1 inch mercury=1.134 feet of water), a plot may be made correlating the two measurements. The slope of the straight line of closest fit through each set of measurements for each well will give the barometric efficiency of each hole. If a relationship is recognized, barometric readings should be made at the same time as water level measurements during the test and the required correction made to the measured water levels.

A day or two before the test, the well should be tested for several hours for yield and drawdown, operation of the discharge measuring equipment, general operating conditions, and approximate best rates of discharge for the test. At the same time, water levels should be measured in the observation wells to assure response.

The measuring point at the pumped well and all observation holes should be selected, marked clearly with paint, and the elevations determined by leveling, if required.

The distance and bearing from the center of the pumping well to the center of all observation wells should be measured to the nearest foot. The distance and bearing to the closest point of any nearby boundary such as a lake or stream or other discharging wells should also be measured.

9–7. Instrumentation and Equipment Required for a Test.—The following items should be available for use in the test:

- An orifice, weir, flowmeter, or other type of water measuring device that will measure accurately in the range of the discharges expected [1,4] (see sec. 9–9).
- Depth-to-water measuring devices as described in section 7–4, excluding air lines which are not sufficiently accurate for use in pumping tests. If electric sounders are used, extra batteries, waterproof electrical tape, and other supplies should be available for servicing them on the job. There should be as many measuring devices available for the test as the number of observation wells and the pumping well, plus one extra for use as a spare. Continuous water stage recorders are very useful if the recording rate is compatible with the drawdown rate.
- A tachometer or revolution counter if an internal-combustion engine is used for power.
- Steel tapes, graduated in hundredths of a foot.
- A thermometer with a range between 32° and 120° F if water temperature is an important factor.
- Synchronized watches for all observers.
- Log-log 3- by 5-cycle and semilog 3-cycle graph paper, rulers, pencils, and forms for recording measurements.
- A gate valve should be installed on the pump discharge pipe to control the discharge. Wells will ordinarily show a slow decline in discharge with time as the drawdown increases. This may be compensated for by limiting the discharge by partially closing the valve at the start of the test and opening it slightly when measurements of discharge show a recognizable decline. The objective is to maintain a constant discharge of the pump throughout the test. A maximum variation of about 5 percent in the discharge should be the goal.
- A barometer or recording barograph if the test is in an artesian aquifer.

- A stopwatch.
- A carpenter's level if an orifice is used.
- Two or more 1-quart water sample bottles.
- A ¾-inch inside-diameter or larger pipe should be installed in the pumping well from above the pump base to the top of the pump bowls. If there is oil on the water, a cork should be inserted in the lower end of the pipe. The cork can be dislodged by air pressure or by pouring water in the pipe. Drawdown measurements are made in the pipe which protects the probe from oil on top of the water and dampens turbulénce caused by vibration of the pump and permits more accurate measurements.

9-8. Running a Pumping Test.—Immediately before starting the pump, the water levels should be measured in all observation wells and in the pumped well to determine the static water levels upon which all drawdowns will be based. These data and the time of measurement should be recorded.

The instant of starting the pump should be recorded as the zero time of the test. The initial discharge may be somewhat in excess of the continuous discharge, so it may be necessary to reduce it by partially closing the valve on the discharge pipe.

The well discharge should be controlled to keep it as constant as possible after the initial excess discharge has been stabilized. This can be done by either regulating the valve (preferable) or, if an internal-combustion engine is used for power, by changing the speed of the engine. The tone or rhythm of an internal-combustion engine provides an aural check of performance. If there is a sudden change in tone, the discharge should be checked immediately and proper adjustments made to the gate valve or to the engine speed if necessary.

During a pumping test to determine aquifer characteristics, water levels in the pumping well and observation holes should be measured to give at least 10 observations of drawdown within each log cycle of time. Adherence to the time schedule should not be at the expense of accuracy in the drawdown measurements. A suggested scheduling measurement is as follows:

0 to 10 minutes: 1, 1.5, 2, 2.5, 3.25, 4, 5, 6.5, 8, and 10 minutes.

10 to 100 minutes: 10, 15, 20, 25, 30, 40, 50 65, 80, and 100 minutes.

100 minutes to completion: 1- to 2-hour intervals.

During the early part of the test, sufficient manpower should be available to have at least one person at each observation well and at the pumping well. After the first 2 hours, two people are usually sufficient to continue the test.

It is important, particularly in the early part of the test, to record with maximum accuracy the time at which readings are taken. The foregoing time schedule should be followed as closely as possible, but if for some reason the schedule is missed, the actual time of taking the reading should be recorded. Estimating drawdown readings to fit the schedule may lead to erroneous results. It is not necessary that readings in the pumping well and observation wells be taken simultaneously as long as the schedule is generally followed and readings recorded at the exact time taken.

9–9. Measurement of Discharge.—Practically any type of device designed for measurement of flow in pressure conduits can be used to measure pump discharge when the discharge pipe is running full. In some instances, it may be necessary to insert an "L" on the pipeline with a riser extending above the elevation of the pump discharge or elevate the discharge end to keep the pipe full at all times. Commercially available flowmeters with totalizing registers are commonly used, particularly for discharges less than about 200 gallons per minute. Such meters should be calibrated and checked for accuracy within the discharge and pressure ranges to which they will be subjected, and they should be installed in conformance with the manufacturer's recommendations. Rate of discharge with such meters is usually determined by measuring with a stopwatch the time required to discharge a given volume; however, an instantaneous flow indicator is preferable.

Reference [1] in the bibliography gives data on venturi, flow nozzle, and orifice meters. Some meters of the required capacity may be available in the field offices or laboratories of the Bureau.

Where discharge is free flowing from the discharge pipe into a ditch or canal, weirs as described in reference [1] can be installed.

The most commonly used device for measuring discharge during a pumping test is probably the free discharge pipe orifice. When used in conjunction with a pipeline, the orifice may be placed at the end of the pipeline or it may discharge into a tank or reservoir which feeds the pipeline. The latter arrangement is usually more convenient because measurement and adjustment of discharge can then be made in the immediate vicinity of the well. Figures 9–1 and 9–2 illustrate pipe orifice arrangement and details.

Numerous combinations of pipe and orifice sizes and applicable tables are available. As a standard, the orifice and tables developed at the Engineering School of Purdue University are recommended [4]. If constructed, set up, and operated as described below, the pipe orifice combinations permit an accuracy of plus or minus 2 percent. Figure 9–3 gives the general discharge equation. Table 9–1 gives the specifications of the various sizes of steel pipe used.

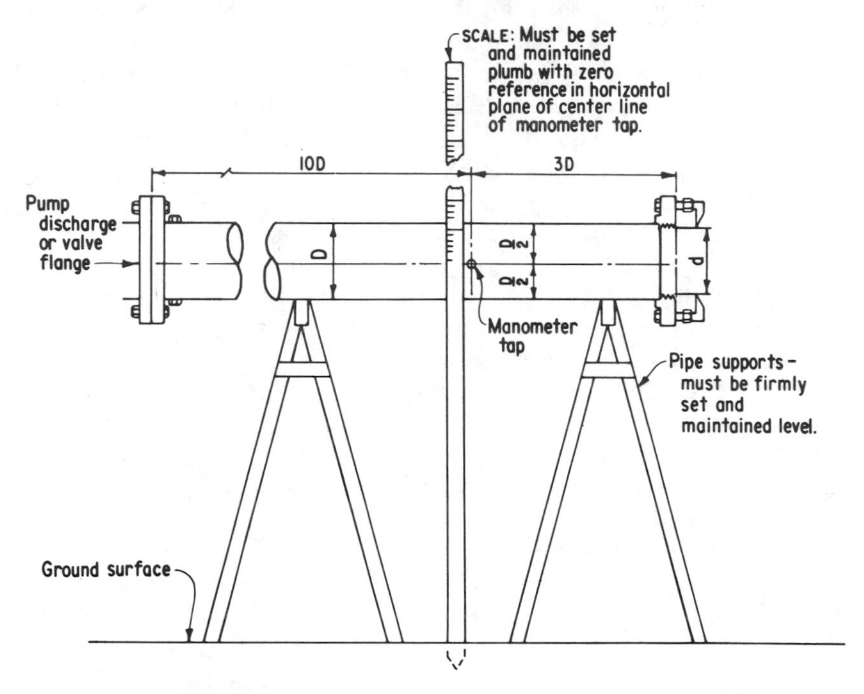

FIGURE 9–1.—Free discharge pipe orifice. 103–D–1466.

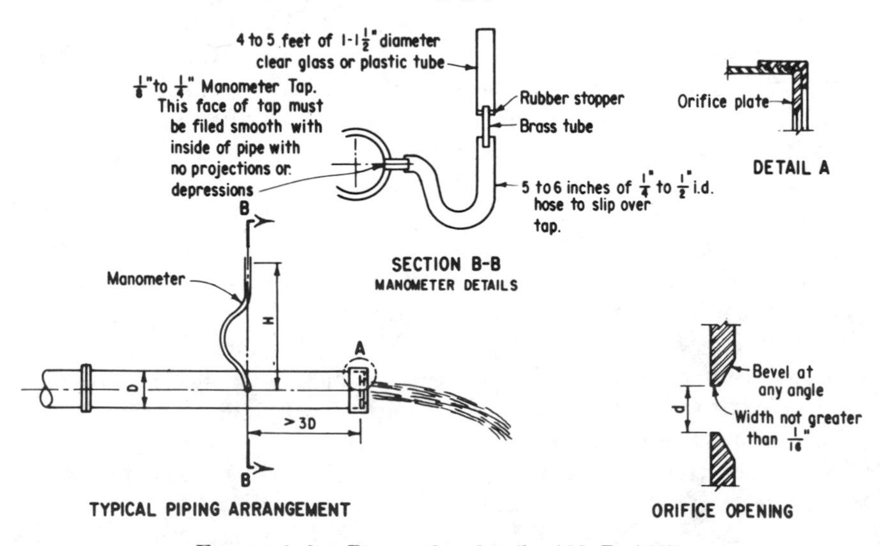

FIGURE 9–2.—Pipe orifice details. 103–D–1467.

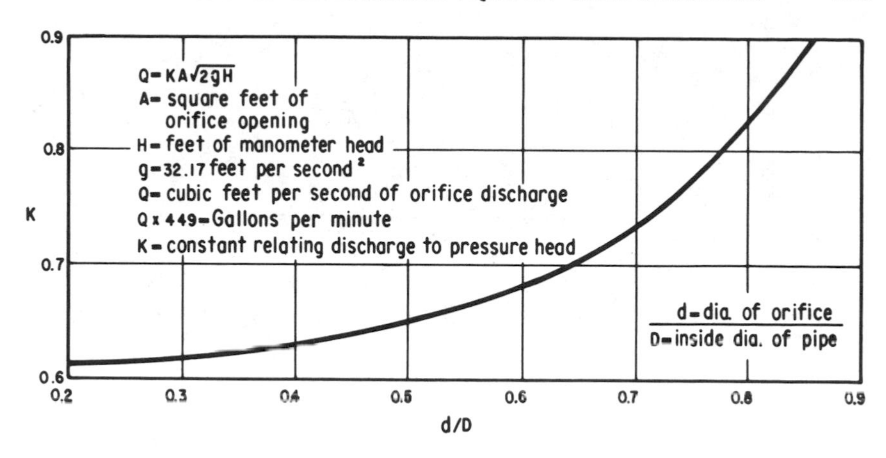

FIGURE 9–3.—Pipe orifice discharge equation. 103–D–1468.

TABLE 9–1.—*Pipe recommended for free discharge orifice use*

Nominal pipe size, inches	Outside diameter, inches	Inside diameter, inches	Wall thickness, inches	Class	Schedule No.	Weight per foot (plain ends), pounds
4	4. 500	4. 026	0. 237	Std.	40	10. 79
6	6. 625	6. 065	. 280	Std.	40	18. 97
8	8. 625	8. 071	. 277	--------	30	24. 70
10	10. 750	10. 192	. 279	--------	--------	31. 20
12	12. 750	12. 090	. 330	--------	30	73. 77

Table 9–2 shows the range of discharges which can be measured with various orifice and pipe combinations.

Orifices may be machined from threaded pipe caps or from 3/16- to 1/4-inch steel plate stock and attached to the pipe by thread protectors or similar devices. The plates should be carefully machined as true circles to automatically center in the pipe when attached. The orifice should be accurately machined to the specified diameter and centered in the plate.

The downstream edge of the orifice should be beveled at an angle of about 45° but leaving a root of uniform width of one-sixteenth of an inch or less on the upstream side. The upstream edge of the orifice should be sharp, clean, and free from rust and any pits or other imperfections. Indices consisting of two lines normal to each other and passing through the center of the orifice are commonly inscribed on the downstream face of the orifice plates to assist in centering them on the end of the pipe.

TABLE 9–2.—*Orifice tables*

[For measurement of water in gallons per minute through pipe orifices with free discharge. The following tables have been compiled by the Engineering Department of Layne and Bowler, Inc., from original calibrations by Purdue University.]

Head in inches	3'' Orifice		4'' Orifice		5'' Orifice		6'' Orifice		7'' Orifice—	8'' Orifice—	9'' Orifice—	10'' Orifice—	Head in inches
	4 in. pipe	6 in. pipe	6 in. pipe	8 in. pipe	6 in. pipe	8 in. pipe	8 in. pipe	10 in. pipe	10 in. pipe	10 in. pipe	12 in. pipe	12 in. pipe	
5.	100	76	145	140	280	220	380	320			825	1100	5
5. 5	104	79	153	145	293	230	394	333			860	1150	5. 5
6	108	82	160	150	305	240	408	345			895	1200	6
6. 5	111	86	167	155	316	250	421	358			930	1250	6. 5
7	115	88	172	160	328	260	433	370			965	1300	7
7. 5	119	91	179	165	339	270	446	383			1000	1350	7. 5
8	122	94	185	170	350	280	458	395	600	935	1032	1400	8
8. 5	125	96	190	175	361	289	471	408	617	963	1065	1440	8. 5
9	128	99	195	180	372	298	483	420	633	992	1093	1480	9
9. 5	130	102	200	185	383	307	495	433	650	1016	1120	1520	9. 5
10	133	104	205	190	393	316	508	445	666	1040	1148	1560	10
10. 5	137	107	210	195	402	324	521	458	682	1060	1172	1600	10. 5
11	140	109	215	200	412	330	533	470	698	1080	1200	1635	11
11. 5	143	111	220	204	421	338	545	480	713	1100	1225	1670	11. 5
12	146	114	225	208	430	346	556	490	728	1120	1250	1705	12
12. 5	149	116	230	212	439	354	567	500	743	1139	1277	1740	12. 5
13	151	118	234	216	448	362	578	510	757	1158	1303	1775	13
13. 5	154	121	239	219	457	369	589	520	771	1176	1328	1810	13. 5
14	157	123	243	224	465	376	599	530	785	1194	1352	1845	14
14. 5	159	126	247	227	473	383	609	540	799	1212	1376	1875	14. 5
15	162	128	250	231	480	390	618	550	812	1230	1400	1905	15
15. 5	164	130	254	234	488	396	627	559	825	1248	1421	1940	15. 5

16	167	132	257	238	495	402	636	568	838	1266	1441	1970	16
16.5	170	134	261	241	503	408	645	577	851	1284	1460	2000	16.5
17	172	136	264	245	510	414	654	586	863	1302	1480	2030	17
17.5	175	138	268	249	517	420	663	595	875	1319	1500	2060	17.5
18	178	140	271	252	524	426	672	604	887	1336	1520	2089	18
18.5	180	142	275	256	530	432	681	612	899	1353	1540	2118	18.5
19	183	144	278	259	536	438	690	620	910	1370	1560	2146	19
19.5	185	146	282	263	542	444	699	628	922	1387	1580	2175	19.5
20	187	148	285	266	548	449	708	636	933	1404	1600	2204	20
20.5	190	150	289	270	554	455	717	643	945	1421	1620	2232	20.5
21	192	152	292	273	560	460	726	650	956	1438	1640	2260	21
21.5	195	154	295	275	566	465	735	657	968	1455	1659	2288	21.5
22	197	156	299	279	572	470	744	664	979	1471	1677	2316	22
22.5	199	158	302	282	578	475	752	671	990	1486	1695	2343	22.5
23	201	160	305	285	584	479	760	678	1001	1500	1714	2360	23
23.5	203	162	307	288	590	484	768	685	1012	1515	1732	2382	23.5
24	205	164	310	291	596	488	776	692	1022	1529	1750	2409	24
24.5	207	165	314	294	602	492	784	699	1033	1543	1767	2435	24.5
25	210	167	317	297	608	496	791	706	1043	1557	1783	2461	25
25.5	212	169	320	300	614	500	798	713	1059	1571	1799	2487	25.5
26	214	171	323	303	620	504	805	720	1065	1585	1815	2513	26
26.5	216	173	326	305	626	508	812	727	1074	1599	1830	2539	26.5
27	219	174	329	308	632	512	818	734	1084	1313	1845	2565	27
27.5	221	176	332	311	638	516	825	741	1094	1327	1860	2590	27.5
28	222	177	335	314	644	520	831	747	1104	1341	1875	2610	28
28.5	224	179	337	317	650	524	838	754	1114	1355	1890	2630	28.5
29	226	180	340	320	656	528	844	760	1124	1369	1905	2650	29
29.5	228	182	343	323	662	532	851	767	1134	1383	1920	2670	29.5
30	230	183	346	325	668	536	857	773	1143	1397	1935	2690	30
30.5	232	185	348	328	674	540	863	780	1153	1711	1950	2713	30.5
31	235	186	351	330	680	544	869	786	1162	1725	1965	2736	31

See footnote at end of table.

TABLE 9-2.—*Orifice tables*—Continued

Head in inches	3'' Orifice		4'' Orifice		5'' Orifice		6'' Orifice		7'' Orifice—	8'' Orifice—	9'' Orifice—	10'' Orifice—	Head in inches
	4 in. pipe	6 in. pipe	6 in. pipe	8 in. pipe	6 in. pipe	8 in. pipe	8 in. pipe	10 in. pipe	10 in. pipe	10 in. pipe	12 in. pipe	12 in. pipe	
31. 5	236	188	354	333	686	548	876	793	1172	1739	1980	2759	31. 5
32	239	189	357	335	692	552	882	799	1181	1753	2005	2782	32
32. 5	240	191	360	338	697	556	889	806	1191	1767	2020	2805	32. 5
33	242	192	363	340	703	560	895	812	1200	1791	2040	2828	33
33. 5	244	194	366	342	709	564	901	818	1209	1795	2050	2850	33. 5
34	246	195	369	345	715	568	907	824	1218	1809	2060	2873	34
34. 5	248	196	372	347	720	572	913	830	1227	1823	2075	2896	34. 5
35	250	197	375	349	726	576	919	836	1235	1837	2090	2919	35
35. 5	252	198	377	351	732	580	925	842	1243	1851	2100	2964	35. 5
36	254	200	380	354	737	584	931	847	1251	1865	2112	2964	36
36. 5	256	201	383	356	743	588	937	852	1259	1879	2124	2980	36. 5
37	257	203	385	358	748	592	943	857	1266	1893	2136	3002	37
37. 5	259	204	388	360	754	596	949	863	1274		2148	3024	37. 5
38	260	205	390	363	759	600	955	867	1281		2160	3046	38
38. 5	262	206	393	365	767	604	961	872	1289		2173	3068	38. 5
39	263	208	396	367	770	608	967	877	1295		2185	3088	39
39. 5	265	209	398	369	776	612	974	882	1304		2197	3110	39. 5
40	266	210	401	371	781	616	979	887	1311		2210	3130	40
40. 5	267	211	403	373	786	620	985	891	1319		2225	3146	40. 5
41	269	212	406	375	790	624	990	896	1326		2233	3160	41
41. 5	271	213	408	378	795	628	996	901	1334		2245	3179	41. 5
42	272	214	411	380	800	631	1001	906	1341		2257	3199	42
42. 5	274	216	413	382	805	635	1007	910	1349		2273	3219	42. 5
43	275	217	415	384	810	638	1012	915	1356		2285	3230	43

43. 5	277	218	418	386	815	642	1018	920	1364	2397	3250	43. 5
44	278	219	420	388	820	645	1023	925	1371	2309	3263	44
44. 5	280	220	422	390	824	649	1029	929	1379	2326	3280	44. 5
45	281	222	425	392	828	652	1034	934	1387	2338	3298	45
45. 5	283	223	427	394	832	656	1040	939	1394	2350	3316	45. 5
46	284	224	429	396	837	659	1045	944	1401	2363	3334	46
46. 5	285	225	432	399	842	663	1051	948	1409	2375	3351	46. 5
47	287	227	434	401	847	666	1056	953	1416	2387	3368	47
47. 5	289	228	437	403	851	669	1062	958	1424	2399	3389	47. 5
48	290	229	440	405	855	672	1067	963	1431	2411	3405	48
48. 5	292	230	442	407	859	676	1073	967	1439	2423	3426	48. 5
49	293	231	444	409	863	679	1078	972	1446	2434	3443	49
49. 5	294	232	446	411	868	683	1084	977	1454	2444	3460	49. 5
50	296	234	448	413	872	686	1089	982	1461	2454	3477	50
50. 5	298	235	450	415	876	690	1095	986	1469	2464	3494	50. 5
51	300	236	453	417	880	693	1100	991	1476	2474	3511	51
51. 5	301	237	455	419	884	697	1105	996	1484	2486	3527	51. 5
52	302	238	457	421	888	700	1110	1000	1491	2498	3544	52
52. 5	303	239	459	423	892	704	1115	1005	1499	2510	3560	52. 5
53	304	240	461	425	896	707	1120	1009	1506	2522	3575	53
53. 5	305	241	463	427	900	711	1125	1014	1513	2534	3591	53. 5
54	307	243	465	429	904	714	1130	1018	1520	2545	3602	54
54. 5	309	244	467	431	908	718	1135	1023	1527	2555	3618	54. 5
55	310	246	469	433	912	721	1140	1027	1534	2565	3634	55
55. 5	311	247	471	435	915	725	1145	1032	1541	2575	3650	55. 5
56	313	248	472	437	919	727	1150	1036	1548	2586	3667	56
56. 5	314	249	474	439	923	730	1155	1040	1554	2597	3684	56. 5
57	315	250	476	441	927	733	1160	1044	1560	2608	3702	57
57. 5	316	251	478	443	930	736	1165	1046	1567	2619	3719	57. 5
58	317	252	480	445	934	739	1170	1052	1574	2630	3736	58
58. 5	319	253	482	447	938	742	1175	1056	1580	2641	3752	58. 5

See footnote at end of table.

TABLE 9–2.—*Orifice tables*—Continued

Head in inches	3'' Orifice		4'' Orifice		5'' Orifice		6'' Orifice		7'' Orifice—10 in. pipe	8'' Orifice—10 in. pipe	9'' Orifice—12 in. pipe	10'' Orifice—12 in. pipe	Head in inches
	4 in. pipe	6 in. pipe	6 in. pipe	8 in. pipe	6 in. pipe	8 in. pipe	8 in. pipe	10 in. pipe					
59	320	254	485	449	942	745	1180	1060	1586		2653	3768	59
59. 5	321	256	487	451	945	748	1185	1064	1592		2665	3784	59. 5
60	323	257	489	453	948	751	1190	1068	1598		2676	3800	60
60. 5	324	258	491	455	951	754	1195	1072					60. 5
61	325	259	492	457	955	757	1200	1076					61
61. 5	326	261	494	459	958	760	1205	1080					61. 5
62	328	262	496	461	961	763	1209	1084					62
62. 5	329	263	498	463	964	766	1214	1088					62. 5
63	330	264	500	465	968	769	1218	1092					63
63. 5	331	265	502	467	971	772	1223	1096					63. 5
64	333	266	504	469	974	775	1227	1099					64
64. 5	334	267	507	471	977	778	1232	1103					64. 5
65	335	268	509	472	981	781	1236	1106					65
65. 5	336	269	511	474	984	784	1241	1110					65. 5
66	338	271	513	475	988	787	1245	1113					66
66. 5	339	272	515	477	991	790	1250	1117					66. 5
67	340	273	517	479	995	793	1254	1120					67
67. 5	341	274	518	481	998	796	1259	1124					67. 5
68	343	275	520	483	1002	799	1263	1127					68
68. 5	344	276	521	485	1005	802	1268	1131					68. 5
69	346	277	523	487	1009	805	1272	1134					69
69. 5	347	278	524	489	1012	808	1276	1137					69. 5
70	349	280	525	491	1016	811	1280	1140					70

NOTE.—Capacities are given in nearest whole numbers.

To assure accurate measurements, the pipe orifice assemblies must be installed as follows:

(a) The position of the manometer tube tap must be at least three pipe diameters from the orifice plate and accurately located on the horizontal diameter of the pipe.

(b) The manometer tube tap must be at least 10 pipe diameters ahead of an elbow, valve, reducer, or similar fitting.

(c) The manometer tap fitting should have an inside diameter of ⅛ to ¼ inch and must be smooth and flush with the inside surface of the pipe.

(d) The pipe must be truly horizontal.

(e) The pipe must be full of water at all times and the water must fall freely from the orifice into the air without any obstruction.

(f) Before each measurement, the bottom of the pipe immediately behind the orifice plate should be cleaned of sand or other debris.

(g) The interior of the pipe should be clean, smooth, and free of grease.

(h) The manometer hose and gage should be free of air bubbles whenever a reading is being made.

(i) Manometer readings should not register less than 1 inch greater than the inside radius of the pipe nor greater than 60 inches. If readings are more or less than these values, the orifice size should be changed.

(j) There should be no leaks between the pump head and the orifice plate.

Usual practice is to securely anchor a 6-foot-long foot scale, reading in inches, in a truly vertical position with the zero point accurately located at the centerline of the manometer tap.

A 5- or 6-foot length of plastic hose or tubing of ¼- to ½-inch inside diameter should be attached to the manometer tap. Clear hose should be used and held against the scale when it is to be read. A recommended method is to use a 1- to 2-foot length of 1-inch-diameter clear glass or plastic tube with a hollow rubber plug with a glass or brass tube at the bottom to which the smaller diameter hose is attached. This arrangement dampens the surging commonly associated with many pumps and permits easier and more accurate readings to be made. If regular surging is evident in the tube, the range of such surging should be noted and the mean taken of the readings.

It is good practice to lower the hose and tube below the manometer tap and allow water to flow through it for a short time to clear all air bubbles and sand from the system before making a reading. Often when lighting conditions are poor, reading the gage may be facilitated by adding a few drops of vegetable coloring or cooking dyes to the clear tube just before making a reading.

When very large discharges are to be measured or the range of discharges to be measured exceeds that of a single orifice, an arrangement using two pipe orifices as shown on figure 9–4 is suggested. This arrangement sometimes permits ready changing of orifice plates by diverting the flow through one pipe without shutting down the pump. The sum of the readings from both orifices can be used to determine larger discharges.

At times the sizes of orifices specified in table 9–2 may not be available and those furnished by the contractor or other sources must be used. The condition and installation of such meters should conform to the requirements previously stated. In addition, the d/D ratio should be between 0.4 and 0.85, where d is the diameter of the orifice and D is the inside diameter of the pipe. The rating curve for the orifice can be checked by using figure 9–3 and the following equation:

$$Q=8.02\ KA\sqrt{h} \text{ or } Q=KA\sqrt{2gh} \quad (2)$$

Where K is taken from figure 9–3 for the applicable d/D ratio and:

A=area of the orifice in square feet,

h=head of water in manometer above the center of the pipe in feet (should be a minimum of 1 inch greater than the inside radius of the pipe), and

Q=discharge from the orifice in cubic feet per second. For discharge in gallons per minute, use 3,600 instead of 8.02 as the constant in the equation.

For measurements of discharge up to 100 gallons per minute, a stopwatch-timed filling of a 55-gallon oil drum with the top removed or a similar container is satisfactory. For very small discharges, a 5-gallon container may be used.

In addition to the previously discussed relatively accurate methods of measuring discharge, a number of trajectory methods such as the California pipe and the coordinate method, which give good to fair approximations of discharge when conditions are favorable, can be found in reference [1]. These methods are not recommended for use where high accuracy is required; however, they are useful in reconnaissance and similar surveys where a general idea of the range of capacities of wells is desired. They require little equipment and are relatively easy to perform.

9–10. Determining Duration of a Test.—The duration of a test is determined by the adequacy of the data in the form of curves obtained from plotting time versus drawdown, distance versus drawdown, or

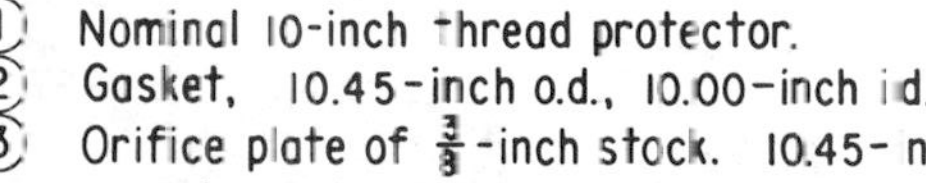

(1) Nominal 10-inch thread protector.
(2) Gasket, 10.45-inch o.d., 10.00-inch i.d.
(3) Orifice plate of $\frac{3}{8}$-inch stock. 10.45-inch o.d. by 8-inch i.d. and a 10.45-inch o.d. by 6-inch i.d.
(4) $\frac{1}{8}$ to $\frac{1}{4}$-inch i.d. tap.
(5) Minimum 9-foot length of $10\frac{3}{4}$-inch o.d. pipe with 0.279-inch wall. Weight per foot (plain ends) of 31.20 lbs. and threaded both ends.
(6) Tapped and threaded 10-inch screw flange.
*(7) 10-inch gate valve.
(8) 10-inch, 90° flanged elbow.
(9) 10-inch flanged tee.
(10) 10-inch flanged discharge from pump.
(11) Nominal 8-inch thread protector.
(12) Gasket, 8.35-inch o.d., 8.00-inch i.d.
(13) Orifice plate of $\frac{3}{8}$-inch stock. 8.35-inch o.d by [illegible]-inch i.d. and a 8.35-inch o.d. by 6-inch i.d.
(14) Minimum 7-foot length of $8\frac{5}{8}$-inch o.d. pipe with 0.277-inch wall. Weight per foot (plain ends) of 24.70 lbs. and threaded both ends.
(15) Tapped and threaded 8-inch screw flange.
*(16) 8-inch gate valve.
(17) 10-inch by $8\frac{5}{8}$-inch flanged reducer taper elbow.

* For convenience, two 10-inch valves could be used between the tee and the elbows rather than at the discharge end of elbows.

FIGURE 9–4.—Double pipe orifice. 103–D–1469.

both relationships, and possibly economic factors related to costs of pumping and monitoring the test.

The time-drawdown graph (see sec. 5–5(b)) is a plot of drawdown against the log of time since pumping began. It is most simply made on semilog paper with drawdown plotted on the arithmetic scale and time plotted on the log scale. There should be a graph made for the pumping well and for each observation well included in the test.

The distance-drawdown graph (see sec. 5–5(a)) is a plot of drawdowns occurring simultaneously in each observation well against the log of the distance from the observation wells to the pumping well. The time selected would usually be the longest available or the last reading taken unless a boundary has been encountered. This graph can be used for determining duration of test only when three or more observation wells are included in the test, since at least three points are needed to establish and verify a straight line.

In a relatively simple test, the time-drawdown plotted points for each observation hole fall initially on a curve which, with time, approximates a straight line within the limits of plotting. When the straight-line condition is attained, continued pumping will result in measured points which will fall on a prolongation of the straight line. The straight-line plot will be attained earliest for the pumping well, then for nearby observation wells, and at later times for more distant wells. When three or more drawdowns, measured at hourly intervals in the most distant hole, fall on the line, the time-drawdown conditions have been met. In an artesian aquifer, the straight-line condition is reached quite rapidly, but in an unconfined aquifer, the condition develops more slowly. Where three or more observation wells are available, a distance-drawdown graph is made as a check before pumping is stopped because sometimes this graph shows that the test should be continued. If pumping has continued long enough, these plotted points will also fall on a straight line within the limits of plotting. When this condition is reached, the test may be stopped. The time for approximate straight-line plotting conditions to be reached may range from 2 hours to as much as 3 weeks, but usually a satisfactory test can be completed within 48 hours. (See also secs. 9–5 and 9–6.)

When a partially penetrating pumping well and observation wells are used, a preliminary estimate of transmissivity and storativity should be made during the test from the straight-line plots of drawdown against the log of time. The test ideally should be continued until the value of:

$$u=\frac{r^2S}{4Tt}, \text{ estimated for each hole is less than } 0.1\,\frac{r^2}{M}$$

where:

r = distance from pumping well to observation hole, ft,
S = storativity, dimensionless,
T = transmissivity, ft^2/t,
t = time, and
M = aquifer thickness, ft.

The measurements from some tests may show an irregular curve or dispersion which does not appear to approach the straight-line condition even after several hours of pumping. If this condition persists for more than 24 hours past the minimum estimated pumping time (see sec. 9–6(d)), pumping may be stopped and recovery measurements made. Study of conditions and results may show that a good test cannot be obtained or that a longer test may be required.

If boundaries are suspected, a test may be run for a longer period to determine whether the boundary effects will become apparent.

9–11. Recovery Test for Transmissivity.—When the pump is stopped after running the pump-out test, the drawdown and time at which it was shut down are recorded. Measurement of water level is immediately initiated in the pumped well and in all observation wells, if any. The same procedure and time pattern is followed as at the beginning of a pumping test (see sec. 9–10). As in the pumping test, the time and depth-to-water are noted for each measurement. The recovery usually will not return to the original static water level within a reasonable length of time, so, when several measurements at 1-hour intervals show less than 0.1-foot difference in recovery, measurements may be discontinued. A good check on the transmissivity value can be made of recovery in the pumped well and of transmissivity and storativity from recovery measurements in the observation wells (see sec. 5–7).

Transmissivity, but not storativity, can be approximated by the following procedure using only a pumping well. The static water level is measured and the time recorded before starting the pump discharging at a uniform rate. Measurements of the drawdown in the well are continued until the plot of drawdown versus log t falls on a straight line over a 3-hour period. The well must fully penetrate the aquifer to obtain a reasonably accurate value for T by this method. The accuracy obtained will be less than from the pump-out test using observation wells. It will be best for homogeneous, isotropic conditions and will lessen as these conditions deteriorate and as boundary conditions increase in effect. The pump is then stopped, and the remainder of the procedure is the same as previously described for a recovery test (sec. 5–7) after the pump is stopped.

9–12. Bibliography.—

[1] "Water Measurement Manual," second edition, Bureau of Reclamation, 1967.

[2] Ferris, J. G., Knowles, D. B., Brown, R. H., and Stallman, R. W., "Theory of Aquifer Tests," U.S. Geological Survey Water-Supply Paper 1536–E, 1962.

[3] Lohman, S. W., "Ground-Water Hydraulics," U.S. Geological Survey Professional Paper 708, 1972.

[4] "Measurement of Water Flow through Pipe Orifice with Free Discharge," Purdue University, Lafayette, Ind., 1949.

Chapter X

PERMEABILITY TESTS IN INDIVIDUAL DRILL HOLES

10–1. General Considerations.—Chapters IV, V, VI, and IX considered various aspects of pumping tests to determine values of transmissivity, storativity, and boundary conditions of aquifers for use in ground-water inventories, well field design, drainage feasibility, and related activities. Such tests are usually large-scale activities that are time consuming, costly, and require significant manpower. Moreover, they are applicable only to saturated materials.

Laboratory permeability tests of subsurface materials usually are not satisfactory. Test specimens from such materials can seldom, if ever, be obtained in an entirely undisturbed state, and a specimen may represent only a limited portion of the material being investigated.

Often, though, the hydraulic characteristics, especially permeability, of unsaturated, slowly permeable rock or unconsolidated materials are sought.

Tests have been devised that are relatively simple and less costly than aquifer pumping tests. These tests are usually conducted in conjunction with exploratory drilling and are designed to obtain data relating to possible or existing seepage, uplift pressures, and similar problems that may occur during the construction and operation of water storage, control, or conveyance facilities.

Exploratory drilling to determine foundation and other conditions is costly and time consuming, but its value is generally recognized. Permeability testing is often an integral part of the operation and the cost of such exploratory drilling. Properly conducted and controlled permeability tests will yield reasonably accurate and reliable data.

The tests described herein show semiquantitative values of permeability. If they are performed properly, however, the values obtained are sufficiently accurate for most engineering purposes. Linear, volumetric, pressure, and time measurements should be made as accurately as available equipment will permit and gages should be checked periodically for accuracy.

The quality of water used in permeability tests is of primary importance. The presence of only a few parts per million of turbidity or air dissolved in water can plug soil and rock voids and cause serious errors in test results. Water should be clear and silt-free. To avoid plugging of the rock pores by air bubbles, the use of water that is a few degrees warmer than the ground temperature of the test section is a desirable practice.

Pumps of up to 250 gallons per minute capacity against a total dynamic head of 160 feet may be required in some tests.

If meters and gages are located in relation to each other as recommended, the arrangement of pipe, hose, etc., will not seriously influence the tests. However, in the interest of pumping efficiency, and other factors, sharp bends in hoses, 90° fittings on pipes, and unnecessary changes in pipe and hose diameters should be avoided.

The equations given for computing permeability are applicable for laminar flow. The velocity at which turbulent flow occurs depends on the grain size of the materials tested and other factors, but a safe average figure below which flow would be considered laminar is about 0.1 foot per second. Therefore, in tests, if the quotient of the water intake in cubic feet per second divided by the open area of the test section in square feet times the estimated porosity of the tested material is greater than 0.10, the various given equations may not be applicable.

In an open hole test, the total open area of the test section is computed as follows:

$$a = \pi d A + \pi r^2 \quad (1)$$

where:

a = total open area of the hole face plus the hole bottom, L^2,
r = radius of the hole, L,
d = diameter of the hole, L, and
A = length of the test section of the hole, L.

In a test using perforated casing, the open area of the perforations is computed as follows:

$$a_p = n a_s \quad (2)$$

where:

n = number of perforations,
a_p = total open area of perforations, L^2, and
a_s = area of each perforation, L^2.

If the bottom of the perforated casing is open, this area must be added to the area of the perforations to obtain the total open area.

Where fabricated well screens are used, estimates of open area may require precise measurement of screen component dimensions and the computations based on these measurements.

For purposes of discussion, permeability tests are divided into three types: pressure tests, constant head gravity tests, and falling head gravity tests. Pressure tests and falling head gravity tests are those in which one or two packers are used to segregate the test section in the hole. In pressure tests, water is forced into the test section through combined applied pressure and gravity head, although the tests can be performed using gravity head only. In falling head tests, gravity head only is used. Constant head gravity tests are those in which no packers are used, and a constant water level is maintained [1].[1]

10–2. Pressure Permeability Tests in Stable Rock.—Pressure permeability tests are run using one or two packers to isolate various zones or lengths of drill hole in stable rock. Hole diameters usually do not exceed 3½ inches, but larger holes can also be tested if suitable equipment is available. The tests may be run in vertical, angled, or horizontal holes and analyzed if the head and zone relationships can be determined. Pressure tests are often the only practical tests to use when it is necessary to determine permeability of streambeds or lakebeds below water.

Compression packers, inflatable packers, leather cups, and similar types of packers have been used for pressure testing. Inflatable packers are usually more economical because they reduce testing time and assure a tighter seal, particularly in rough-walled or out-of-round holes. The packers are inflated through tubes extending to a cylinder of air or nitrogen at the surface. If a pressure sensing instrument is included, pressure in the test section is sensed by the instrument and is transmitted to the surface by an electrical circuit where it is either read from a register at the surface or is recorded on a chart. Although this arrangement permits an accurate determination of test pressures, other observations outlined in this section should still be made to permit an estimate of permeability should the pressure sensor fail. This double packer arrangement permits successive tests at different depths in a completed hole without having to remove the packer between each test. The pressure sensor can also be adapted where a single packer is used.

(a) *Methods of Testing.*—The most common testing practice by the Bureau is the drilling of about 10 feet of hole and pressure testing the newly drilled section. In rock that tends to ravel or bridge the hole and which must be cemented to permit continuation of drilling, this is the only practical method of testing. In such rock, long lengths of open hole are impractical and the test sections must be kept short because the tests must be made before the hole is cemented if good test data are to be obtained.

Where the rock is stable and does not require cementing, the following method of testing may offer distinct advantages. The hole

---

[1] Numbers in brackets refer to items in the bibliography, section 10–5.

is drilled to the total depth without testing. Two inflatable packers 5 to 10 feet apart are mounted near the bottom of the rod or pipe used for making the test. The bottom of the rod or pipe is sealed, and the section between the packers is perforated. The perforations should be at least one quarter of an inch in diameter, and the total area of all perforations should be greater than two times the inside cross-sectional area of the pipe or rod. Tests are made beginning at the bottom of the hole. After each test, the packers are raised the length of the test section and another test made. This procedure is followed until the entire length of the hole has been tested.

(b) *Cleaning Test Sections Before Testing.*—Before each test, the test section should be surged with clear water and bailed out to clean cuttings and drilling fluid from the face of the hole. If the test section is above the water table and will not hold water, water should be poured into the hole during the surging, then bailed out as rapidly as possible. When a completed hole is tested using two packers, the entire hole can be cleaned in one operation. Cleaning the hole is frequently omitted from testing procedures; however, this omission may result in a permeable rock appearing to be impermeable because the hole face is sealed by cuttings or drilling fluid. In such cases, the computed permeability will be lower than the true permeability.

Alternative methods to surging and bailing a drill hole in consolidated formations before pressure testing include the use of a rotating, stiffly bristled brush while washing and jetting with water. An average jet velocity of 150 feet per second is desirable. This is approximated by a rate of pumping equal to 1.4 gallons per minute per $\frac{1}{16}$-inch-diameter hole in the rod. On completion of jetting, the hole should be blown or bailed out to the bottom, if possible.

(c) *Length of Test Section.*—The length of the test section is governed by the character of the rock, but generally a length of 10 feet is desirable. At times, a good seal cannot be obtained for the packer at the planned elevation because of bridging, raveling, or the presence of fractures. Under these circumstances, the test section length should be increased or decreased or test sections overlapped to assure that the test is made with well-seated packers. On some tests, a 10-foot section will take more water than the pump can deliver; hence, no back pressure can be developed. When this occurs, the length of the test section should be shortened until back pressure can be developed, or the falling head test (sec. 10–4) might be tried.

The test sections should never be shortened to where the ratio $\frac{A}{D}$ is less than 5, where D is the diameter of the hole and A is the length of the test section. Under no circumstances should a packer be set inside the casing when making a test unless the casing has been grouted in the hole. Except under the most adverse conditions, the use of test sections greater than 20 feet in length is inadvisable. Longer test sections may not permit sufficient localization of permeable zones and may complicate computations.

(d) *Size of Rod or Pipe to Use in Tests.*—Drill rods are commonly used as intake pipes to make pressure and permeability tests. NX and NW rods can be used for this purpose without seriously affecting the reliability of the test data, if the intake of the test section does not exceed 12 to 15 gallons per minute and the depth to the top of the test section does not exceed 50 feet. For general use, 1¼-inch or larger pipe is more satisfactory. Figures 10–1 through 10–4 show head losses per 10-foot section at various deliveries of water for different sizes of drill rod and 1¼-inch pipe. These figures were compiled from experimental tests. The desirability of using the 1¼-inch pipe, particularly where holes 50 feet or more in depth are to be tested, is obvious from study of the graphs. The couplings on the 1¼-inch pipe must be turned down to an outside diameter of 1.8 inches for use in AX holes.

(e) *Pumping Equipment.*—Tests are commonly run using a mud pump for pumping the water. Such pumps are generally of the multiple cylinder type with a uniform fluctuation in pressure. Many of these pumps have a maximum capacity of about 25 gallons per minute and, if not in good condition, the capacities may be as small as 17 to 18 gallons per minute. Tests are often difficult, if not impossible, to analyze because such pumps do not have sufficient capacity to develop back pressure in the length of hole being tested. When this happens, the tests are generally reported: "took capacity of pump, no pressure developed." This result does not permit determination of permeability of the material tested other than it is probably high. The fluctuating pressures of multiple cylinder pumps, even when an air chamber is used, are often difficult to read accurately because the high and low readings must be averaged to determine the approximate true effective pressure, a difficulty which may be a source of error. In addition, such pumps occasionally develop instantaneous excessively high pressures which may fracture the rock or blow out a packer.

Permeability tests made in drill holes ideally should be performed using centrifugal pumps having sufficient capacity to develop back pressure. A pump with a capacity of up to 250 gallons per minute against a total head of 160 feet would be adequate for most testing. Head and discharge of such pumps are easily controlled by changing engine speed or with a control valve on the discharge.

(f) *Swivels for Use in Tests.*—Swivels used for testing should be selected for minimum head losses.

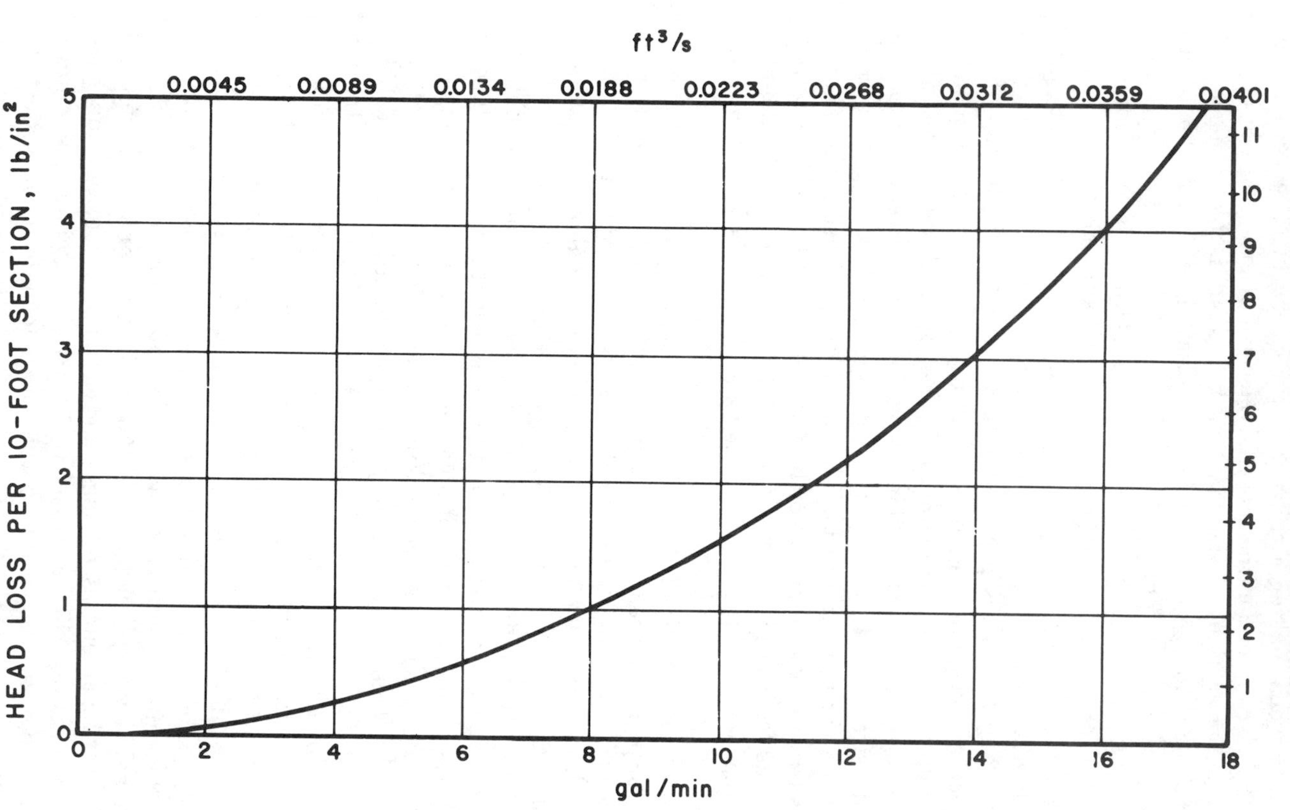

FIGURE 10-1.—Head loss in a 10-foot section of AX drill rod. 103-D-1470.

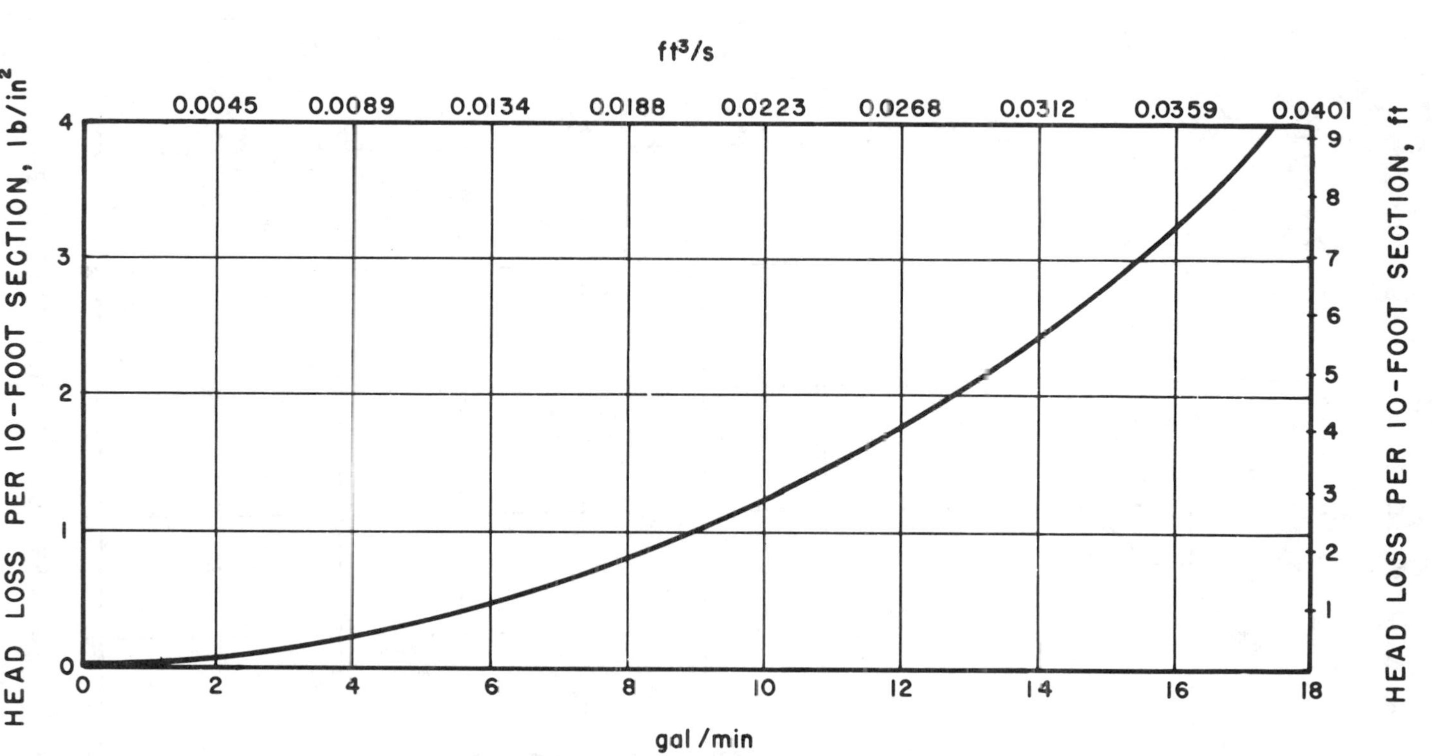

FIGURE 10-2.—Head loss in a 10-foot section of BX drill rod. 103-D-1471.

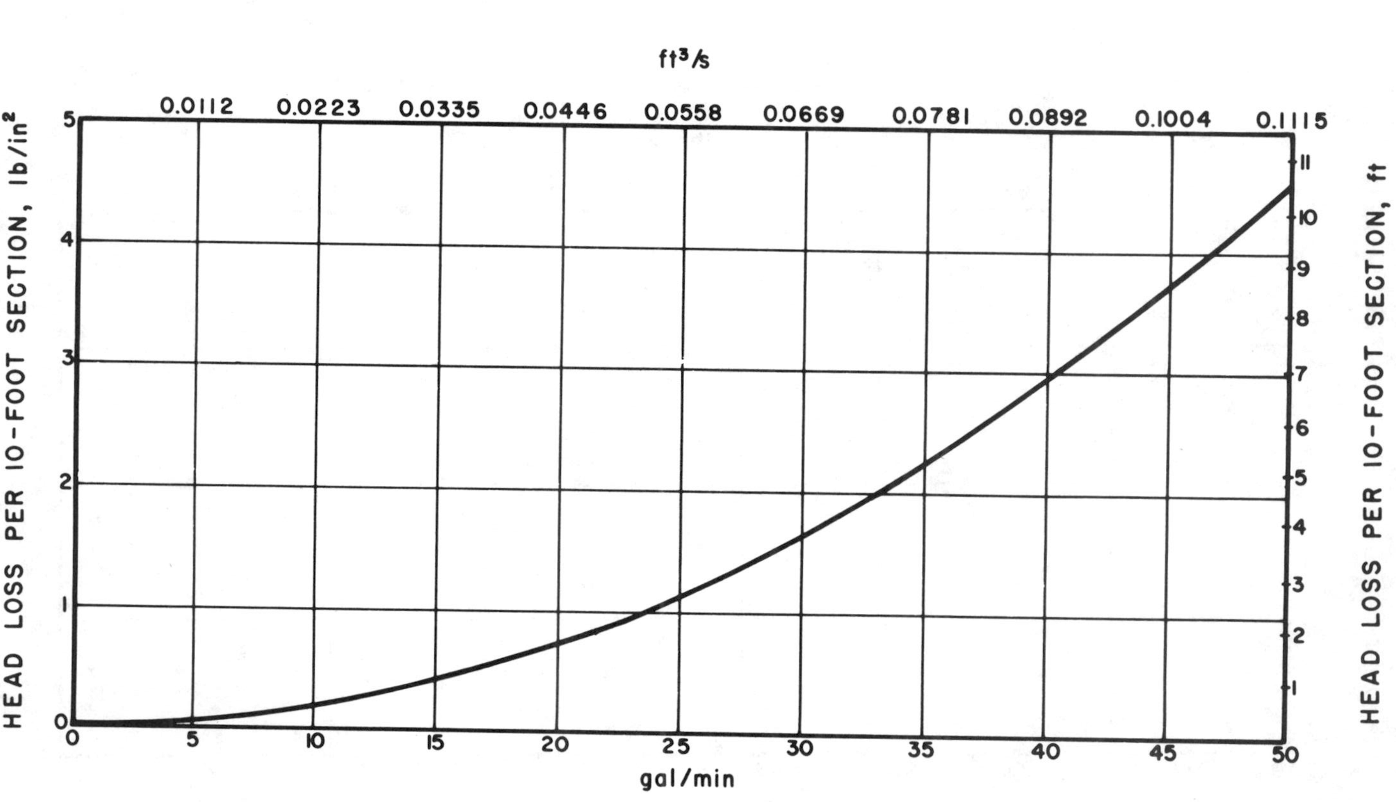

FIGURE 10-3.—Head loss in a 10-foot section of NX drill rod. 103-D-1472.

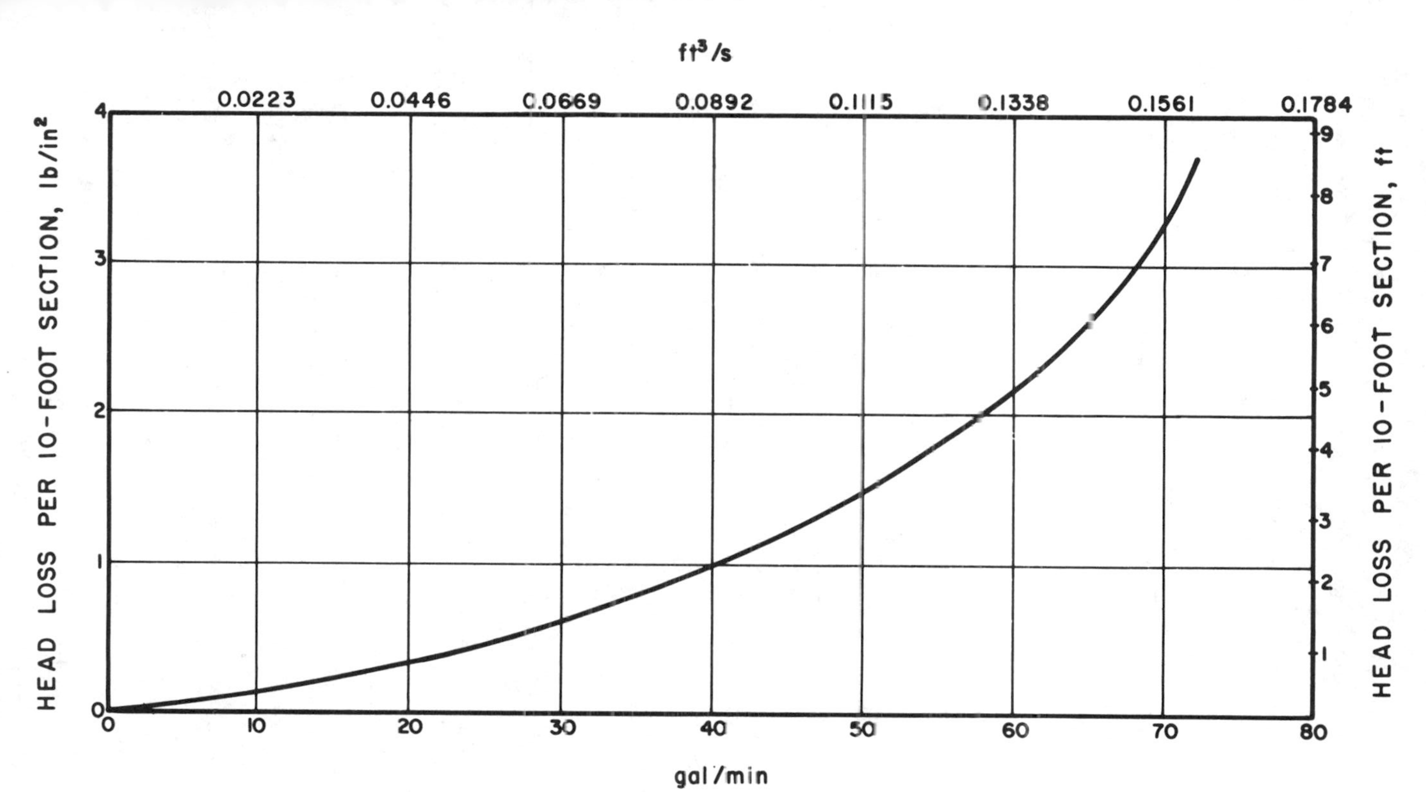

FIGURE 10-4.—Head loss in a 10-foot section of 1¼-inch iron pipe 103-D-1473.

(g) *Location of Pressure Gage in Tests.*—The ideal location for a pressure gage is near the well head, preferably between the packer and the swivel.

(h) *Recommended Watermeters.*—Required water deliveries in pressure tests may range from less than 1 gallon per minute to as much as 400 gallons per minute. No one meter is sufficiently accurate at all ranges to be reliably used. Therefore, two meters are recommended: (1) a 4-inch propeller or impeller-type meter to measure flows between 50 and 350 gallons per minute; and (2) a 1-inch disk-type meter for flows between 1 and 50 gallons per minute. Ideally, each meter should be equipped with an instantaneous flow indicator as well as a totalizer. Watermeters should be tested frequently to assure reliability.

Inlet pipe adapters should be available for each meter to minimize turbulent inflow. The adapters should be at least 10 times as long as the diameter of the rated size of the meter.

(i) *Length of Time for Tests.*—The minimum length of time to run a test depends upon the nature of the material tested. Tests should be run until stabilization occurs; that is, until three or more readings of water intake and pressure taken at 5-minute intervals are essentially equal. In tests above the water table, water should be pumped into the test section at the desired pressure for about 10 minutes in coarse materials or 20 minutes in fine-grained materials before making measurements.

Stability is obtained more rapidly in tests below the water table than in unsaturated material. When multiple pressure tests are made, each pressure theoretically should be maintained until stabilization occurs. This is not practicable in some cases, however, but good practice requires that each pressure stop be maintained for at least 20 minutes with intake and pressure readings made at 5-minute intervals as the pressure is increased and for 5 minutes as pressure is decreased.

(j) *Pressures to be Used in Testing.*—Where subsurface conditions for proposed reservoirs or other water-impounding or storage facilities are being investigated, the theoretical minimum pressure used in the test section should equal the head imposed by the maximum reservoir level. However, when tests are made in locations where the ground surface is below the proposed maximum pool level, the use of such test pressures may be impractical because of the danger of blowouts or fracturing the hole face. Under these conditions, a safe pressure in consolidated rock is 0.5 pound per square inch, or 1.15 feet of water per foot of depth from the ground surface to the top of the test section. In all other locations, the same criterion is a good rule-of-thumb guide. Tables 10–1 and 10–2 are provided for converting pounds per square inch to feet of water and vice versa, respectively.

TABLE 10–1.—*Conversion of heads of water in feet to hydrostatic pressures in pounds per square inch.* 103–D–1486.

Weight of Water - 62.4 pounds per cubic foot

HEAD IN FEET	ADDITIONAL UNITS									
	0	1	2	3	4	5	6	7	8	9
0	——	0.43	0.87	1.30	1.73	2.17	2.60	3.03	3.47	3.90
10	4.33	4.77	5.20	5.64	6.07	6.50	6.94	7.37	7.80	8.24
20	8.67	9.10	9.54	9.97	10.40	10.84	11.27	11.70	12.14	12.57
30	13.00	13.44	13.87	14.31	14.74	15.17	15.61	16.04	16.47	16.91
40	17.34	17.77	18.21	18.64	19.07	19.51	19.94	20.37	20.81	21.24
50	21.67	22.11	22.54	22.98	23.41	23.84	24.28	24.71	25.14	25.58
60	26.01	26.44	26.88	27.31	27.74	28.18	28.61	29.04	29.48	29.91
70	30.34	30.78	31.21	31.65	32.08	32.51	32.95	33.38	33.81	34.25
80	34.68	35.11	35.55	35.98	36.41	36.85	37.28	37.71	38.15	38.58
90	39.01	39.45	39.88	40.32	40.75	41.18	41.62	42.05	42.48	42.92
100	43.35	43.78	44.22	44.65	45.08	45.52	45.95	46.38	46.82	47.25
110	47.68	48.12	48.55	48.99	49.42	49.85	50.29	50.72	51.15	51.59
120	52.02	52.45	52.89	53.32	53.75	54.19	54.62	55.05	55.49	55.92
130	56.36	56.79	57.22	57.66	58.09	58.52	58.96	59.39	59.82	60.26
140	60.69	61.12	61.56	61.99	62.42	62.86	63.29	63.72	64.16	64.59
150	65.02	65.46	65.89	66.33	66.76	67.19	67.63	68.06	68.49	68.93
160	69.36	69.79	70.23	70.66	71.09	71.53	71.96	72.39	72.83	73.26
170	73.69	74.13	74.56	75.00	75.43	75.86	76.30	76.73	77.16	77.60
180	78.03	78.46	78.90	79.33	79.76	80.20	80.63	81.06	81.50	81.93
190	82.36	82.80	83.23	83.67	84.10	84.53	84.97	85.40	85.83	86.27
200	86.70	87.13	87.57	88.00	88.43	88.87	89.30	89.73	90.17	90.60

(k) *Arrangement of Equipment.*—Recommended arrangement of test equipment starting at the source of water is as follows:

> Source of water, suction line, pump, waterline to settling and storage tank or basin, suction line, centrifugal test pump, line to watermeter adapter (if required) or to watermeter, short length of pipe, plug valve, waterline to swivel, sub for gage, and pipe or rod to packer. All connections should be kept as short and straight as possible with a minimum number of changes in diameter of hose, pipe, etc.

All joints, connections, and hose between the watermeter and the packer or casing should be tight so no water loss occurs between the meter and the test section.

(l) *Pressure Permeability Test Methods.*—A schematic drawing of the following two methods is shown on figure 10–5.

TABLE 10-2.—*Conversion of hydrostatic pressures in pounds per square inch to heads of water in feet.* 103-D-1487.

POUNDS PER SQ. INCH	ADDITIONAL UNITS									
	0	1	2	3	4	5	6	7	8	9
0	——	2.31	4.62	6.93	9.24	11.55	13.86	16.17	18.48	20.79
10	23.10	25.41	27.72	30.03	32.34	34.65	36.96	39.27	41.58	43.89
20	46.20	48.51	50.82	53.13	55.44	57.75	60.06	62.37	64.68	66.99
30	69.30	71.61	73.92	76.23	78.54	80.85	83.16	85.47	87.78	90.09
40	92.40	94.71	97.02	99.33	101.64	103.95	106.26	108.57	110.88	113.19
50	115.50	117.81	120.12	122.43	124.74	127.05	129.36	131.67	133.98	136.29
60	138.60	140.91	143.22	145.53	147.84	150.15	152.46	154.77	157.08	159.39
70	161.70	164.01	166.32	168.63	170.94	173.25	175.56	177.87	180.18	182.49
80	184.80	187.11	189.42	191.73	194.04	196.35	198.66	200.97	203.28	205.59
90	207.90	210.21	212.52	214.83	217.14	219.45	221.76	224.07	226.38	228.69
100	231.00	233.31	235.62	237.93	240.24	242.55	244.86	247.17	249.48	251.79
110	254.10	256.41	258.72	261.03	263.34	265.65	267.96	270.27	272.58	274.89
120	277.20	279.51	281.82	284.13	286.44	288.75	291.06	293.37	295.68	297.99
130	300.30	302.61	304.92	307.23	309.54	311.85	314.16	316.47	318.78	321.09
140	323.40	325.71	328.02	330.33	332.64	334.95	337.26	339.57	341.88	344.19
150	346.50	348.81	351.12	353.43	355.74	358.05	360.36	362.67	364.98	367.29
160	369.60	371.91	374.22	376.53	378.84	381.15	383.46	385.77	388.08	390.39
170	392.70	395.01	397.32	399.63	401.94	404.25	406.56	408.87	411.18	413.49
180	415.80	418.11	420.42	422.73	425.04	427.35	429.66	431.97	434.28	436.59
190	438.90	441.21	443.52	445.83	448.14	450.45	452.76	455.07	457.38	459.69
200	462.00	464.31	466.62	468.93	471.24	473.55	475.86	478.17	480.48	482.79

Method 1: This method is primarily applicable to testing in consolidated rock which requires casing as the hole is drilled, although it may be used in stable material if desired. The hole is drilled, the tools removed, a packer seated a given distance above the bottom of the hole, water under pressure is pumped into the test section, and the readings recorded. The packer is then removed, the hole drilled deeper, the packer is inserted so as to leave the full length of the newly drilled hole within the test section, and the test repeated.

Method 2: This method is applicable in consolidated rock which is stable and does not require cementing. The hole is drilled to the final depth, cleaned, and blown out or bailed. Two packers, spaced on pipe or drill stem to isolate the desired test section, are used. Tests should be started at the bottom of the hole. After each test, the pipe is lifted a distance equal to the A dimension shown on figure 10-5 and the test repeated until the entire hole is tested.

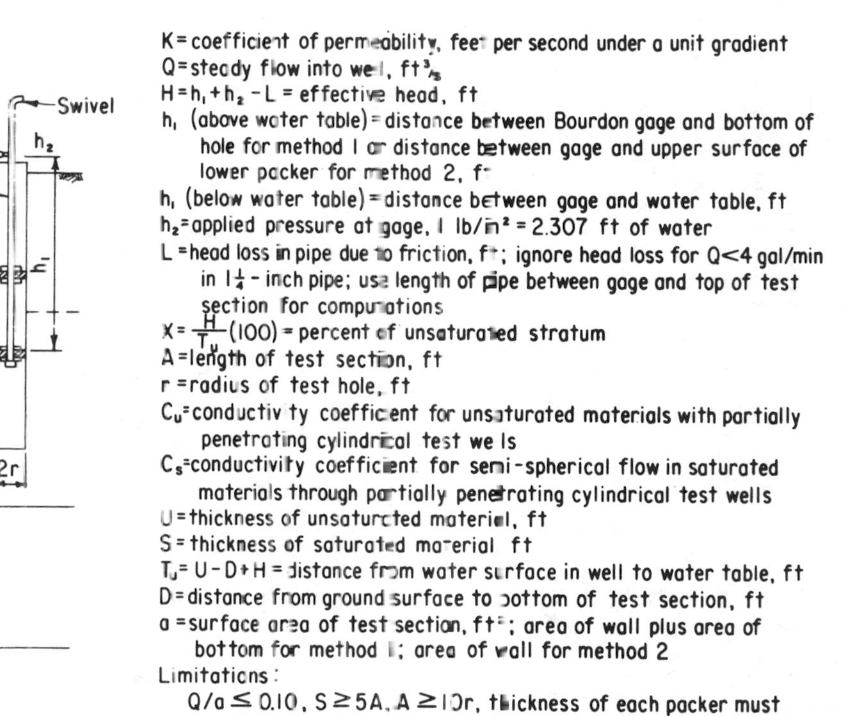

FIGURE 10–5.—Permeability test for use in saturated or unsaturated consolidated rock [4] 103-D-1474.

Data required for computing the permeability may not be available until the hole has encountered the water table or a relatively impermeable bed. The required data for each test include:

1. Radius r of the hole, in feet.
2. Length of test section A, the distance between the packer and the bottom of the hole, method 1, or between the packers, method 2, in feet.
3. Depth h_1 from pressure gage to bottom of the hole, method 1, or from gage to upper surface of lower packer in method 2, in feet. If a pressure sensor is used, substitute the pressure recorded in test section prior to pumping for the h_1 value.
4. Applied pressure h_2 at the gage, in feet, or the pressure recorded during pumping in test section if a sensor is used.
5. Steady flow Q into well at 5-minute intervals, in ft^3/s.
6. Nominal diameter in inches and length of intake pipe in feet between the gage and upper packer.
7. Thickness U of unsaturated material above water table, in feet.
8. Thickness S of saturated material above a relatively impermeable bed, in feet.
9. Distance D from the ground surface to the bottom of the test section, in feet.
10. The time that the test is started and time measurements are made.
11. If tests are made in streambeds or lakebeds below water, the effective head is the difference in feet between the elevation of the free water surface in the pipe and the elevation of the gage plus the applied pressure.
12. If a pressure sensor is used, the effective head in the test section is the difference in pressure before water is pumped into the test section and the pressure readings made during the test.

The following examples show some typical calculations using methods 1 and 2 in the different zones shown on figure 10–5. Figure 10–6 shows the location of the zone 1 lower boundary for use in unsaturated materials.

PRESSURE PERMEABILITY TESTS EXAMPLES OF METHODS 1 AND 2

Example 1:

Zone 1—*Method* 1

Given: $U=75$ ft, $D=25$ ft, $A=10$ ft, $r=0.5$ ft,
$h_1=32$ ft, $h_2=25$ lb/in^2$=57.8$ ft, and
$Q=20$ gal/min$=0.045$ ft^3/s.

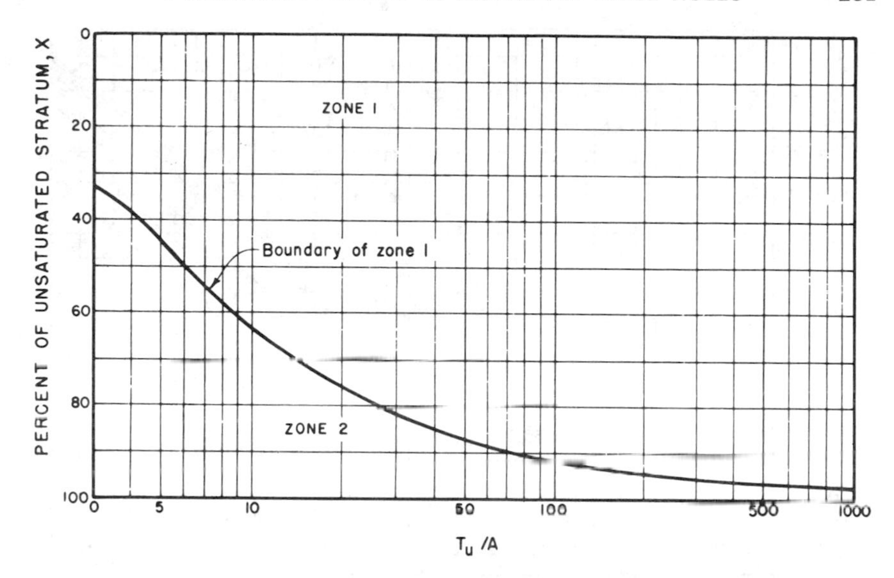

FIGURE 10–6.—Location of zone 1 lower boundary, for use in unsaturated materials [4]. 103-D-1475.

From figure 10–4: Head loss L for a 1¼-inch pipe at 20 gal/min is 0.76 foot per 10-foot section. If the distance from the Bourdon gage to the bottom of the pipe is 22 ft, the total L is (2.2)(0.76)=1.7 ft.

$H=h_1+h_2-L=32+57.8-1.7=88.1$ ft of effective head.
$T_u=U-D+H=75-25+88.1=138.1$ ft

$$X=\frac{H}{T_u}(100)=\frac{88.1}{138.1}(100)=63.8 \text{ percent}$$

$$\frac{T_u}{A}=\frac{138.1}{10}=13.8$$

The values for X and $\frac{T_u}{A}$ lie in zone 1 (fig. 10–6). To determine the conductivity coefficient C_u from figure 10–7:

$$\frac{H}{r}=\frac{88.1}{0.5}=176.2$$

$$\frac{A}{H}=\frac{10}{88.1}=0.11 \text{ and } C_u=62$$

Then,

$$K=\frac{Q}{C_u rH}=\frac{0.045}{(62)(0.5)(88.1)}=0.000016 \text{ ft/s}$$

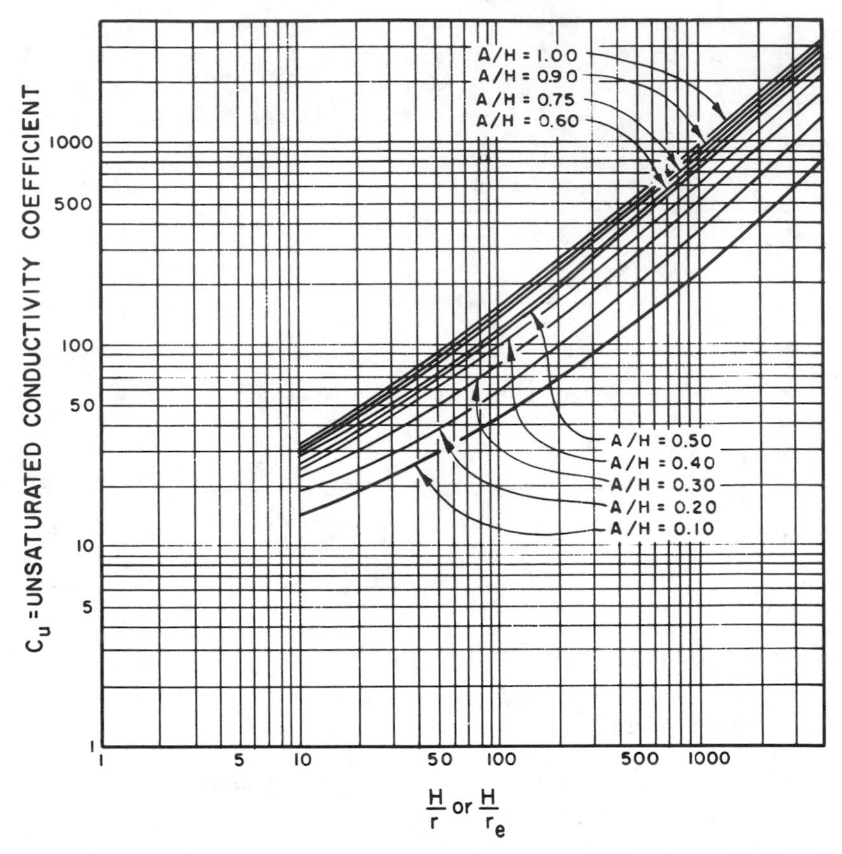

FIGURE 10–7.— Conductivity coefficients for permeability determination in unsaturated materials with partially penetrating cylindrical test wells [4]. 103-D-1476.

Example 2:

Zone 2

Given: U, A, r, h_2, Q, and L are as given in example 1.
$D=65$ ft, and $h_1=72$ ft.

If the distance from the Bourdon gage to the bottom of the intake pipe is 62 ft, the total L is (6.2)(0.76)=4.7 ft.

$H=72+57.8-4.7=125.1$ ft
$T_u=75-65+125.1=135.1$ ft

$$X=\frac{125.1}{135.1}(100)=92.6 \text{ percent and } \frac{T_u}{A}=\frac{135.1}{10}=13.5$$

The test section is in zone 2 (fig. 10–6). To determine the conductivity coefficient C_s from figure 10–8:

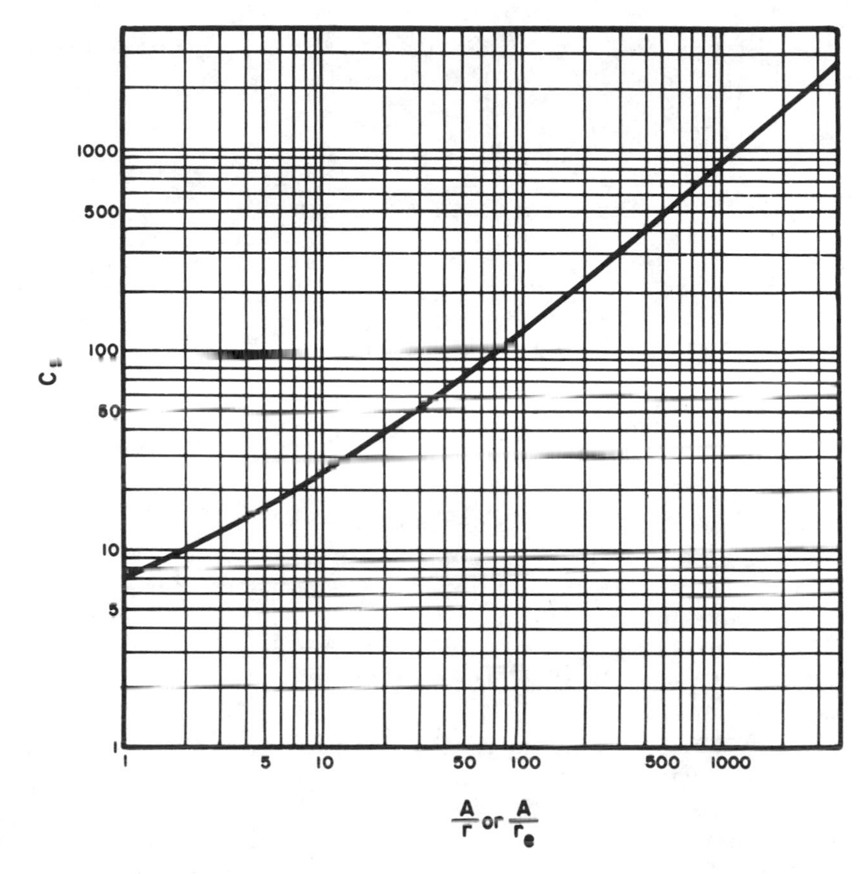

FIGURE 10-8.—Conductivity coefficients for semispherical flow in saturated materials through partially penetrating cylindrical test wells [4]. 103-D-1477.

$$\frac{A}{r}=\frac{10}{0.5}=20 \text{ and } C_s=39.5$$

Method 1

$$K=\frac{2Q}{(C_s+4)r(T_u+H-A)}=\frac{(2)(0.045)}{(39.5+4)(0.5)(135.1+125.1-10)}$$

$$=0.000016 \text{ ft/s}$$

Method 2

$$K=\frac{2Q}{(C_s r)(T_u+H-A)}=\frac{(2)(0.045)}{(39.5)(0.5)(135.1+125.1-10)}=0.000018 \text{ ft/s}$$

Example 3:

Zone 3

Given: U, A, r, h_2, Q, and L are as given in example 1.
$D=100$ ft, $h_1=82$ ft, and $S=60$ ft.

If the distance from the Bourdon gage to the bottom of the intake pipe is 97 ft, the total L is $(9.7)(0.76)=7.4$ ft.

$H=82+57.8-7.4=132.4$ ft

$$\frac{A}{r}=\frac{10}{0.5}=20 \text{ and } C_s=39.5 \text{ from figure 10–8}$$

Method 1

$$K=\frac{Q}{(C_s+4)rH}=\frac{0.045}{(39.5+4)\ (0.5)\ (132.4)}=0.000016 \text{ ft/s}$$

Method 2

$$K=\frac{Q}{C_s rH}=\frac{0.045}{(39.5)\ (0.5)\ (132.4)}=0.000017 \text{ ft/s}$$

(m) *Multiple Pressure Tests.*—These tests are run in the same manner as other pressure permeability tests except that the pressure is applied in three or more approximately equal steps. For example, if the allowable maximum differential pressure is 90 pounds per square inch or 210 feet of water, the test would be run at pressures of about 30, 60, and 90 pounds per square inch (70, 140, and 210 ft).

Each pressure step should be maintained for 20 minutes and intake readings made at 5-minute intervals. The pressure is then raised to the next step. On completion of the highest step, the process should then be reversed with the pressure being maintained for 5 minutes at approximately the same middle and the lowest pressure steps. A plot of intake against pressure for the five steps in a multiple pressure test may be useful in assessing hydraulic conditions.

Graphs of synthetic test results of multiple pressure tests are plotted on figure 10–9. Synthetic plots have been used in the interest of uniform spacing for illustrative purposes, but the curves are typical of those most often encountered. The results should be analyzed on the basis of confined flow hydraulic principles combined with data obtained from the core or hole logs.

Probable conditions represented by the circled numbers on figure 10–9 are:

① Probably very narrow, clean fractures. Flow is laminar and permeability is low with discharge directly proportional to head.

② Firm, practically impermeable material; fractures are tight. Little or no intake regardless of pressure.

③ Highly permeable, relatively large open fractures indicated by high rates of water intake and no back pressure. Pressure shown on gage due entirely to pipe resistance.

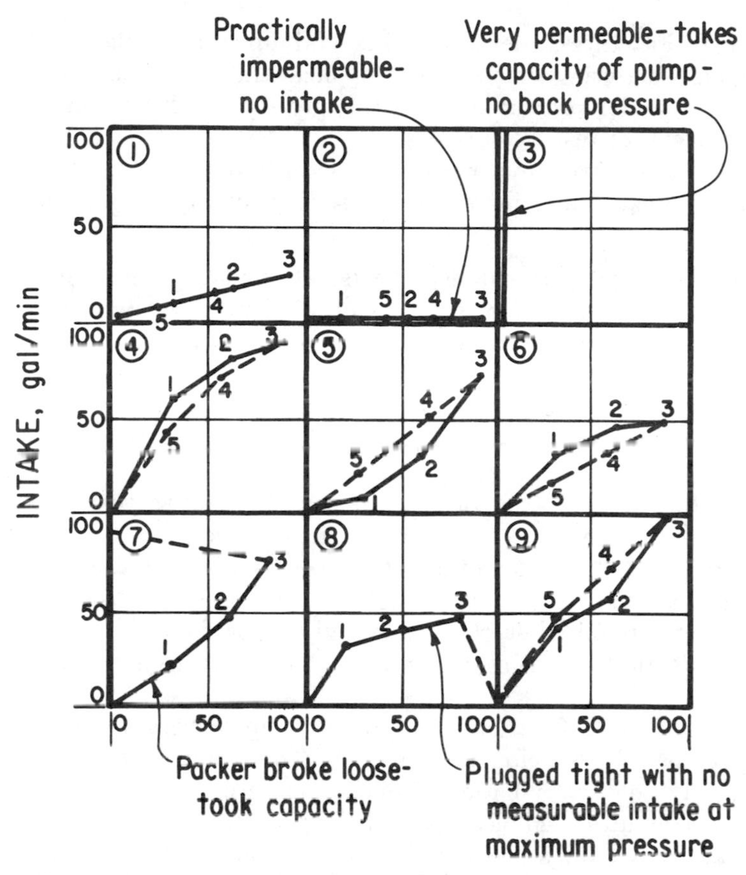

FIGURE 10-9.—Plots of simulated, multiple pressure, permeability tests. 103-D-1478.

④ Permeability high with fractures that are relatively open and permeable, but contain filling material which tends to expand on wetting or dislodges and tends to collect in traps that retard flow. Flow is turbulent.

⑤ Permeability high, with fracture filling material which washes out, increasing permeability with time. Fractures probably are relatively large. Flow is turbulent.

⑥ Similar to ④ but fractures are tighter and flow is laminar.

⑦ Packer failed or fractures are large, flow is turbulent. Fractures have been washed clean; highly permeable. Test takes capacity of pump with little or no back pressure.

⑧ Fractures are fairly wide and open but filled with clay gouge material which tends to pack and seal when subject to water under pressure. Takes full pressure with no water intake near end of test.

⑨ Open fractures with filling which tend to first block and then break under increased pressure. Probably permeable. Flow is turbulent.

10–3. Gravity Permeability Tests.—Gravity permeability tests are intended primarily for use in unconsolidated or unstable materials and are usually made in larger diameter holes than pressure tests. Gravity tests can be run only in vertical or near-vertical holes. A normal test section length is 5 feet; however, if the material is stable, will stand without caving or sloughing, and is relatively uniform, sections up to 10-foot long may be tested. Shorter test sections may be used if the length of the water column in the test section is at least five times the diameter of the hole. After each test, the casing for open hole tests is driven to the bottom of the hole and a new test section opened below it. If perforated casing is used, the pipe may be driven to the required depth and cleaned out, or the hole drilled to the required depth, then casing driven to the bottom of the hole and cleaned out.

(a) *Cleaning Test Sections.*—Each newly opened test section should be developed by surging and bailing. This should be done slowly and gently so that a large volume of loosely packed material will not be drawn into the hole, but so that compaction due to drilling will be broken down and some fines will be removed from the formation.

(b) *Measurement of Water Levels Through Protective Pipe.*—In making gravity tests, insertion of a small diameter perforated pipe (¾- to 1½-inch) in the hole is helpful in dampening wave or ripple action on the water surface due to the inflow of water and also permits more accurate water level measurements.

In an uncased test section in friable materials liable to wash, the end of the pipe should rest on a 4- to 6-inch cushion of coarse gravel at the bottom of the hole. In more stable material, the pipe may be suspended above the bottom of the hole, but the bottom of the pipe should be at least 2 feet below the top of the water surface maintained in the hole.

Water may also be introduced through the pipe and measurements of water level made in the annular space between the pipe and the casing.

(c) *Pumping Equipment and Controls.*—Pressure is not required in the test section which normally has a free water surface. However, pump capacity should be adequate.

A problem on many gravity tests is accurate control of the flow of water into the casing. The intake of the test section necessary to maintain a constant head is sometimes so small that it is impossible to limit the inflow sufficiently using a conventional arrangement. A source of error is that many meters register inaccurately at very low flows. To overcome this difficulty, a constant head tank has been developed as shown on figure 10–10.

This tank will deliver 0.05 to 25 gallons per minute of controlled flow. The materials used in its construction are available on most projects or are readily procured from plumbing supply houses, and the tank can be easily assembled by a welder. When the tank is completed, a rating curve must be prepared for it.

A hose delivering 20 to 25 gallons per minute of water is placed in the tank until overflow begins. The valve is opened to each gradation in turn and the gallons per minute discharge at each opening measured. Three measurements should be made at each gradation. The chart is prepared plotting the gallons per minute discharge against the gage opening. In the field, a close approximation of the discharge can be obtained by reference to the chart when the gage is opened between gradations. A sample chart, figure 10–11, was prepared for a tank assembled in the research laboratory in Denver.

Some precautions must be observed in using the tank. The hose transmitting water from the tank to the casing should never be attached directly to the valve outlet except as noted in the following paragraph. Water discharging from the valve should fall freely into a hose two or more times the diameter of the valve outlet or into an open tank from which a hose leads to the test hole. Such arrangements give an erratic flow in the hole for a few minutes, but it quickly stabilizes. The crest of the overflow trough must be level to ensure accurate discharge.

For some gravity tests in which the test sections take a very small amount of water, a plug valve is used in the waterline from the meter to the casing. A bypass line around the plug valve contains a ½-inch needle valve. The entire setup is connected directly to the outlet valve of the constant head tank. A 1-inch watermeter will not measure accurately the low flows used under such conditions, and after stabilization is obtained, actual flow is determined by the time required to fill a container of known volume.

(d) *Recommended Watermeters.*—The watermeter recommendations given in section 10–2(h) are equally satisfactory for use in gravity tests.

(e) *Length of Time for Tests.*—As in pressure tests, establishment of stabilized conditions is of primary importance if good results are to be obtained from gravity tests. Depending on the type of test per-

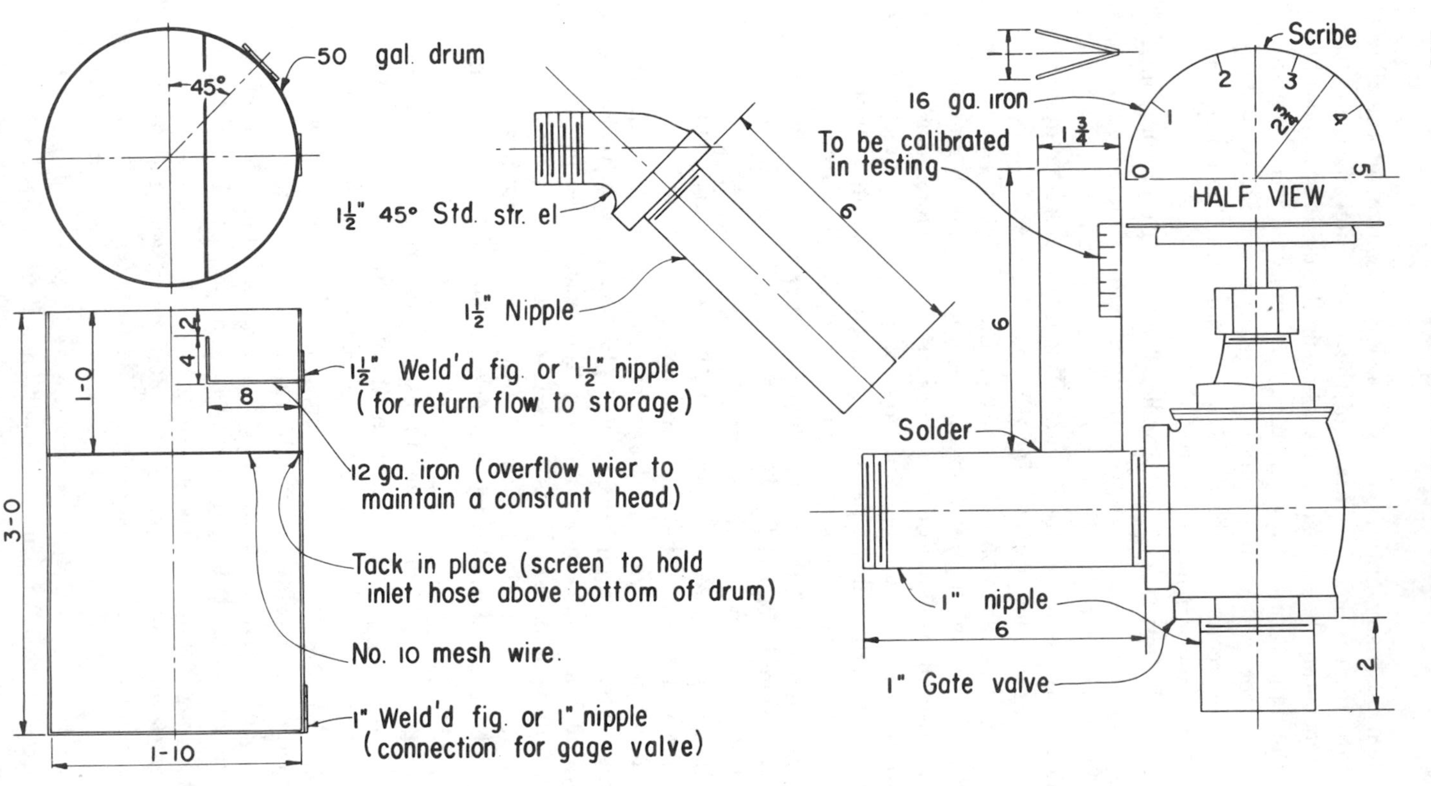

FIGURE 10-10.—Supply tank for constant head permeability tests. 103-D-1479.

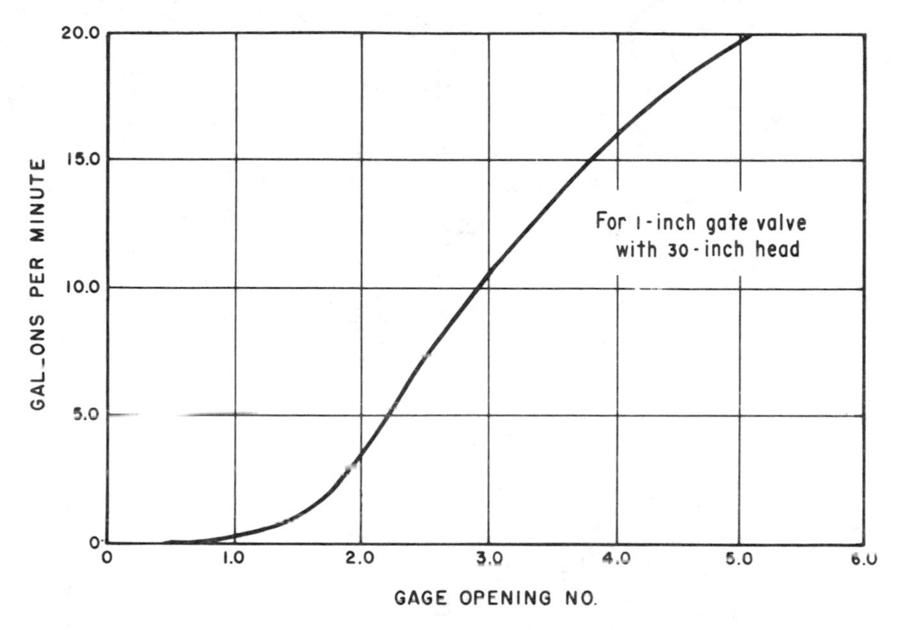

FIGURE 10-11.—Discharge curve for constant head supply tank. 103-D-1480.

formed, one of two methods is used. In one method, the inflow of water is controlled until a uniform inflow results in stabilization of water level at a predetermined level. In the other method, a uniform flow of water is introduced into the hole until the water level stabilizes.

(f) *Arrangement of Equipment.*—The recommendations given for equipment arrangement in section 10-2(k) are suitable for use in gravity tests. If a constant head tank is used, it should be placed so that the flow is directly into the casing.

(g) *Gravity Permeability Test—Method* 1.—For tests in unsaturated and unstable material using only one drill hole, method 1, figure 10-12, is the most accurate available. Because of mechanical difficulties, this test cannot be economically carried out to a depth greater than about 40 feet when it is necessary to use a gravel fill in the hole. In performing the test, care should be exercised (after the observation and intake pipes are set) to add gravel in small increments as the casing is pulled back; otherwise, the pipes may become sandlocked in the casing. For tests in unsaturated and unstable material at depths greater than about 40 feet, method 2 should be used.

Hole conditions

a. *Unconsolidated Material.*—A 6-inch or larger hole is drilled or augered to the depth at which the test is to be made and then developed gently. A cushion of coarse gravel is placed at the

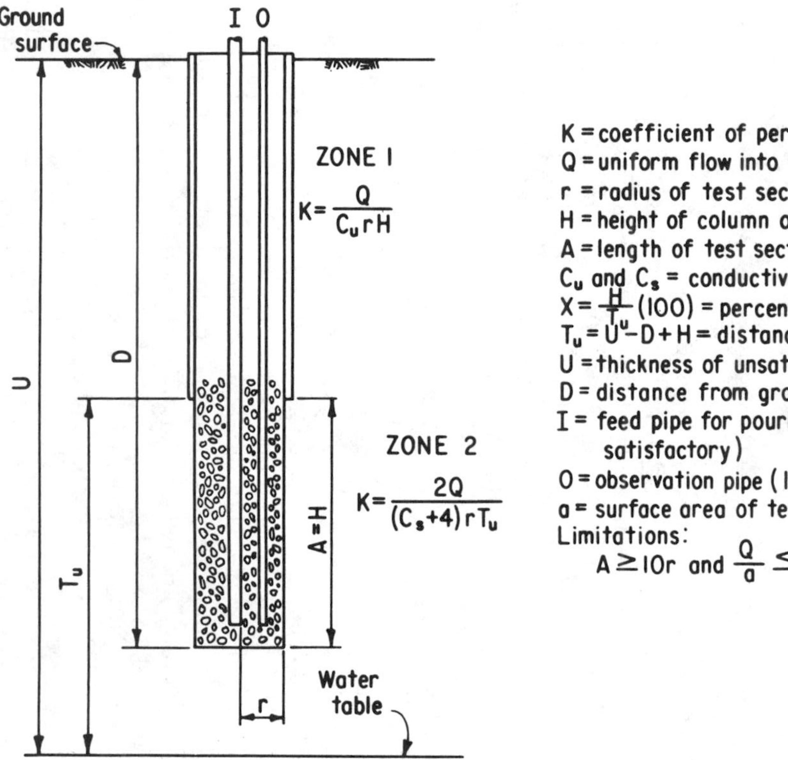

FIGURE 10–12.—Gravity permeability test—Method 1 [4]. 103-D-1481.

bottom of the hole and the feed pipe (I) and the observation pipe (O) are set in position (fig. 10–12). After the pipes are in position, the hole is filled to a depth at least five times the diameter of the hole with medium gravel. If the material will not stand without support, the hole must be cased to the bottom. After casing, the gravel cushion and pipes are put in and the casing is pulled back slowly as medium gravel is fed into the hole. The casing should be pulled back only enough to assure that the water surface to be maintained in the hole will be below the bottom of the casing. It is advisable to have a few inches of the gravel fill protrude into the casing.

A metered supply of water is poured into the feed pipe until three or more successive measurements, taken at 5-minute intervals, of the water level through the observation pipe are within plus or minus 0.2 foot. The water supply should be controlled so that the stabilized water level will not be within the casing, but is more than five times the hole diameter above the bottom of the hole. It generally is necessary to adjust the flow of water to obtain the desired conditions.

b. *Consolidated Materials.*—In consolidated material, or unconsolidated material which will stand without support even when saturated, the gravelfill and casing may be omitted. The use of the coarse gravel cushion is advisable. In all other respects, the test is carried out as in unconsolidated, unstable materials.

c. *General.*—Tests should be made at the successive depths selected so that the water level in each test is at or above the bottom of the hole in the preceding test.

The conductivity coefficients within the limits ordinarily employed in the field can be obtained from figures 10–7 and 10–8. The zone in which the test is made and applicable equations can be found on figures 10–6 and 10–12, respectively.

Data required for computing the permeability may not be available until the hole has penetrated the water table. The required data include:

1. Radius of hole r, in feet.
2. Depth of hole D, in feet.
3. Depth to bottom of casing, in feet.
4. Depth of water in hole H, in feet.
5. Depth to top of gravel in hole, in feet.
6. Length of test section A, in feet.
7. Depth to water table T_u, in feet.

8. Steady flow Q introduced into the hole to maintain a uniform water level, in ft^3/s.
9. Time test is started and time each measurement is made.

Some examples using method 1 are as follows:

GRAVITY PERMEABILITY TESTS EXAMPLES OF METHOD 1

Example 4:

Zone 1—*Method* 1

Given: $H=A=5$ ft, $r=0.5$ ft, $D=15$ ft, $U=50$ ft, and $Q=0.10$ ft^3/s.

$$T_u=U-D+H=50-15+5=40 \text{ ft and } \frac{T_u}{A}=\frac{40}{5}=8$$

$$X=\frac{H}{T_u}(100)=\frac{5}{40}(100)=12.5 \text{ percent}$$

The values for X and $\frac{T_u}{A}$ lie in zone 1 (fig. 10–6). To determine the conductivity coefficient C_u from figure 10–7:

$$\frac{H}{r}=\frac{5}{0.5}=10, \ \frac{A}{H}=\frac{5}{5}=1, \text{ and } C_u=32$$

From figure 10–12:

$$K=\frac{Q}{C_u rH}=\frac{0.10}{(32)\ (0.5)\ (5)}=0.00125 \text{ ft/s}$$

Example 5:

Zone 2—*Method* 1

Given: H, A, r, U, and Q are as given in example 1. $D=45$ ft

$$T_u=50-45+5=10 \text{ ft and } \frac{T_u}{A}=\frac{10}{5}=2$$

$$X=\frac{5}{10}(100)=50 \text{ percent}$$

Point $\frac{T_u}{A}$ and X lie in zone 2 on figure 10–6.

To determine the conductivity coefficient C_s from figure 10–8:

$$\frac{A}{r}=\frac{5}{0.5}=10 \text{ and } C_s=25.5$$

From figure 10–12:

$$K=\frac{2Q}{(C_s+4)rT_u}=\frac{(2)\ (0.10)}{(25.5+4)\ (0.5)\ (10)}=0.00136 \text{ ft/s}$$

(h) *Gravity Permeability Test—Method* 2.—This method may give erroneous results when used in unconsolidated material because of several uncontrollable factors. However, it is the best of the available pump-in tests for the conditions involved. In most instances, if performed with care, the results obtained by its use are adequate. When it is necessary to determine permeabilities in streambeds or lakebeds below water, method 2 is the only practical gravity test available.

A 5-foot length of 3- to 6-inch casing is perforated in a uniform pattern. The maximum number of perforations possible, without seriously affecting the strength of the casing, is desirable. The bottom of the perforated section of casing should be beveled on the inside and case hardened so that a cutting edge can be made.

The casing is sunk by drilling or jetting and driving—whichever method will give the tightest fit of the casing in the hole. In poorly compacted material and soils with a nonuniform grain size, development by filling the casing with water to a few feet above the perforations and gently surging and bailing is advisable before making the test. A 6-inch coarse gravel cushion is poured into the casing and the observation pipe set on the cushion.

A uniform flow of water sufficient to maintain the water level in the casing above the top of the perforations is poured into the well. Depth of water measurements are made at 5-minute intervals until three or more measurements are within plus or minus 0.2 foot. The water should be poured through a pipe and measurements made between the pipe and casing. This procedure may be reversed if necessary.

When a test is completed, the casing is sunk an additional 5 feet and the test repeated.

The test may be run in consolidated material using an open hole for the test section. This practice is not recommended because the bottom of the casing under such conditions is seldom tightly fitted in the hole and considerable error may result from seepage upward through the annular space between the casing and the wall of the hole.

Measurements should be made to the nearest 0.01 foot. The values of C_u and C_s within the limits ordinarily employed in the field can be obtained from figures 10–7 and 10–8. The zone in which test is made and applicable equations can be found on figures 10–6 and 10–13, respectively.

In making tests in a hole, some data required for computing the permeability are not available until the hole has encountered the water table. The recorded data are supplemented by this information as it is determined. The data recorded in each test are as follows:

1. Outside radius of casing, in feet.
2. Length of perforated section of casing A, in feet.

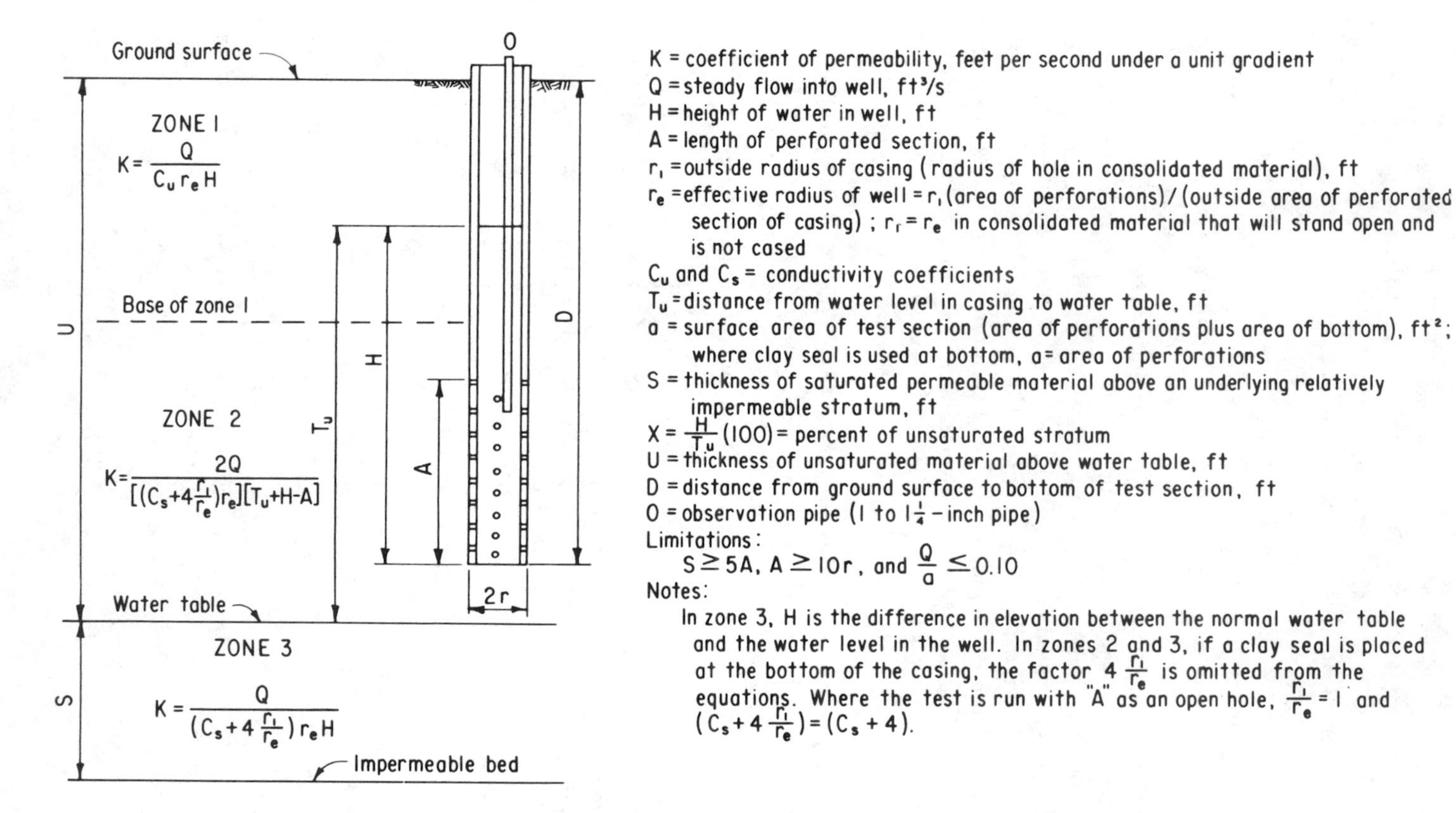

FIGURE 10–13.—Gravity permeability test—Method 2 [4]. 103-D-1482.

3. Number and diameter of perforations in length A.
4. Depth to bottom of hole D, in feet.
5. Depth to water surface in hole, in feet.
6. Depth of water in hole H, in feet.
7. Depth to water table U, in feet.
8. Thickness of saturated permeable material above underlying relatively impermeable bed S, in feet.
9. Steady flow into well to maintain a constant water level in hole Q, in ft³/s.
10. Time test is started and time each measurement is made.

Some examples using method 2 are as follows:

GRAVITY PERMEABILITY TESTS EXAMPLES OF METHOD 2

Example 6:

Zone 1—Method 2

Given: $H=10$ ft, $A=5$ ft, $r_1=0.25$ ft, $D=20$ ft, $U=50$ ft, $Q=0.10$ ft³/s, and 128 perforations of ½-inch-diameter holes.

Area of perforations$=\pi(0.25)^2(128)=25.13$ in$^2=0.174$ ft^2

Area of perforated section$=2\pi r_1 A=2\pi(0.25)(5)=7.854$ ft^2

$$r_e=\frac{0.174}{7.854}(0.25)=0.00554 \text{ ft}$$

$$T_u=U-D+H=50-20+10=40 \text{ ft and } \frac{T_u}{A}=\frac{40}{5}=8$$

$$X=\frac{H}{T_u}(100)=\frac{10}{40}(100)=25 \text{ percent}$$

Point $\frac{T_u}{A}$ and X lie in zone 1 on figure 10–6.

Find C_u from figure 10–7:

$$\frac{H}{r_e}=10/0.00554=1{,}805 \text{ and } \frac{A}{H}=\frac{5}{10}=0.5$$

Then, $C_u=1{,}200$

From figure 10–13:

$$K=\frac{Q}{C_u r_e H}=\frac{0.10}{(1{,}200)(0.00554)(10)}=0.0015 \text{ ft/s}$$

Example 7:

Zone 2—*Method* 2

Given: Q, H, A, r_1, r_e, U, A/H, and H/r_e same as example 6.
D=40 ft

$$T_u=50-40+10=20 \text{ ft and } \frac{T_u}{A}=\frac{20}{5}=4$$

$$X=\frac{10}{20}(100)=50 \text{ percent}$$

Point $\frac{T_u}{A}$ and X lie in zone 2 on figure 10–6.

Find C_s from figure 10–8:

$$\frac{A}{r_e}=\frac{5}{0.00554}=902 \text{ and } C_s=800$$

From figure 10–13:

$$K=\frac{2Q}{\left[\left(C_s+4\frac{r_1}{r_e}\right)r_e\right](T_u+H-A)}=\frac{0.20}{(5.43)(20+10-5)}=0.0015 \text{ ft/s}$$

Example 8:

Zone 3—*Method* 2

Given: Q, H, A, r_1, r_e, A/H, H/r_e, U, C_s, and A/r_e are as given in example 7. In addition, S=60 ft.

From figure 10–13:

$$K=\frac{Q}{\left(C_s+4\frac{r_1}{r_e}\right)r_eH}=\frac{0.10}{(980.5)(0.00554)(10)}=0.0018 \text{ ft/s}$$

(i) *Gravity Permeability Test—Method* 3.—This method is a combination of Gravity Permeability Test—Methods 1 and 2 that was developed to permit testing under difficult conditions. It is the least accurate method of tests, but is the only one available for use under circumstances where method 2 cannot be used because of the nature of the material, see figure 10–14.

In some areas, the material to be tested will be of such character that a casing that is beveled and casehardened at the bottom will not stand up under the driving necessary to sink it. This is particularly true in gravelly materials where the particle size is greater than about 1 inch. Under such conditions, method 3 would probably not be satisfactory because a drive shoe must be used with this method. Use of a drive shoe causes excessive compaction of the materials and forms an annular space about the casing, introducing an opportunity for error in the results. The size of this error is unknown, but if reasonable care is taken in performing the test, the results will probably approximate the correct magnitude for the material tested.

On completion of each test, a 3- to 6-inch perforated casing is advanced a distance of 5 feet or more by drilling and driving. After each new test section is developed by surging and bailing, a 6-inch gravel cushion should be placed on the bottom to support the observation pipe. A uniform flow of water sufficient to maintain the water level in the casing just at the top of the perforations is then poured into the well. The water is poured directly into the casing and measurements are made through the 1¼-inch observation pipe. The test should be run until three or more measurements taken at 5-minute intervals show the water level in the casing to be within plus or minus 0.2 foot of the top of the perforations.

The values of C_u and C_s, within the limits ordinarily employed in the field, can be obtained from figures 10–7 and 10–8. The zone in which the test is made and applicable equations can be found on figures 10–6 and 10–14, respectively.

The data recorded in each test are as follows:

1. Outside radius of casing r_1, in feet.
2. Length of perforated section of casing A, in feet.
3. Number and diameter of perforations in length, A.
4. Depth to bottom of hole D, in feet.
5. Depth to water surface in hole, in feet.
6. Depth of water in hole, in feet.
7. Depth to water table, in feet.
8. Steady flow into well to maintain a constant water level in hole Q, in ft^3/s.
9. Time test is started and time each measurement is made.

GRAVITY PERMEABILITY TESTS
EXAMPLES OF METHOD 3

Example 9:

Zone 1—*Method* 3

Given: $Q=10.1$ gal/min$=0.023$ ft^3/s, $H=A=5$ ft, $D=22$ ft, $U=71$ ft, $T_u=54.5$ ft, $r_e=0.008$ ft, and $r_1=1.75$ in$=0.146$ ft (nominal 3-inch casing).

$$\frac{T_u}{A}=\frac{54.5}{5}=10.9 \text{ and } X=\frac{H}{T_u}(100)=\frac{5}{54.5}(100)=9.2 \text{ percent.}$$

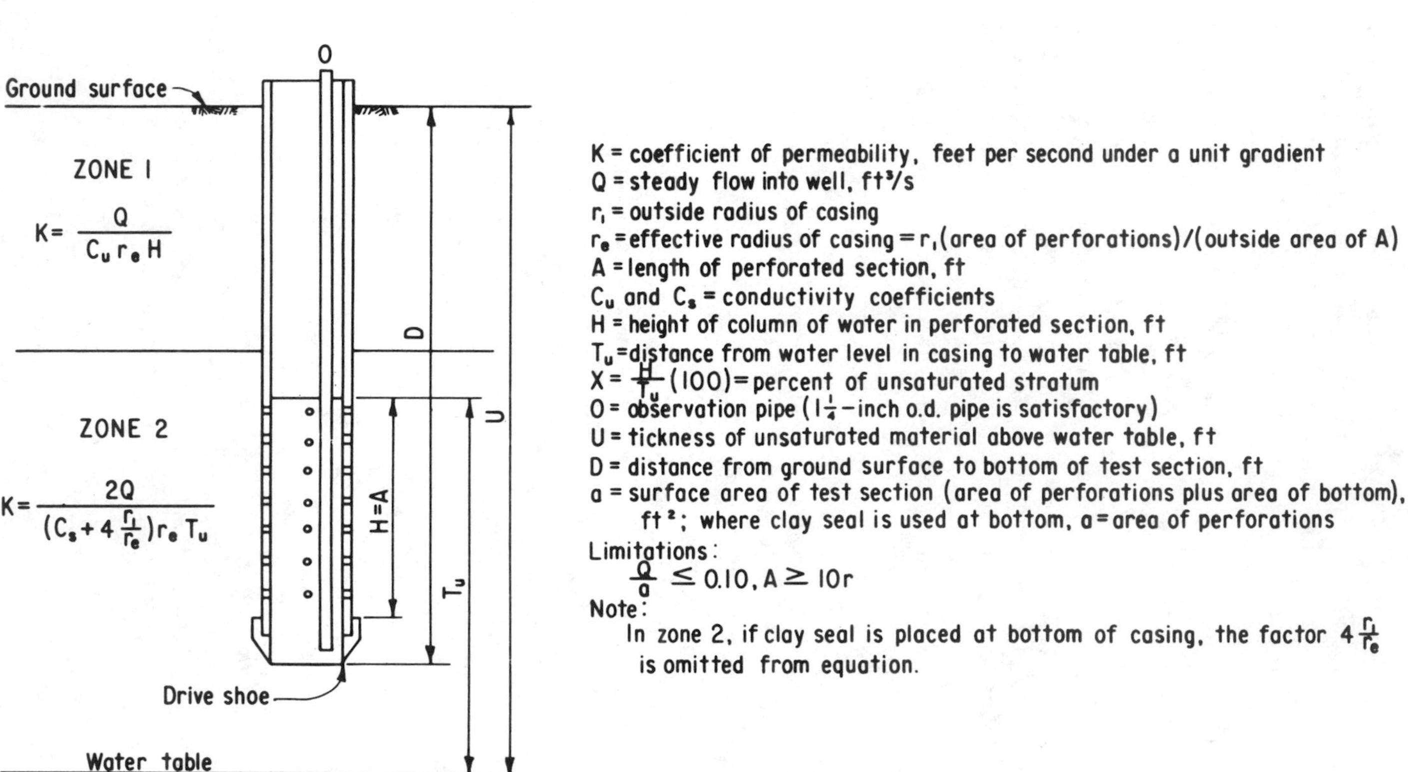

FIGURE 10–14.—Gravity permeability test—Method 3 [4]. 103-D-1483.

These points lie in zone 1 (fig. 10–6).
Find C_u from figure 10–7:

$$\frac{H}{r_e}=5/0.008=625 \text{ and } \frac{A}{H}=1$$

Then, $C_u=640$

From figure 10–14:

$$K=\frac{Q}{C_u r_e H}=\frac{0.023}{(640)(0.008)(5)}=0.0009 \text{ ft/s}$$

Example 10:

Zone 2—*Method* 3

Given: Q, H, A, U, r_e, and r_1 are as given in example 9.
$D=66$ ft and $T_u=10$ ft.

$$\frac{T_u}{A}=\frac{10}{5}=2 \text{ and } X=\frac{5}{10}\,(100)=50 \text{ percent.}$$

These points lie in zone 2 (fig. 10–6).
Find C_s from figure 10–8:

$$\frac{A}{r_e}=\frac{5}{0.008}=625 \text{ and } C_s=595$$

From figure 10–14:

$$K=\frac{2Q}{\left(C_s+4\,\frac{r_1}{r_e}\right)r_e T_u}=\frac{(2)(0.023)}{(668)(0.008)(10)}=0.00086 \text{ ft/s}$$

(j) *Gravity Permeability Test—Method* 4.—Under some circumstances, method 4 can be used to advantage in determining the overall average permeability of unsaturated materials above a widespread impermeable layer. However, it does not permit determination of relative permeability variations with depth. The method is actually an application of the steady state pumping test theory discussed in section 5–2.

An intake well (preferably 6 inches or larger) is drilled to a relatively impermeable layer of wide areal extent or to the water table. The well is uncased in consolidated material, but in unconsolidated material a perforated casing or screen should be set from the bottom to about 5 feet below the ground surface. The well should be developed by pouring water into it while surging and bailing prior to the testing for intake capacity.

Before the observation wells are drilled, a test run of the intake well should be made to determine the maximum height H of the column of water in the intake pipe above the top of the impermeable stratum that is possible to maintain with available pumping equipment (see fig. 10–15). The spacing of the observation wells can be determined from this test run. A 1- to 1¼-inch observation pipe should be inserted near the bottom of the intake well to facilitate water level measurements.

A minimum of three observation wells should be installed, by jetting or some other method, to the top of the impermeable stratum. Suitable pipe, perforated for the bottom 10 to 15 feet, should be set to the bottom of these wells. The observation wells should be set from the intake well at distances equal to multiples of one-half the

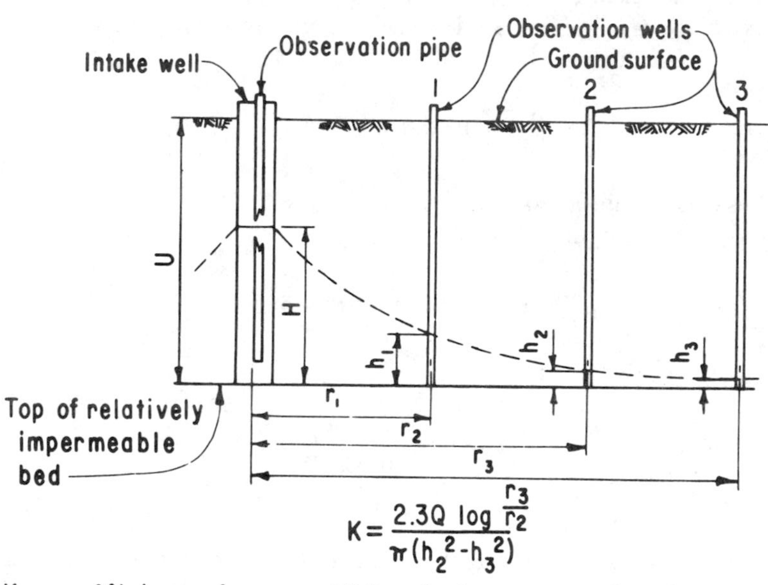

K = coefficient of permeability, feet per second under a unit gradient

Q = uniform flow into intake well, ft³/s

r_1, r_2, and r_3 = distance from intake well to observation holes, ft

h_1, h_2, and h_3 = height of water in observation holes r_1, r_2, and r_3 respectively, above elevation of top of impermeable layer, ft

H = height of column of water in intake pipe above top of impermeable stratum, ft

U = distance from ground surface to impermeable bed, ft

FIGURE 10–15.—Gravity permeability test—Method 4 [4]. 103-D-1484.

height H of the water column which it is possible to maintain in the intake well.

The elevations of the top of the impermeable strata or water table in each well are determined and the test started. After water has been poured into the intake at a constant rate for an hour, measurements are made of water levels in the observation wells. Measurements are made at 15-minute intervals thereafter, and each set of measurements plotted on semilog paper with the square of the height of the water level above the top of the impermeable bed h^2 against the distance from the intake well to the observation holes r for each hole (see fig. 10–16). When the plot of a set of measurements permits a straight line to be drawn through the points within the limits of plotting, stable conditions prevail and the permeability may be computed.

The data recorded in each test are as follows:

1. Ground elevations at sites of intake well and observation wells.
2. Elevations of reference points at intake well and observation wells, in feet.
3. Distances from center of observation wells to center of intake well, r_1, r_2, and r_3, in feet.
4. Elevation of top of impermeable bed at intake well and observation wells, in feet.
5. Depths of water below reference point in intake well and observation wells at 15-minute intervals, in feet.
6. Uniform flow of water Q, introduced into well, in ft^3/s.
7. Time pumping is started and time each measurement is made.

GRAVITY PERMEABILITY TESTS
EXAMPLE OF METHOD 4

Example 11:

Method 4

Given: U=50 ft, Q=1 ft^3/s, H=30 ft, r_1=15 ft, r_2=30 ft, r_3=60 ft, h_1=23.24 ft, h_2=17.89 ft, and h_3=10.0 ft.

$$h_1^2=540 \text{ ft}^2, h_2^2=320 \text{ ft}^2, \text{ and } h_3^2=100 \text{ ft}^2.$$

A plot of r against h^2, as shown on figure 10–16, shows that a straight line can be drawn through the plotted points which means that stable conditions exist and the permeability may be computed. From figure 10–15:

$$K=\frac{2.3Q \log \frac{r_3}{r_2}}{\pi(h_2^2-h_3^2)}$$

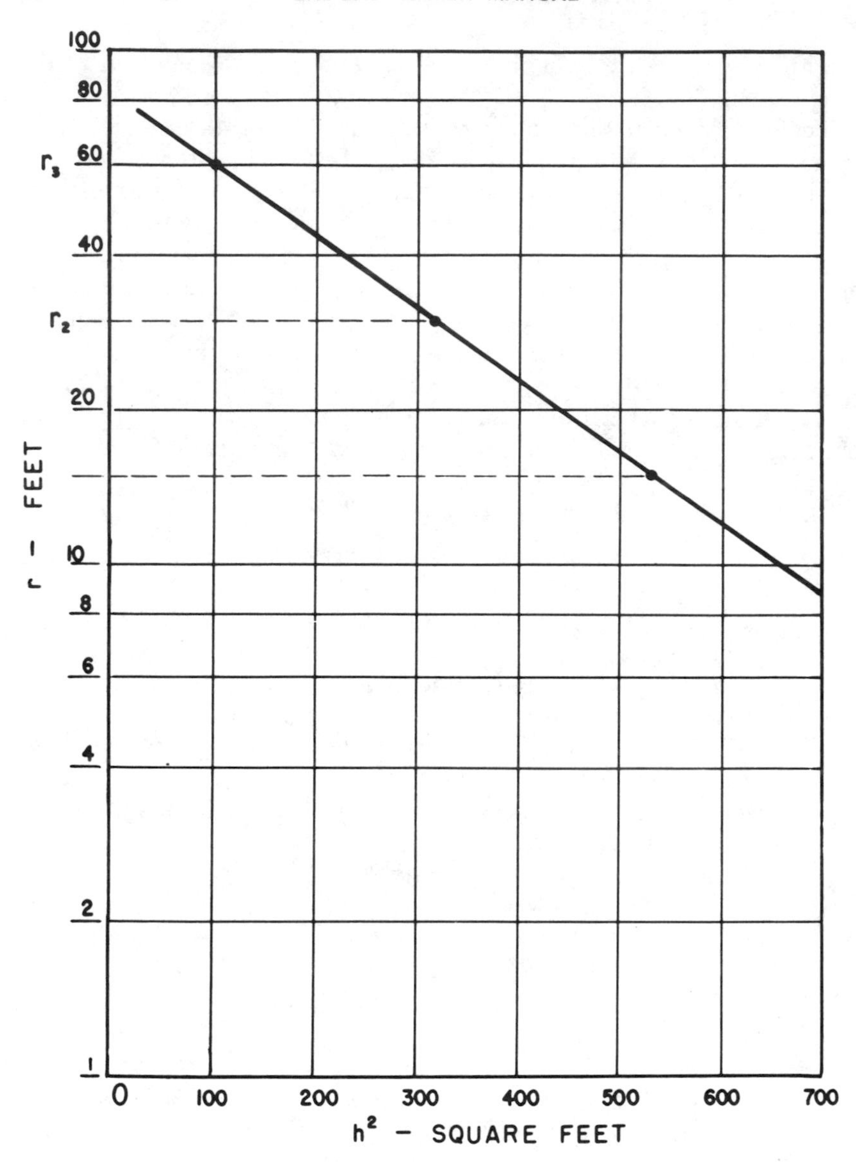

FIGURE 10-16.—Plot of h^2 versus r for gravity permeability test—Method 4. 103-D-1485.

$$\log \frac{r_3}{r_2}=0.3010, \log \frac{r_3}{r_1}=0.6021, \text{ and } \log \frac{r_2}{r_1}=0.3010$$

$$K=\frac{2.3Q \log \frac{r_3}{r_2}}{\pi(h_2^2-h_3^2)}=\frac{2.3Q \log \frac{r_3}{r_1}}{\pi(h_1^2-h_3^2)}=\frac{2.3Q \log \frac{r_2}{r_1}}{\pi(h_1^2-h_2^2)}$$

$$K=\frac{(2.3)(1)(0.3010)}{\pi(220)}=\frac{(2.3)(1)(0.6021)}{\pi(440)}=\frac{(2.3)(1)(0.3010)}{\pi(220)}=0.001 \text{ ft/s}$$

10–4. Falling Head Tests.—Falling head tests are used primarily in open holes in consolidated rock. They are performed using inflatable packers identical to those described under pressure testing (sec. 10–2) and can be used as an alternate method under some circumstances when the pressure transducer or other instrumentation fails. The method of cleaning the hole is the same as that described under pressure testing.

(a) *Tests Below the Static Water Level in the Hole.*—

(1) Use inflatable straddle packers with 10-foot isolated intervals on a 1¼-inch standard size drop pipe (i.d.=1.38 inch). Set initially at the bottom of the hole and inflate packers to 300 pounds per square inch differential pressure.

(2) After packers are inflated, measure water level in drop pipe three or more times at 5-minute intervals until water level stabilizes. Stabilized level will be at the static level of water in the test section.

(3) After stabilization of level, pour in 2 or more gallons of water as rapidly as possible into the drop pipe. One gallon of water will raise the water level in a 1¼-inch pipe 12.9 feet if the section is tight.

(4) Measure water level as soon as possible after water is poured in. Measure initial depth to water and time measurement as soon as possible and at two 5-minute intervals thereafter. If rate of decline exceeds 15 feet in 13 minutes, the transmissivity of a 10-foot test section is greater than 200 square feet per year or an average permeability greater than 20 feet per year.

(5) The value of transmissivity so determined is only an approximation but is sufficiently accurate for many engineering purposes.

The test is based on an adaptation of the slug method by Ferris and Knowles [2]. The equation for analysis is:

$$T=\frac{V}{2\pi s\Delta t} \qquad (3)$$

where:

- T=transmissivity of test section, ft²/s,
- V=volume of water entering test section in period Δt, ft³ (1-foot decline in 1¼-inch pipe=0.01 ft³),
- s=decline in water level in period Δt, ft, and
- Δt=period of time, seconds, between successive water level measurements; that is, t_1-t_0, t_2-t_1, etc.

(6) If the log indicates the test section is uniform without obvious points of probable concentrated leakage, the average permeability of the test section in feet per second can be estimated from $K=T/A$ where A is the length of the test section in feet. If the log indicates a predominantly impervious test section but with a zone or zones of probable concentrated flow, the average K of the zones can be estimated from $K=T/A'$ where A' is the thickness of the permeable zone or zones in feet.

(7) After each test, deflate packers, raise test string 10 feet, and repeat until entire hole below the static water level has been tested.

(b) *Tests in Unsaturated Materials Above the Water Table.*—Tests above the water table require somewhat different procedures and analyses than tests in the saturated zone. Tests made in sections straddling the water table or slightly above it will give computed values somewhat too high if the equations in section 10–4(a) are used and somewhat low if the following equations (4) and (5) are used. For tests above the water table, the following procedure is used:

(1) Install a 10-foot straddle packer at bottom of hole if hole is dry or with the top of bottom packer at the water table if it contains water. Inflate the packer.

(2) Fill drop pipe with water to surface if possible, otherwise to level permitted by capacity of the pump.

(3) Measure water level in drop pipe and record with time of measurement. Make two or more similar measurements while water table declines.

(4) On completion of a test, raise packer 10 feet and repeat procedure until all uncased or uncemented hole is tested.

(5) The equation for analysis of each test section is an adaptation of one derived by Jarvis [3]:

$$K=\frac{r_1^2}{2A\Delta t}\left[\frac{\sinh^{-1}\frac{A}{r_e}}{2}\log_e\left(\frac{2H_1-A}{2H_2-A}\right)-\log_e\left(\frac{2H_1H_2-AH_2}{2H_1H_2-AH_1}\right)\right] \tag{4}$$

where:

K=average permeability of the test section, ft/s,
A=length of test section, ft,
r_1=inside radius of drop pipe, ft (0.0575 ft for 1¼-inch pipe),
r_e=effective radius of test section, ft (0.125 ft for 3-inch hole),
Δt=time intervals (t_1—t_0, t_2—t_1), seconds,
$\sinh^{-1}$=inverse hyperbolic sine,
$\log_e$=natural logarithm, and
H=length of water column from bottom of test interval to water surface in standpipe, ft (H_0, H_1, H_2 lengths at time of measurements t_0, t_1, t_2, etc.)

(6) For the particular equipment specified, and a 10-foot test section, equation (4) may be simplified as follows:

$$K=\frac{1.653\times10^{-4}}{\Delta t}\left[2.5\log_e\left(\frac{H_1-5}{H_2-5}\right)-\log_e\left(\frac{H_1H_2-5H_2}{H_1H_2-5H_1}\right)\right] \quad (5)$$

10-5. Bibliography.—

[1] Ahrens, T. P., and Barlow, A. C., "Permeability Tests Using Drill Holes and Wells, including Comments Regarding Equipment, etc.," Bureau of Reclamation Geology Report No. G-97, January 5, 1951.
[2] Ferris, J. G., Knowles, D. B., Brown, R. H., and Stallman, R. W., "Theory of Aquifer Tests," U.S. Geological Survey Water-Supply Paper 1536-E, pp. 104-105, 1962.
[3] Jarvis, D. H., "Theory of Falling Head Permeameter in Unsaturated Material," Memorandum to C. W. Jones, Bureau of Reclamation, May 5, 1949.
[4] Zanger, C. N., "Theory and Problems of Water Percolation," Engineering Monograph No. 8, Bureau of Reclamation, April 1953.

«Chapter XI

COMPONENTS OF A WELL AND PARTICULARS OF DESIGN

11-1. General.—Chapter III discussed relationships and conditions in ideal isotropic, homogeneous aquifers of uniform thickness and infinite areal extent and the response of such aquifers to the discharge of ideal fully penetrating wells. However, ideal aquifer conditions are never encountered and wells must be designed with practical aspects in mind to economically utilize an aquifer. The more important of these practical aspects are discussed in this chapter.

A generally accepted, all-inclusive standard for water well design is not available. The American Water Works Association Publication, AWWA-A100-66, Standards for Deep Wells [3][1], describes a number of commonly used design standards, but it is impossible to describe all the different standards or even a majority of those in common use. Water well design involves the consideration of many factors including: desired yield and pump size; casing diameters, wall thicknesses, and lengths; screen diameters; slot sizes; percentages of open area and length; the type and characteristics of the aquifer or aquifers to be developed; sanitation; control of corrosion and incrustation; local well drilling designs and practices; and State and local statutes and regulations. These factors might be assumed to preclude standardization; however, many of them are basic to all wells and these common factors make possible a degree of design standardization. Figures 11-1 through 11-8 illustrate this degree of standardization for drilled wells involving different types of construction.

The basic information desired before undertaking the design of a well are:

- Thickness, character, and sequence of materials above the water table or at the top of a confined aquifer.
- Thickness, character, and sequence of the aquifer(s), nature of the permeability (interstitial or resulting from secondary voids), and the degree of confinement of the aquifer.
- Size and gradation of aquifer materials.
- Transmissivity and storativity of the aquifer.
- Water level conditions and trends.
- Quality of water.

---

[1] Numbers in brackets refer to items in the bibliography, section 11-14.

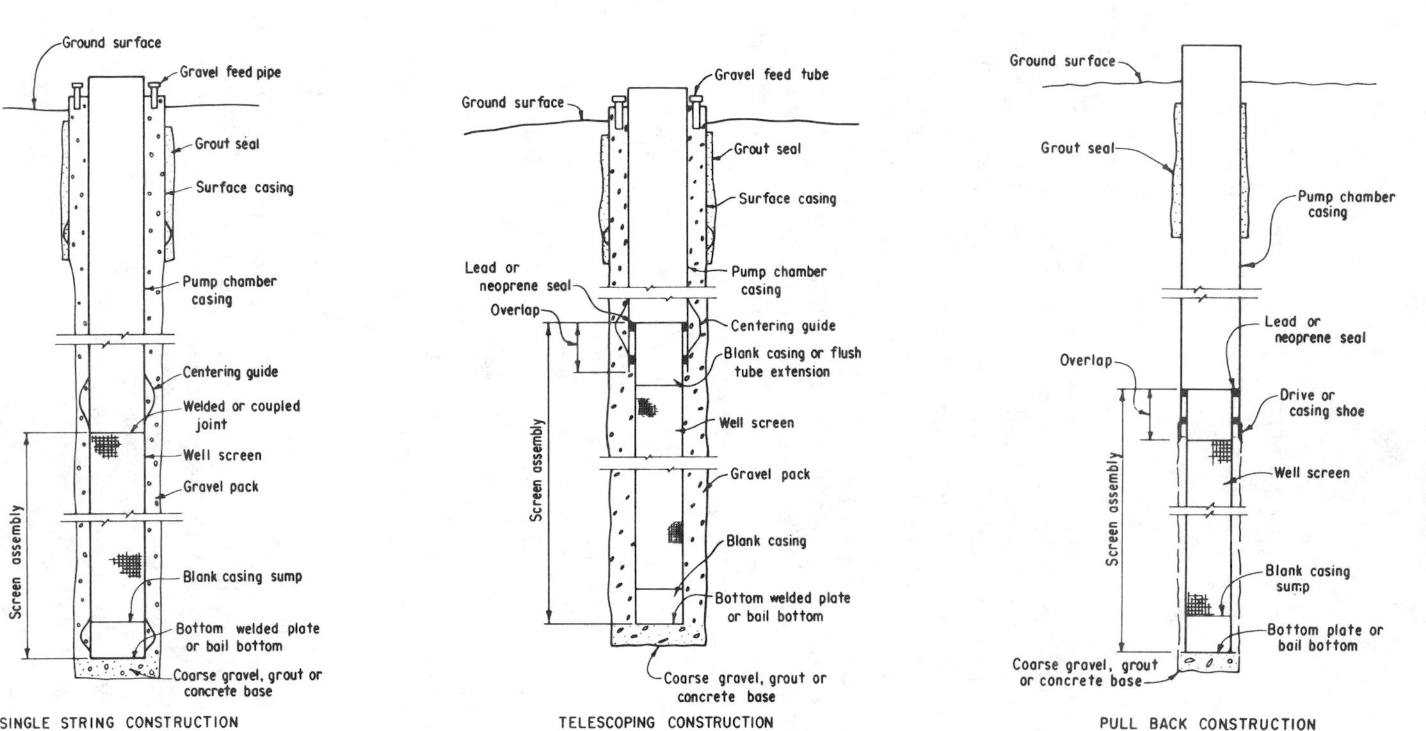

FIGURE 11-1.—Gravel packed, rotary drilled well for single string construction. 103-D-1488.

FIGURE 11-2.—Gravel packed, rotary drilled well for telescoping construction. 103-D-1489.

FIGURE 11-3.—Straight wall, cable tool drilled well for pull back construction. 103-D-1490.

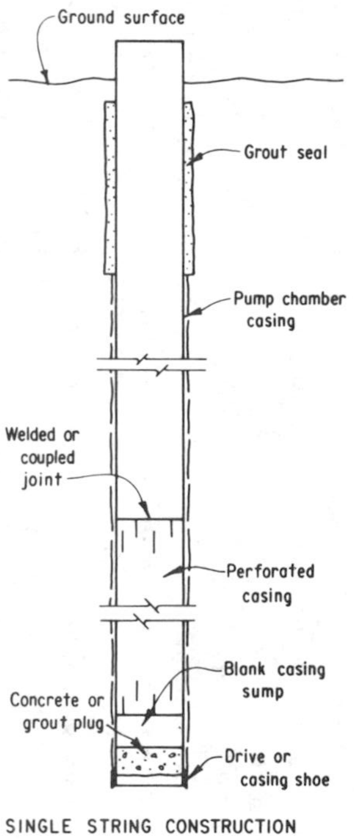

Ground surface
Grout seal
Pump chamber casing
Welded or coupled joint
Perforated casing
Blank casing sump
Concrete or grout plug
Drive or casing shoe
SINGLE STRING CONSTRUCTION

FIGURE 11–4.—Straight wall, cable tool drilled well for single string construction. 103–D–1491.

Ground surface
Grout seal
Pump chamber casing
Lead seal
Overlap
Drive or casing shoe
Well screen
Screen Assembly
Blank casing sump
Coarse gravel, grout or concrete base
Bottom plate or bail bottom
MULTIPLE TELESCOPING CONSTRUCTION

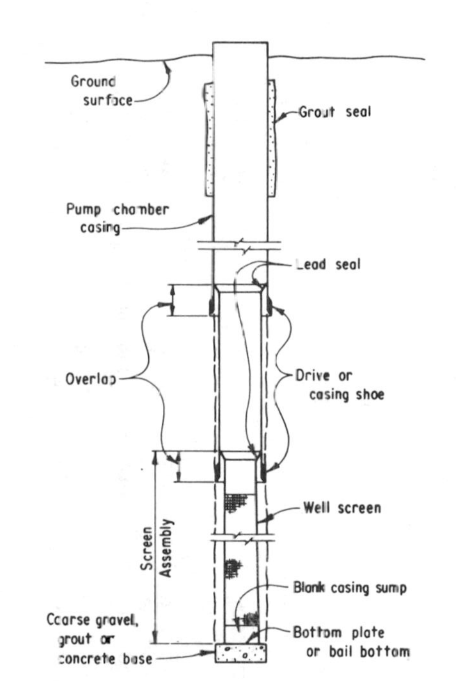

FIGURE 11–5.—Straight wall, cable tool drilled well for multiple telescoping construction. 103–D–1492.

Ground surface
Grout seal
Pump chamber casing
Drive or casing shoe set in grout
Open hole
WELL IN CONSOLIDATED MATERIAL (ROCK)

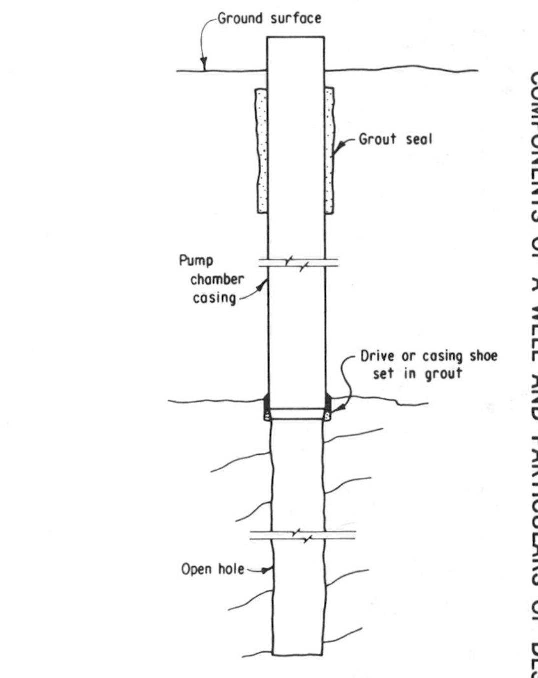

FIGURE 11–6.—Straight wall, cable tool drilled well in consolidated material (rock). 103–D–1493.

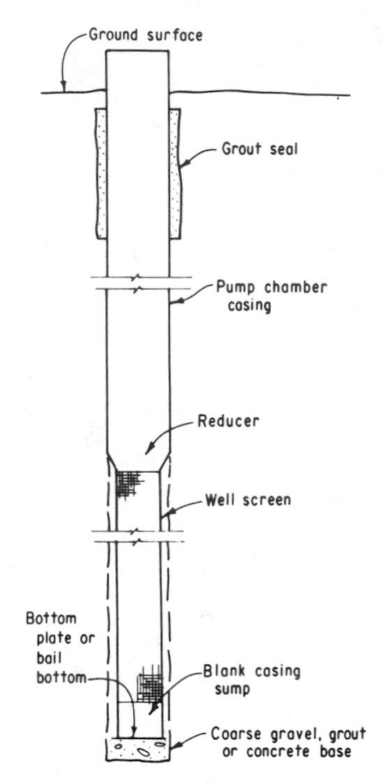

FIGURE 11-7.—Straight wall, rotary drilled well for single string construction using a reducer. 103-D-1494.

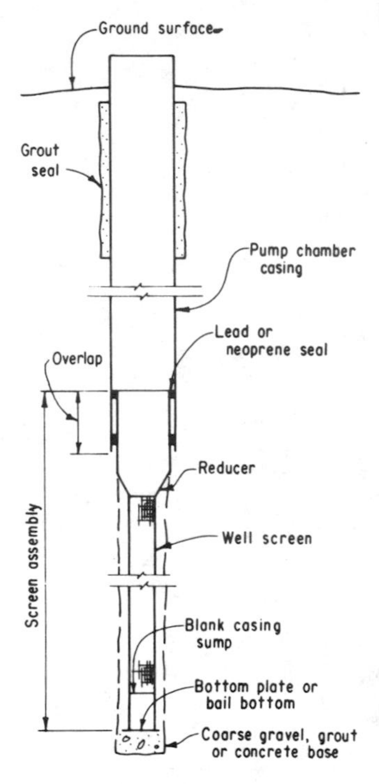

FIGURE 11-8.—Straight wall, rotary drilled well for telescoping construction using a reducer. 103-D-1495.

- Design and construction features of wells previously constructed in the area.
- Operation and maintenance history of previously constructed wells.
- Purpose and desired yield of the proposed well.

Unfortunately, all the information desired is seldom available, even in a developed area. In an undeveloped area, all that may be known initially is the approximate location of the proposed well and the desired yield.

For major wells (yields greater than 100 to 125 gallons per minute), all available information on existing wells should be obtained and, so far as possible, aquifer transmissivity characteristics as described in chapters V and IX should be determined. These data should be supplemented by a geological investigation and the drilling of a pilot hole. A properly drilled and sampled pilot hole will furnish an accurate lithologic log, aquifer samples for mechanical analyses, information on static water levels, water samples for analyses, and the type of aquifer present. The site, if found to be inadequate for any reason, can be abandoned without the major cost of drilling a producing well. If the site is found to be satisfactory, a design can be prepared and specifications written. Pilot hole data permit the preparation of firm design and specifications features which minimize the unknown factors and the risks during construction. Consequently, the contractor can foresee many problems, assemble necessary equipment and materials, and more accurately schedule his program. Contract savings resulting from data gained in drilling a pilot hole may balance the cost of the pilot hole.

For minor wells, the well drilling costs may be about the same as for a pilot hole; thus, the latter may not be economical. The recommended procedure for the design of such wells is to make a preliminary design based on the desired yield, readily available information, and evaluation of existing geological knowledge of the area. Additions or changes in the preliminary design such as in pump chamber depth and screen slot size and setting can be made on the basis of information obtained during the drilling.

The designations used for various components of a well are not standardized. Various terms are used for similar components in different publications and in different parts of the country. In the following sections, the terminology favored by the Bureau is used, but other terms commonly encountered are also noted.

11-2. Surface Casing (soil casing, conductor casing, outer casing).—

(a) *Description and Purpose.*—Surface casing may or may not be used, depending upon local conditions and established drilling practices. The casing may be temporary and removed when completing the well, or it may be a permanent part of the structure. Surface casing is installed, where possible, from near the ground surface through unstable, unconsolidated, or fractured materials a short distance into a firm, stable, or massive and, if possible, relatively impermeable material. The purpose of surface casing is to: (1) simplify and facilitate drilling a well by supporting unstable materials so they will not cave and fall into the hole; (2) minimize washing and erosion of the side of the hole by drilling fluids and tools; (3) reduce loss of drilling fluids; (4) facilitate installing or pulling back other casing; (5) facilitate placing a sanitary seal; and (6) serve as a reservoir for gravel pack and under some circumstances to provide a degree of protection against upward caving around the well [1,3,4,7,9,10,12].

Table 11-1 gives the recommended minimum diameters of surface casing for various pump chamber casing diameters. Temporary surface casing normally is pulled from the hole on completion of the well.

TABLE 11–1.—*Minimum pump chamber and permanent surface casing diameters*

Well yield, gal/min	Nominal pump chamber casing diameter, inches [1]	Surface casing diameter, inches	
		Naturally developed wells [2]	Gravel packed wells
Up to 90	6*	8–10	18
50 to 150	8**	10–12	20
100 to 500	10**	12–14	22
300 to 1,500	12**	16–18	24
500 to 2,000	16**	16–18	26
1,500 to 3,000	16**	18–20	28
2,000 to 5,000	20**	20–22	30
3,000 to 5,000	24***	24–26	34
4,000 to 8,000	28***	26–28	36

[1] Based on use of deep well turbine pumps with the following nominal rotation rates:
*3,600 revolutions per minute (r/min)
**1,800 r/min
***1,200 r/min

[2] Larger values should be used if nominal ¾- to 1-inch pipe will be inserted into the annular space for grouting the pump chamber casing as the temporary surface casing is withdrawn.

Concurrent with pulling the temporary casing, the annular space about the permanent casing is normally filled with grout as described in section 15–3 in chapter XV.

When permanent surface casing is installed, the first operation in construction of the well is to drill an oversized hole and to install, center, and grout the surface casing. Table 11–2 shows the minimum hole diameter for different casing sizes. After the surface casing has been installed and the grout has set, the well is drilled deeper through the bottom of the surface casing. The hole drilled is usually about 2 inches smaller in diameter than the outside diameter of the surface casing.

In gravel-packed wells of the depth and diameter of most Bureau of Reclamation wells, the gravel pack is usually placed through nominal 2- to 4-inch coupled tremie pipes. To permit insertion of these pipes, the annular space between the wall of the hole and pump chamber casing must be adequate to permit passage of the pipe couplings.

Schedule 10 or 20 weight pipe is usually adequate for permanent surface casing to depths shown in table 11–3. For setting depths greater than those shown in table 11–3, heavier casing should be used.

In areas where unstable material extends to depths beyond normal surface casing placement, the casing may have to be supported at the ground surface. I–beams welded to the casing may provide adequate support as shown on figure 11–9.

TABLE 11-2.—*Minimum hole diameter required for adequate grout thickness around different sizes of casing*

Nominal pump chamber casing diameter, inches	[1] Hole diameter for casing with couplings, inches	Hole diameter for casing with welded joints, inches
6	10⅜	9⅝
8	12⅝	11⅝
10	------------------------	13¾
12	------------------------	15¾
14	------------------------	17
16	------------------------	19
18	------------------------	21
20	------------------------	23
22	------------------------	25
24	------------------------	27
26	------------------------	29
28	------------------------	31
30	------------------------	33
32	------------------------	35
34	------------------------	37
36	------------------------	39

[1] Coupled pipes are not normally available or used.

TABLE 11-3.—*Permanent surface casing—wall thickness, diameter, and maximum depth of setting*

Nominal diameter, inches	ASA Schedule No. or class	Wall thickness, inches	Weight per foot—plain end, pounds	Maximum depth of setting, feet
8	20	0. 250	23. 36	420
10	20	. 250	28. 04	235
12	20	. 250	33. 38	140
14	10	. 250	36. 71	105
16	10	. 250	42. 05	70
18	10	. 250	47. 39	50
20	20	. 375	78. 60	120
22	20	. 375	86. 61	95
24	20	. 375	94. 62	70
26	Std.	. 375	102. 63	55
28	20	. 500	146. 85	105
30	20	. 500	157. 53	85
32	20	. 500	168. 21	70
34	20	. 500	178. 89	60
36	20	. 500	189. 57	50

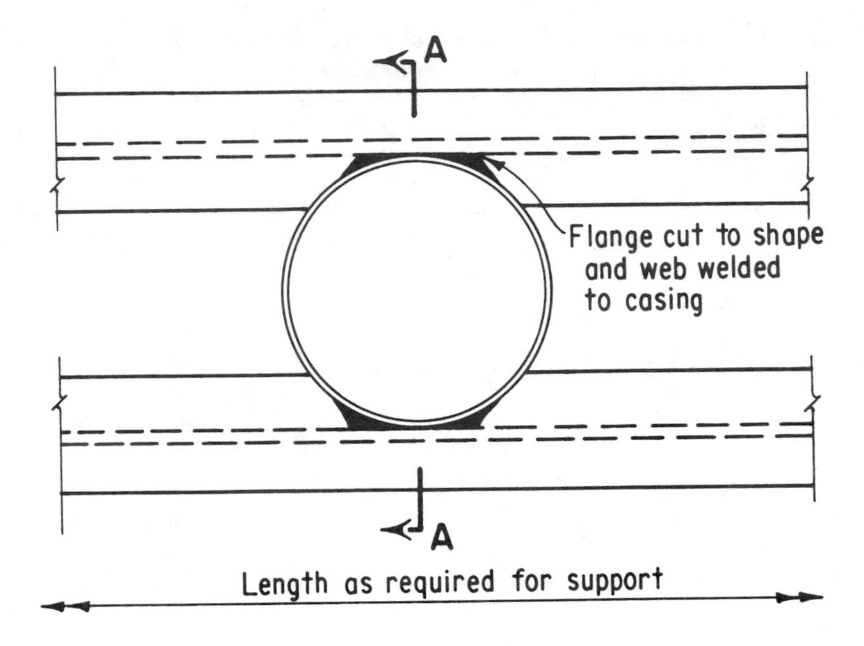

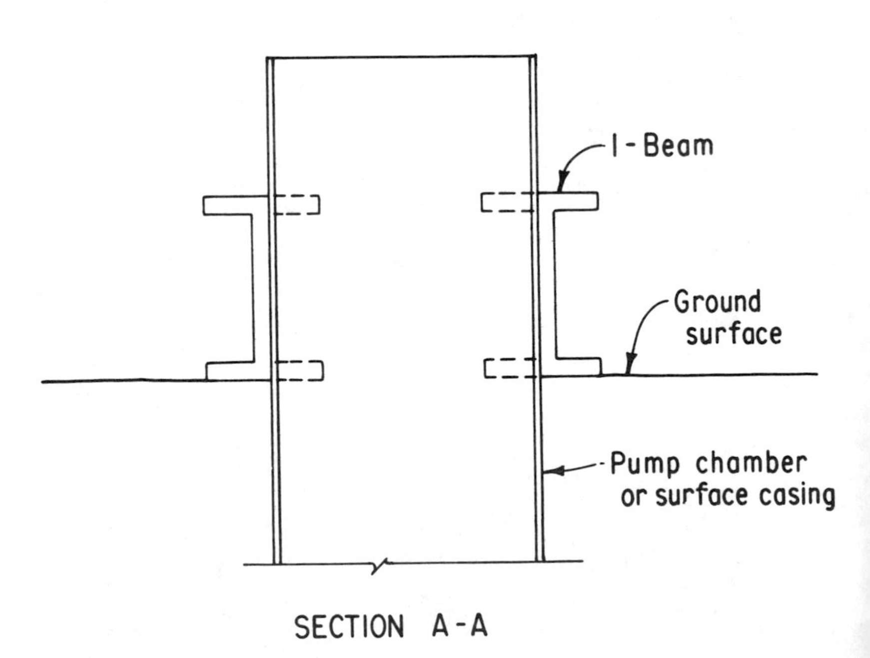

FIGURE 11-9.—Support of a casing by I-beams. 103-D-1496.

(b) *Design Particulars for Surface Casing.*—The wall thickness, depth of setting, and weight of temporary surface casing is usually left to the discretion of the contractor whose responsibility includes its installation and removal. However, the minimum diameter must be large enough not only to give a minimum 1½-inch-thick grout seal about the pump housing casing but also to permit the installation of the tremie pipe. Table 11–4 shows suggested minimum temporary surface casing diameters to use with various nominal pipe sizes of pump chamber casing and nominal 1-inch-diameter tremie pipe.

Permanent surface casing is usually installed with a minimum stick-up of 1 foot above the original ground surface. However, the design of surface facilities may require more or less stickup.

When drilling with rotary rigs using water or other drilling fluid, a hole is normally cut in the surface casing to permit fluid flow between the slush pit and the well. Grouting is, therefore, usually stopped when the grout approaches this hole. On completion of the well, a patch of the same weight material as the casing should be welded over the hole. When the foundation is poured, the unfilled space about the top of the casing is then filled with concrete.

Minimum depth of setting, diameter, wall thickness, and weight of permanent surface casing should be given in the specifications. In some instances, to facilitate his operation the contractor may elect to install, at his own expense, casing exceeding the minimum requirements.

11–3. Pump Chamber Casing (pump housing casing, working casing, inner casing, flow pipe protective casing).—(a) *Description*

TABLE 11–4.—*Minimum temporary surface casing diameters to permit grout around pump housing casing with 1-inch grout pipes*

Nominal pump chamber casing diameter, inches	Temporary surface casing diameter (welded joints on pump chamber casing), [1] inches
6	10
8	12
10	14
12	16
14	18
16	20
18	22
20	24
22	26
24	28

[1] Add 2 inches to the diameter to accommodate pump chamber casing with coupled joints.

and Purpose.—Pump chamber casing is an essential part of every well. In single string construction of uniform diameter, it comprises all casing above the screen. In other types of construction, it is the casing within which the pump bowls will be set.

The pump chamber casing furnishes a direct connection between the surface and the aquifer and when permanent surface casing is not used, it seals out undesirable surface or shallow ground water and supports the side of the hole.

(b) *Design Particulars for Pump Chamber Casing.*—Pipe conforming to API Standard 5L—March 1963 [2] either Grade A or B is commonly used for pump chamber casing. Pipe conforming to ASA Standard for Wrought Steel Pipe, ASA B36.10–59 Schedules 30 and 40 or to standard weight pipe in regard to weight and wall thickness are also used for well casing. A wall thickness of less than 0.250 inch is not recommended because of corrosion possibilities. Where hard driving, deep setting in unconsolidated materials, or aggressive corrosion attack is anticipated, heavier casing should be used. Lighter weight pipe may be used for temporary installations in shallow wells and when pipe is set in an open hole in stable consolidated rock. Table 11–5 gives data on commonly used pipe of various diameters and weights and suggested maximum setting depths. Couplings may be available on casing up to 12 inches in diameter, but most casing joints are welded. Plain ends beveled for welding and threaded ends should, if possible, conform to API Standard 5L, Sections 7 and 8 [2]. Other threads and couplings can be used, but the API Standards have desirable design characteristics and are recommended. Threads on all pipe and couplings should be undamaged and brushed clean and doped before joining. They should be joined without cross-threading and tightened to the maximum possible to ensure watertight joints. Welds should be multiple pass, fully penetrating, continuous running welds, and the weld should be brushed with a steel brush and peened to remove slag and stresses between each pass. When stainless steel or other alloys are used, welding rods and procedures recommended by the manufacturer should be used. Welding, braising, or sweating of joints between dissimilar metals or alloys such as stainless and low carbon steel or brass should be avoided. Such materials should be insulated from each other by a dielectric coupling to avoid galvanic corrosion.

In areas of the southwestern United States, two-ply casing composed of 6- to 12-gage sheet steel in diameters between 8 and 30 inches is commonly used in wells drilled with mud scows, see table 11–6. This casing is manufactured in 5-foot lengths, each of which telescopes one-half of its length into the next section. The casing is forced down the hole by hydraulic jacks, and the outer overlap is welded to the telescoping casing. The casing is usually perforated in

TABLE 11–5.—*Data on standard and line pipe commonly used for water well casing*

Nominal size, inches	Outside diameter, inches	Outside diameter couplings, inches	Schedule or class [1]	Wall thickness, inches	Weight per foot-plain end, pounds	Inside diameter, inches	[2] Suggested maximum setting, feet
4	4. 500	5. 200	--------	0. 219	10. 10	4. 062	1, 190
			40	. 237	10. 79	4. 026	1, 060
6	6. 625	7. 390	--------	. 250	17. 02	6. 125	705
			40(S)	. 280	18. 97	6. 065	850
8	8. 625	9. 625	20	. 250	22. 36	8. 125	420
			30	. 277	24. 70	8. 071	525
			40 (S)	. 322	28. 55	7. 981	695
10	10. 750	11. 750	20	. 250	28. 04	10. 250	235
			30	. 307	34. 24	10. 136	410
			40 (S)	. 365	40. 48	10. 020	580
12	12. 750	14. 000	20	. 250	33. 38	12. 250	140
			30	. 330	43. 77	12. 090	320
			S	. 375	49. 56	12. 000	435
			40*	. 406	53. 56	11. 938	515
14	14. 000	15. 000	10	. 250	36. 71	13. 500	105
			20	. 312	45. 68	13. 376	195
			30 (S)	. 375	54. 57	13. 250	350
			40	. 438	63. 37	13. 124	495
16	16. 000	17. 000	10	. 250	42. 05	15. 500	70
			20	. 312	52. 36	15. 376	140
			30(S)	. 375	62. 58	15. 250	240
			40	. 500	82. 77	15. 000	495
18	18. 000	19. 000	10	. 250	47. 39	17. 500	50
			20	. 312	59. 03	17. 376	100
			S	. 375	70. 59	17. 250	170
			30	. 438	82. 06	17. 124	270
			40	. 562	104. 76	16. 876	495
20	20. 000	21. 000	10	. 250	52. 73	19. 500	35
			20(S)	. 375	78. 60	19. 250	125
			30	. 500	104. 13	19. 000	295
			40*	. 594	123. 06	18. 802	445

[1] See footnotes at end of table.

TABLE 11–5.—*Data on standard and line pipe commonly used for water well casing*—Continued

Nominal size, inches	Outside diameter, inches	Outside diameter couplings, inches	Schedule or class [1]	Wall thickness, inches	Weight per foot-plain end, pounds	Inside diameter, inches	[2] Suggested maximum setting, feet
22	22. 000	----------	10	. 250	58. 07	21. 500	30
			20(S)	. 375	86. 61	21. 250	95
			30	. 500	114. 81	21. 000	220
24	24. 000	----------	10	. 250	63. 41	23. 500	20
			20(S)	. 375	94. 62	23. 250	70
			30	. 562	140. 80	22. 876	240
			40	. 688	171. 17	22. 624	410
26	26. 000	----------	10	. 312	85. 73	25. 376	30
			S	. 375	102. 63	25. 250	55
			20	. 500	136. 17	25. 000	135
28	28. 000	----------	10*	. 312	92. 41	27. 376	25
			(S)	. 375	110. 41	27. 250	45
			20	. 500	146. 85	27. 000	105
			30	. 625	182. 73	26. 750	210
30	30. 000	----------	10*	. 312	99. 08	29. 376	20
			(S)	. 375	118. 65	29. 250	35
			20	. 500	157. 53	29. 000	85
			30	. 625	196. 08	28. 750	170
32	32. 000	----------	10*	. 312	105. 76	31. 376	20
			(S)	. 375	126. 66	31. 250	30
			20	. 500	168. 21	33. 000	70
			30	. 625	209. 43	32. 750	140
34	34. 000	----------	10*	. 312	112. 43	33. 376	15
			S	. 375	134. 67	33. 250	25
			20	. 500	178. 89	33. 000	60
			30	. 625	222. 78	32. 750	115
36	36. 000	----------	10*	. 312	119. 11	35. 376	10
			(S)	. 375	142. 68	35. 250	20
			20	. 500	189. 57	35. 000	50
			30	. 625	236. 13	34. 750	100

[1] ASA Standard B36.10 schedule numbers (S) indicates standard weight pipe; * indicates a non-API standard.

[2] Maximum settings were estimated for the worst possible conditions in unconsolidated formation. A design factor of approximately 1.5 was used for steel with yield strength less than 40,000 lb/in². A 50-percent increase in depth of setting beyond those given is considered safe under favorable conditions.

TABLE 11–6.—*Recommended maximum depth of setting for California stovepipe and similar sheet steel and steel-plate fabricated casing, in feet*

Diameter, inches	Gage [1]						Thickness, inches			
	12		10		8	6	3/16	1/4	5/16	3/8
	D	S	D	S	D	D				
8	340	125	750	260	X	X	X	X	X	X
10	150	60	390	135	X	X	320	750	X	X
12	100	35	225	75	390	X	180	435	875	X
14	60	20	140	45	250	X	115	270	530	X
16	40	15	90	30	165	275	75	180	360	630
18	30		65	20	115	190	55	125	260	445
20	20		45		85	140	35	90	180	320
22			35		60	105	X	X	X	X
24			25		45	80	20	50	100	185
26			20		35	60	X	X	X	X
30			10		25	40	10	25	50	95

[1] U.S. Standard Gage.
D=Telescoping.
S=Single thickness.
X=Not commonly made in these sizes.

place. Numerous successful wells have been constructed with such casing. It is not generally accepted in other parts of the country, however, nor is it recommended for use in wells of a permanent nature.

Aggressive corrosion, heavy incrustation, or both, often suggest use of special alloys, nonmetallic pipes, or special coatings of low carbon steel pipes for pump housing and other casing. Coatings are not recommended. It is practically impossible to install a casing without breaking the coating on the outside of the casing or to install a pump without breaking the coating on the inside of the casing. Where the coating is broken, a point of aggressive, concentrated corrosion attack may result.

Ceramic-clay and concrete pipes are used as casing. Ceramic and clay pipe is resistant to corrosion but even type V cement is subject to some attack and is unsatisfactory if a well must be acidized. All of these materials are heavy, difficult to install and keep in alinement. Also, tightly sealed joints are difficult to make with these materials. Strings of such pipe should be surrounded by a grout seal for their entire length. Because of their weight and strength characteristics, deep settings may result in compressive failure of the pipe at the bottom.

Asbestos-cement pipe is lighter, somewhat more resistant than concrete to corrosion, available in longer lengths, and special joint seals are available; however, this pipe presents many of the problems associated with concrete pipe.

Plastic pipe such as polyvinyl chloride (PVC), polyethylene, acrylonitrile-butadiene-styrene (ABS), or rubber modified plastics offer many advantages such as lightweight, ease of installation, corrosion resistance, and low price. Pipe of diameters up to 6 inches have been used successfully in water wells in firm, consolidated rock to depths in excess of 800 feet. However, these materials lack the tensile, yield, and impact strengths; elasticity; and the ease of joining of low carbon steel pipe. Plastic pipe is manufactured of suitable wall thicknesses and in diameters up to about 10 or 12 inches for a setting depth of about 150 feet in unconsolidated formations.

Fiberglass reinforced plastic pipe up to 10 inches in diameter with 0.180-to 0.200-inch wall thickness has been used extensively in water wells in some areas to depths of about 300 feet. However, the conditions are exceptional, and collapse under normal development procedures has been reported to be a problem. The pipe offers all the advantages of plastic pipe plus greater collapse strength but is still not comparable to steel.

Stainless steel and various copper alloys are satisfactory from nearly all standpoints as well casing. However, they are expensive and justifiable only in permanent wells in very corrosive environments.

The collapse resistance of pipe increases with the wall thickness, the elasticity, and the yield strength of the material and decreases with an increase in pipe diameter. From all standpoints, except corrosion resistance, low carbon steel pipe is the most satisfactory material. Pennington [8] determined that the corrosion resistance follows an exponential curve such that doubling the wall thickness extends the life of the pipe about four times.

In view of this, the low carbon steel casing with increased wall thickness is in a majority of cases the most satisfactory and economical material to use in areas of moderate corrosion. However, this criterion does not apply to perforated casing.

The pump chamber casing should have a nominal diameter at least 2 inches larger than the nominal diameter of the pump bowls of the required capacity. Table 11–1 shows recommended minimum diameters for various desired yields using deep well turbine pumps of standard manufacture. The diameter should be increased to 4 inches for deep settings of the larger pumps.

The top of pump housing casing should be set at least 12 inches above the proposed top elevation of the pump foundation. Any excess may be cut off when the permanent pump is installed.

Depth of setting of a pump, hence depth of the pump chamber casing, is determined by estimating projected pumping levels (sec. 19–6) and considering the following factors:

- Present static water level.
- Minimum static water level of record in area.
- Longtime water level trends in the area.
- Probable drawdown at desired yield.
- Possible interference by other wells or boundary conditions.
- Required pump submergence.
- Ten pipe diameters between the suction-cone of the pump and the top of any reduction in pipe size (desirable but not essential).
- The presence of any telescoping overlap.

In some areas, one or more of these factors may not be in harmony and compromise will be necessary. The controlling factors in permanent well installations are usually sanitation, stability, and an estimated minimum usable well life of 25 years.

Pump housing casing should always be installed plumb and straight. Deviation from the vertical should not exceed two-thirds of the inside diameter of the casing per 100 feet of depth. A 40-foot-long dolly, one-half inch smaller than the inside diameter, should pass without binding (see sec. 16–7). If surface casing is used, it should be installed plumb and straight and the pump housing casing should be centered in such casing. If the pump chamber casing is longer than about 50 feet and the annular space is greater than 1 inch, the casing should have centering guides at the bottom and at 40-foot intervals to the surface where it should be firmly anchored until the foundation is placed and set. Where hard driving is required in setting casing such as with a cable tool rig, several reductions in size of the pump chamber casing may be necessary so that the desired diameter can be set at the required depth by telescoping. With the exception of telescoping nominal 12- into nominal 14-inch pipe, all standard weight and thinner walled pipes with couplings and drive shoes will telescope into the next larger size. Twelve-inch pipe with welded joints and no drive shoe will telescope into some 14-inch pipe.

Pump chamber casing should be grouted in except when it is run inside a grouted-in surface casing. This may be done as the temporary surface casing is withdrawn if such casing is used, or an oversized hole may be drilled with rotary equipment.

In areas of thick, unstable materials, a pump chamber casing may have to be supported from the surface in a manner similar to surface casing (see sec. 11–2(a) and fig. 11–9).

Pump chamber casing is sometimes perforated, or screened sections are included in it, above the pump bowls. This practice should be avoided if possible in permanent installations. If the drawdown increases to depths below the screened section, cascading results in entrained air in the discharge which encourages pump cavitation and may result in other adverse effects. Also, periodic exposure of a screen to the atmosphere is conducive not only to corrosion but also to growth of organisms which may result in plugging of the screen.

11-4. Screen Assembly.—(a) *Description and Purpose.*—The purposes of a screen are to: (1) stabilize the sides of the hole, (2) keep sand out of the well, and (3) facilitate flow into and within the well. The screen proper may range from pipe perforated inplace to carefully fabricated cage-type wire-wound screen with accurately sized slot openings [1,3,4,5,7,9,10,12]. The screen assembly may consist of only a screen or perforated section; or of a screen, associated blank casing, bottom seals, etc., in other well designs.

Formerly, most drilled water wells were constructed with cable tools and finished by perforating the casing inplace with a Mills knife or other similar tool. However, this practice is rapidly declining. The size of the openings resulting from inplace perforating cannot be accurately controlled and the sizes range from one-eighth inch to one-half inch wide and from 1 inch to 2 inches long. The perforations are large and irregular and have rough, ragged edges that encourage corrosion attack and incrustation. On heavier wall casing the perforators sometimes do not make an opening, but only dimple the pipe. Maximum percentage of open area is about 3 to 4 percent. Good development is almost impossible, and unless the aquifer is relatively clean, coarse sand and gravel, wells so perforated are usually sand pumpers.

Perforated casing made by sawing, machining, or torch cutting the slots in the casing is commonly used in wells. Slot openings range from about 0.010 to 0.250 inch and the maximum percentage of open area is about 12 percent for the larger slots. Sawed and machine-cut slots are satisfactory if properly sized and entrance velocity limits can be met. In some instances, entrance velocity limits can be met by enlarging the diameter of the well and screen or increasing the depth of the well and the length of the screen. Torch-cut slots usually have rough edges and slag remnants adhering to them. The finest possible slot is about 0.125 inch. Wells screened with perforated pipe of any kind are usually more difficult to develop than continuous slot or louvre screens, and if slots are not accurately sized to the aquifer, the wells may be sand pumpers.

There are numerous screens manufactured with punched or stamped perforations. Slots range from 0.060 to 0.250 inch and they often have rough and ragged edges. Maximum percentage of open area is about 20 percent. Some of these screens are made of 8-gage and lighter stock and are not suitable for settings at depths much greater than 100 to 150 feet, depending upon diameter.

A number of louvre-type screens are manufactured of 7-gage or heavier materials and are available in six or eight slot sizes ranging from about 0.030 to 0.150 inch. The slots are usually accurately sized and wire brushed to remove roughness or irregularities. Open areas range from about 3 percent for the smaller slot sizes to about 20 percent for the larger.

Some types of screen are made by winding wire around a perforated pipe base with or without spacers between the wires and the pipe. Almost any slot sizes can be readily obtained and open area in those made with spacers compares favorably with that of louvre-type screen of the same slot size. Such screens are satisfactory in clean, rather coarse materials with few or no fines. However, where aquifers contain a large percentage of fines, the channels between the spacers between the pipe and wire wrapping may become clogged and seriously reduce the entrance area. The pipe base is often made by winding stainless steel wire on a low carbon steel pipe. In corrosive waters, this combination may result in rapid corrosion of the pipe base. Such screens should always be made of a single metal or alloy. The only real advantage of pipe-base screens is their superior tensile and collapse strengths.

Cage-type wire-wound screens consist of a continuous winding of round or specially shaped wire on a cage of vertical rods. The wire is attached to the rods by welding or using dovetailed connections. Almost any slot size is readily available, usually in increments of about 0.005 inch from 0.006 inch to 0.250 inch. Screens are made in both telescoping and pipe sizes. The former will just telescope through a casing of the same nominal size of the screen, whereas the latter has the same inside diameter as the casing and may be joined to it by welding or coupling. Cage-type wire-wound screens are the most efficient available. Open areas are the largest obtainable and slot sizes can be closely matched to aquifer gradations. Although such screens are more expensive initially than other types of screen, they are usually more economical, especially when used in thin but highly productive aquifers.

Most screens are made in lengths ranging from 5 to 20 feet which can be joined by welding or couplings to give almost any length of screen and combination of slot sizes desired. Couplings and welding materials should be composed of the same materials as the screen.

Screens are available in diameters ranging from 1¼ to 60 inches. Screen diameters should be selected on the basis of the desired yield from the well and thickness of the aquifer. Table 11–7 gives the recommended minimum diameters for various well capacities. These diameters may be increased to obtain acceptable entrance velocities if necessary, and smaller diameters are sometimes specified in the interest of economy. With smaller diameters, the initial cost is lower but well efficiency is also lowered. Smaller diameters are not recommended for permanent installations. However, there is evidence that equal efficiency as well as certain other advantages may result when installing 30 feet or more of screen by increasing the diameter toward the top of the well, provided that a satisfactory average entrance velocity results. For example, the screen might consist of 10-foot lengths of 8-, 10-, and 12-inch diameters connected by reducers, with the topmost screen of the minimum diameter recommended in table 11–7 for the desired yield.

In uniform aquifers a continuous length of screen is normally installed. In thick, nonuniform aquifers, however, usual practice is to set the screen opposite only the best aquifer materials and set blank pipe opposite the poorer materials between the screen sections. The blank sections between screens should be pipe size where pipe size screens are used or flush tubing where telescoping size screens are used. Blank pipe or flush tube extensions may extend from the top of the screen up to the pump chamber casing to which it may be attached by welding, a coupling, a reducer, or the extensions may telescope for 5 feet or more into the casing. Where the extension is relatively short, it is sometimes referred to as a flush tube extension or overlap pipe;

TABLE 11–7.—*Recommended minimum screen assembly diameters*

	Minimum nominal screen assembly diameter, inches
Discharge, gal/min:	
Up to 50	2
50 to 125	4
125 to 350	6
350 to 800	8
800 to 1,400	10
1,400 to 2,500	12
2,500 to 3,500	14
3,500 to 5,000	16
5,000 to 7,000	18
7,000 to 9,000	20

otherwise, it is called a riser pipe. Installation of a bottom sump consisting of 5 to 10 feet of blank casing or a flush tube extension below the bottom of the lowest screened section in the well is recommended. Use of the bottom sump in well design is not common, but it offers several advantages. During development, materials drawn into the well settle in the sump and do not encroach on the screen. In addition, sand enters all wells in unconsolidated materials when they are pumped. The sump provides a storage space for sand which settles to the bottom and prolongs the effective operation of the total screen. The sump also provides a suitable place for the attachment of centering guides at the bottom of the screen assembly.

Screen assemblies set in unconsolidated materials should have a bottom seal. This seal may consist of a steel plate welded to the bottom, a bail bottom which is welded or coupled to the bottom of the assembly and has a bail on the upper surface to facilitate installation, any of a variety of float shoes, bail down shoes, self-sealing jets, and other special fixtures, or a concrete plug. The bottom seal not only precludes heaving of materials up into the well under certain circumstances but also provides a bearing area for support of the screen assembly.

Screens may be made of many different metals and metal alloys, plastics, concrete, asbestos-cement pipe, glass reinforced epoxy, coated base metals, and wood. The least expensive and most commonly available screens are fabricated of low carbon steel. Those made from the nonferrous metals and alloys, plastics, and exotic materials are used in areas of aggressive corrosion and incrustation to prolong well life and efficiency or where permanence and continuous service are essential.

A well screen is particularly susceptible to corrosion attack and incrustation by mineral deposits. The many perforations expose more surface area to a reactive environment than a pipe of similar size. In addition, water flowing to and through a screen brings a constantly renewed supply of reactive materials into contact with it and at the same time removes protective coatings or corrosive products which otherwise would offer some protection against further attack. Minimum hydrostatic pressure and maximum water velocity occur at the well face which may result in release of carbon dioxide and other dissolved gases, some of which are aggressive corrosion agents. These and related factors also upset the chemical balance of the water so that carbonates of calcium, magnesium, iron, and other minerals may deposit on the screen and in the formation adjacent to it, blocking the slots and reducing the open area. Incrustation is commonly remedied by acidizing the well (see sec. 21–4). However, even when inhibited acids are used, such treatment results in some corrosion of the screen.

Concrete and asbestos-cement are particularly susceptible to both incrustation and acid attack, whereas vitreous clay pipe is practically immune to acid attack. Screens made of these materials have a low percentage of open area and may (unless the contractor has special equipment and experience) present difficulties in construction, except in shallow wells.

Plastic, wood, and fiberglass-reinforced epoxy screens are practically immune to corrosion attack. Incrustation, while it cannot be avoided, is reported to be less troublesome and can be removed without damage to the screen. However, these screens commonly have relatively low percentages of open area. Nonreinforced plastics are subject to creep under sustained load with resultant changes in slot sizes. The collapse resistance of plastic screens in unconsolidated materials is questionable, particularly in wells deeper than about 150 feet, unless wall thicknesses are properly sized to resist stresses. Increasing the wall thickness increases the cost to where stainless steel or other similar alloys may be cost competitive and more satisfactory.

Coated steel or other base metal screens are not recommended. The coating, to be effective, would have to be applied after the perforations were made. This would change the slot sizes an indeterminate amount. Moreover, it is practically impossible to install a coated screen in a well without some of the coating being scratched or otherwise damaged. These spots or holidays then become points for aggressive, concentrated corrosion attack.

When all factors are considered, the most satisfactory screen materials, except under unusual conditions, are steel, stainless steel, or some of the metal alloys.

(b) *Design Particulars of Screen Assemblies.*—Achieving a satisfactory screen design generally is not possible until a pilot hole or the well has been drilled, logged, sampled, mechanical analyses made of the formation samples, and chemical analyses made of water samples. In addition, if other wells are present in the area, the depth, design, and history of such wells should be examined for data on corrosion and incrustation experience, sand pumping, and drawdown.

Casing in a typical well normally consists of low carbon steel pipe of adequate wall thickness not only to support the hole but also to give a satisfactory life span consistent with the corrosion potential. Screens are a different matter, however, because of their construction. Corrosion resistance cannot readily be accomplished by increasing the weight of the screen as in casing because of the critical role of slot width. Instead, corrosion resistance for screens is increased by using a suitable corrosion-resistant material.

Table 11–8 shows the more commonly used metallic screen materials in the order of increasing costs. As mentioned previously, plastic, glass

TABLE 11–8.—*Well screen material*

Material [1]	Acid resistance	Corrosion resistance in normal ground water
Low carbon steel	Poor.	Poor. [2,3]
Toncan and Armco iron	Poor.	Fair. [2,3]
Admiralty red brass	Good.	Good.[3]
Silicon red brass	Good.	Good.[3]
304 stainless steel	Good.	Very good.
Everdure bronze	Very good.	Very good.[4]
Monel metal	Very good.	Very good.[4]
Super nickel	Very good.	Very good.[4]

[1] Other materials are available for use under special situations such as installations in aquifers containing high-temperature corrosive brines.
[2] Not recommended for permanent installations where incrustation is a serious problem.
[3] Not recommended for permanent installations where sulfate-reducing or similar bacteria are present or where water contains more than 60 p/m SO_4.
[4] Recommended only in areas where corrosion is very aggressive.

reinforced epoxy, vitreous clay title, and wooden screens have high corrosion resistance, but their use should be limited to relatively shallow, small diameter wells of low capacity or in unusual applications where other materials are unsuitable.

Where exceptionally deep settings or unstable ground is encountered which would require extra column strength and collapse resistance, some of the wire-wound screens are available in extra strength designs. These screens have a somewhat reduced percentage of open area compared to the standard designs but are still superior to other types. Where perforated casing is used, extra strength can be gained by increasing the wall thickness of the casing.

The hydraulic advantages of various patterns and types of slot or perforations are discussed in section 3–8. While these factors are significant, the most important characteristics of a screen are the slot size and amount of open area. Slot size is determined from the mechanical analyses of the formation samples obtained from the pilot hole or well.

If the uniformity coefficient of a sample for a naturally developed well is 5 or less, a slot size should be selected which will retain from 40 to 50 percent of the aquifer and allow 40 to 50 percent to pass through the screen during development. If representation of the sample is questionable, or if corrosion may be a problem, a slot size which will retain 40 to 45 percent of the aquifer should be used. If the sample is representative and corrosion is not a problem but incrustation is anticipated, a slot size is selected which will retain from 45 to 50 percent of the aquifer.

If the uniformity coefficient of a sample is greater than 5, a slot size which will retain from 30 to 50 percent of the aquifer should be selected. Thirty to 40 percent retention is used if the sample is representative, corrosion is not a problem, or if incrustation is anticipated. Forty to 50 percent is used if there is doubt that the sample is representative or if corrosion is a problem. The upper limit of retention in each case is the extreme and if screens are not available in a standard slot size, the next smaller standard size should be selected. Slot sizes are commonly described in 1/1,000 inch; thus, a No. 60 slot has a 0.060-inch slot width. Figure 11–10 shows slot sizes for some representative screens.

Most aquifers are not homogeneous and uniform but consist of layers containing granular materials of different gradation, size, and uniformity coefficient. Consequently, a single slot size is seldom suitable for use throughout a well. A suggested method of treating this usual occurrence in large capacity wells is to arrange a table based on the log and mechanical analyses of the samples in descending order as they were encountered in the hole. Blank casing is set opposite zones consisting of clay or silt or in which more than 20 percent of the grains pass the 100-mesh screen. The selected slot size is then entered on the table adjacent to the interval represented by each usable aquifer sample. If coarser materials in beds 5 feet or less in thickness are interlayered with finer aquifer material, it is seldom worthwhile to attempt to screen them separately, so screen suitable for the finest is used throughout.

Aquifers pack and settle during development of a well and a fine sand may migrate downward to a point opposite a screen which is too coarse. Therefore, where finer material overlies coarser material, the finer screen should be extended downward into the coarser material at least 10 percent of the thickness of the coarser material.

The diameter of the pump chamber is determined by the size of the pump required to discharge the desired yield. The diameter of the screen, however, is determined by the desired yield. Table 11–7 indicates the recommended minimum diameters of screen for a range of discharges. In Bureau wells, pump chamber diameter is usually 2 or more inches larger than the screen diameter, and pipe size screen assemblies are generally used, since they will telescope through the larger diameter pump chamber casing.

When the minimum diameter and slot sizes of the screen have been determined, the average entrance velocity can be estimated by dividing the desired yield in cubic feet per second by the total open area of the screen in square feet.

Table 11–9 gives minimum open areas of some representative screens of various slot sizes and diameter in square feet per foot of

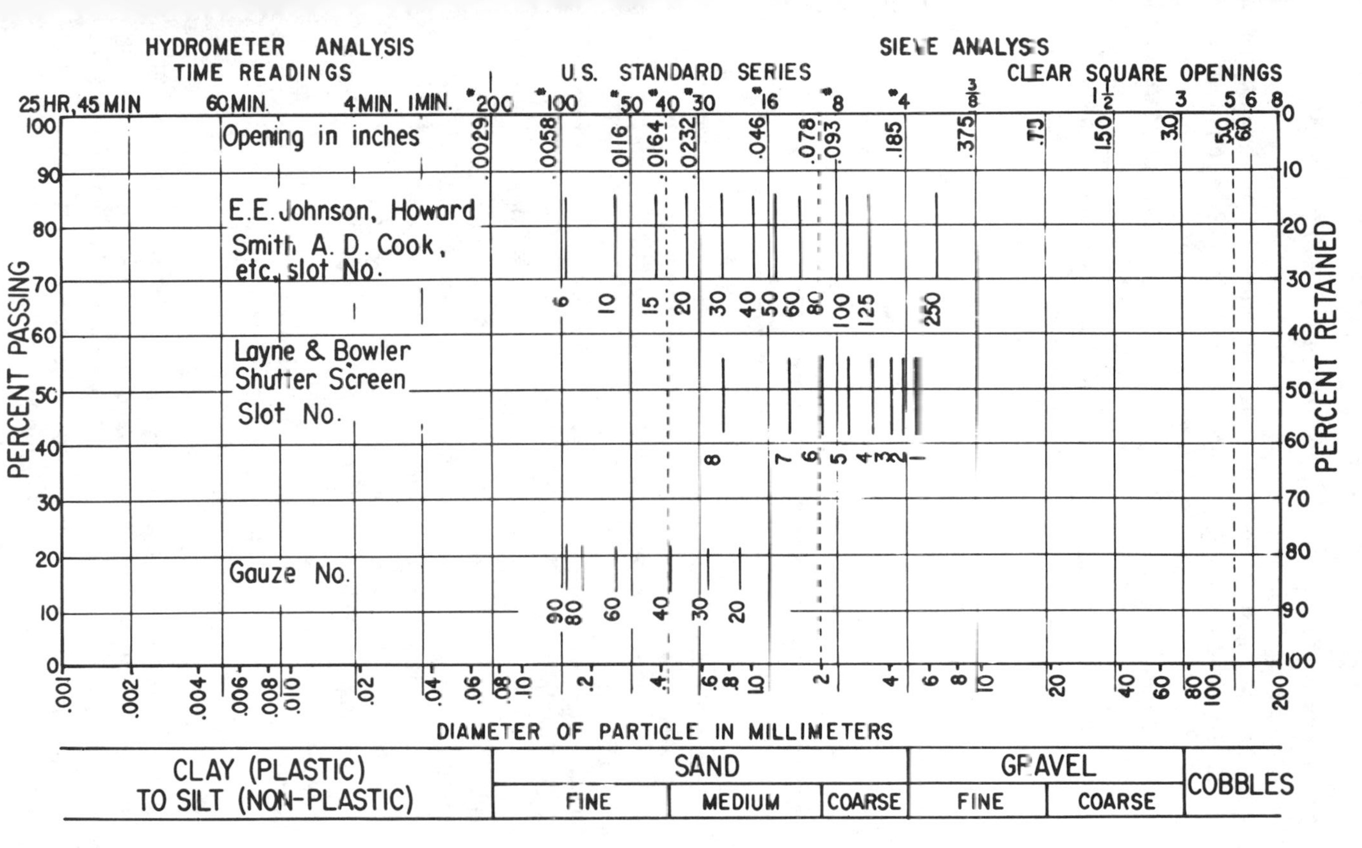

FIGURE 11-10.—Slot sizes for representative well screens. 103-D-1497.

length. Estimates based on interpolation are usually adequate for combinations between those shown in the table.

The values given in table 11–9 were computed from the manufacturer's published data. The square feet and percentage open area for louvre or shutter screens, perforated screens, and slotted pipe are believed to be accurate within a few percent. On wire-wound screen, the open area may be more than 30 percent less than shown because of the manufacturer's method of stating open area and lack of data on vertical wire sizes used. In selecting diameters and open areas for cage-type wire-wound screen, a smaller diameter or slot size may actually have more open area because of changes in wire size to maintain strength. Some savings in cost may be realized in some instances by selecting a smaller diameter screen or slot size and at the same time obtaining an equal or greater open area.

The average entrance velocity through the screens, neglecting blockage by aquifer or gravel pack material, should be 0.1 foot per second or less. If this is exceeded by more than about 0.05 foot per second, the screen diameter should be increased or the screen lengthened if possible to give the desired maximum entrance velocity. It is obvious that the greater the percentage of screen open area, the shorter the length of screen required to obtain an acceptable entrance velocity. Consideration should also be given to the effect of percentage of open hole as discussed in section 3–7. In confined aquifers, full penetration and maximum percentage of open hole are recommended where aquifer depth and thickness make such construction economically feasible. Where aquifers are deep and thick, judgment should be used to determine the most economical combination of penetration and open hole. In unconfined aquifers, full penetration and a 35 to 50 percent open hole at the bottom of the well are recommended, depending upon the thickness, stratigraphy, productivity of the aquifer, and the economy of such construction. Where the aquifer is deep and thick, judgment is again necessary to determine the most economical combination. Basically, the screen should always be placed at the bottom of the well and the screen length should not be less than 35 percent of the estimated thickness of the aquifer penetrated by the well.

Screen components such as blank sections of pipe or flush tube extensions, bottom plates, float-down or jetting shoes, centering guides, and other accessories should be fabricated of the same material as the screen; otherwise, wood, plastic, or concrete components should be used. Dissimilar metals should never be incorporated in the screen assembly. When the screen assembly is fabricated of nonferrous metals, the assembly should be separated from the low carbon steel

TABLE 11–9.—*Minimum open areas of screens in square feet per linear foot and the percentage of open area* [1]

[Cage-type wire-wound screen—telescoping sizes (from Johnson Division, UOP Inc.)]

Screen size, inches	Slot sizes, thousandth of an inch						
	10	20	40	60	80	100	150
4	0. 139	0. 243	0. 389	0. 493	0. 555	0. 514	0. 611
	14. 1	24. 7	39. 6	50. 2	56. 5	52. 3	[illegible]
4.5	0. 160	0. 278	0. 444	0. 555	0. 632	0. 583	0. 694
	14. 3	25. 0	39. 9	49. 9	56. 8	52. 4	62. 3
5	0. 180	0. 312	0. 493	0. 618	0. 701	0. 752	0. 777
	14. 4	25. 0	39. 6	49. 6	56. 3	52. 4	62. 4
5.625	0. 194	0. 347	0. 548	0. 687	0. 784	0. 722	0. 861
	14. 1	25. 2	39. 8	40. 0	57. 0	52. 5	62. 0
6	0. 208	0. 368	0. 451	0. 590	0. 694	0. 777	0. 916
	14. 1	24. 9	30. 6	40. 0	47. 1	52. 7	62. 1
8	0. 194	0. 354	0. 604	0. 784	0. 916	1. 027	1. 110
	9. 9	18. 0	30. 7	39. 9	46. 6	52. 2	56. 5
10	0. 249	0. 451	0. 763	0. 992	1. 166	1. 312	1. 409
	10. 0	18. 1	30. 6	39. 8	46. 8	52. 7	56. 6
12	0. 291	0. 534	0. 902	0. 999	1. 200	1. 367	1. 666
	9. 8	18. 1	30. 6	33. 9	40. 7	46. 3	56. 5
14	0. 264	0. 493	0. 847	1. 110	1. 339	1. 513	1. 853
	8. 2	15. 3	26. 4	34. 5	41. 7	47. 1	57. 7
16	0. 298	0. 555	0. 964	1. 263	1. 527	1. 735	2. 110
	7. 8	14. 6	25. 3	33. 2	40. 2	45. 6	55. 5
16 sp	0. 305	0. 576	0. 992	1. 298	1. 568	1. 776	2. 165
	7. 9	15. 0	25. 8	33. 8	40. 9	46. 3	56. 5
18	0. 340	0. 638	1. 110	1. 450	1. 749	1. 985	2. 415
	7. 9	14. 9	26. 0	33. 9	40. 9	46. 5	56. 6
18 sp	0. 347	0. 645	1. 117	1. 464	1. 763	1. 999	2. 443
	8. 0	14. 9	25. 8	33. 8	40. 7	46. 2	56. 6
20	0. 263	0. 479	0. 881	1. 221	1. 499	1. 721	2. 193
	5. 5	10. 0	18. 4	25. 5	31. 3	36. 0	45. 8
24	0. 319	0. 596	1. 097	1. 513	1. 874	2. 138	2. 727
	5. 3	10. 0	18. 5	25. 5	31. 6	36. 0	46. 0

[1] Top value shown is ft^2/ft and bottom value is percent.

TABLE 11–9.—*Minimum open areas of screens in square feet per linear foot and the percentage of open area* [1]—Continued

[Cage-type wire-wound screen—telescoping sizes (from Cook Well Strainer Company)]

Sscreen size, inches	Wire size, inches	Slot size, thousandth of an inch										
		10	20	25	30	40	50	60	70	80	100	125
4	0. 09	0. 097	0. 178	0. 213	0. 245	0. 302	0. 350	0. 392	0. 430	0. 461	0. 516	---------
		10. 0	18. 2	21. 7	25. 0	30. 7	35. 7	40. 0	43. 8	47. 0	52. 6	---------
6		0. 136	0. 247	0. 295	0. 339	0. 418	0. 485	0. 542	0. 593	0. 637	0. 722	---------
		10. 0	18. 2	21. 7	25. 0	30. 8	35. 7	40. 0	43. 8	47. 0	52. 6	---------
8		0. 196	0. 356	---------	0. 490	0. 604	0. 700	0. 784	0. 858	0. 923	1. 032	---------
		10. 0	18. 2	---------	24. 9	30. 8	35. 7	39. 9	43. 8	47. 0	52. 6	---------
8	. 1467	0. 125	0. 236	0	0. 334	0. 421	0. 500	0. 571	0. 625	0. 694	0. 797	0. 904
		6. 4	12. 0	---------	17. 0	21. 5	25. 5	29. 9	32. 4	35. 4	40. 6	46. 1
10	. 09	0. 248	0. 447	0. 553	0. 621	0. 765	0. 869	0. 993	1. 101	1. 168	1. 307	1. 440
		10. 0	18. 7	21. 7	25. 0	30. 8	35. 7	39. 9	43. 6	47. 0	52. 6	58. 0
10	. 1467	0. 159	0. 299	---------	0. 423	0. 533	0. 633	0. 723	0. 805	0. 879	1. 010	1. 146
		0. 64	12. 0	---------	17. 0	21. 5	25. 5	29. 1	32. 4	35. 4	40. 6	46. 1
10	. 1875	0. 126	0. 241	0. 294	0. 345	0. 440	0. 528	0. 607	0. 682	0. 749	0. 872	1. 003
		5. 0	9. 6	11. 7	13. 7	17. 5	21. 0	24. 2	27. 1	29. 8	34. 7	40. 0

12--------	.09	0.300	0.546	---------	0.751	0.926	1.115	1.208	1.316	1.414	1.582	1.054
		10.0	18.1	---------	24.9	30.8	35.7	39.9	43.7	47.0	52.6	58.0
12--------	.1467	0.192	0.362	---------	0.512	0.646	0.767	0.876	0.974	1.064	1.222	1.387
		6.4	12.0	---------	17.0	21.4	25.5	29.1	32.4	35.3	40.5	46.1
12--------	.1875	0.153	0.292	0.356	0.417	0.532	0.638	0.733	0.824	0.905	1.053	1.213
		5.0	9.7	11.7	13.7	17.5	21.0	24.2	27.1	29.8	34.7	40.0
14--------	.1875	0.180	0.345	---------	0.492	0.630	0.755	0.874	0.962	1.071	1.248	1.435
		5.0	9.7	---------	11.9	13.7	17.5	21.0	24.2	29.8	34.7	40.0
16--------	1.469	0.245	0.460	---------	0.652	0.821	0.975	1.117	1.239	1.353	1.555	1.764
		6.4	12.0	---------	17.0	21.4	25.5	29.1	32.4	35.3	40.5	46.1
16--------	.1875	0.191	0.367	---------	0.447	0.523	0.669	0.803	0.925	1.138	1.325	1.527
		5.0	9.7	---------	11.7	13.7	17.5	21.0	24.2	29.8	34.7	40.0
18--------	.1875	0.219	0.416	0.510	0.596	0.762	0.913	1.050	---------	1.296	1.505	1.735
		5.0	9.7	11.7	13.7	17.5	21.0	24.2	---------	29.8	34.7	40.0
20--------	.1875	0.246	0.468	0.571	0.669	0.854	1.027	1.177	---------	1.458	1.690	1.945
		5.0	9.7	11.7	13.7	17.5	21.0	24.2	---------	29.8	34.7	40.0
24--------	.1875	0.298	0.567	0.692	0.811	1.034	1.240	1.426	1.599	1.760	2.047	2.180
		5.0	9.6	11.7	13.7	17.5	21.0	24.2	27.1	29.9	34.7	40.0

[1] Top value shown is ft^2/ft and bottom value is percent.

TABLE 11–9.—*Minimum open areas of screens in square feet per linear foot and the percentage of open area* [1]—Continued

[Cage-type wire-wound screen—telescoping sizes (from Howard Smith Company)]

Screen size, inches	Slot size, thousandth of an inch								
	8	10	12	14	16	20	30	40	50
4	0. 104	0. 125	0. 145	0. 167	0. 187	0. 229	0. 305	0. 382	0. 437
	9. 0	11. 2	13. 0	15. 0	16. 8	20. 6	27. 4	34. 3	39. 2
6	0. 118	0. 146	0. 174	0. 194	0. 222	0. 271	0. 368	0. 451	0. 437
	8. 0	9. 9	11. 8	13. 2	15. 1	18. 4	25. 0	30. 6	29. 6
8	0. 160	0. 194	0. 229	0. 264	0. 298	0. 354	0. 493	0. 604	0. 700
	8. 1	9. 8	11. 6	13. 4	15. 2	18. 0	25. 1	30. 7	35. 6
10	0. 201	0. 250	0. 291	0. 333	0. 375	0. 451	0. 625	0. 763	0. 888
	8. 1	10. 1	11. 7	13. 3	15. 1	18. 1	25. 1	30. 6	35. 6
12	0. 243	0. 291	0. 347	0. 396	0. 444	0. 534	0. 736	0. 902	1. 047
	8. 3	9. 9	11. 8	13. 4	15. 1	18. 2	25. 0	30. 7	35. 6
14	0. 215	0. 263	0. 312	0. 360	0. 403	0. 492	0. 687	0. 854	1. 006
	6. 3	7. 8	9. 2	10. 6	11. 9	14. 6	20. 3	25. 3	29. 8
16	0. 194	0. 243	0. 284	0. 333	0. 375	0. 479	0. 680	0. 853	1. 006
	5. 2	6. 5	7. 6	8. 9	10. 0	12. 8	18. 2	22. 9	26. 9

[1] Top value shown is ft^2/ft and bottom value is percent.

TABLE 11-9.—*Minimum open areas of screens in square feet per linear foot and the percentage of open area* [1]—Continued

[Cage-type wire-wound screen—pipe size screen (from Johnson Division, UOP Inc.)]

Screen size, inches	Slot size, thousandth of an inch						
	10	20	40	60	80	100	150
4	0. 174	0. 305	0. 472	0. 597	0. 680	0. 639	0. 756
	14. 3	25. 2	38. 9	49. 2	56. 1	52. 7	62. 4
6	0. 174	0. 319	0. 534	0. 694	0. 812	0. 916	0. 986
	5. 7	18. 4	30. 8	40. 0	46. 8	52. 8	56. 8
8	0. 222	0. 410	0. 694	0. 902	1. 055	1. 187	1. 277
	9. 8	18. 1	30. 7	39. 9	46. 7	52. 5	56. 5
10	0. 285	0. 514	0. 868	0. 958	1. 152	1. 305	1. 596
	10. 1	18. 2	30. 8	34. 0	40. 9	46. 3	56. 7
12	0. 264	0. 500	0. 868	1. 131	1. 367	1. 548	1. 888
	7. 8	14. 9	25. 9	33. 8	40. 9	46. 3	56. 5
14	0. 298	0. 555	0. 965	1. 263	1. 527	1. 735	2. 110
	8. 1	15. 1	26. 3	34. 4	41. 6	47. 3	47. 5
16	0. 340	0. 639	1. 110	1. 450	1. 749	1. 985	2. 415
	8. 1	15. 2	26. 4	34. 6	41. 7	47. 3	57. 6

[1] Top value shown is ft²/ft and bottom value is percent.

TABLE 11-9.—*Minimum open areas of screens in square feet per linear foot and the percentage of open area*—Continued

[Cage-type wire-wound screen—pipe size irrigator screen (from Johnson Division, UOP Inc.)]

Screen size, inches	Slot size, thousandth of an inch								
	30	40	50	60	70	80	90	100	125
8	0. 21	0. 27	0. 32	0. 37	0. 41	0. 45	0. 49	0. 52	0. 60
10	. 26	. 33	. 39	. 45	. 50	. 55	. 59	. 63	. 72
12	. 31	. 39	. 46	. 53	. 59	. 65	. 70	. 75	. 80
14	. 35	. 45	. 53	. 61	. 68	. 75	. 81	. 87	. 99
16	. 40	. 51	. 60	. 69	. 78	. 85	. 92	. 98	1. 12
Approximate percent open area	9. 2	11. 7	13. 9	16. 0	17. 9	19. 5	21. 2	22. 8	26. 0

TABLE 11-9.—*Minimum open areas of screens in square feet per linear foot and the percentage of open area* [1]—Continued

[Cage-type wire-wound screen—pipe size screen (from Cook Well Strainer Company)]

Screen size, inches	Wire size, inches	Slot size, thousandth of an inch									
		10	20	25	30	40	50	60	80	100	125
6	0.09	0.171	0.312	0.372	0.429	0.528	0.612	0.686	0.800	0.902	----------
		10.0	18.2	21.7	25.0	30.8	35.7	40.0	46.7	52.6	----------
6	.1467	0.111	0.208	0.252	0.295	0.372	0.442	0.504	0.614	0.705	0.799
		6.4	12.0	14.5	17.0	21.4	25.4	29.0	35.4	40.5	46.0
6	.1875	0.191	0.117	0.214	0.249	0.319	0.384	0.477	0.547	0.635	0.732
		5.0	9.7	11.7	13.7	17.5	21.0	24.2	29.9	34.7	40.0
8	.1875	0.118	0.225	0.274	0.321	0.410	0.492	0.567	0.707	0.814	0.938
		5.1	9.6	11.7	13.7	17.5	21.0	24.2	29.9	34.7	40.0
10	.1875	0.145	0.276	0.336	0.394	0.503	0.693	0.694	0.857	0.997	1.140
		5.0	9.7	11.7	13.7	17.5	21.0	24.2	29.9	34.7	40.0
12	.1875	0.171	0.326	0.398	0.467	0.595	0.709	0.821	1.013	1.179	1.356
		5.0	9.7	11.7	13.7	17.5	21.0	24.2	29.8	34.7	40.0
18	.1875	0.250	0.477	0.582	0.683	0.871	1.043	1.200	1.482	1.724	1.984
		5.0	9.7	11.7	13.7	17.5	21.0	24.2	29.8	34.7	40.0
20	.1875	0.277	0.528	0.644	0.755	0.963	1.154	1.327	1.637	1.905	2.194
		5.0	9.7	11.7	13.7	17.5	21.0	24.2	29.8	34.7	----------

[1] Top value shown is ft^2/ft and bottom value is percent.

TABLE 11–9.—*Minimum open areas of screens in square feet per linear foot and the percentage of open area* [1]—Continued

[Cage-type wire-wound screen—pipe size screen (from Howard Smith Company)]

Screen size, inches	Slot size, thousandth of an inch								
	8	10	12	14	16	20	30	40	50
4	0. 104	0. 125	0. 146	0. 167	0. 187	0. 229	0. 305	0. 382	0. 437
	8. 8	10. 6	12. 4	14. 2	15. 9	19. 4	25. 9	32. 4	0. 371
6	0. 139	0. 174	0. 201	0. 236	0. 264	0. 312	0. 430	0. 534	0. 618
	8. 0	10. 0	11. 6	13. 6	15. 2	18. 0	24. 8	30. 8	35. 6
8	0. 187	0. 229	0. 264	0. 305	0. 340	0. 408	0. 562	0. 694	0. 805
	8. 3	10. 1	11. 7	13. 5	15. 0	18. 1	24. 8	30. 7	35. 6
10	0. 229	0. 278	0. 333	0. 382	0. 423	0. 514	0. 701	0. 861	0. 979
	8. 1	9. 9	11. 8	13. 6	15. 0	18. 2	24. 8	30. 6	35. 6
12	0. 222	0. 271	0. 319	0. 368	0. 416	0. 500	0. 701	0. 874	1. 020
	6. 6	8. 1	9. 5	11. 0	12. 4	14. 9	20. 9	26. 1	30. 5
14	0. 243	0. 298	0. 353	0. 403	0. 451	0. 548	0. 770	0. 958	1. 124
	6. 6	8. 1	9. 6	11. 0	12. 3	14. 9	21. 0	26. 1	30. 7
16	0. 278	0. 340	0. 402	0. 465	0. 521	0. 632	0. 881	1. 096	1. 284
	6. 6	8. 1	9. 6	11. 1	12. 5	15. 1	21. 0	26. 2	30. 7

[1] Top value shown is ft^2/ft and bottom value is percent.

TABLE 11–9.—*Minimum open areas of screens in square feet per linear foot and the percentage of open area*[1]—Continued

[Wire-wound on pipe base (from Howard Smith Company)]

Screen size, inches	Pipe perfora-tions	Slot size, thousandth of an inch								
		8	10	12	14	16	20	30	40	50
4	0. 208	0. 083	0. 115	0. 135	0. 155	0. 174	0. 210	0. 291		
	17. 6	6. 4	9. 0	10. 6	12. 1	13. 6	16. 4	22. 8		
6	0. 310	0. 103	0. 126	0. 150	0. 172	0. 194	0. 236	0. 333	0. 418	0. 496
	17. 9	5. 6	6. 8	8. 1	9. 3	10. 5	12. 8	18. 0	22. 6	26. 9
8	0. 375	0. 133	0. 163	0. 193	0. 222	0. 251	0. 305	0. 430	0. 541	0. 638
	16. 6	5. 6	6. 9	8. 1	9. 3	10. 5	12. 8	18. 1	22. 7	26. 9
10	0. 408	0. 160	0. 202	0. 239	0. 275	0. 311	0. 379	0. 532	0. 670	0. 793
	14. 5	5. 5	6. 9	8. 1	9. 4	10. 6	12. 9	18. 1	22. 8	27. 0
12	0. 491	0. 193	0. 237	0. 281	0. 323	0. 366	0. 445	0. 626	0. 780	0. 931
	14. 7	5. 6	6. 8	8. 1	9. 3	10. 6	12. 9	18. 1	22. 7	26. 9
14	0. 525	0. 212	0. 260	0. 308	0. 355	0. 401	0. 487	0. 685	0. 862	1. 021
	14. 3	4. 9	6. 9	8. 1	9. 4	15. 3	12. 9	18. 0	22. 7	27. 0
16	0. 624	0. 211	0. 296	0. 341	0. 404	0. 457	0. 555	0. 799	0. 980	1. 162
	14. 9	4. 9	6. 9	7. 9	9. 4	10. 6	12. 8	18. 1	22. 7	27. 0
18	0. 691	0. 269	0. 332	0. 393	0. 452	0. 511	0. 622	0. 874	1. 083	1. 301
	14. 6	5. 6	6. 9	8. 1	9. 4	10. 6	12. 9	18. 1	22. 4	
20	0. 708	0. 300	0. 369	0. 436	0. 502	0. 566	0. 690	0. 970	1. 221	1. 416
	13. 5	5. 6	6. 9	8. 2	9. 4	10. 6	12. 9	18. 1	22. 8	26. 5

[1] Top value shown is ft^2/ft and bottom value is percent.

TABLE 11–9.—*Minimum open areas of screens in square feet per linear foot and the percentage of open area* [1]—Continued

[Louvre or shutter-type screen—standard (3/16- to 1/4-inch wall) (from Roscoe Moss Company)]

Screen size, inches	Slot size, inch					
	1/16	3/32	1/8	5/32	3/16	1/4
6	0. 017	0. 025	0. 039	0. 042	0. 050	0. 068
	0. 9	1. 4	1. 9	2. 4	2. 8	3. 9
8	0. 025	0. 038	0. 050	0. 063	0. 076	0. 101
	1. 1	1. 6	2. 2	2. 7	3. 3	4. 4
10	0. 027	0. 040	0. 055	0. 069	0. 083	0. 111
	0. 0	1. 1	1. 0	2. 1	2. 0	3. 0
12	0. 036	0. 055	0. 073	0. 092	0. 111	0. 147
	1. 0	1. 6	2. 1	2. 7	3. 3	4. 3
14	0. 036	0. 055	0. 073	0. 092	0. 111	0. 147
	0. 9	1. 4	1. 9	2. 4	2. 9	3. 8
16	0. 046	0. 069	0. 092	0. 115	0. 138	0. 183
	1. 0	1. 5	2. 1	2. 6	3. 1	4. 1
18	0. 046	0. 069	0. 092	0. 115	0. 138	0. 183
	0. 9	1. 4	1. 8	2. 3	2. 8	3. 7
20	0. 055	0. 083	0. 111	0. 138	0. 165	0. 222
	1. 0	1. 5	2. 0	2. 5	3. 0	4. 1

[1] Top value shown is ft²/ft and bottom value is percent.

TABLE 11–9.—*Minimum open areas of screens in square feet per linear foot and the percentage of open area* [1]—Continued

[Full flow screen (3/16- to 1/4-inch wall) (from Roscoe Moss Company)]

Screen size, inches	Slot size, inch					
	1/16	3/32	1/8	5/32	3/16	1/4
6	0. 050	0. 076	0. 101	0. 127	0. 151	0. 202
	2. 8	4. 3	5. 8	7. 3	8. 7	11. 6
8	0. 067	0. 101	0. 135	0. 169	0. 202	0. 235
	2. 9	4. 4	5. 9	7. 4	8. 9	10. 3
10	0. 106	0. 165	0. 222	0. 282	0. 346	0. 472
	3. 7	5. 8	7. 8	10. 0	12. 2	16. 7
12	0. 132	0. 206	0. 278	0. 353	0. 432	0. 589
	3. 9	6. 1	8. 3	10. 5	12. 9	17. 6
14	0. 132	0. 206	0. 278	0. 353	0. 432	0. 589
	3. 4	5. 4	7. 3	9. 2	11. 3	15. 4
16	0. 158	0. 247	0. 333	0. 424	0. 519	0. 707
	3. 6	5. 6	7. 6	9. 7	11. 8	16. 1
18	0. 184	0. 289	0. 389	0. 494	0. 605	0. 825
	3. 7	5. 9	7. 9	10. 1	12. 4	16. 9
20	0. 212	0. 331	0. 444	0. 564	0. 691	0. 944
	3. 9	6. 1	8. 2	10. 4	12. 7	17. 4

[1] Top value shown is ft²/ft and bottom value is percent.

TABLE 11–9.—*Minimim open areas of screens in square feet per linear foot and the percentage of open area* [1]—Continued

[134 shutter screen—3 gage (0.25-inch) wall (from the Layne and Bowler Company)]

Screen size, inches	Slot size, thousandth of an inch				
	30	55	80	105	130
4	0. 039	0. 072	0. 104	0. 138	0. 168
	3. 3	6. 1	8. 8	11. 7	14. 2
6	0. 074	0. 133	0. 196	0. 256	0. 318
	4. 3	7. 8	11. 5	15. 1	18. 7
8	0. 094	0. 172	0. 251	0. 329	0. 410
	4. 2	7. 7	11. 2	14. 7	18. 3
10	0. 126	0. 231	0. 336	0. 441	0. 545
	4. 5	8. 4	12. 2	16. 1	19. 8
12	0. 147	0. 270	0. 392	0. 514	0. 637
	4. 5	8. 3	12. 0	15. 7	19. 5
16	0. 200	0. 364	0. 532	0. 696	0. 863
	4. 8	8. 7	12. 7	16. 6	20. 6
20	0. 220	0. 403	0. 586	0. 770	0. 952
	4. 7	8. 5	12. 4	16. 3	20. 2
24	0. 283	0. 518	0. 754	0. 990	1. 225
	4. 5	8. 2	11. 9	15. 7	19. 5

[1] Top value shown in ft²/ft and bottom value is percent.

TABLE 11–9.—*Minimim open areas of screens in square feet per linear foot and the percentage of open area* [1]—Continued

[Punched screens—gravel guard well screen, 0.25-inch wall (from Doerr Metal Products)]

Screen size, inches	Slot size, inch			
	1/32	1/16	1/8	3/16
8	0. 054	0. 120	0. 263	0. 410
	2. 5	5. 7	12. 5	19. 5
10	0. 069	0. 153	0. 335	0. 522
	2. 6	5. 8	12. 8	19. 9
12	0. 084	0. 185	0. 407	0. 634
	2. 7	5. 9	12. 9	20. 1
14	0. 098	0. 218	0. 478	0. 746
	2. 7	5. 9	13. 0	20. 2
16	0. 111	0. 245	0. 538	0. 839
	2. 7	5. 9	12. 8	20. 0
18	0. 126	0. 278	0. 610	0. 951
	2. 7	5. 9	12. 9	20. 2
24	0. 160	0. 352	0. 773	1. 21
	2. 5	5. 6	12. 3	19. 2

[1] Top value shown is ft²/ft and bottom value is percent.

TABLE 11–9.—*Minimum open areas of screens in square feet per linear foot and the percentage of open area* [1]—Continued

[Slotted pipe [2]—horizontally slotted casing]

Pipe size, inches	Slot size, inch			
	1/8	5/32	3/16	1/4
10	0. 061	0. 076	0. 090	0. 120
	2. 1	2. 7	3. 2	4. 3
12	0. 074	0. 092	0. 109	0. 145
	2. 2	2. 8	3. 3	4. 3
14	0. 085	0. 106	0. 127	0. 167
	2. 3	2. 9	3. 5	4. 6
16	0. 098	0. 122	0. 145	0. 192
	2. 3	2. 9	3. 5	4. 6
18	0. 109	0. 136	0. 163	0. 216
	2. 3	2. 9	3. 5	4. 6
20	0. 115	0. 144	0. 173	0. 228
	2. 2	2. 8	3. 3	4. 3

[1] Top value shown is ft²/ft and bottom value is percent.
[2] Slots are 1.5 inches long on 6⅜-inch centers on a plane around the pipe or on 1¼-inch centers vertically with each horizontal row staggered.

TABLE 11–9.—*Minimim open areas of screens in square feet per linear foot and the percentage of open area* [1]—Continued

[Oil field milled slotted casing [2]]

Pipe size, inches	Slot size, thousandth of an inch					
	100	120	140	180	200	250
6	0. 017	0. 020	0. 023	0. 030	0. 033	0. 042
	0. 98	1. 0	1. 0	2. 0	2. 5	3. 2
8	0. 022	0. 027	0. 031	0. 040	0. 044	0. 056
	0. 97	1. 2	1. 4	1. 8	1. 9	2. 5
10	0. 028	0. 033	0. 039	0. 050	0. 056	0. 069
	1. 0	1. 0	1. 1	1. 8	2. 0	2. 4
12	0. 033	0. 040	0. 047	0. 060	0. 067	0. 083
	0. 99	1. 2	1. 4	1. 8	2. 0	2. 5
14	0. 039	0. 047	0. 054	0. 070	0. 078	0. 097
	1. 1	1. 3	1. 5	1. 9	2. 1	2. 6
16	0. 044	0. 053	0. 062	0. 080	0. 089	0. 111
	1. 1	1. 3	1. 5	1. 9	2. 1	2. 7
18	0. 050	0. 060	0. 070	0. 090	0. 100	0. 125
	1. 0	1. 2	1. 4	1. 8	2. 1	2. 5
20	0. 056	0. 067	0. 078	0. 100	0. 111	0. 139
	1. 0	1. 2	1. 4	1. 8	2. 1	2. 5

[1] Top value shown is ft^2/ft and bottom value is percent.
[2] Vertical 2-inch-long slots spaced at two diameter centers on staggered horizontal rows around the pipe. Vertical spacing or horizontal rows on 3-inch centers.

pump housing casing by neoprene, plastic, cement, other nonmetallic materials, or couplings.

In straight wall wells where the diameter of the hole results in an annual space about the screen greater than about 2 inches, a formation stabilizer should be used (sec. 11–11(b)).

A common misconception holds that straightness and plumbness are unimportant in the screen assembly since the pump is not set in the assembly. The hole should be straight enough, however, to permit installing the screen without having to force or drive the screen down. If installed crooked or too much out of plumb, the screen is subject to bending stresses that may cause slot enlargement or collapse. Plumbness of the screen therefore should meet the same criteria as the casing in not deviating from the vertical more than two-thirds of the inside diameter of the screen per 100 feet and the axis of the screen assembly should coincide with that of the pump housing casing or riser pipe in the vicinity of their junction.

11–5. Drive Shoes.—When casing is driven into place, particularly in cobbly or bouldery materials, the bottom of the casing should be reinforced with a hardened steel ring or drive shoe which is screwed or welded to the casing. The drive shoe should have an outside diameter and beveled cutting edge that are about the same as that of couplings for the diameter of pipe being used. The cutting edge shaves irregularities off the side of the hole as the casing is driven and will split or force large rocks into the side of the hole, thus preventing the bottom of the casing from collapsing. Commercially available drive shoes come in two patterns, regular and Texas. The Texas pattern is longer and somewhat more rugged than the regular pattern and is used where driving is particularly difficult. The selection of the type of shoe to be used is usually left to the discretion of the contractor.

11–6. Reducers and Overlaps.—(a) *Description and Purpose.*—When drilling with cable tools, a point is reached where it is no longer possible, due to skin friction and other factors, to drive the casing further. When this occurs, a smaller casing is telescoped into the installed casing and drilling is continued using a smaller bit. On completion of the drilling, the smaller casing may be cut off at a point some distance above the bottom of the larger casing. On some deep holes there may be six or more such reductions. The starter casing should always be of suitable diameter so that the diameter of the pump chamber casing at the depth of pump setting will be adequate.

In other designs, particularly when the hole is drilled uncased as with most rotary rigs, the pump chamber casing may be directly attached to the screen assembly riser pipe or extension by a coupling or reducer. The entire casing and screen assembly is then lowered into the well as a continuous string of pipe with additional lengths being added at the surface as the string is lowered. If the string is allowed to rest on the bottom, the weight of the entire string is carried by the screen, which is the weakest section; thus, the possibility of buckling or collapse of the screen is increased. Because of this hazard, single string construction should provide for maintaining the casing and screen string in tension until the well is developed and permanently anchored at the surface. This is particularly important in wells exceeding about a 100-foot depth. Designs in which the screen assembly is telescoped into place offer some advantage since the screen does not support the entire column weight. If necessary, the screen can be withdrawn and replaced, an operation which is impossible with a solidly connected line of pipe (sec. 11–2(a)).

(b) *Design Particulars of Reducers and Overlaps.*—A commonly used length of overlap between casing and screen assembly is 5 to 10 feet. In extremely deep wells or where the possibility of settlement of the screen assembly is present, more overlap may be necessary [1,9,10]. An overlap should always be sealed as described in section 11–7.

Reducers used between the casing and the screen assembly are, in many instances, fabricated by the contractor or local shops, both of which have a tendency to use flat conical sections. From the standpoint of hydraulic efficiency and strength, the upper straight end and conical section of reducers should be fabricated of the same weight or wall thickness and material as the larger pipe to which they will be attached. The conical section of the reducer should have a length at least 10 times the difference in diameter of the two pipes which it will connect.

11–7. Seals.—(a) *Description and Purpose.*—The grout seal commonly placed around the permanent surface casing or the pump chamber casing is primarily a sanitary seal which should be of sufficient thickness, depth, and imperviousness to prevent any surface water or poor quality ground water from entering the well. Native clay, bentonite, and other materials are also used as grout and may be satisfactory from the sealing standpoint. However, a good cement-based grout mixed with a proper amount of bentonite or aluminum powder will produce a better seal as well as serve other useful functions. Such a mixture protects the casing against corrosion attack, and if the casing is removed by corrosion, the grout serves as a concrete casing. If correctly installed, grout forms a bond between the casing and the soil, stabilizes the soil about the well, and sometimes acts as a keystone to limit the extent of upward caving. From the standpoint of a sanitary seal, grouting the full length of casing is probably unnecessary; but in view of the other functions the grout performs, the practice is recommended. Once the placement equipment is installed, additional amounts of grout are relatively inexpensive.

Where casings or screen assemblies are telescoped down the hole, a seal should be placed at the top of each telescoped section. The most commonly used seal is a commercial, swaged lead seal which fits on the top of the smaller casing and is swaged out with a special tool against the inside of the larger casing. A correctly installed swaged lead seal is practically leakproof so far as permitting the entrance of water from outside the casing. In addition, the seal prevents sand and gravel from being carried up the annular space and into the well when the well is pumped. A swaged lead seal also permits casing and screen assemblies to be pulled out of a well if necessary.

A recent development is neoprene rubber seals which are vulcanized around the smaller casing at the factory. The outside diameter of the seals, which have flexible lips, is slightly larger than the inside diameter of the larger casing. If the inside of the casing is wet when the smaller casing is telescoped inplace, the seals slide down readily. When inplace, they form a tight seal which keeps out undesirable water and stops upward movement of sand or gravel in the annular space. In addition, seals act as insulation, separating dissimilar materials which otherwise would cause galvanic corrosion. They permit easier removal of casing or screen assemblies, if necessary, than other seals. Where insulation as well as sealing is desired, usual practice is to use two or more of the seals spaced about 1½ to 3 feet apart to assure separation of the dissimilar metals in the overlap.

Where the difference in diameter of the overlapping casings is sufficient to permit insertion of a ½-inch or larger pipe into the annular space, a neat cement-bentonite grout seal may be placed. The grout acts also as an insulation. It has sufficient strength to resist the flow of water but is readily broken if the lower casing needs to be pulled.

Designs may be encountered in which a grout seal is placed in the annular space between a surface casing and the pump chamber casing and the surface casing is not grouted in. The theory is that caving of unstable materials will create a positive seal about the surface casing. This cannot be depended upon, so permanent surface casing should always be grouted in. However, the top of the annular space between the surface casing and the pump chamber casing should always be tightly sealed with an expanding packer, a concrete plug, or a steel ring welded to the wall of the pump chamber casing and the top of the surface casing [1,9,10].

(b) *Design Particulars for Seals.*—The grout seal around surface or pump chamber casing should have a minimum thickness of 1½ inches about the pipe or about the couplings, if they are used. The annular space should be flushed with water before commencing the grouting operations. The grout should be introduced at the bottom of the space to be grouted and should be placed in a continuous operation. If cement grout is used, it should be entirely placed before the occurrence of initial set. AWWA-A100-66, Standard for Deep Wells, Section A1-8.4 outlines various acceptable methods used for placing grout. For most water wells, however, placement through a tremie pipe is acceptable. Section 15-3 gives specifications for acceptable grout materials.

11-8. Gravel or Concrete Base.—When a well is bottomed in fine sands, plastic clay, or other similar soft or unstable material, a recommended practice is to overdrill the well 3 to 4 feet. This interval should be filled with coarse gravel or concrete to provide a firm base for the casing and screen.

11-9. Centering Guides.—Where casing or screen assemblies over 40 feet long are installed in holes having nominal diameters 2 inches or larger than the outside diameter of the casing, centering guides should be installed. Such guides hold the casing in the center of the hole as well as offer support against bending and buckling because of axial and unbalanced horizontal loads. Centering guides are essential for centering casing and screens for grouting and gravel packing. The guides should be placed at the bottom and at about 40- to 50-foot intervals up the hole. In gravel packed wells, care should be exercised to keep the centering guides on approximately straight lines from top to bottom so as not to interfere with the insertion of tremie pipes. Centering guides should not be welded directly to the screen proper if avoidable. Preferably, a short section of blank casing to which the centering guides can be welded should be inserted in the screen at approximately the desired intervals. Centering guides may be of wood, strap steel, or alloy. Metallic guides should always be of the same alloy as the casing or screen assembly to which they are attached. The guides are set in a plane around the casing at 90° or 120° intervals. Figures 11-11 and 11-12 show designs of acceptable centering guides for various conditions which may be encountered [1,10].

11-10. Tremie Pipes.—In gravel packed wells, the pack material is generally installed through one or two temporary tremie pipes which are withdrawn in stages as the pack is placed. Tremie pipes consist of nominal 2- to 4-inch coupled steel pipe. The diameter depends on the pack, grain size, clearance, and other factors. The design of the well must provide for adequate annular space to permit passage of the tremie pipes including couplings. Many different designs are used for accommodating gravel packs in wells. Most designs provide for the gravel pack to be extended for 20 or more feet above the top of the uppermost screen. In some designs, the annular space above the pack is filled with grout and no provision is made for the addition of gravel pack if required.

Prudent design calls for both adequate and protected storage for the pack such as partially within a surface casing. Also, a means of permanent replenishment of the pack from the surface should be provided. Such features minimize the possibility of direct aquifer contact with the screen in the event of excessive pack settlement. Storage can be provided between the pump chamber and surface casings. Replenishment can be accomplished through permanent tremie pipes installed when the well is completed. One or two permanent tremie pipes are installed between the surface and pump chamber casing and extending to the desired depth. If required, a concrete seal is then placed around them. The top of permanent tremie pipes should always be threaded and sealed with a screw cap

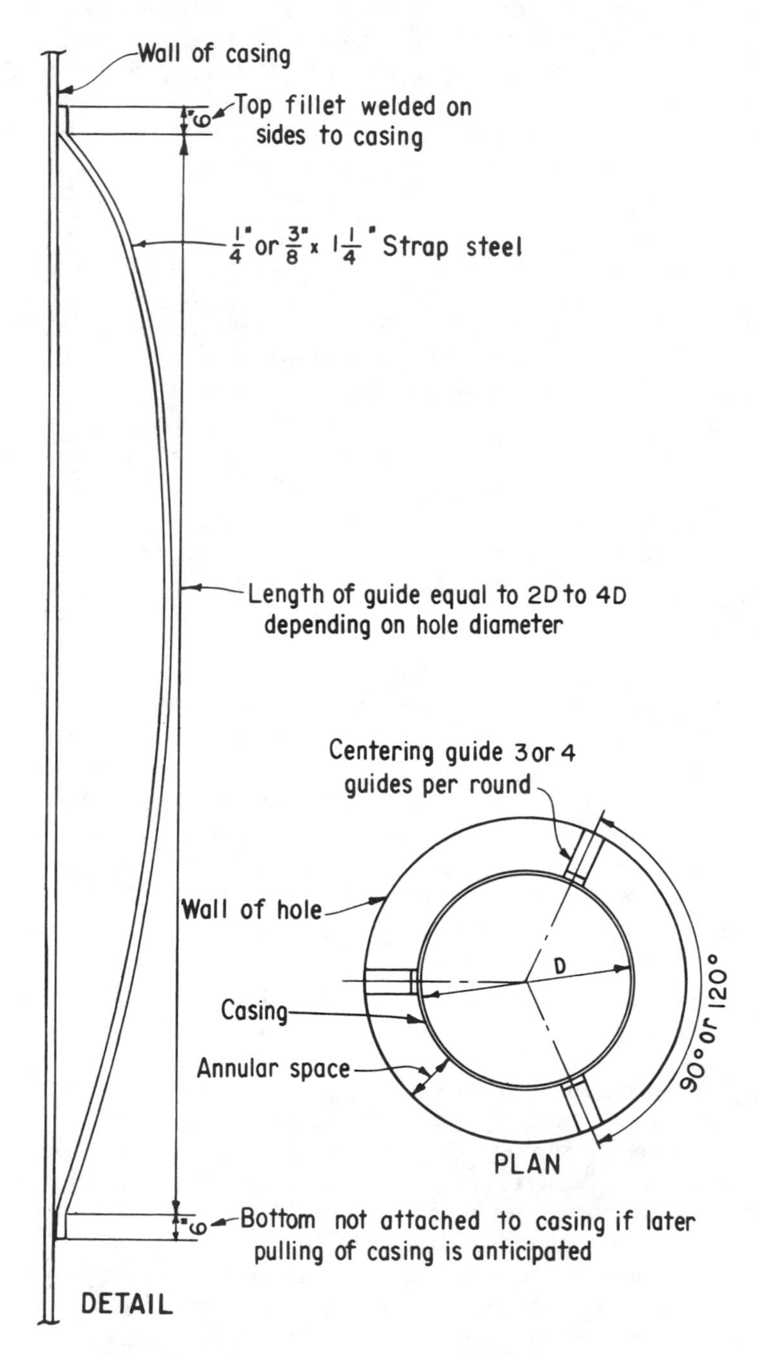

FIGURE 11-11.—Steel strap centering guide. 103-D-1498.

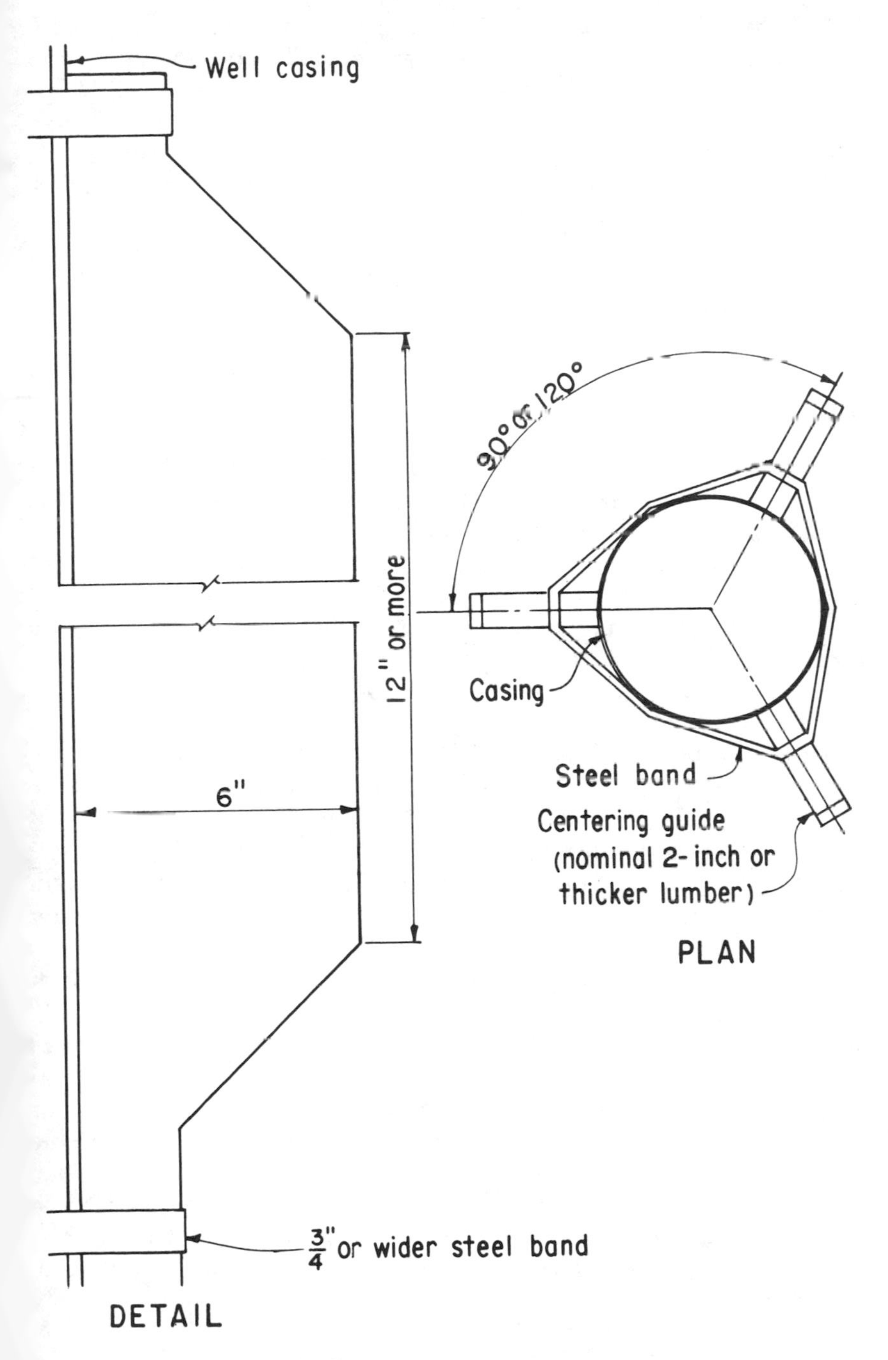

FIGURE 11–12.—Wood block centering guide. 103–D–1499.

[1,10]. Sizes of temporary tremie pipe may be left to the discretion of the contractor, but permanent tremie pipe should be of adequate size and weight for permanent service.

11-11. Gravel Packs and Formation Stabilizers.—(a) *Description and Purpose.*—Where casing and screen is set in an oversized hole where the annular space is larger than 2 inches but gravel pack construction is not intended, a formation stabilizer should be placed in the annular space. The stabilizer does not need to be carefully selected in regard to gradation as long as the smaller grains are larger than the screen slot size and the largest are three-eighths of an inch or less. The purpose of the formation stabilizer is to support the pipe against unbalanced forces which might arise during development of the well and to facilitate development of the well [10].

The principal functions of a gravel pack are: (1) to stabilize the aquifer and minimize sand pumping, (2) to permit use of the largest possible screen slot with resultant maximum open area, and (3) to provide an annular zone of high permeability, thus increasing the effective radius of the well and the yield. A gravel pack normally should not be used unless there are conditions that make pack use mandatory or at least advantageous. Use of a pack usually increases the cost and difficulty of construction of a well.

Gravel packs[2] should be designed to have a small coefficient of uniformity, and grain sizes should be carefully selected to match the aquifer material. Screens should then be selected to pass not more than 5 percent of the pack material. Maximum grain size of a pack should not exceed three-eighths of an inch if placed through a nominal 4-inch tremie pipe. Minimum design thickness depends on the ability to place the pack. A ½-inch-thick pack is theoretically adequate. Maximum design thickness should not exceed 8 inches because of the difficulty of development through a thicker pack [1,4,6,9,10,12].

Conditions which especially favor the use of a pack include: (1) presence of a fine, uniform sand aquifer; (2) presence of a layered aquifer with alternating sand and clay layers; (3) a requirement for maximum yield from a marginal aquifer; and (4) presence of a friable sandstone or similar aquifer.

(b) *Placement Procedures and Design of Gravel Packs.*—If the design permits, the screen assembly should be supported at the surface and kept under tension while the gravel pack is being placed.

Gravel pack should be placed in a manner which assures complete filling of the annular space and minimizes bridging and segregation. In wells drilled to depths of up to 500 feet, gravel pack is best placed

[2] Although referred to as gravel packs, material is most commonly in the sand size range.

through two tremie pipes placed 180° apart and extending initially to within about 5 feet from the bottom of the hole. The inside diameter of the tremie pipes should be at least 12 times the diameter of the coarsest pack material if placed by gravity and 10 times if pumped. As the pack is being placed, the tremie pipes are raised in such a manner that the free fall of pack material below the bottom of the pipe does not exceed 5 feet. The placement of gravel should be continued at a uniform rate until completed. Gravel may be poured or shoveled into the tremie pipes dry or may be washed or pumped in. If washed in, a constant flow of gravel and water is fed into each tremie pipe. If pumped in, the gravel-water ratio should be 1 yard of gravel to 1,800 to 3,000 gallons of water. The gravel-water ratio will depend upon the stability of the hole walls, the grain size of the gravel pack, and the type and size of pump available that will permit a uniform constant pumping rate without causing caving of the wall of the hole.

In rotary drilled holes, the fluid in the hole should be circulated and the viscosity reduced by dilution with water until the Marsh Funnel velocity is less than 30 seconds before gravel is introduced into the well. A desirable practice, if well design and available equipment permit, is to pump fluid from the bottom of the well as the gravel pack is pumped in. To avoid collapse of the hole after the drilling fluid has been thinned, the pumping rate should be adjusted to maintain the water level in the well above the static water level. In some cases, however, the aquifer may be too permeable to permit building up an adequate head above the static water level.

Other methods have been devised to place gravel pack in deep rotary drilled wells where use of tremie pipes would be impracticable. Such equipment as the cross-over tool permits pumping of gravel through the drill pipe and into the annulus.

Numerous formulas based on standard mechanical analyses (MA) of grain size have been developed for the selection of gravel pack gradations. None is entirely satisfactory, but those described in this discussion are usually adequate. A number of terms commonly used in the literature must be carefully examined in regard to their meaning. In most ground-water literature, the grain size terms, D_{10}, D_{60}, D_{100}, etc., refer to the percent retained sizes. Bureau practice is to refer to them as the percent passing or the percent smaller than. For example, the uniformity coefficient C_u in Bureau terminology is the $\frac{D_{60}}{D_{10}}$ size ratio, while in most other literature it is the $\frac{D_{40}}{D_{90}}$ size ratio.

The terminology used in most ground-water literature is used in the following discussion and all references to an MA plot refer to the percent retained values on the right side of Bureau form 7–1415 (fig. 7–10). The criteria for (1) and (2) below generally have been taken from Kruse [6] but have been modified to conform to Bureau field experience. Criteria for (3) have been taken from [10].

(1) Where the uniformity coefficient C_u of aquifer material is less than 2.5:

a. It is preferable to use gravel pack with C_u between 1 and 2.5 and with the 50-percent size a maximum of 6 times the 50-percent size of the aquifer material.

b. If uniform pack material is not readily available, it is acceptable to use gravel pack material with a C_u between 2.5 and 5. Select the gravel pack to have a 50-percent size not greater than 9 times the 50-percent size of the aquifer.

c. Normally, the screen slots should not pass more than 10 percent of the pack material. However, under certain conditions, this may be exceeded.

(2) Where uniformity coefficient C_u of aquifer material is between 2.5 and 5:

a. It is preferable to use gravel pack material with C_u between 1 and 2.5 and with the 50-percent size of pack material not more than 9 times the 50-percent size of the formation.

b. An acceptable but less desirable criterion is to use gravel pack material with C_u between 2.5 and 5 and the 50-percent size of the pack material not more than 12 times the 50-percent size of the formation.

c. The screen slots should not pass more than 10 percent of the pack material unless conditions permit.

(1)a. and (2)a. are the most efficient packs and most readily developed, but pack material with a low uniformity coefficient is sometimes not readily available and may be costly.

(3) Where uniformity coefficient C_u of the formation is greater than 5:

a. Multiply the 70-percent retained size of the formation by 6 and 9 and locate the points on the graph.

b. Through these points draw two parallel lines representing materials having uniformity coefficients of 2.5 or less.

c. Prepare specifications for gravel pack material falling between the two lines.

d. Select a screen slot size which will retain 90 percent or more of the pack material.

Regardless of the criteria used in selecting the gravel pack, the gravel should be washed, screened, rounded where possible, abrasive-resistant, dense, and of siliceous materials with less than 5 percent flat grains. The pack should contain not more than 5 percent earthy or soft materials such as clay, shale, or anhydrite or readily soluble materials such as limestone or gypsum.

A newly introduced product consists of a well screen surrounded by a prefabricated pack of epoxy-cemented sand grains. The usefulness of this method has not been determined at the time of this writing.

A mechanical analysis should be run on three random samples of the pack material. Each sample should be taken from a different part of the shipment to assure conformance to the gravel pack gradation requirements. The gravel is acceptable if 95 percent passes the coarsest designated screen, plus or minus 8 percent of that designated is held on smaller screens, and not more than 10 percent passes the finest designated screen.

Bureau form 7–1415 (fig. 7–10), while based on the U.S. Standard Series (fourth root of two ratio), does not include all the screen sizes available. Uniformity coefficients of materials plotting between two of the adjacent solid lines on form 7–1415 will be around 1.55 to 1.60 and between two alternate solid lines between about 2 and 2.8. At times, closer approximations may be desirable in selecting slot sizes and similar values. Table 11–10 lists the approximate volume of the annulus between various sizes of holes and casings.

11–12. Pump Foundations.—Surface-mounted pumps must be supported on foundations capable of resisting all loads. Separate pedestals may also be necessary to elevate the pump above potential surface water levels. Supporting pumps by mounting them directly onto well casings is not recommended. Foundations for vertical shaft pumps installed over wells should be tied to the surface or pump chamber casing and should support the pump as closely as possible at the vertical centerline of the pump column, provide support for the full area of the head base, and minimize all possible deflections of the pump base. Foundations generally should be constructed of minimum 3,750 pounds per square inch concrete placed on solid ground. A typical schematic section of a pump foundation is shown on figure 11–13. The type of drilled well and local government regulations may affect some features of the foundation, but recommended foundations have a minimum depth and soil bearing area as determined by:

$$B=\frac{144W}{K}+\frac{\pi d^2}{4}, \text{ in inches} \qquad (1)$$

$$T=12+\frac{B-A}{2}\left(\frac{K}{2{,}000}\right), \text{ in inches} \qquad (2)$$

TABLE 11–10.—*Volume of annulus between casing or screen and hole for grout and gravel pack*

Casing outside diameter, inches	Hole diameter, inches	Annulus, cubic yards per 100 ft
2⅜	4	0.2
	6	0.6
	8	1.2
4½	8	0.9
	10	1.6
	12	2.5
6⅝	10	1.1
	12	2.6
	14	3.1
	16	4.3
	18	5.7
8⅝	12	1.4
	14	2.5
	16	3.7
	18	5.0
	20	6.6
10¾	14	1.6
	16	2.8
	18	4.2
	20	5.7
	22	7.4
	24	9.3
12¾	16	1.9
	18	3.2
	20	4.8
	22	6.5
	24	8.4
	26	10.8
14	18	2.6
	20	4.1
	22	5.6
	24	7.7
	26	9.7
	28	11.9
16	20	2.9
	22	4.6
	24	6.4
	26	8.5
	28	10.7
	30	13.0
18	22	3.2
	24	5.1
	26	7.1
	28	9.3
	30	11.7
	32	14.1
20	24	3.5
	26	5.6
	28	7.8
	30	10.1
	32	12.6
	34	15.3
22	26	3.9
	28	6.1
	30	8.4
24	28	4.2
	30	6.6
	32	9.0

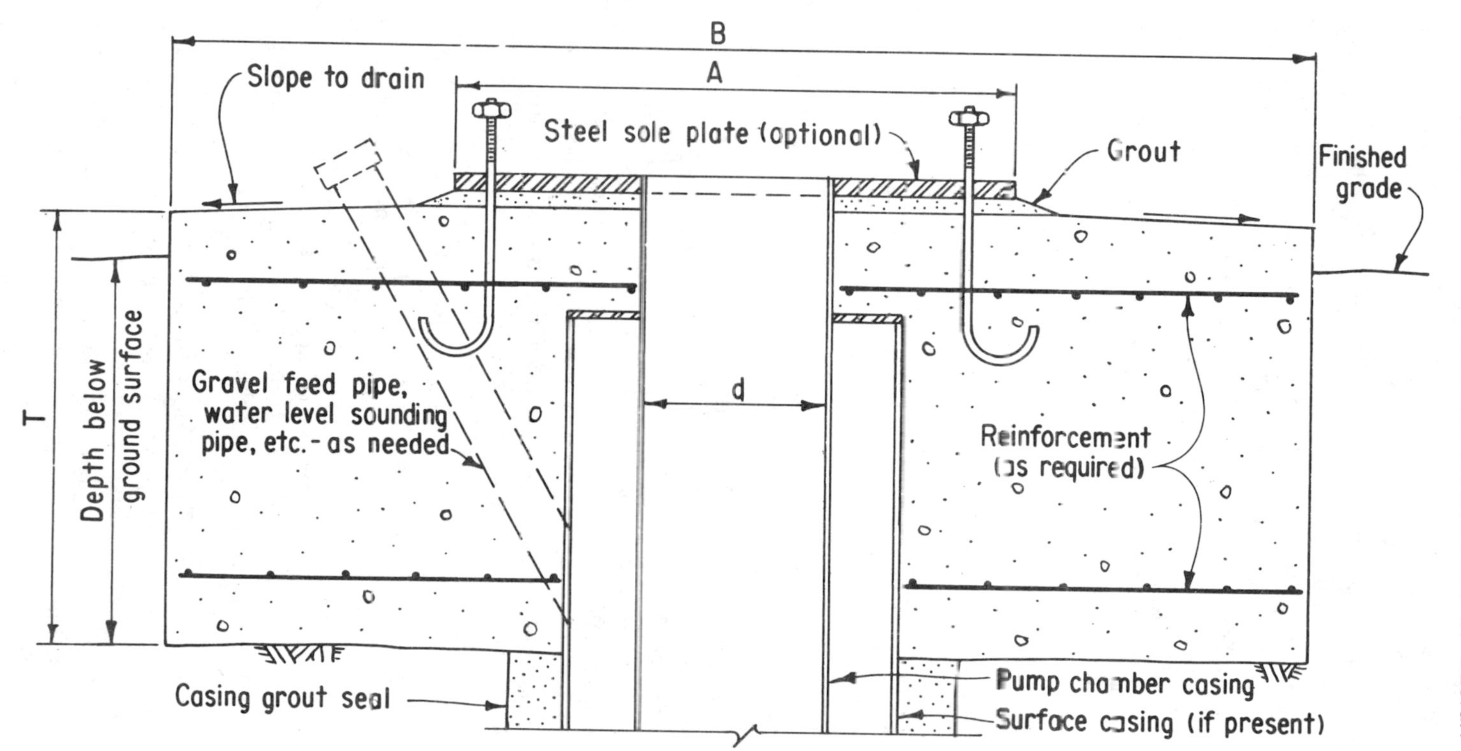

FIGURE 11-13.—Schematic section of a typical concrete pump foundation. 103-D-1500.

where:

B=minimum width to support the pump, in inches for a square foundation base,

W=total weight in pounds to be supported by the foundation (includes, as a minimum, the total weight of the driver and of the pump with a full column of water [11]),

d=inside diameter of pump housing casing or surface casing, in inches,

T=minimum required thickness of foundation, inches,

A=width or diameter of discharge head base, inches, and

K=recommended soil bearing factor in lb/ft^2 for various soil types taken from the following tabulation:

Estimated Bearing Capacities of Various Soil Types

Type of soil:	K, lb/ft^2 *of area*
Clay	2, 000
Packed gravel	16, 000
Compacted sand	8, 000
Dry sand	4, 000
Alluvial topsoil	1, 000

Steel reinforcement of the concrete foundation is recommended where T is greater than 2 feet and where B–A is greater than T/2. The pump should be secured to the foundation with anchor bolts. Setting plans indicating head base dimensions are available from pump manufacturers. Where necessary, a pump foundation pedestal should be constructed to raise the pump base above the elevation of any possible floodwaters or surface runoff which might inundate the area. If a foundation pad is constructed about the pedestal, it should be an integral part of the pedestal and its surface should slope gently away from the pedestal so water will not accumulate around it. Pump pits are restricted in most States especially for domestic or municipal water supplies and are not recommended for Bureau installations. If protection of the pump is required, a surface pumphouse is preferred. Pitless adaptors, however, if approved by the Water Systems Council or the National Sanitation Foundation, are satisfactory particularly for small capacity, submersible pump installations.

11–13. Special Well Types.—(a) *Drainage Wells.*—Drainage wells are usually conventional ground-water wells designed for the special purpose of relieving and controlling a high water table. For drainage purposes, wells are designed to prevent the water table from encroaching within a certain depth below the land surface and are deliberately located to interfere with each other to accomplish this purpose.

For an area to be susceptible to drainage by wells, a suitable aquifer must be present and the soil lying between the root zone and the aquifer must have adequate vertical permeability to permit deep percolation.

There is little basic difference in the design of a well for drainage and of a well for the production of water. However, there is a distinct difference in the criteria used in the design. In drainage wells, the basic purpose is to lower and maintain the water level to a given depth within a given period of time. A given volume of water must be removed to accomplish this purpose. Three factors—volume, drawdown, and time—are the parameters that, along with the aquifer characteristics, will give the most economical installation from the standpoint of initial and operational costs.

(b) *Inverted Wells.*—A special type of well, called an inverted, recharge, or injection well, is used to return surplus or unwanted surface water to an aquifer. Such wells usually operate by gravity and have been used, where geological conditions are favorable, to dispose of irrigation waste. Other applications include injection of freshwater to build a barrier against intrusion of saltwater, disposal of industrial wastes and treated sewage effluent, and recharge of ground water.

The design and construction methods used for inverted wells that operate by gravity are similar to those used for pumping wells. Where injection pressures are used, however, special design features may be necessary to control such pressures. Sometimes inverted wells may be designed with backwashing and flushing features to permit cleaning the wells if they become plugged by silt or other foreign matter being carried into them with the recharge water. Usually when large volumes of surface water must be recharged through inverted wells, settling ponds and filters are used to reduce the sediment load. In some instances, the recharge water may be chlorinated or otherwise treated. Without such precautions, use of inverted wells may promote serious ground-water contamination.

(c) *Pressure Relief Wells.*—Another special purpose well, called a pressure relief well, is used to reduce and control excessive artesian head. As the name implies, the purpose of this type of well is to reduce the pressure in artesian aquifers, thereby reducing the upward leakage of ground water through the overlying materials or reduce hydraulic uplift. Such wells have been used to drain agricultural lands and reduce pressure under engineering structures such as dams and powerplants and unstable earth masses such as landslides.

(d) *Collector Wells.*—In many areas, aquifers are too thin, contain poor quality water, or for other reasons cannot furnish water in the desired quantity or quality to standard wells. Under such circumstances, a collector well may be a solution. A collector well commonly consists of a concrete caisson 6 to 15 feet in diameter which is sunk to an adequate depth to directly intercept a thin aquifer or to permit horizontal screens to be extended radially into such an aquifer [4,12].

Drawdown resulting from such a well is spread over a relatively large area and is less than that which would result from a single well pumping the same volume. Also, temperature and quality of the water may be subject to a degree of control. The collecting-type well requires intensive local exploration and testing to determine conditions and data for design purposes. Construction, which may take 10 months to a year, requires special skill, knowledge, and equipment and is usually expensive. However, many installations have been constructed and operated economically where conditions were such that other types of development were impracticable.

11–14. Bibliography.—

[1] Ahrens, T. P., "Basic Considerations of Well Design," Water Well Journal, vol. 24, No. 4, 5, 6, and 8, 1970.

[2] "Specification for Line Pipe," API Std 5L, American Petroleum Institute Division of Production, Dallas, Tex., March 1963.

[3] "AWWA Standard for Deep Wells," American Water Works Association, AWWA–A100–66, New York, 1967.

[4] Campbell, M. D., and Lehr, J. H., "Water Well Technology," McGraw-Hill, New York, 1973.

[5] "Water Well Standards—State of California," California Department of Water Resources Bulletin 74, Sacramento, 1968.

[6] Kruse, E. G., "Selection of Gravel Packs for Wells in Unconsolidated Aquifers," Colorado State University, Agricultural Experiment Station, Technical Bulletin No. 66, Fort Collins, March 1960.

[7] "Manual of Water Well Construction Practices for the State of Oregon," second edition, Oregon Drilling Association, Salem, 1968.

[8] Pennington, W. A., "Corrosion of Some Ferrous Metals in Soil with Emphasis on Mild Steel and on Gray and Ductile Cast Irons," Bureau of Reclamation, Chemical Engineering Branch Report No. ChE–26, March 9, 1965.

[9] Reinke, J. W., and Kill, D. L., "Modern Design Techniques for Efficient High Capacity Irrigation Wells," Paper No. 70–732, presented at American Society of Agricultural Engineers Winter Meeting, December 8–11, 1970.

[10] "Ground Water and Wells," UOP Johnson Division, St. Paul, Minn., 1966.

[11] "Turbine Pump Facts," Vertical Turbine Pump Association, Pasadena, Calif., 1962.

[12] Walton, W. C., "Ground Water Resources Evaluation," McGraw Hill, New York, 1970.

Chapter XII

INFILTRATION GALLERIES

12–1. Purpose of Infiltration Galleries.—In many areas, subsurface conditions preclude ground-water development using normal well construction. Such conditions include the presence of thin aquifers or a thin freshwater layer underlain by saline water. In such areas, an infiltration gallery may permit adequate development of ground water.

An infiltration gallery may be considered a horizontal well or subsurface drain that intercepts underflow in permeable materials or infiltration of surface water.

12–2. Basic Components of an Infiltration Gallery.—Infiltration galleries usually are constructed to discharge into a sump whose bottom is some distance below the invert of the gallery screen and casing. The sump may be of almost any dimensions but commonly is a circular or square structure 4 to 8 feet in diameter or on a side. Depth should be adequate to permit the pump bowls to be set with adequate submergence and clearance. The sump is usually cased with concrete or corrugated culvert pipe, although wood shoring, brick, or concrete block have been used.

The bottom is sealed with concrete or a metal plate. The top is finished with a reinforced concrete slab incorporating an inspection and cleanout manhole and a hole through which the pump is installed, see figure 12–1. All points of possible leakage through the top are sealed with mastic, compression seals, or other waterstops. The top of the sump should be at an elevation which will preclude its submergence by any flood flows or surface runoff.

The conductor pipe, usually fabricated of corrugated pipe, leads from the sump, to which it is tightly sealed, to the screen or manifold of the gallery. It may be horizontal or slope slightly downward toward the sump. Where more than one screen is used, the conductor pipe may connect to a manifold from which two or more screens, possibly of somewhat smaller diameter, extend into the permeable material. If only one screen is used, it should have the same diameter as that of the conductor pipe and may be normal to it or an extension of it. Depending upon conditions and purpose, the screen may be perforated corrugated culvert, cage-type wire-wound screen, slotted concrete, asbestos-cement pipe, or other perforated materials.

Unless placed in very clean, permeable gravels, the screen should always be gravel packed. The same criteria are used in selecting the

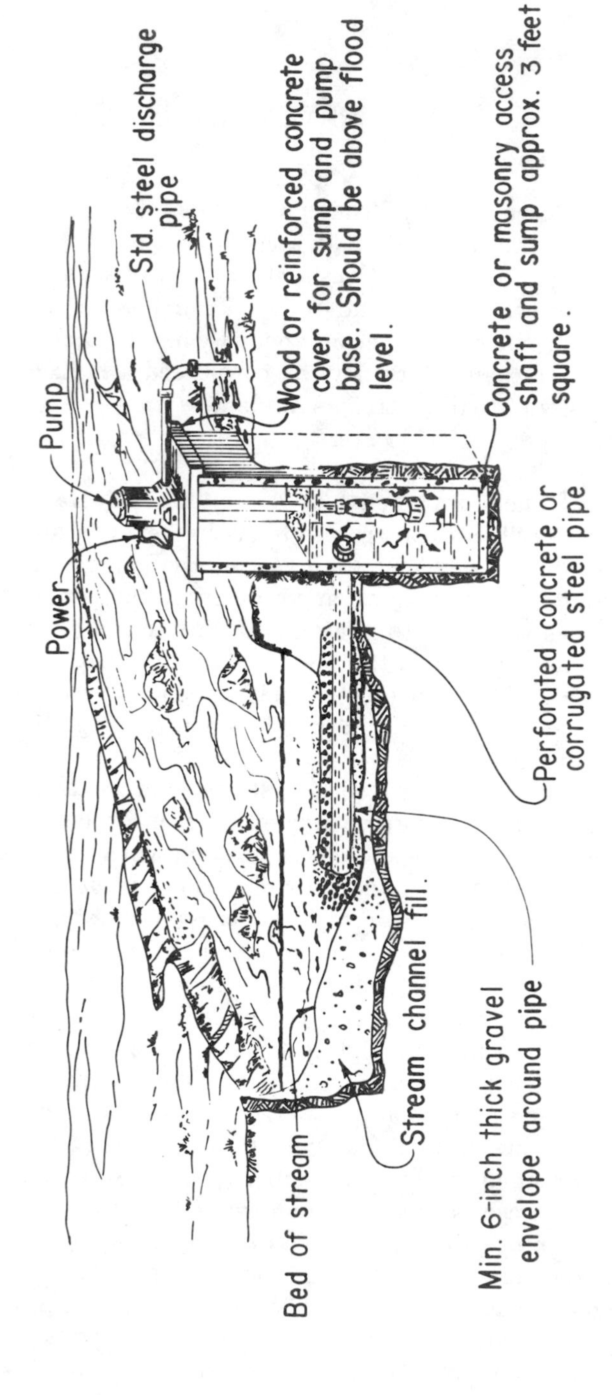

FIGURE 12-1.—Schematic section of an infiltration gallery constructed to intercept infiltration from a stream. 103-D-1501.

gravel pack and screen slot sizes as are used for gravel packed wells (sec. 11–11). Other criteria are used in determining the thickness of the pack.

Under some circumstances, the gravel pack is protected by an overlying layer or a diversion wall of riprap.

Provision should be made to permit back flushing of the gallery from a fairly large capacity storage tank or by direct pumping of clear stream or lake water into the sump.

12–3. Types of Infiltration Gallery Installations.—In some areas, lakes or perennial streams may be present which are not suitable for simple intake structure pumping directly from the open water body because of silt load, shoreline slope, presence of inert and physiologically harmless contaminants, rapid and unpredictable changes in water level, excessive wave action, or danger from ice pressure.

In any of these or similar cases, if there is a predictable minimum water depth, an infiltration gallery may be a source of an adequate and suitable water supply. The type of installation will depend upon the nature and permeability of the streambed or lakebed involved. In the case of a relatively impermeable bed, the gallery is placed beneath the channel or lakebed in a suitable excavation which is backfilled to completely surround and cover the gallery with selected gravel pack which serves as an artificial aquifer.

Where aquifer materials have good permeability but a relatively thin freshwater lens that is underlain by salt or brackish water, as is common near many coastal areas, the gallery is placed normal to the direction of flow of the ground water and at a depth which will preclude drawdowns great enough for saltwater to come up into the gallery.

Many ephemeral or intermittent streams are dry during a significant part of the year but the channels are underlain by sand and gravel containing a significant and perennial underflow which may be captured more readily and completely with an infiltration gallery than with wells. These variable conditions dictate to a large extent the type of structure that will be required.

12–4. Design of Infiltration Galleries [3].[1]—In any infiltration gallery, the design should provide for an average entrance velocity of 0.1 foot per second or less. The smaller the entrance velocity, within economical and physical limits, the better.

Depth beneath the minimum water surface, whether surface or ground water, should be as great as physically possible and economically practical.

---

[1] Numbers in brackets refer to items in the bibliography, section 12–5.

Permeability and storativity of a natural aquifer should be determined by pumpout tests, if possible. Where such tests are not feasible, permeability can sometimes be determined by pump-in tests (see sec. 10–2) and storativity approximated on a judgment basis from a number of representative samples of the material. Where galleries are placed in relatively impermeable streambed or lakebed material, the permeability of the gravel pack should be determined by laboratory tests on two or more representative samples.

Recommended minimum diameter of screen and conductor pipe is 18 inches, although smaller sizes have been used successfully for domestic supplies in some areas. When installed in impermeable lakebeds or streambeds, or other places where a gravel pack is used, or where there is a relatively impermeable base to the gravel or sand, the invert of the screen should be a distance equal to at least one diameter of the screen above the bottom of the excavation.

Since the sealing by silt and the resultant decline in yield are common occurrences, infiltration galleries are commonly overdesigned by increasing the screen diameter or length to compensate for decline in yield.

Estimates of yield are made with one of the following applicable equations:

(a) *For a gallery in slowly permeable material with minimum depth of water above streambed, channel, or lakebed.*—Under this condition, it is assumed that the river or lake has direct access to the gravel pack or backfill. The flow moves directly downward from the water body into the pack and then into the pipe [3]. The equation for determining the length of screen necessary to yield a given volume is:

$$L=\frac{Qd}{KHB} \quad (1)$$

where:

L=length of screen required, ft,
Q=desired discharge, ft^3/s,
d=vertical distance between riverbed and center of screen, ft,
K=coefficient of permeability of the gravel backfill, ft/s,
H=head acting on the center of the pipe (distance between minimum water surface elevation and the center of the pipe), ft, and
B=average width of the trench backfilled with gravel.

The axis of the gallery is commonly placed normal to the bank or shore, but there is no reason why it could not be placed at any angle between parallel and normal so long as the minimum water level requirements are met.

(b) *For a gallery in permeable riverbed or lakebed with minimum depth of water above the bed* [3] (see fig. 12–2).—The equation for computing the length of screen required is:

$$L=\frac{Q\ln\frac{2d}{r}}{2\pi\ KH} \tag{2}$$

where:

L=computed length of screen required to yield the desired rate discharge, ft,
Q=desired rate of discharge, ft³/s,
ln=natural log,
d=distance from bed of river to center of the pipe, ft,
r=radius of the pipe, ft,
K=permeability of the channel fill or lake bottom material, ft/s, and
H=minimum depth of water above the lake or channel bed, ft.

Depending upon the gradation of the sand and gravel, a gravel pack may or may not be required. When not required, the pipe is covered and the excavation backfilled with the materials removed from it. When the pack is required, the same criteria are used in selecting the gravel pack as were described for wells (see sec. 11–11).

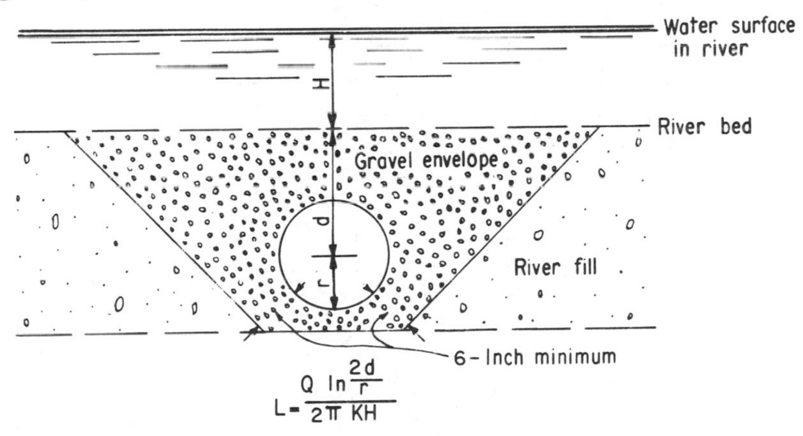

$$L=\frac{Q\ \ln\frac{2d}{r}}{2\pi\ KH}$$

L- computed length of screen to yield desired discharge, ft
Q- desired discharge, ft³/s
ln- natural logarithm
r - radius of pipe, ft
K - coefficient of permeability of channel fill or lake bottom material, ft/s
H - depth of water in the river at low flow, ft
d - distance from river bed to center of pipe, ft

FIGURE 12–2.—Factors influencing flows to an infiltration gallery screen in a permeable streambed. 103–D–1502.

(c) *For a gallery in an ephemeral or intermittent stream channel filled with permeable material through which a perennial underflow is moving* [1].—Under such conditions, the gallery is located normal to the axis of the channel and as far beneath the water table as is practical and still leave a distance equal to at least one screen diameter thickness of permeable material beneath the invert of the screen. The yield per foot of length of the gallery is estimated using equation (4) which has been rearranged from equation (3) derived by Moody and Ribbens [1].

$$s(r,t)=\frac{q}{2K}\left\{\sqrt{\frac{4Kt}{\pi MS}}\exp\left(-\frac{r^2S}{4Tt}\right)+\frac{r}{M}\operatorname{erf}\sqrt{\frac{r^2S}{4Tt}} -\frac{2}{\pi}\ln\left[\exp\left(\frac{\pi r}{2M}\right)-\exp\left(-\frac{\pi r}{2M}\right)\right]\right\} \quad (3)$$

$$q=\frac{2Ks}{\sqrt{\frac{4Kt}{\pi MS}}\exp\left(-\frac{r^2S}{4Tt}\right)+\frac{r}{M}\operatorname{erf}\sqrt{\frac{r^2S}{4Tt}}-\frac{2}{\pi}\ln\left[\exp\left(\frac{\pi r}{2M}\right)-\exp\left(-\frac{\pi r}{2M}\right)\right]} \quad (4)$$

where:

s = distance between the water table and the top of the screen, ft,

q = yield for a linear foot of gallery, ft^3/s,

K = permeability of the aquifer, ft/s,

t = time since pumping began on any pumping schedule, seconds,

exp = exponential function, $\exp x = e^x$,

erf = error or probability function,

$$\operatorname{erf} x=\frac{2}{\sqrt{\pi}}\int_0^x e^{-x^2}dx;$$

tables for this function can be found in most standard mathematical tables publications,

M = undisturbed saturated thickness of the aquifer, ft,

r = radius of the screen, ft,

S = storativity of the aquifer, dimensionless,

$T=KM$ = transmissivity of the aquifer, ft^2/s,

$\pi=3.1416$, $e=2.71828$, and ln = natural log,

limits: $s\leq 0.1M$

Since the equation gives the yield q per foot of length, the required length for a desired Q is $\frac{Q}{q}$, where Q is the desired or required yield.

If s is larger than about 10 percent of M, Q will be somewhat smaller than estimated.

If the width and average depth of the saturated channel fill, the gradient of the top of the saturated underflow zone, and the permeability of the fill are known, the volume of underflow can be estimated from:

$$Q_t = KA \frac{h_1 - h_2}{L} \tag{5}$$

where:

Q_t = total underflow, ft²/s,
K = permeability, ft/s,
A = cross sectional area, ft²,
$h_1 - h_2$ = elevations of the tops at the saturated section on lines normal to the direction of flow, ft, and
L = distance between h_1 and h_2, ft.

The total estimated underflow would be the maximum recoverable, but it is seldom that more than 60 to 75 percent of the total underflow can be intercepted.

The value of r has a small influence on the value of s and Q. Also, s and Q decrease with time until $Q = Q_t$ or 60 to 75 percent of Q at which time the system stabilizes. When using the equation to design a gallery, three or four different combinations of s, t, and Q may be required to estimate a satisfactory relationship necessary to intercept the desired or available discharge.

(d) When a gallery is installed to skim freshwater from on top of a saltwater wedge or lens (see fig. 12–3), the gallery should be placed normal to the direction of flow, and equation (4) is used except M represents the thickness of the freshwater layer. However, s must be limited so that saltwater does not come up to the screen. The Ghyben-Herzberg principle is usually adequate to determine the maximum value of s [2]:

$$h_s = \frac{d_s}{d_s - d_f} h_f \tag{6}$$

where:

h_s = distance below mean sea level of the saltwater point head at the freshwater-saltwater interface,
h_f = distance from top of the freshwater table to mean sea level or the point saltwater head,
d_s = density of the saltwater, and
d_f = density of the freshwater.

For sea water of average density of 1.027 and freshwater of average density of 1.000:

$$h_s = \frac{1.027}{1.027 - 1.000} h_f = 38.0\, h_f$$

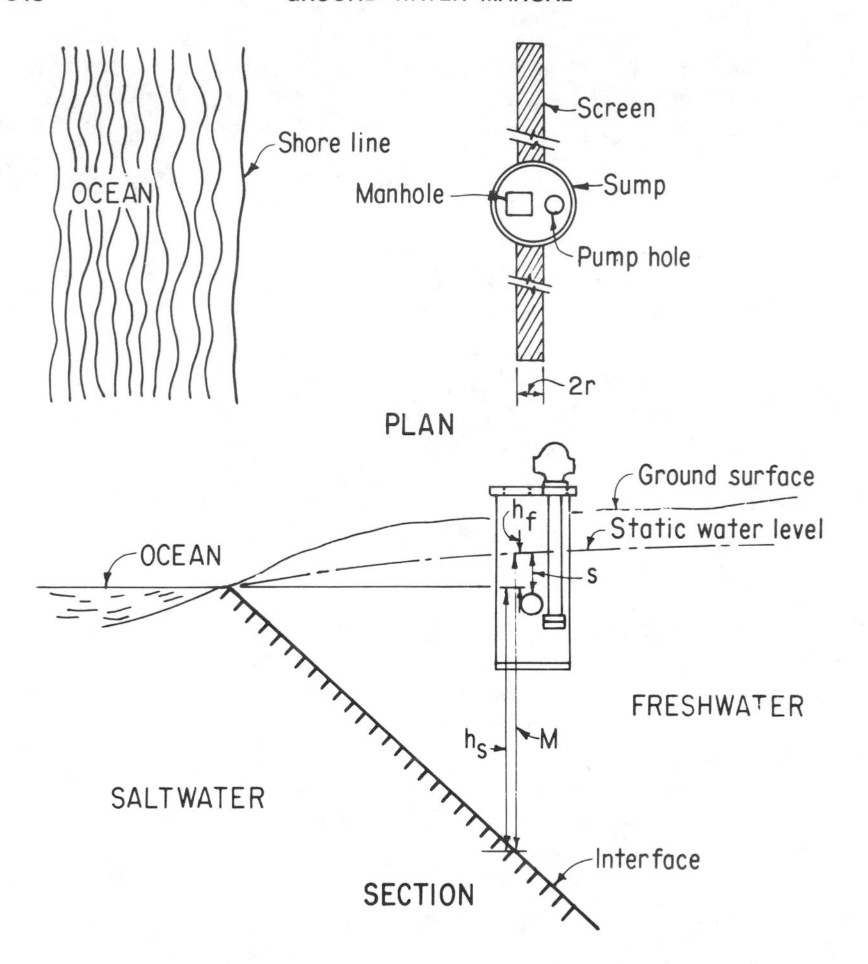

FIGURE 12–3.—Schematic section of an infiltration gallery constructed to obtain freshwater near a seacoast. 103–D–1503.

The depth to the interface is about 38 times the distance between the water table and mean sea level or the saltwater piezometric surface at the point of measurement. For example, if h_f was 3 feet, the interface would be about 114 feet below the water table. If an infiltration gallery lowered the water table 2.5 feet, h_f would be 0.5 foot and the saltwater could come up to only 19 feet below the drawdown.

The saltwater-freshwater interface is not a sharp contact but a zone of brackish water between the fresh and saltwater bodies. Approximate saltwater point head h_s may be determined by installing a piezometer to below the brackish zone; slowly pump or bail water out of the casing until it is completely filled with saltwater, and permit the level to stabilize. The water table elevation can be determined in a shallow observation well. The shallow freshwater table elevation minus the elevation of the point head of saltwater equals h_f.

12-5. Bibliography.—

[1] Moody, W. T., and Ribbens, R. W., "Ground Water—Tehama-Colusa Canal Reach No. 3, Sacramento Canals Unit, Central Valley Project, California," Memorandum to Chief, Canals Branch, Bureau of Reclamation, December 29, 1965.

[2] Walton, W. C., "Groundwater Resource Evaluation," McGraw-Hill, New York, p. 194, 1970.

[3] Zanger, C. N., "Determination of Perforated Pipe Length Embedded in Permeable Soil Necessary to Supply 55 cfs for the N-Bar-N Pumping Plant, Montana Pumping Unit, MBBP," Memorandum to R. E. Kruger, Bureau of Reclamation, November 3, 1948.

«Chapter XIII

DEWATERING SYSTEMS

13–1. Purposes of Dewatering Systems.—Excavating below the water table requires lowering the water table to promote stability; prevent raveling and sloughing of the slopes, heaving, or boiling of material in the bottom of the excavation, and to bring about dry, firm working conditions. Where an excavation lies below the piezometric surface, water may not be encountered, but blowouts may develop due to excessive uplift pressure. A properly designed and constructed dewatering system will lower the water table or piezometric surface adequately to permit safe and dry construction. The finer grained the material, the greater the desired depth to the lowered zone of saturation or pressure must be to avoid upward pumping of water due to vibrations resulting from work activities.

13–2. Methods of Dewatering and Soil Stabilization.—In some small excavations, dewatering is done by alternate stage excavation and subsequent gravity drainage to ditches and sumps in the excavation (fig. 13–1). In some instances, this is supplemented by sheet or similar types of piling driven at the edge of the excavation and, if possible, into an impermeable underlying bed (fig. 13–2). Similar methods have been used in the past on larger excavations. Over several decades, improvements and developments in well points, pumps, well construction techniques, and increased knowledge of ground-water hydraulics have led to the design of better, more efficient methods using well points, deep well systems, horizontal drains, sand drains, and vacuum and electro-osmosis techniques.

In massive hardrock, interception and disposal of water is the principal requirement of a dewatering system. This can usually be controlled by sumps, well points, or wells. Where the permeability of the rock consists primarily of fractures and other similar openings, grouting may be required in conjunction with the dewatering procedures [1].[1]

(a) *Well Point Systems.*—A well point is a 1½- to 3½-inch-diameter well screen, usually 18 to 40 inches long (figs. 13–3 and 13–4). Well points are made in a number of designs and may be jetted, driven, gravel packed, installed in open holes, or otherwise emplaced. In some types of installation, regular 4- to 6-inch-diameter well screens

[1] Numbers in brackets refer to items in the bibliography, section 13–8.

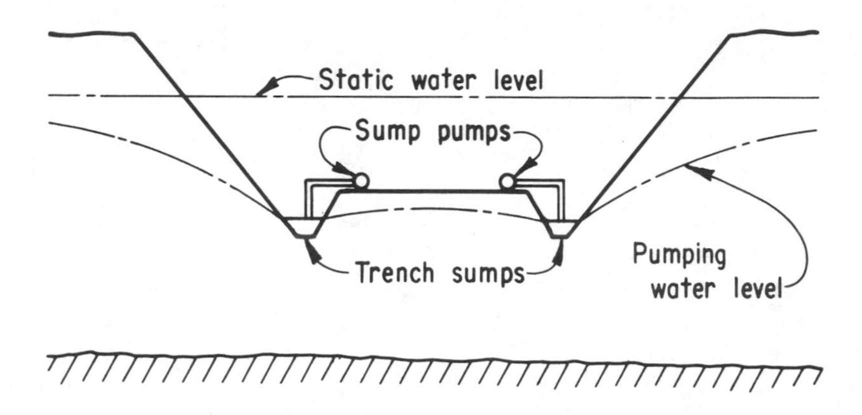

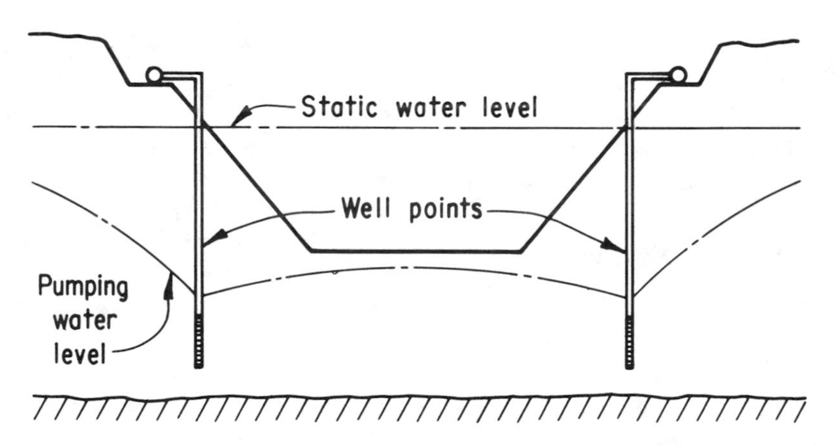

FIGURE 13–1.—Schematic sections showing dewatering by pumping from sumps and well point systems. 103–D–1504.

and casing may be set. The well point is usually attached to a riser pipe which may be 1½- to 3 inches in diameter. Well point systems for dewatering consist of a number of well points placed at 2- to 6-foot intervals along a line. The riser pipes are connected to a header usually 6 to 8 inches in diameter (fig. 13–5) with a swing connection (fig. 13–6) which can rotate through 360° horizontally and about 270° vertically.

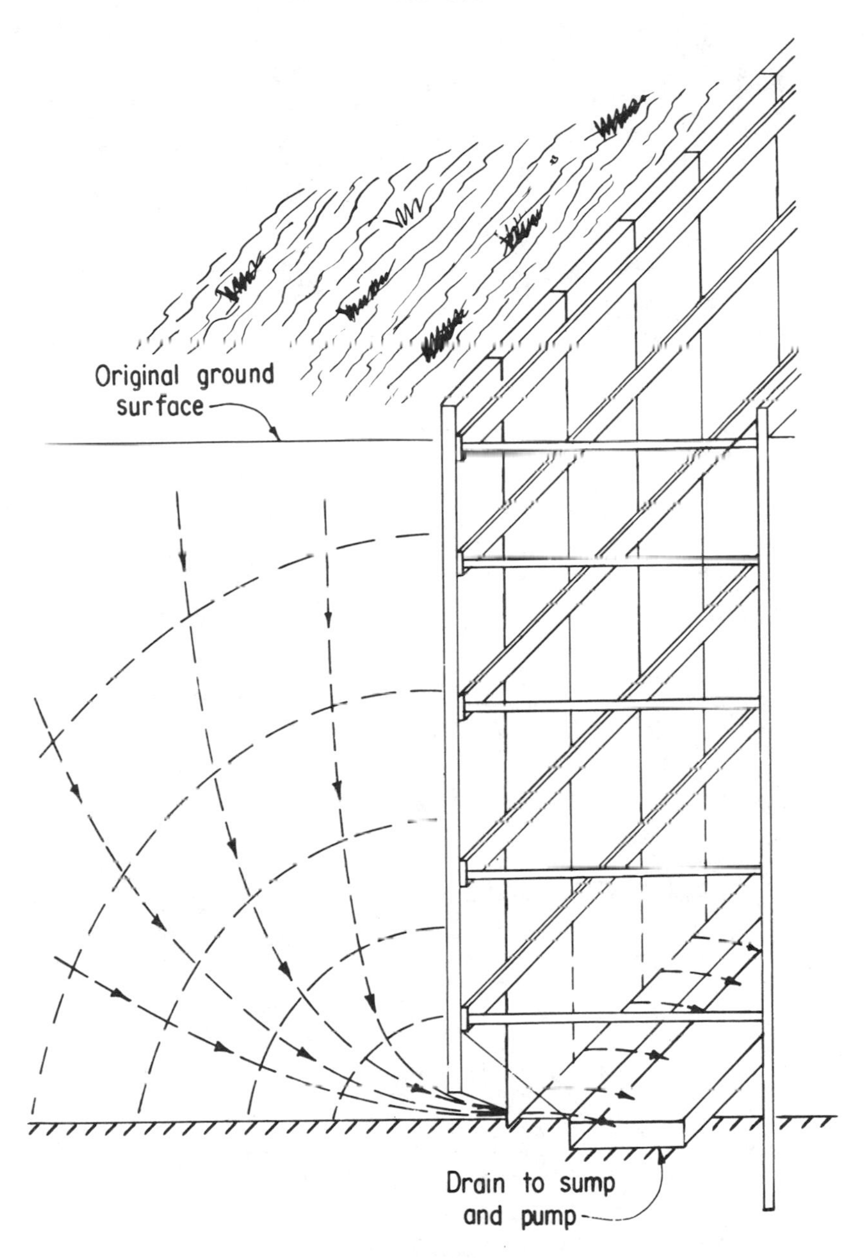

FIGURE 13-2.—Dewatering by drainage under or through shoring to a sump 103-D-1505.

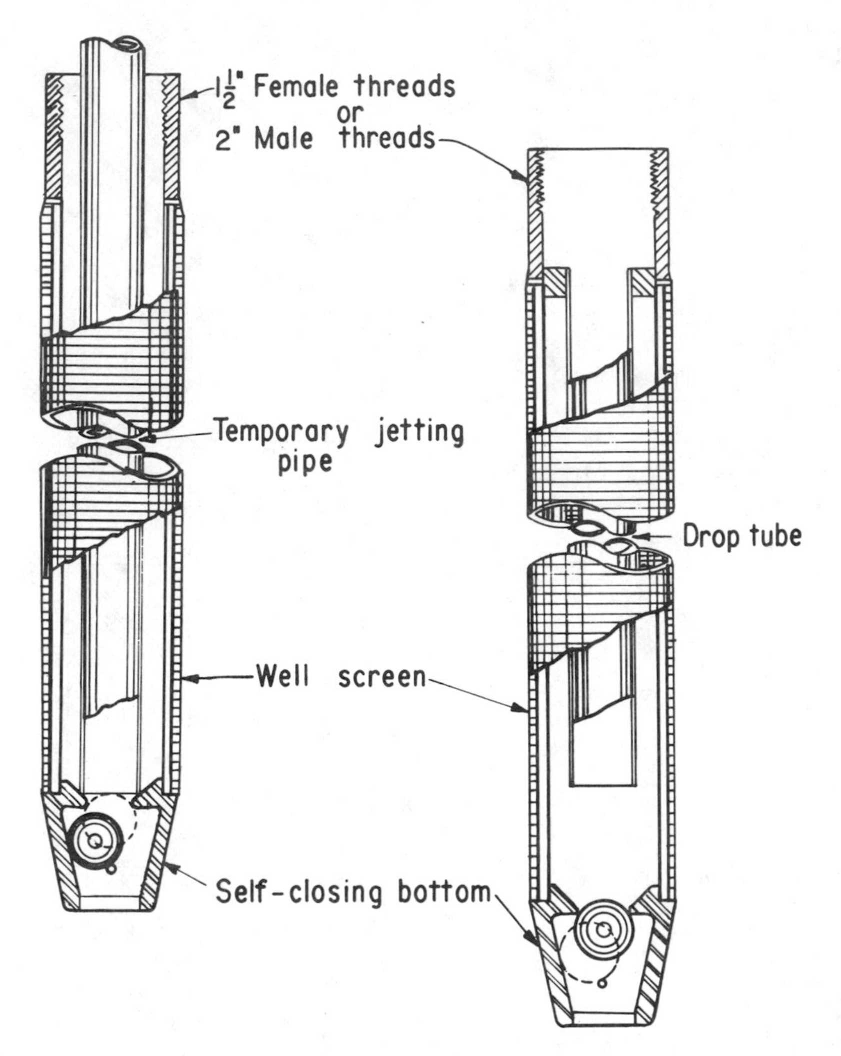

FIGURE 13-3.—Typical well points equipped with jetting tips. From [5]. 103-D-1506.

The swing connector contains a valve that controls water withdrawal from the well point. The yield may vary between well points in any system; consequently, the discharge of each well point must be controlled by adjusting the valve so that drawdown is not large enough to expose the top of the screen and draw air into the system. The header pipe may be up to 600 feet long and is placed as straight and level as possible. The pipe is usually connected to a centrifugal pump

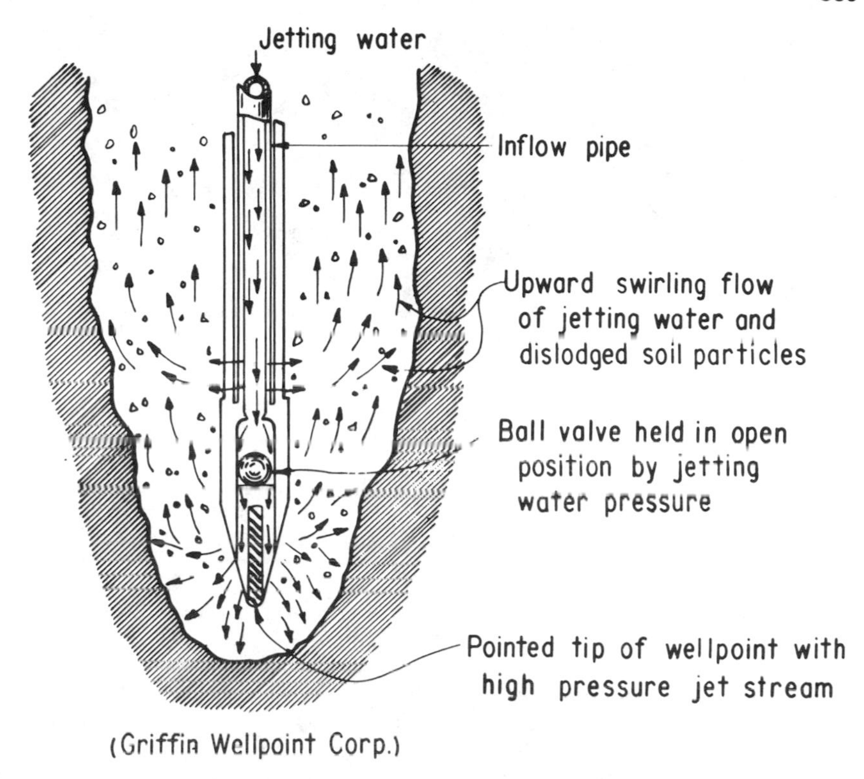

FIGURE 13-4.—Installation of a well point by jetting. 103-D-1507.

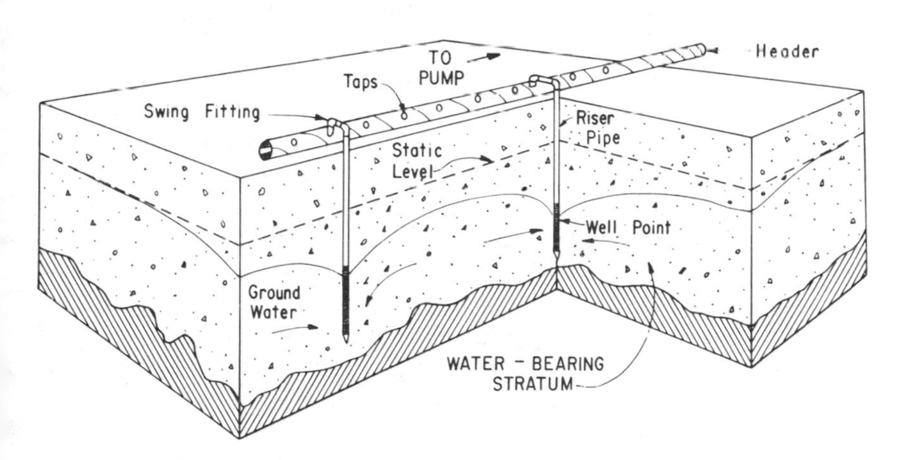

FIGURE 13-5.—Schematic section of a portion of a well point dewatering system. From [5]. 103-D-1508.

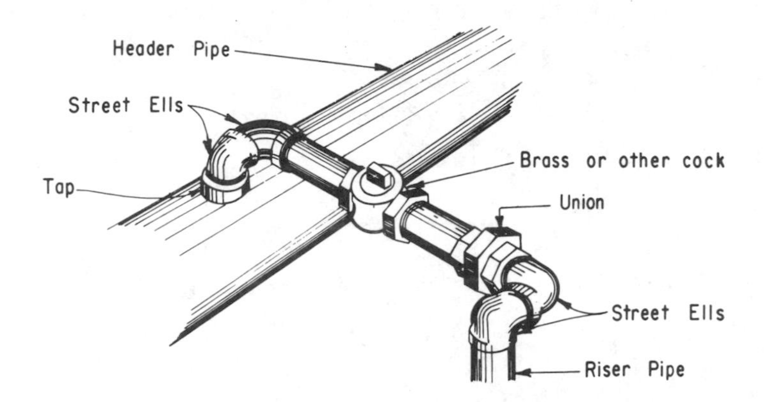

Details of connecting fittings from riser pipe to header pipe. Made up of 1-inch to $2\frac{1}{2}$-inch pipe fittings, the connection may be swung horizontally and vertically to meet the top of the riser pipe. This flexibility is important since, in practice, the well point and its riser pipe will always be a few inches off the calculated position.

FIGURE 13–6.—Swing connection fittings for riser pipe to header pipe connection From [5]. 103–D–1509.

located in about the middle of the header and which can develop from 15 to 22 feet of suction lift. Well points operating by suction lift are placed with the top of the screen at depths 15 to 18 feet below the static water level. The maximum effective lift in the well points is about 15 feet.

An excavation is usually dug to within a few inches of the water table before installing a well point system so that maximum advantage may be made with the suction lift available, see figures 13–1 and 13–7. Where the excavation is to be carried down more than 12 to 15 feet below the water table, well points may be installed in stages with each stage about 15 feet lower than the previous one, see figures 13–7 and 13–8. When more than two such stages are involved, supplemental wells and well points may be required to control certain unstable spots in the excavation. Larger well points using jet ejector (eductor) or submersible pumps in place of suction lift pumps are commonly used. Jet ejector pumps can lift water from depths of 100 feet or more. However, this type pump requires an additional header system to bring operating water to each pump which results in a more costly operation than suction lift pumps. Submersible pump lifts are almost unlimited, but the pumps are more complex and costly than suction lift and ejector pumps. Operation and maintenance (O&M) costs are larger for ejector and submersible pumps than for a single-stage suction installation. However, such pumps may be more economical than multiple stage installations (fig. 13–9) [1,5].

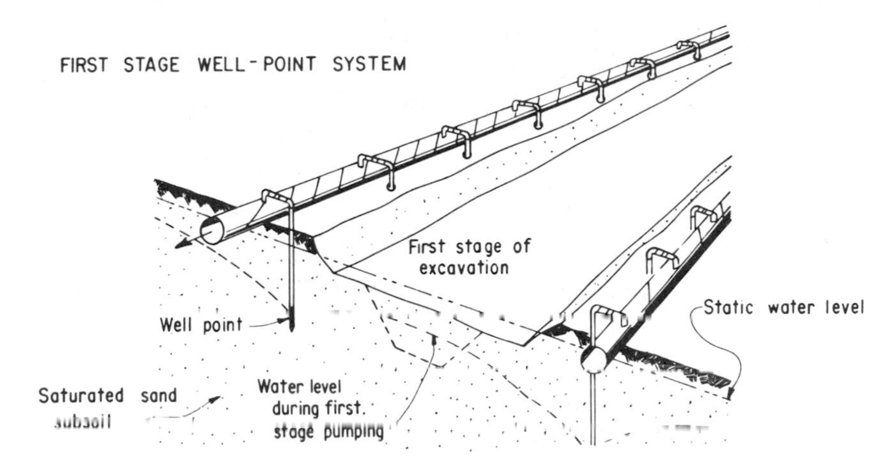

FIGURE 13-7.—Single stage well point system. From [5]. 103-D-1510.

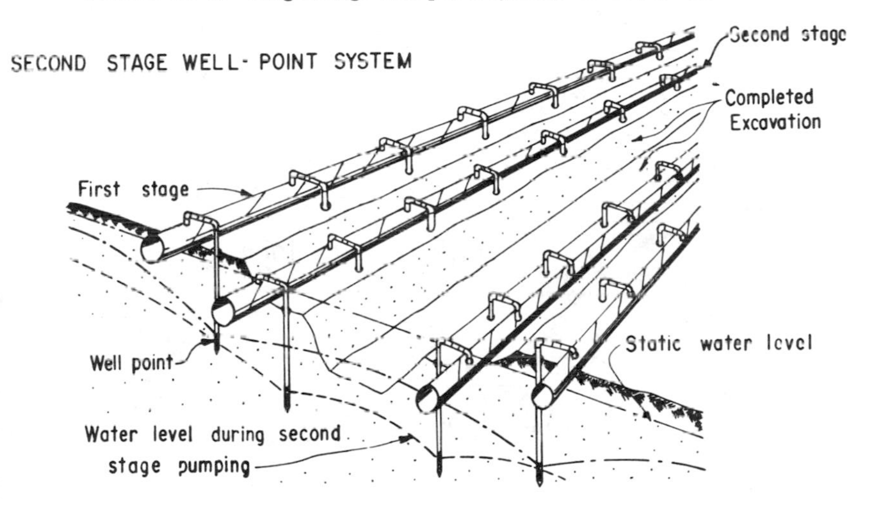

FIGURE 13-8.—Two stage well point system. From [5]. 103-D-1511.

Fine silts and other slowly permeable materials cannot be readily drained by gravity methods alone because of their high specific retention. Such soils can, however, be partially drained and stabilized by means of vacuum wells or well point systems which create negative pore pressure or tension in the soil. A vacuum dewatering well or well point (fig. 13-10) should be gravel packed from the bottom of the hole to within a few feet of the surface of the poorly permeable material. The remainder of the hole, the top of the poorly permeable material, should be sealed with bentonite or other impermeable soil. By maintaining a vacuum in the well screen and pack, flow towards the well or well points is increased, particularly in stratified soils. The

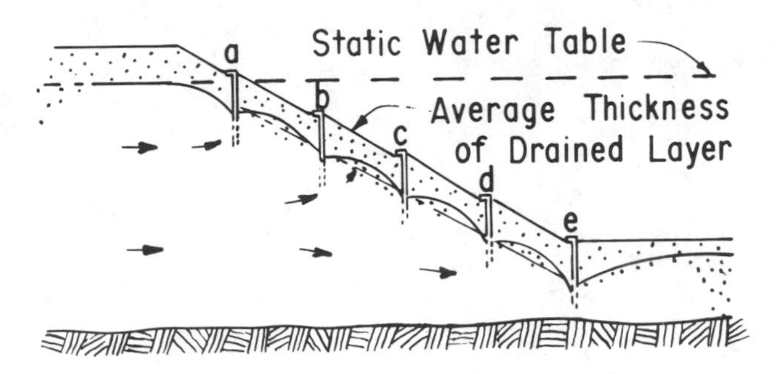

MULTISTAGE WELL POINT SYSTEM
(Stages a through e)

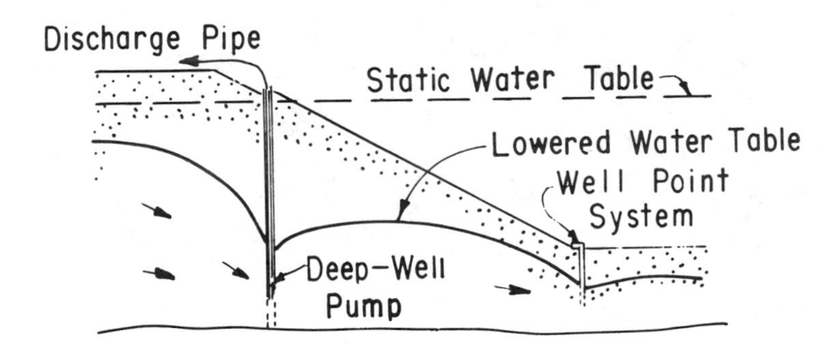

COMBINED DEEP WELL AND
WELL POINT SYSTEMS

FIGURE 13-9.—Multistage well point and combined systems. A system combining a deep-well pump and a well point system that may be used as an alternative to a multistage well point system. From [1]. 103-D-1512.

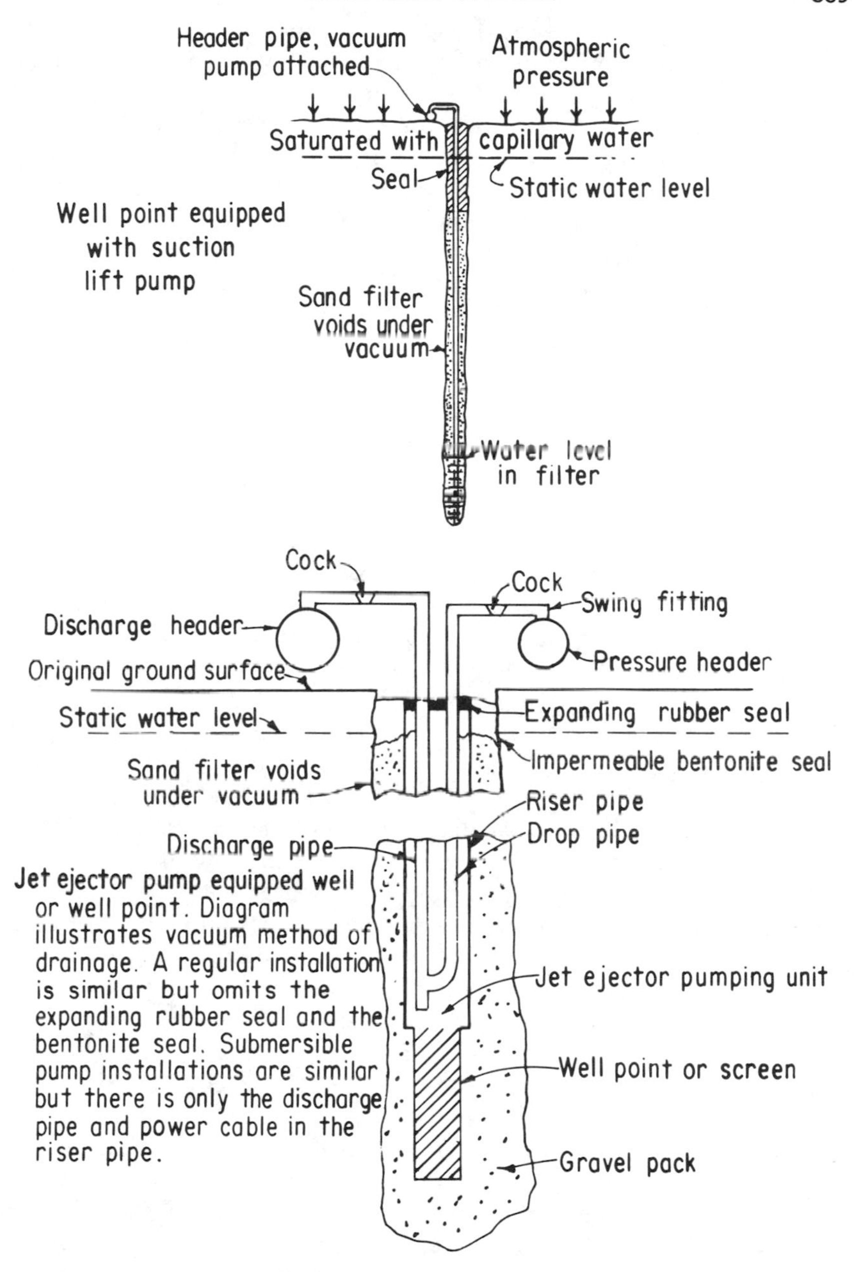

FIGURE 13-10.—Use of a vacuum to increase well point efficiency. 103-D-1513.

system normally requires use of closely spaced well points and pumping capacity is usually small. Vacuum booster pumps may be required on the headers or individual wells for effective operation.

(b) *Deep Wells for Pumps.*—When artesian pressures are involved or where the soil to be dewatered is part of a relatively uniform thick and permeable aquifer, dewatering can frequently be accomplished with a few deep, high-capacity wells equipped with deep well turbine or other similar pumps. If the transmissivity and storativity of the aquifer can be determined, application of the principle of superposition (sec. 2–20) can be used to determine approximate well capacity, depth, and spacing required to accomplish the dewatering. An advantage of deep wells rather than well points is that they often may be located outside the boundaries of the excavation and, hence, offer little or no interference with the excavation operations. See figures 13–9 and 13–11.

(c) *Horizontal Drains.*—Where excavations are located on slopes or where landslides have occurred, such excavations may be dewatered and stabilized with horizontal wells or drains. These drains can be installed with drills especially designed for horizontal drilling. The first horizontal drain should be installed near an existing observation well in which the influence and effect of the drawdown can be observed. Using the principle of superposition, approximate drain spacing to achieve an adequate drawdown can be estimated. Methods are available for gravel packing such drains if conditions make such treatment desirable. Dewatering can usually be accomplished by gravity flow or, if necessary, pumping and vacuum drainage can be used [3].

(d) *Electro-osmosis.*—Electro-osmosis has been used to drain and stabilize saturated fine-grained soils. The installation and O&M costs are usually higher than with a well point system, but the electro-osmosis system may accomplish the purpose where a well point system will not. An electro-osmosis system consists of negative and positive electrodes installed at spacings of 10 to 100 feet. The cathodes (negative electrodes) consist of smaller diameter well points since the rate of discharge is usually small and intermittent. A small suction pump is usually used on the manifold. The anodes (positive electrodes) may be iron or steel pipe, rails, or other available conductors placed midway between the cathodes, or there may be separate lines of cathodes and anodes placed 15 to 20 feet apart (fig. 13–12). The current requirement may be relatively high since, even with spacing of about 20 feet, 25 or more amperes are required to produce 35 to 40 gallons per day per well point. Detailed design procedures are not included herein since the system is seldom used [1,3].

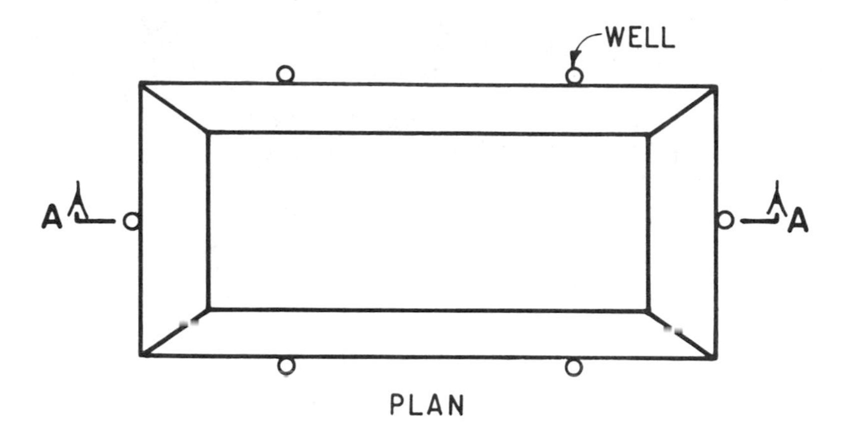

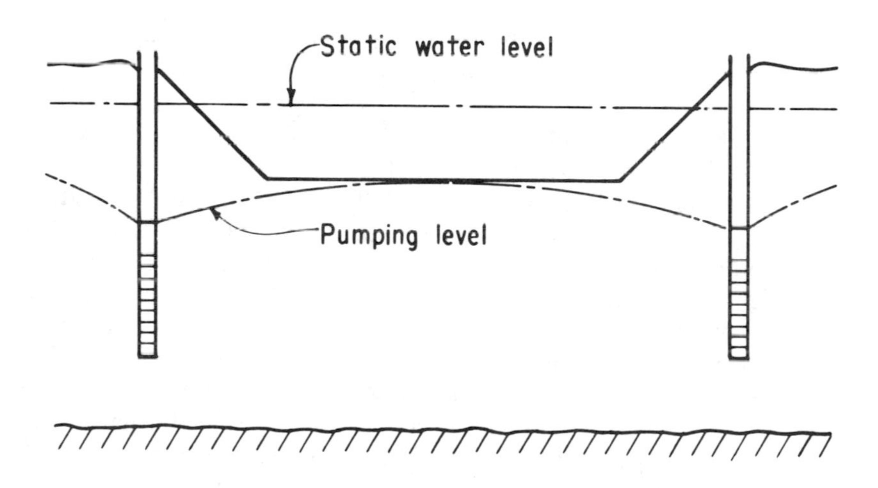

FIGURE 13–11.—Use of deep wells and high lift pumps for dewatering. 103–D–1514.

(e) *Vertical Sand Drains.*—Vertical sand drains have been used in conjunction with well points to facilitate drainage in stratified soils. The usual installation calls for 16- to 20-inch-diameter sand drains to be installed on 6- to 10-foot centers through the materials to be stabilized and extended to the underlying permeable layer where the well points or other wells are placed. Under such conditions, the well points or wells are usually gravel packed and may be designed for vacuum drainage as well. The sand drain backfill is selected according to the same criteria used in selecting a gravel pack (see sec. 11–11) [1].

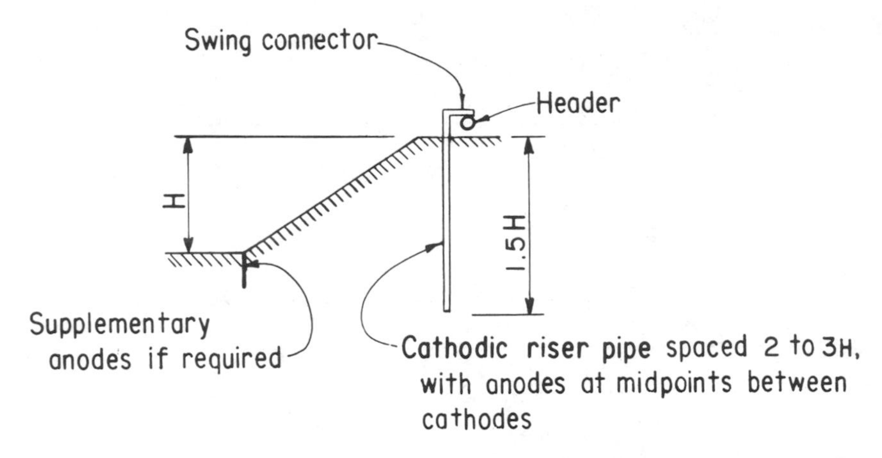

FIGURE 13–12.—Use of the electro-osmosis technique for dewatering. 103–D–1515.

13–3. Field Investigation for Dewatering Systems.—The amount of investigation necessary to permit design of a satisfactory dewatering system depends upon the geologic and hydrologic complexity of the area and the depth and areal extent of the excavation. The field procedures have been described and discussed in chapters X and XI and analyses of the results in chapter V. The principles and procedures discussed in these chapters are directed primarily toward design, maintenance, and operation of permanent production wells. In dewatering an unconfined aquifer, the equations and formula previously discussed are not always applicable. Drawdowns are usually in excess of 65 percent of the unconfined aquifer thicknesses. Furthermore, well points are closely spaced and estimates of yield, drawdown, interference, and similar factors do not conform to equations usually used. Much depends upon the judgment and experience of the technician, use of empirical equations, and the results of specific investigations conducted prior to design and installation of a dewatering system. If possible, such investigations should be initiated a year or more before construction is undertaken so that data may be obtained on seasonal and annual variation of ground-water levels and the response of such levels to the variation in water stages of adjacent lakes and streams and to precipitation. Such data are obtained by periodic measurement of observation holes and piezometers and from water records of surface water levels and precipitation. Drilling, completion of observation wells, and piezometers are discussed in sections 7–7 and 7–8.

In and adjacent to the area of excavation, exploration holes should be drilled to a depth 1½ times the proposed depth of the excavation. The holes should be carefully logged and, ideally, samples should be taken at 5-foot intervals or at each change in formation, whichever is less. Mechanical analyses should be made on samples consisting of silt or coarser materials. When more than one aquifer is encountered in the holes, static water levels should be measured in each aquifer by means of piezometers. Pumpout aquifer tests to determine transmissivity and storativity and boundary conditions should be made of the aquifers present.

Where use of well points is planned, a test installation may be desirable to determine response to the pumping of one or more well points.

On completion of the exploration program, a three-dimensional diagram of the area of excavation and adjacent land, based on available geologic and hydrologic data, should be prepared and studied as a basis for design of a dewatering program and possible location of additional observation wells. Key exploration holes, both in and adjacent to the proposed excavation, should be completed as observation wells and piezometers (fig. 13–13). These should be measured periodically prior to construction. As many observation wells as possible should be preserved during construction and measured frequently during construction as a check on the performance of the dewatering system.

13–4. Design of Well Points and Similar Dewatering Systems.— (a) *Suction Lift Well Point Systems.*—Reference [1] gives theoretical and empirical steady-state equations for use in the design of suction lift well point systems. This reference should be consulted along with [3,4,5] for an understanding of the principles involved. It is seldom

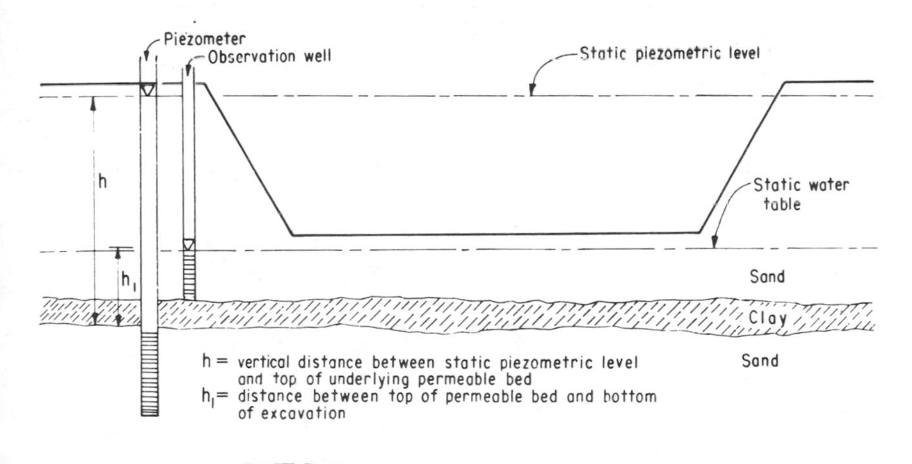

FIGURE 13–13.—Use of piezometers and observation wells to delineate complex ground-water conditions. 103–D–1516.

practical to calculate the size and shape of the area of influence of this type of well point system. The cost of making tests to measure controlling factors cannot be justified because aquifers are usually unconfined. Drawdowns are usually too great for the usual equations to be applicable and partial penetration and anisotropy may further complicate the problem.

Well point spacing is selected primarily on the basis of judgment and operating experience. In silt and fine sands, 2 to 2½ feet is the usual spacing and as the coarseness of the materials increases, spacing may be increased to about 6 feet. Thickness of the aquifer and percent of penetration are also controlling factors. In thin aquifers of less than 15 feet and when the percent of penetration in thicker aquifers is less than about 25 percent, spacing is also small (2 to 2½ feet). Spacing may be increased as the transmissivity and percent of penetration increases.

For silt and other fine-grained materials, well points with diameters of 1½ inches generally are satisfactory. The diameter should be increased for more permeable materials. One-inch-diameter riser pipes are suitable for the smaller diameter well points and should be increased to 2 to 2½ inches for well point diameters up to 3½ inches on suction lift systems.

The well points of the first stage should be set 3 to 5 feet below the bottom of the proposed excavation with maximum drawdown at about 15 feet below the water table. Following dewatering by the first stage, the excavation is carried to within about 1 foot of the water table and the second stage is then installed. Theoretically, the procedure could be carried to almost any depth in a thick homogeneous aquifer, but the dewatered thickness is relatively thin on the side of the slope relative to the adjacent saturated thickness (fig. 13–8). When three or more stages are required, seepage pressures may cause slope instability. Under such conditions, supplemental deep wells, a deep well system alone, or a supplemental well point system should be used.

Where an excavation is underlain by a relatively impermeable bed which limits the drawdown, sufficient dewatering may sometimes be obtained by drilling 12- to 14-inch-diameter relatively shallow holes into the impermeable bed, setting the well points into the holes in the impermeable material, and gravel packing around the well points (fig. 13–14).

An excavation may be underlain by a confining layer of relatively impermeable material which, in turn, is underlain by an aquifer. Water in this aquifer may be under similar or higher head than water above the confining layer. If the material above the layer is dewatered and the bottom of the excavation is lowered, a point may be reached where either the bottom of the excavation will heave or blow out or boils will occur in areas where the confining layer is thin. Under such conditions, relief wells or well points should be installed in the underlying aquifer to reduce the pressure and stabilize the bottom of the excavation (fig. 13–15). If the upward hydrostatic pressure is greater than the pressure on the confining layer, a heave or blow may occur [3].

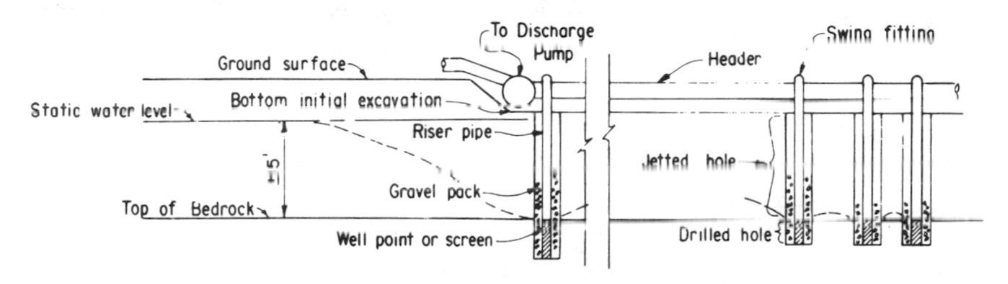

FIGURE 13–14. Dewatering a thin aquifer overlying an impermeable material. 103–D–1517.

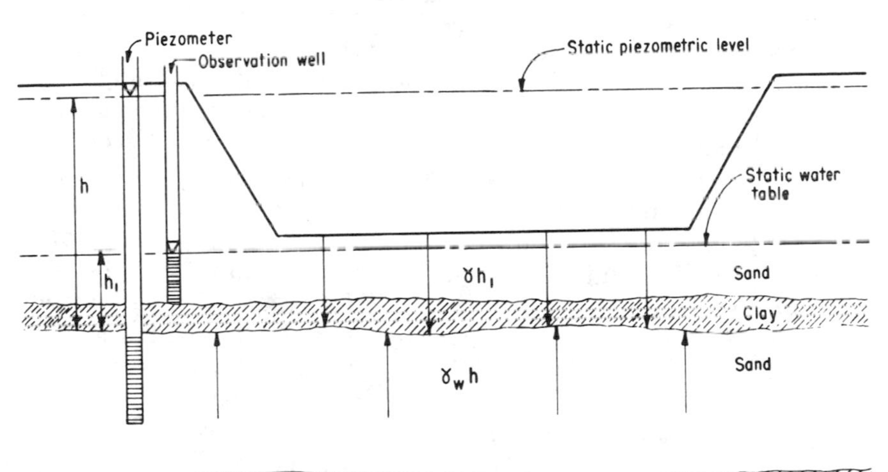

FIGURE 13–15.—Factors contributing to blowouts or boils in an excavation. 103–D–1518.

The maximum suction lift which can be obtained with available pumps is about 20 to 25 feet; however, friction losses in the system may reduce this to about 15 to 18 feet. To keep the loss of suction head to a minimum, well points, riser pipes, swing connections, and header pipes should be sized generously. In addition, all joints in the system should be made airtight.

The pumping and other tests recommended previously in this section should permit an estimate of the probable discharge which will be obtained for well points. From this, required pipe diameters to keep friction losses at a minimum and pump capacities can be estimated.

The basic installation is probably that of a single line of well points necessary to dewater a relatively shallow trench for a pipeline or similar installation. The top of the well points are usually set a minimum of 4 feet below the bottom of the excavation and as close to the edge of the trench as possible without interfering with the work. In some very permeable materials, a line of well points on each side of the excavation may be required to provide adequate dewatering.

Where an excavation has considerable lateral extent in all directions, it is usually outlined by a dewatering system.

In relatively slowly permeable soils such as silty clays, silts, and fine sands or where such materials overlie a more permeable layer, the well points should be gravel packed except for 4 or 5 feet at the top of the hole which is tightly sealed. This permits the development of a vacuum in the deeper portions of the hole which encourages horizontal migration of water to the gravel pack and then down the hole to the well point. The amount of water removed may be small, but in many instances it is sufficient to stabilize an otherwise unstable material (fig. 13–10). Gravel packing may also be used where clay beds are irregularly distributed through the saturated mass of material to be drained. The selection of screen slot sizes, gravel-packed gradations, etc., is based on the same criteria outlined in section 11–11.

(b) *Jet Ejector Well Point Systems.*—In some cases, well point systems are installed using a jet ejector pump in each riser pipe rather than pumping by suction lift. Jet ejector pumps operate by an induced suction created by flow of water through a venturi in addition to normal suction from centrifugal and vacuum pumps. Jet ejector well points have only about a 25- to 35-percent efficiency, but they can lift water from depths of 60 to 100 feet and develop a vacuum of 18 to 20 feet of head in each hole. The discharge is usually in the range of 12 to 15 gallons per minute. Well point spacing is usually 4 to 10 feet.

Two headers are employed. One delivers water under pressure to the venturi in each pump and the other provides discharge.

The riser pipes are usually 3½ inches in diameter or larger to permit installation of the jet ejector pumps whose intake is usually a few feet above the top of the screen or well point. Well points or screens may be as long as 10 to 15 feet and as much as 6 inches in diameter. At times, particularly in smaller installations, submersible pumps may be used rather than jet ejector pumps.

If the thickness, permeability, and storativity of an aquifer are known, a line of closely spaced well points may be considered as a drain and the equations given in section 12–4 used to compute the time required for dewatering [?] Each line of well points is considered a drain or collector.

(c) *Deep Well Dewatering Systems.* Deep wells can be used for dewatering thick homogeneous aquifers and to lower heads in underlying artesian aquifers which might give rise to heaving or boils in the floor of an excavation. The design and installation of the wells generally are the same as described in chapter XI except that lower design standards may be adequate because of the temporary nature of the wells. If storativity, transmissivity, boundary conditions, and aquifer thickness are known, wells can usually be located outside the excavation limits.

13–5. Installation of Dewatering Wells and Well Points.— Numerous types of well points are manufactured. Some use wire screen of various types that cover perforated pipe while others are wire wound, slotted, or perforated. The percentage of open area is variable and the slots are usually fixed to a relatively small number of sizes. The wire-wound type offers the greatest flexibility in slot size as well as the greatest percentage of open area and most efficient distribution of slots.

Well points are made to be driven inplace, jetted down, or to be installed in open holes. For dewatering purposes, the points are seldom driven. A more common practice is to jet the well point down to the desired depth, flush out the fines leaving the coarser fraction of material to collect in the bottom of the hole, and then drive the point into this coarser material. Additional gravel may be added to fill the hole near or to the ground surface or a seal may be placed over the pack in the upper 2 to 3 feet.

A method used in some unstable material consists of jetting down or otherwise sinking temporary casing into which the well point and riser pipe are installed. As the casing is pulled, gravel may be placed around the well point after which the pipe is pulled.

Major dewatering contractors have developed equipment and techniques for mass installation of dewatering wells of all types.

It is desirable, regardless of the method of installation, to develop a well point by pumping, surging, or other means prior to use.

13–6. Pumps for Dewatering Systems.—Pumps for suction lift well point systems should have adequate air handling capacity and be capable of producing a high vacuum. Well point system pumps are usually self-priming centrifugal pumps with an auxiliary vacuum pump which permits developing a vacuum of 20 to 25 feet of water. The intake of the pump should be as close to the bottom of the excavation as practicable [1].

Submersible pumps and jet ejector pumps are usually standard off-the-shelf items selected for the estimated yields and lifts involved.

Good engineering practice requires excess capacity in the pumps and standby units in the event of failure of a pump.

13–7. Artificial Ground-Water Barriers.—Natural ground-water dams caused by faulting, dikes, and similar features are frequently encountered in ground-water investigations. Similar barriers may be constructed artificially for the purposes of ground-water control, such as seepage under dams, protection of excavation, and the raising or maintenance of ground-water levels.

(a) *Sheet Piling.*—Sheet piling is a commonly used method, but of questionable effectiveness in some aspects, for construction of ground-water barriers. If completely interlocking and driven to an impermeable barrier, sheet piling is an effective means of controlling piping and instability but not of stopping subsurface flow. The interlocking of sheet piling does not give a watertight joint and if 10 percent open area is present, about 70 percent of the water will flow through the piling. While such barriers may have relatively small effect in decreasing the flow of water, they do cause an increase or dispersal of the exit area with a consequent decrease in the possibility of the formation of boils and development of piping.

Where boulders and cobbles are present and sheet piling is driven as a cutoff, the piles may be interlocking and vertical at the surface, but the joints may be broken and the piles diverted in a random manner below the surface so that little or no damming occurs. This condition may be conducive to the formation of boils and development of piping. In such locations, piling might better be jetted or otherwise installed to insure the integrity of the wall. In some instances, this is not feasible and concrete soldier piles may have to be set in preexcavated holes.

(b) *Cutoff Walls.*—Cutoff walls are often placed beneath dams located on permeable foundations. There is seldom a problem regarding the integrity of such cutoff walls, since they are commonly placed in open cuts dewatered by wells or well points and protected by cofferdams. To be entirely effective, cutoff walls should provide for 100 percent closure, but this may be economically or physically impossible in some instances. Under such conditions, a careful study of soil conditions, permeabilities, etc., should be made, usually in combination with flow net analyses, to determine the best design from the standpoint of leakage and stability.

The recent development of slurry trenches has provided a safer and less expensive method of installing ground-water barriers in unconsolidated materials. Draglines or backhoes, depending upon the depth and size of the excavation, are used to excavate the trench which is kept full of a bentonite-based fluid. This fluid, which has a weight of 9½ to 10 pounds per gallon, develops a filter cake on the side of the trench which reduces seepage of the fluid away from the excavation and exerts hydraulic pressure on the side of the trench which minimizes caving. Excavations by such a process have been carried to depths of up to 100 feet. When reaching the top of bedrock, it may be cleaned of pockets of sand by operation of an airlift pump.

The completed excavation can be filled with concrete placed through tremie pipes extending to the bottom. As the concrete is pumped in, the bentonite-based fluid is displaced and flows out at the surface. The trench may also be excavated and backfilled in sections with the bentonite being diverted to the next section as the completed section is being backfilled.

Another procedure is to displace the bentonite in the trench with the sand and coarser fractions of the material excavated, thus creating a clayey-sand mixture. This results in a watertight and permanent ground-water barrier. In using the slurry trench method, an engineer experienced in the handling of mud should be available to advise on proper use and treatment of the slurry.

(c) *Grouting.*—The formation of impermeable barriers to water seepage by grouting is an established practice. Grouting consists of introducing sealing compounds or mixtures, usually under pressure, into rock and soil to fill fractures and voids with stable insoluble materials. Grouting has usually been successful in fractured hard rocks; however, the results in unconsolidated materials have been variable. Native clays, bentonite, silts, and sand are natural materials which have been used for grouting. Cement, various chemicals such as combined sodium silicate and calcium chloride, tar, asphalt, and various epoxies have also been used.

The nature of the openings, their size and continuity, the velocity of water flow through the openings, and other factors influence the type of materials to be used and the method of injection.

Successful grouting is both an art and a science and should be undertaken only under the direction of an experienced and knowledgeable engineer.

13-8. Bibliography.—

[1] Mansur, C. I., and Kaufman, R. I., "Dewatering," Chapter 3 in "Foundation Engineering," (edited by G. A. Leonards), McGraw-Hill, New York, pp. 241–350, 1962.

[2] Moody, W. T., and Ribbens, R. W., "Ground Water—Tehama-Colusa Canal Reach No. 3, Sacramento Canals Unit, Central Valley Project, California," Memorandum to Chief, Canals Branch, Bureau of Reclamation, December 29, 1965.

[3] Terzaghi, K., and Peck, R. B., "Soil Mechanics in Engineering Practice," second edition, John Wiley & Sons, New York, 1967.

[4] "Ground Water and Wells," UOP Johnson Division, St. Paul, Minn., 1966.

[5] "Well Point Systems," UOP Johnson Division, Bulletin No. 467–D, St. Paul, Minn., 1969.

Chapter XIV

CORROSION AND INCRUSTATION

14–1. Corrosion.—Metal screens, casings, and pumps used in wells may be subjected to deterioration and eventual destruction by reaction with the environment in which they are placed. This destructive process is called corrosion. For purposes of this discussion, corrosion of well components can be considered as either chemical or galvanic. Chemical corrosion results from chemical reaction of the metals with components of the soil or water in which they are placed. Chemical corrosion usually results in the corroded metal going into solution and being carried away from the point of attack. Galvanic corrosion arises from the action of electrolytic cells formed between dissimilar metals or surface conditions in the well, which results in attack and corrosion of metals at anodic points, and frequently the deposition of the corrosion products at the cathodes. Organism-induced corrosion is initiated, accelerated, and aggravated by certain algae, fungi, and bacteria. These organisms do not attack the metal directly, but cause environmental changes which, in conjunction with the byproducts of their metabolism, cause chemical and galvanic corrosion [2,3,4]. [1]

Corrosion, to some degree, is inevitable in any well installation. It is difficult to impossible to accurately predict the type of corrosion and the intensity of attack to be expected. Where wells are to be constructed in locations where there are existing wells, discussion with well owners, well drillers, and well service and supply companies will usually establish the presence or absence of prevailing aggressive corrosion conditions in an area and possibly the cause. In areas where there is no experience record, the designer has little to guide him. Analysis and interpretation of water samples generally will indicate whether the water will cause chemical corrosion or be conducive to galvanic corrosion but may give little information in regard to the intensity or severity of the effects [2,3].

As a general guide, chemical corrosion can be anticipated when CO_2, HCO_3, O_2, H_2S, HCl, Cl_2, H_2SO_4 or salts of these substances are present in excess of 5 to 10 p/m, the pH of the water is below 7, and the Ryznar Index is above 9 [5]. If the Ryznar Index is above 9, the pH below 7.8, and the total dissolved solids content above about 300 p/m, galvanic corrosion may occur. With galvanic corrosion, the products of the attack on the anodic areas are frequently deposited

[1] Numbers in brackets refer to items in the bibliography, section 14–3.

on the cathodic areas, causing blocking of screen slots and water channels in pumps.

Bacterial corrosion is more difficult to predict or anticipate. Whether the bacteria are everywhere present in an aquifer or whether they are introduced into a well during drilling or servicing is not known, but there is substantial evidence that the latter does occur. Some so-called iron bacteria not only foster corrosion but also block screens and give a disagreeable taste and odor to the water. Other types of bacteria are not so readily recognized and can be identified only by special sampling, incubation, and study techniques, which are expensive and time consuming. Determinations, therefore, are not usually made unless experience in an area shows that such investigations are necessary. Since so little is known of the origin or control of bacterial corrosion, it is good practice to sterilize every well on completion of drilling and at the time the permanent pump is installed [1, 3].

In the following galvanic series tabulation, metals and alloys are arranged in order of increasing resistance to corrosion. Galvanic corrosion between two adjacent metals or alloys in the tabulation is low. The farther two metals are apart in the list, the greater the potential which will develop between them and the more aggravated the corrosion attack will be on the anodic material. Since surface area is also important if two dissimilar materials must be used, the more cathodic material should have as small a surface area as possible and the anodic material should have as large a surface area as possible to diffuse and spread out the attack.

When corrosive environments are known to be present, consideration should be given to the use of corrosive-resistant metals in the screen and in various parts of the pump.

Except under unusual conditions, it generally is more economical to use less costly metals and to increase the wall thickness of the casing and column pipe than to use the more costly metals. The increased wall thickness provides more metal to resist corrosion and thus increases the life of the casing and column pipe [4].

When the well screen is installed by telescoping, it may be possible to remove and replace the screen. In this situation, the choice of metal for the screen becomes a problem of balancing the cost of the more expensive metal against the replacement cost of the less expensive metal. When the well screen is attached directly to the casing, a dielectric coupling should be provided between the two components if they are of dissimilar metals to provide corrosion protection to both.

The use of corrosive-resistant metals in the manufacture of different parts of the pump to be installed in the well must also be considered. Here the problem is additionally complicated because not only are the

Galvanic Series

(Arranged in order of increasing resistance to corrosion)

Anodic—Corroded End

Magnesium
Magnesium alloys
Zinc
Aluminum 2 S
Cadmium
Aluminum 17ST
Steel, iron, or cast iron
Chromium-iron (active)
Ni-Resist
Chromium-nickel-iron (active)
Lead, tin, or lead-tin solders
Nickel (active)
Inconel (active)
Brass or copper
Bronze
Monel
Silver solder
Nickel (passive)
Inconel (passive)
Chromium-iron (passive)
Chromium-nickel-iron (passive)
Silver
Gold or platinum

Cathodic—Protective End

individual parts attacked by chemical corrosion, but also dissimilar metals in contact with each other are conducive to galvanic corrosion. When corrosion is known to be present, the metals of the pump parts must be chosen with regard to their location in the previous tabulation as related to the intensity of corrosion expected.

14–2. Incrustation.—Incrustation may be considered the opposite of corrosion insofar as it is characterized by the accumulation of minerals deposited primarily in and about openings in the screen and in the voids of the formations surrounding the well. This accumulation hinders water from entering the well, eventually the efficiency of the well declines, the drawdown increases, and the discharge decreases. The rate of mineral deposition depends upon the character and quality of the water involved and apparently increases with increased drawdown and the entrance velocity of the water. In some wells, a screen may be completely blocked within a few months or a year, whereas in others the deposition is so slow that the effects are not recognizable over many years. As far as can be determined, the materials of which the well and screen are constructed have little influence upon the rate or character of the deposition unless galvanic corrosion is associated with the incrustation.

A total dissolved solids value in excess of 150 p/m, high calcium and iron bicarbonate content, a Ryznar Index below 7, and a pH above 7.5 are indicative of possible incrustation problems [5].

The most common forms of mineral incrustation encountered in wells are summarized as follows:

(a) Precipitation of iron and calcium carbonates in the form of a hard, brittle, cement-like material adhering to the screen and frequently cementing the gravel pack or aquifer particles for some distance from the screen.

(b) An accumulation of iron and manganese hydroxides or hydrated oxides on the screen or in the formation immediately adjacent to the screen. The hydroxides are insoluble, jelly-like masses unless oxygen is present, in which case they are oxidized into oxides or hydrated oxides having a black, brown, or reddish granular appearance.

(c) Occasionally, it will be found that silts and clays in suspension are deposited about the screen and reduce the entrance of water. This is a relatively common occurrence in some types of gravel-pack construction, particularly when the pack is too coarse or development of the well has been inadequate.

(d) In some parts of the Western United States where lignite beds are prevalent, decomposition of the lignite results in the formation of a slimy black or brown viscous material about the screen and in the adjacent aquifer.

While there is no known way of avoiding mineral incrustation entirely in an area where conditions are favorable for its formation, there are some practices which can retard incrustation. The simplest practice is to assure maximum inlet area in the screen and consequently minimum inlet velocity of the water. In addition to having maximum inlet area, the screen should have uniform openings over the entire surface; these openings should be clean cut without burrs or rough edges which encourage deposition.

The second important factor is proper development of the well to ensure that as many fines as possible are withdrawn from the aquifer in the immediate vicinity of the screen. If a gravel pack is used, it should have a permeability sufficiently higher than the aquifer so that a reduction in velocity occurs as the water passes through the pack. This practice ensures minimum velocity and maximum open area in the formation and the gravel pack.

When an aggravated mineral incrustation problem has been encountered in an existing well, it can often be overcome by reducing the rate of pumping and, consequently, the entrance velocity of the water flowing into the well. This may be accomplished by installing a smaller pump and operating the well longer to obtain a required quantity. Where this practice is not possible, the rate of deposition may be decreased by installing several wells and pumping them at a relatively low rate to obtain the needed volume of water.

Since incrustation cannot be entirely avoided, rehabilitation of permanent wells should be anticipated in many cases. It is good practice to carry out such rehabilitation at stated intervals before the problem becomes too acute at any one well. The rehabilitation of wells is discussed in chapters XVIII and XXI.

Mineral incrustation caused by bacteria often is a factor in the decline of yield of a well. Crenothrix or similar organisms form a slimy gelatinous mass which accumulates on the screen or other metal parts in the well. This may appear in the well water as fine, short, reddish-brown filaments in a gelatinous matrix. The mass not only blocks the screen but gives a disagreeable taste and odor to the water and fosters aggravated corrosion of the ferrous metal parts. Where experience has shown that such organisms infect wells in an area, newly drilled and serviced wells and pump installations should be thoroughly sterilized as a preventive measure. Similarly, sterilization of infected wells may, in some cases, destroy the organisms and alleviate the problem.

14-3. Bibliography.—

[1] Grange, J. W., and Lund, E., "Quick Culturing and Control of Iron Bacteria," Journal of the American Water Works Association, vol. 6, No. 5, pp. 242–245, May 1969.

[2] Kuhlman, F. W., "Corrosion of Iron in Aqueous Media," Canadian Mining and Metallurgical Bulletin No. 52, pp. 713–729, November 1959.

[3] Moehrl, K. E., "Corrosion Attack in Water Wells," Corrosion, vol. 17, No. 2, pp. 26–27, February 1961.

[4] Pennington, W. A., "Corrosion of Some Ferrous Metals in Soil with Emphasis on Mild Steel and On Gray and Ductile Cast Iron," Bureau of Reclamation Chemical Engineering Branch Report No. ChE–26, March 9, 1965.

[5] Ryznar, J. W., "A New Index for Determining the Amount of Calcium Carbonate Scale Formed by Water," Journal of the American Water Works Association, April 1944.

«Chapter XV

WELL SPECIFICATIONS

15–1. Purposes of Well Specifications and Available Standards.— Well specifications should provide that: (1) the completed well will be constructed of material which will be compatible with the environment and which will give an adequate well life, (2) workmanship will be of an acceptable standard, and (3) the well will be of adequate size and design to yield the desired discharge with acceptable minimum efficiency. In addition, specifications should: (4) be reasonably flexible, (5) foster competitive bidding, (6) permit the preparation of cost estimates, and (7) be a guide to those responsible for supervising construction of the well.

(a) *National Standards.*—Two national standards are available: the American Water Works Association-National Water Well Association (AWWA–NWWA) "AWWA Standard for Deep Wells, AWWA A100-66" [2] [1] and the American Society of Agricultural Engineers (ASAE) "Recommended Practices for Designing and Constructing Water Wells for Irrigation" [1].

The former is an excellent general guide for the construction of most types of wells. In using the AWWA–NWWA Standard as a guide, the designer should be familiar with the fundamentals of well design and construction.

The ASAE recommended practices are a great improvement over the design and construction practices that have been followed in many parts of the country. However, they are designed primarily for low initial construction cost with little or no consideration for well life or efficiency.

(b) *Local, Regional, and State Standards.*—Many water wells have been constructed in the past by drillers and engineers with local attitudes, interests, and experience. During this development period, little was known about well efficiency, well life, sanitation, and related factors. If the initial cost of a well was low and the desired quantity of water was obtained, the well was considered acceptable.

With the introduction of the deep well turbine pump, larger and more efficient drilling equipment, and the increased importance of ground water, as surface water supplies become inadequate or unsatisfactory, the situation has changed. Economic and sanitary conditions are given more consideration, and research and engineering have become an integral part of the industry.

[1] Numbers in brackets refer to items in the bibliography, section 15–4.

Many State and regional drilling contractors' organizations [3] have issued minimum standards for the guidance of their members and for the protection of the public.

Many States and counties have also established minimum standards covering such aspects of water well construction as: location with respect to possible pollution, minimum casing types and weights, sanitary seals, the protection of aquifers, and the sterilization of wells [4,5,6,7,8,9,10,11,12,13,14,16,17]. As would be expected, these are usually more stringent than those of the well drilling contractors' organizations. To foster and guide further action in this regard, the U.S. Department of Health, Education, and Welfare has issued a model law governing the construction of water wells which it is urging the various State legislatures to enact [15]. The model law was developed after consultation with Federal, State, and local officials; engineers; water well drillers; and equipment suppliers. The enactment by most of the States within the next few years of legislation based on the model law and requirements of the Environmental Protection Agency is anticipated.

Most of the referenced standards, statutes, and regulations require the construction of an acceptable well to a minimum standard. They are not written to give the best construction from all aspects nor is such construction warranted on all wells.

Consultation with Federal, State, and local officials before any water development activities is advisable.

Well design and construction should equal or exceed local or State requirements so long as they are reasonable and are of good engineering practice. Local practices and procedures should also be considered and, if engineeringly sound, included as possible alternatives in specifications. Recognition of local practices will often result in lower bids, but they should not be specified as the only method or the specifications may then prove to be too restrictive.

15-2. Content of Technical Paragraphs.—Variations in geological and hydrological conditions, water requirements, and purposes of the installation preclude the writing of standard technical specifications paragraphs for well specifications.

Usually one of two conditions will dictate to a large extent the format and content of the specifications. Where a pilot hole has been drilled and a log and mechanical analysis of samples are available, a firm design can be made. From this, the amount and nature of drilling, materials, and related items can be specified. Lump-sum bids can be requested for definite work and unit price bids requested for the remainder of the work. Where such specifications can be prepared, bids are usually lower and more realistic since the contractor knows the drilling condition he will encounter, the equipment he must have, and the amount of material he must order. The amount of risk involved in well drilling, which is usually reflected in the bids, may be measurably reduced by proper preparation of the specifications.

Depending on time schedules, the number of wells involved, and information already available on an area, specifications may be written to provide for the drilling of pilot holes and subsequent construction of a producing well in one contract. However, where more than two or three wells are involved, the use of two separate contracts may be more desirable.

The other condition involves the drilling of small-diameter, low-capacity wells in virgin or geologically complex areas. A pilot hole cannot be justified for such wells because the cost of the pilot hole may approach the cost of the well. In such areas, specifications should provide for careful logging and sampling of materials during drilling. Estimates should be made of the probable number of units of work and materials which will be involved for the purpose of comparing bids, but provision should be made that if more or less work or material is involved, the unit prices will prevail. An estimate for standby time should be included to permit time for the determination of screen type, slot size, diameter and length, and delivery of screen.

In the interest of competition and nonrestrictive bidding, the method of drilling generally is not specified in Bureau well specifications. At times this may involve the inclusion of alternatives or apparently redundant paragraphs and bid items. However, circumstances which may require the specification of a particular drilling method include: impracticability of other methods due to adverse subsurface conditions, requirement for a particular sampling procedure or type of sample, or similar factors which limit the acceptable method of drilling or other activity. When such requirements occur, the specifications should be carefully written to assure that the bidder will furnish the type of operation desired, and there must be a defensible and convincing reason for the provisions.

15–3. Grout and Concrete for Water Well Construction.—Sufficient data have been furnished in chapter XI for writing specifications for wells except in regard to cement grout seals and concrete. The following subsections, (a) through (h), were prepared by the Concrete and Structural Branch of the Division of General Research, E&R Center, Denver, to cover these items.

(a) *General.*—Concrete may be required in the bottom of the well for sealing purposes and in pump foundations and pads. The concrete should consist of portland cement, sand, coarse aggregate, air-entraining agent, and water. In addition, a water-reducing, set-controlling admixture (WRA) should be used in seal concrete and may be used in concrete for pump foundations and pads.

Grout should be placed in the annular space between the casing and the wall of the drilled hole for the full depth of the space to be grouted. The grout should consist of portland cement, aluminum powder, water, and WRA.

Concrete and grout shall be well mixed and brought to the proper consistency.

(b) *Materials.*—

(1) *Cement.*—Portland cement should conform to the requirements of Federal Specification SS–C–192g for Type V, low-alkali cement and shall meet the false-set limitation specified therein. The cement shall be free from lumps when used in concrete and grout. Other types of cement may be specified for unusual conditions.

(2) *Sand and coarse aggregate.*—Sand and coarse aggregate shall be clean, hard, dense, and durable; consisting of uncoated rock particles and fragments, respectively; and shall not contain injurious amounts of dirt, organic matter, or other deleterious substances. The sand shall be natural sand and shall all pass a standard 3/16- or 1/4-inch screen. The coarse aggregate shall be reasonably well graded from 3/16 or 1/4 inch to 1/2 inch.

(3) *Aluminum powder.*—Aluminum powder shall be finely ground and shall contain no polishing agents such as stearates, palmitates, or fatty acids.

(4) *Air-entraining agent.*—The air-entraining agent shall be an approved agent conforming to ASTM Designation: C 260.

(5) *Water-reducing, set-controlling admixture* (*WRA*).—The WRA should be a hydroxylated carboxylic acid.

(6) *Water.*—Water used in concrete and grout, and for curing concrete, should be free from objectionable quantities of silt, organic matter, alkalies, salts, or other impurities.

(7) *Sealing compound.*—The sealing compound, if used for curing concrete, should be an approved white-pigmented compound conforming to Bureau of Reclamation "Specifications for Sealing Compound for Curing Concrete," dated June 1, 1961.

(8) *Reinforcing bars.*—Reinforcing bars, when required for reinforcing concrete in pump foundations and pads, should conform to Federal Specification QQ–S–632b, Type II, Class B40, C40, B50, C50, or R50.

(c) *Concrete Mixtures.*—Concrete for seal concrete, and for pump foundations and pads, should be proportioned to contain one part cement to two parts sand and two parts coarse aggregate by volume. (One sack of cement, 94 pounds net per sack, equals 1 cubic foot.) The concrete should contain 1 fluid ounce of single-strength air-entraining agent per sack of cement. In addition, 6 fluid ounces of WRA per sack of cement should be used in seal concrete. The contractor may use WRA in concrete for pump foundations and pads, in which case 4 fluid ounces of WRA per sack of cement should be used.

The quantity of water used in each concrete mixture should be such that the slump shall not exceed 3 inches for concrete to be placed in pump foundations and pads and should not exceed 8 inches for seal concrete. Each batch of concrete should be mixed for not less than 1½ minutes.

(d) *Grout Mixture.*—The grout mixture should be proportioned on the basis of sacks of portland cement to which ½ teaspoon of aluminum powder per sack of cement has been added. The aluminum powder should be thoroughly preblended with the cement. The cement-aluminum powder mixture should be mixed with not less than 5 nor more than 5½ gallons of water per sack of mixture. The grout mixture should contain WRA, and the quantity to be used should be in accordance with the manufacturer's recommendations and should be such that would delay the set of the grout for not less than 3 hours but not more than 12 hours. A trial mix may be necessary to test and determine quantities of WRA and water required to obtain an acceptable grout composition suitable for placing and meeting the specified setting time requirements.

(e) *Placing Grout.*—Grout should be placed from the bottom to the top of the space to be grouted in one continuous operation and shall be entirely placed before the occurrence of initial set in accordance with the provisions of AWWA–A 100–66, American Water Works Association Standard for Deep Wells, Sections 1–5.2 and AL–8.4. Grout returning to the surface through the annular space should be wasted until it shows little or no contamination with drilling fluid or formation material, as determined by the contracting officer. After the grout has been placed, no further work should be performed on the well for at least 48 hours.

(f) *Forming, Placing, Finishing, and Curing Concrete.*—Seal concrete should be placed with a dart valve or dump bailer or other suitable equipment that will properly place concrete down the hole; through water when encountered; to the required depth; and provide a dense, homogeneous concrete seal at the bottom of the hole.

Concrete in pump foundations and pads should be formed where and as required to shape the concrete to the required lines. The concrete shall be thoroughly consolidated. Exposed unformed surfaces of concrete should be brought to uniform surfaces and given a reasonably smooth wood float or steel trowel finish, as directed. The concrete should be water cured by keeping the exposed surfaces continually moist by sprinkling or spraying for at least 14 days after being placed or by spraying one coat of sealing compound on the exposed surfaces to provide a continuous uniform membrane over all exposed areas. The concrete, when being placed, should have a temperature of not more than 90°F, should be maintained at a temperature not lower than 50°F for at least 72 hours after placement, should be protected from freezing for not less than 3 days following the 72-hour protection at plus 50°F, and should be protected against injury until final acceptance by the Government.

(g) *Placing Reinforcing Bars.*—Steel reinforcing bars should be placed in foundation and pad concrete where and as shown on the drawings. The reinforcing bars should be cleaned of heavy flaky rust, loose mill scale, dirt, grease, or other foreign substances and shall be accurately placed and secured in position.

(h) *Measurement and Payment.*—Measurement, for payment, of concrete in pump foundations and pads should be on the basis of concrete having the dimensions shown on the drawings or prescribed by the contracting officer. Measurement, for payment, of seal concrete should be on the basis of the number of cubic yards placed as determined by the number of sacks of cement used in the concrete actually placed, each seven sacks of cement being considered as 1 cubic yard of concrete. Measurement, for payment, of grouting should be made on the basis of the actual number of sacks of cement used in grout placed in the well and used in the test batches of grout.

15–4. Bibliography.—

[1] "Designing and Constructing Water Wells for Irrigation," American Society of Agricultural Engineers, St. Joseph, Mo., December 1964.

[2] "AWWA Standard for Deep Wells," American Water Works Association, AWWA–A100–66, New York, 1967.

[3] "Recommended Standards for Preparation of Water Well Construction Specifications," Associated Drilling Contractors of California, Sacramento, September 17, 1960.

[4] "Water Well Standards—State of California," California Department of Water Resources Bulletin 74, Sacramento, 1968.

[5] "Rules and Regulations and Water Well Drilling Pump Installation Contractors' Law," Colorado Division of Water Resources, Denver, 1967.

[6] "Idaho Minimum Well Construction Standards," Idaho Department of Reclamation, Boise, July 1968.

[7] "Water Resources Regulation 2.3, Well Drillers and Well Construction," Maryland Water Resources Commission and Department of Water Resources, Annapolis, 1969.

[8] "Ground-Water Quality Control," Michigan Department of Public Health, Act 294PA 1965 and Rules, Lansing, 1966.

[9] "Nebraska Minimum Standards for Gravel Packed Irrigation Wells," University of Nebraska Extension Service, College of Agriculture, E.C. 57–702, Lincoln, August 1957.

[10] "Rules and Regulations for Drilling Wells and Other Related Material," Nevada Department of Conservation and Natural Resources, Division of Water Resources, Carson City, 1969.

[11] "Rules and Regulations Governing Drilling of Wells and Appropriation and Use of Ground-Water in New Mexico," New Mexico State Engineer, Santa Fe, 1966.

[12] "Rules and Regulations Prescribing General Standards for the Construction and Maintenance of Water Wells in Oregon," Oregon State Engineer, (Preliminary Draft Subject to Revision), Salem, 1972.

[13] "Ohio Water Well Construction Code," Ohio Water Resources Board, Columbus, October 1946.

[14] "Construction Standards Individual Water Supplies," Pennsylvania Department of Health, Harrisburg, 1961.

[15] "Recommended State Legislation and Regulations, Urban Water Supply, Water Well Construction and Individual Sewage," U.S. Department of Health, Education, and Welfare, Public Health Service, Washington, D.C., 1965.

[16] "Well Construction and Pump Installation," Wisconsin Administrative Code, Chapter NR112, Wisconsin Department of Natural Resources, Madison, 1972.

[17] "Regulations and Instructions, Water Well Minimum Construction Standards," Wyoming State Engineer, Cheyenne, 1971.

«Chapter XVI

WATER WELL DRILLING

16-1. Introduction.—Most wells are drilled by mechanically powered equipment normally referred to as drill rigs. This chapter is intended to acquaint the reader with the major types of drill rigs and the capabilities and limitations of such rigs.

16-2. Drilling and Sampling with Cable Tool Rigs and Variations.—(a) *Drilling Methods.*—The cable tool method of drilling, often referred to as the standard method, churn drill, percussion method, or facetiously as the yo-yo, is one of the oldest, most versatile, and simplest drilling devices [6].[1]

The cable tool drills by lifting and dropping a string of tools suspended on a cable. A bit at the bottom of the tool string strikes the bottom of the hole, crushing, breaking, and mixing the cuttings. A string of tools in ascending order consists of a bit, a drill stem, jars, and a swivel socket, which is attached to the cable.

In stable rock, an open hole can be drilled, but in unconsolidated or raveling formations, casing must be driven down the hole during the drilling. Above the water table or in otherwise dry formations, water is added to the hole to form a slurry of the cuttings so they may be readily removed by a bailer. The bottom of the casing is usually fitted with a drive shoe.

As casing is driven in unconsolidated formations, the vibration causes the sides of the hole to collapse against it and compact. Frictional forces increase until it is no longer possible to drive the casing. When this occurs, a smaller diameter casing is telescoped inside of the casing already in place, and drilling is continued using a smaller diameter bit. On deep holes, as many as four or five reductions in casing may be required.

Cable tool rigs are generally limited to drilling maximum hole diameters of 24 to 30 inches and to depths of less than 2,000 feet.

The cable tool rig is probably the most versatile of all rigs in its ability to drill satisfactorily under a wide range of conditions. Its major drawback, compared to some other rigs, is its slower rate of progress and depth limitations.

The initial cost of a cable tool rig complete with tools is one-half to two-thirds that of a rotary rig of equivalent capacity. The rigs

[1] Numbers in brackets refer to items in the bibliography, section 16-8.

are usually compact, require less accessory equipment than other types, and are more readily moved in rugged terrain. The simplicity of design, ruggedness, and ease of maintenance and repair of the rigs and tools are particularly advantageous in isolated areas. They generally require less skilled operators and a smaller crew than other rigs of similar capacity. The low horsepower requirements are reflected in lower fuel consumption, an important aspect where fuel costs are high or sources of fuel are remote. While somewhat slower than other rigs in drilling some formations, they can usually drill through boulders and fractured, fissured, broken, or cavernous rocks which often are beyond the capabilities of other rigs. In addition, much less water is required for drilling than with most other commonly used rigs, an important consideration in arid and semiarid zones. Also, sampling and formation logging are simpler and more accurate with the cable tool rig. The cuttings bailed from each drilled interval usually represent about a 5-foot zone. When casing is used, there is little chance of contamination of the sample. A skilled driller can usually recognize a change in formation by the response of the rig to the changed drilling condition and can then take samples at a shorter interval. In some unconsolidated formations, casing can be sunk by merely bailing and driving so that samples are relatively unbroken and representative.

The samples are not greatly contaminated by drilling mud and clay; shale and silt fractions are less likely to be lost by dispersion in the drilling fluid. Cuttings of unconsolidated formations are usually not too finely pulverized, and some are usually of sufficient size to permit ready identification and description. A more reliable method of sampling involves the use of a drive barrel sampler driven by drilling jars. This provides a means of obtaining representative to undisturbed samples from moderate to great depths. When promising aquifer materials are encountered, they are readily tested for yield and quality of water by bailing or, if of sufficient importance, by pumping.

The disadvantages of a relatively slow rate of progress and the economical and physical limitations on depth and diameter have been mentioned previously. A further disadvantage of the cable tool rig is the necessity of driving casing coincident with drilling in unconsolidated materials. This precludes the use of electric logs, which are desirable in many instances. However, gamma logs may be taken inside a casing. The driving of casing necessitates a heavier wall pipe than would otherwise be required in some installations. Also, screens often must be set by pullback or bail-down methods. The pullback method in deep or large diameter wells is sometimes extremely difficult, and the bail-down method may give rise to problems of alinement.

Mud scow drilling [3,7,8] uses a heavy scow in place of a bit for drilling large diameter holes in gravel and finer materials. The scow is a heavy pipe, 10 feet or more in length, fitted at the bottom with a heavy shoe somewhat similar to a drive shoe and generally with a heavy steel knife blade welded across the diameter of the shoe. When drilling with a mud scow, the casing used is usually California double-walled stovepipe (see sec. 11–3) in 3- to 5-foot lengths which are jacked down as drilling proceeds. The method is peculiar to the southwestern part of the United States.

The cable tool rig is readily adapted to drilling 2- to 4-inch-diameter holes with jet or hollow-rod tools in soft formations such as clay or sand. Jet drilling [1,4,5,7,8] is basically a percussion method combined with a pressure pump. The drill pipe is lifted and dropped, which chops up material at the bottom of the hole. The water helps to jet the broken material loose and carries the cuttings up the hole where they are discharged into the pit. The method is useful in installing observation holes and small capacity water wells.

Hollow-rod drilling [1,4,5,7,8] is somewhat similar to jet drilling except that a check valve is used above the bit instead of a pump. It is suitable for drilling small wells and observation holes of 2- to 4-inch diameter to about 100-foot depths in soft unconsolidated material.

Both the hollow-rod and jetting methods have the disadvantages of the direct circulation rotary rig discussed below in regard to sampling of formations and water and the measurement of static water levels.

(b) *Sampling.*—The accessories and equipment for cable tool rigs are fairly uniform and standard, although the sampling procedures of different drillers are variable. To ensure obtaining good samples meeting a minimum standard, Bureau specifications usually require that they be taken at each 3- to 5-foot interval or at each change in material, whichever is less. To assure reliable sampling, the Bureau requires samples to be deposited in a sample box. One such box contains four different sample compartments. Each sample batch is mixed and quartered until a 2-quart representative sample remains. This is placed in two separate 1-quart containers, each marked with the well designation, the date, and the drilled interval it represents. After the sample is obtained, the compartment is thoroughly cleaned and flushed before another sample is placed in it. A typical sample box is shown in figure 16–1.

16–3. Drilling and Sampling with Direct Circulation Rotary Rigs and Variations.—(a) *Drilling Methods.*—The rotary rig drills by turning a fishtail, toothed cone, or similar bit at the bottom of a string of drill pipe. The typical string consists of a bit which scrapes, grinds,

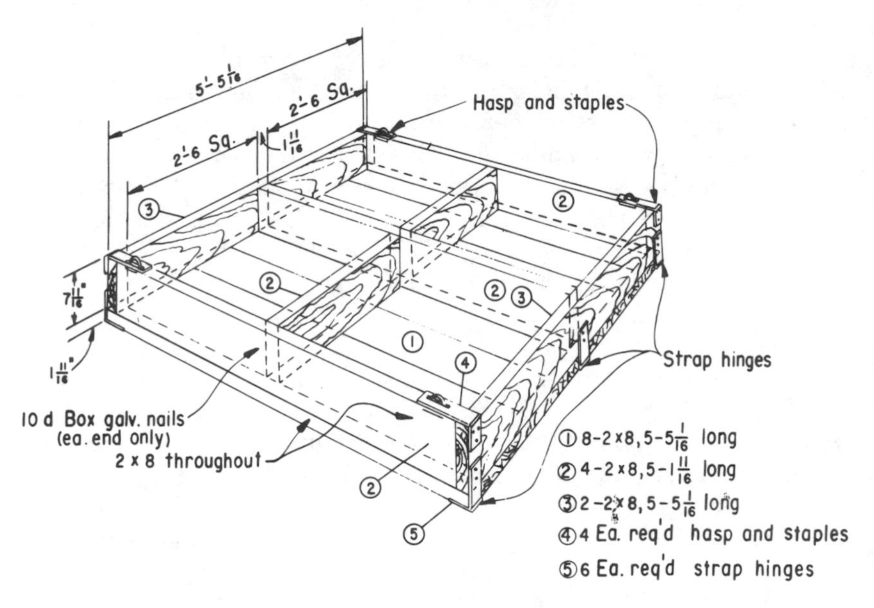

FIGURE 16–1.—Compartmented cuttings sample box for use with a cable tool rig. 103–D–1519.

fractures, or otherwise breaks the formation drilled; a drill collar of heavy walled pipe which adds weight to the bit and helps to maintain a straight hole; and a drill pipe which extends to a kelly (shaft) near the surface which imparts rotation. As the bit is turned, drilling fluid (mud) is pumped down the pipe to lubricate and cool the bit, jet material from the bottom of the hole, and to clean the hole by transporting the cuttings to the surface in the annular space between the hole wall and the drill pipe. The drilling fluid also forms a thin layer of mud on the wall of the hole which reduces seepage losses and, together with the hydrostatic head exerted by the mud column, holds the hole open [1,2,3,4,5,7,8].

The selection of the correct mud and the maintenance of mud weight, viscosity, jelling strength, and a low percentage of suspended solids, together with a suitable uphole velocity, contribute to rapid, trouble-free drilling. Numerous drilling fluids are used, but for water wells, a suspension of bentonite or similar clays in water is most commonly used.

The cost of a rotary rig is considerably greater than that of a cable tool rig of equal capacity. Operation of the rotary rig requires much more training and skill than a cable tool rig and a larger crew is needed. Maintenance and repair are more complex and more water is required for drilling when using drilling fluids. Also, if the permeability of the aquifer is much greater than about 50 feet per day, mud may invade the aquifer and jell at some distance from the wall of the hole. While chemicals are available to break down the mud, it is frequently impossible to fully develop a water well in which mud has invaded the aquifer.

Clayey materials are frequently mixed and incorporated in the drilling fluid and may not be recognized. Cuttings are usually fine, and because of a variable rate of travel in the mud stream, become mixed and separated and are not always representative. Because of the mud-filled hole and dispersal of cutting in the return flow, possible aquifer materials may be overlooked in the drilling. Static water levels, water samples, and pumping tests are not readily obtained from aquifers without special equipment. An electric log is usually desirable in conjunction with the driller's formation log when interpreting the results of rotary drilling for well design purposes.

Despite the above disadvantages, the direct circulation rotary rig offers relatively rapid drilling in most formations, greater depth capacities, and an open hole which simplifies installation of casing, screen, and grout and permits the use of most geophysical well-logging equipment.

Although plain water is often used as a drilling fluid, the fluid usually consists of a suspension of native clay, bentonite, or organic thickeners in water. Native clays are seldom a satisfactory mud base. Bentonite is far more effective and efficient, particularly when the better grades are used. The desirable properties obtainable with bentonite are high viscosity, jelling strength, and a relatively low solids content in the mud. Organic bases have little or no jell strength but excellent viscous properties. They degrade with time instead of jelling and permit more rapid and thorough development of a well. For water wells, mud weight is usually 9 to 9½ pounds per gallon and the viscosity is from 32 to 36 seconds from a marsh funnel. The sand content of the drilling fluid, where it is picked up by the mud pump, should not exceed about 2 percent. Drilling fluid should be tested at about 4-hour intervals for weight and viscosity. If it cannot be treated to obtain the desired properties, the old mud should be discarded and a new batch mixed.

Where a pilot hole has been drilled and suitable aquifers found, the pilot hole is normally reamed to the desired diameter.

(b) *Sampling.*—The method of sampling required by the Bureau is to drill 3 to 5 feet, raise the bit from the bottom of the hole, and continue circulation until all cuttings from the sample interval are cleared from the hole and caught in a sample catcher. Drilling is then resumed for another 3 to 5 feet. The sample is then removed from the catcher and the sample catcher thoroughly cleaned prior to drilling the next interval. The sample is mixed and quartered until about a 2-quart representative sample remains. This is placed in a pail or drum to which about 5 gallons of clear water are added, stirred, and permitted to settle for about 20 minutes. The muddy water is then decanted and the sample placed in two 1-quart containers, each of which is clearly marked to show the well designation, the depth interval represented, and the date the sample was taken. Sampling by rotary methods using clay-based muds is not recommended as a basis for the design of wells in granular materials. An exception is drive sampling. Figure 16–2 shows a typical sample catcher. A convenient arrangement which saves considerable time is to use two sample catchers in parallel with a diversion gate at the end of the ditch, which permits the return flow to be diverted into either catcher. The cleaning of one catcher is then possible while the other is being used.

16–4. Air Rotary Drilling.—Air rotary drilling [1,7,8] was developed primarily in response to the need for a rapid drilling technique in hard rock in arid areas. The rig, bits, etc., are essentially the same as for direct circulation rotary drilling except the fluid channels in the bit are of uniform diameter rather than jets, and the mud pump is replaced by an air compressor. Air is circulated down the drill string to cool the bit and to blow the cuttings to the surface.

When initially developed, the air rotary method was used for relatively small diameter holes in hard rock. Larger holes have become possible through use of foams and other air additives, and diameters up to 36 inches have been drilled successfully.

Sampling of cuttings is not adequate or practical for well design because there is some delay in bringing cuttings to the surface. During

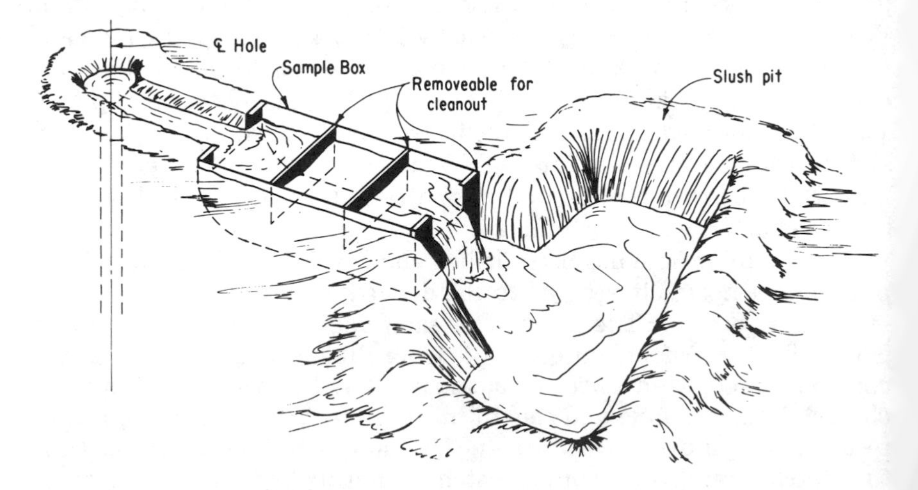

FIGURE 16–2.—Cuttings sample catcher for use with a direct circulation rotary rig. 103–D–1520.

this delay, considerable mixing of cuttings from various depths occurs. The rigs are mostly applicable to hard rock terrains where water is encountered in fractures or similar openings and wells are completed as open holes.

Shortly after the development of air rotary drilling, the down-the-hole hammer bit was developed. This consists of a maximum 8-inch-diameter bit working on the principle of the jackhammer which replaces the conventional bit at the bottom of the drill string. This arrangement efficiently combines some of the advantages of the cable tool and rotary rig. The air used to activate the bit either blows the cuttings to the surface or lifts them by the principles of the airlift pump when drilling below the water table. The bit is particularly applicable to rapid drilling of hard rock.

Both rotary air and down-the-hole drilling in saturated materials are limited in depth by the available air pressure which must be greater than that exerted by the column of water in the hole if the rig is to function.

16–5. Drilling and Sampling with Reverse Circulation Rotary Drills.—(a) *Drilling Methods.*—The reverse circulation rotary rig operates essentially the same as a direct circulation rotary rig except that the water is pumped up through the drill pipe rather than down through it. Large capacity centrifugal or jet pumps similar to those used on gravel dredges are used. The discharge is directed into a large pit in which the cuttings settle out. The water then runs through a ditch and into the hole so that the water level in the hole is maintained at the ground surface [1,2,3,7,8].

Velocity of water down the hole cannot exceed a few feet per minute to avoid erosion of the side of the hole at the restricted annulus around each flange joint. Consequently, the minimum hole diameter is about 16 inches. The drag bits generally used range in diameter from about 16 to 72 inches. When boulders or cobbles too large to pass through the drill pipe are encountered, the bit is pulled from the hole and the larger rocks removed with an orangepeel bucket. Recently, compound bits consisting of combinations of cones similar to the bits used in direct circulation rotary drilling and the use of drill collars have made drilling in rock more feasible.

The water velocity up the drill pipe is usually in excess of 400 feet per minute and separation of cuttings is at a minimum. Samples caught are representative of the formation being drilled within depths of a few inches at most. Because of the large volume and high velocity of the water, however, special sample catchers such as the Cope sampler are required to obtain good samples (see fig. 16–3).

To maintain a stable hole, a differential head of 8 to 13 feet is required. If the static water level is less than this range, an arrangement to increase the head must be devised.

The reverse circulation rig is probably the most rapid drilling equipment available for unconsolidated formations, but it requires a large volume of water which must be constantly replenished since drilling mud is seldom used.

Where the water table is in excess of 20 feet, surface casing should be installed and grouted to minimize loss of water. The column of water in the hole acts to keep the hole open in a manner similar to the drilling fluid used in a direct circulation rotary rig.

Because of the large hole diameters, reverse circulation drilled holes are usually gravel packed.

The normally equipped reverse circulation rig can drill to a depth of approximately 450 feet at sea level. Deeper drilling is usually not possible because friction losses in the drill pipe and the weight of the cuttings-charged column of water become too great for the suction lift of the pump. However, by introducing air into the lower third of the drill pipe and using airlift pumping rather than centrifugal pumping, wells have been drilled to more than 1,200 feet at elevations of over 5,000 feet using this method.

(b) *Sampling.*—A common method of obtaining samples with a reverse circulation rig is to catch them either in a buckct or with a screen at the end of the discharge pipe. However, such samples are

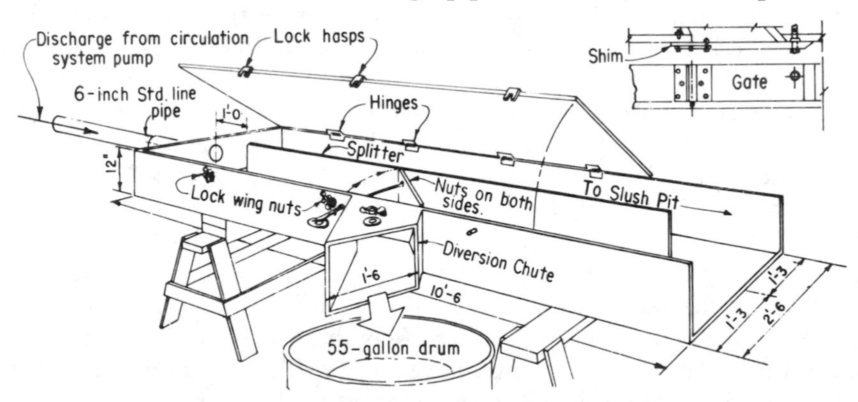

Original model was made of welded $\frac{3}{16}$ inch steel plate; however, if general measurements are followed, model could be made from wood. Discharge hose from Reverse Circulation Pump is connented to 6-inch pipe so all materials go through sampler. Materials can be observed in open 2-foot discharge without cover. Sampler is mounted on heavy sawhorses or similar supports with about a 6-inch slope towards the slush pit into which material is discharged. To obtain sample, control gate is thrown open against splitter to divert sample through chute and into 55-gal. oil drum.

FIGURE 16-3.—Cope cuttings sample catcher for use with a reverse circulation rotary rig. 103-D-1521.

never representative. The large volume of water discharging at a high velocity tends to wash fines over the edge of the bucket or through the screen mesh. To overcome this problem, the sampler shown on figure 16–3 was developed by the Cope Drilling Company of Idaho Falls, Idaho. This sampler permits the catching of representative samples without loss of fines.

Several drums are used so that samples may be caught at frequent intervals when drilling in a thick aquifer. Each sample is allowed to stand for about 10 minutes to permit the fines to settle to the bottom. The water is then decanted, the sample dumped on a clean plywood panel or similar surface, and mixed and quartered until a 2-quart representative sample remains. The sample is placed in two 1-quart containers which are clearly marked with the well designation, the depth represented, and the date.

16–6. Other Drilling Methods [8].—Numerous other methods such as auger and chilled shot are used, but most of them have limited depth capacities, only special applications, small diameters, or are slow and costly.

16–7. Plumbness and Alinement Tests.—Each new well should be checked for conformance to the specifications regarding plumbness and alinement or straightness.

The measurements made are of the plumbness and straightness of the cased hole. Thus, an oversized hole may be out of line or plumb, but the casing may fall within the limits of the specifications. The casing should not be permitted to excessively encroach on the annulus and hinder placement of grout or gravel pack.

The usual standard for plumbness requires that the axis of the well casing not deviate from the vertical in excess of two-thirds the inside diameter of the casing per 100 feet of depth and that the deviation be reasonably consistent regarding direction. This requirement applies to both the casing and the screen [9].

The usual standard for alinement or straightness requires that a 40-foot-long dolly can be passed freely through the pump housing casing without hanging. The dolly should be rigid and fitted with 1-foot-wide rings which have a ½-inch smaller outside diameter than the inside diameter of the casing or screen being tested. The rings are placed at each end and in the center of the dolly [9].

The dolly is hung so it is centered at the top of the well with the supporting cable attached at the exact center of the dolly. The cable sheave or support should be adjusted and firmly fixed so that the cable is vertical between the support and top of the dolly. The dolly is then lowered in 5-foot increments and the deviation of the cable from the center of the casing measured for amount and direction at each 5-foot interval. During lowering, the cable should be watched to detect any deviation from the general direction of displacement or other sudden deflection.

The deviation of the well from the vertical at any depth can be computed by the equation:

$$X = \frac{D(H+h)}{h} \tag{1}$$

where:

X=well deviation at any given depth, in inches,
D=distance the cable departs from the center of the casing, in inches,
H=distance to the top of the dolly that is below the top of the casing, in feet, and
h=distance from the suspension point of the cable to the top of the casing, in feet.

If the dolly passes freely through the casing and deviations from the vertical are within acceptable limits, the well is satisfactory. If trouble is encountered, it can be checked with a cage (see fig. 16-4).

The cage should be at least 1-foot long and have a minimum outside diameter ½-inch smaller than the inside diameter of the casing. The cage is first set in the top of the casing and centered. Deviations of the casing from the vertical and the direction of deviation can be determined by measuring the distance and direction of movement of the cable from the center of the casing and applying equation (1). A special template as shown in figure 16-5 can be used if desired to simplify measurement of the direction and amount of movement of the cable. The computed deviation can then be plotted on graph paper to determine conformance to the specifications or the location of any difficulty encountered.

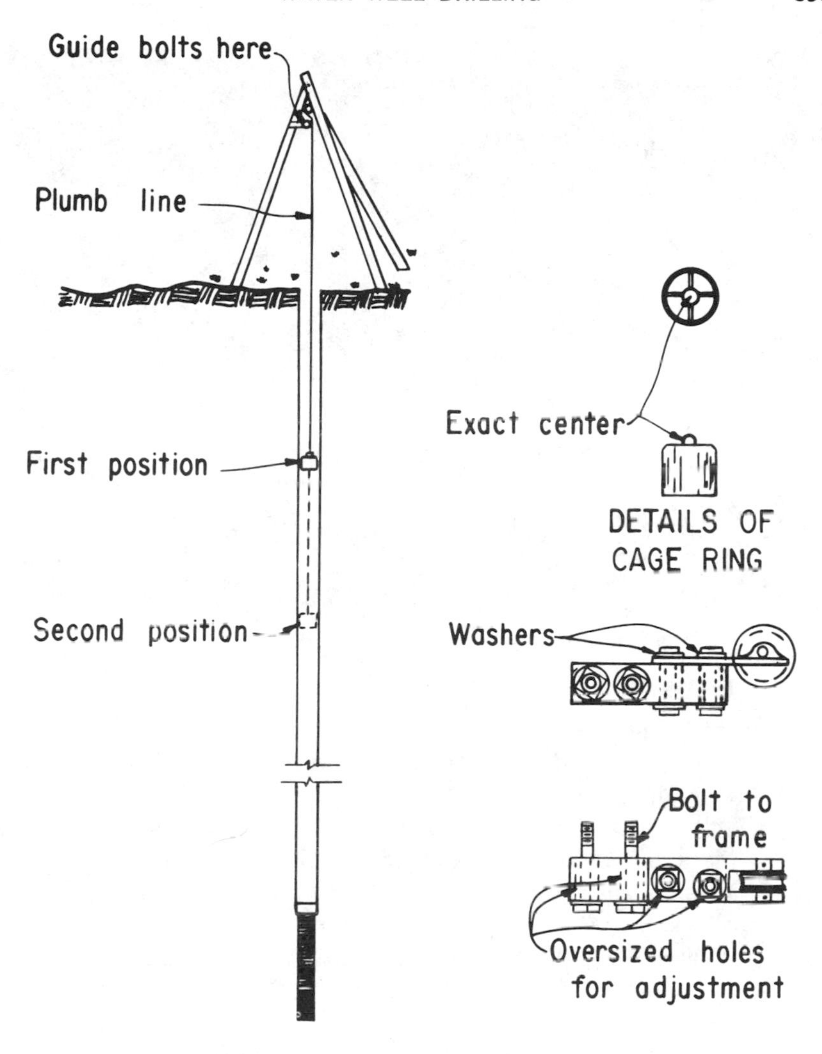

FIGURE 16-4.—Cable suspended cage for checking straightness and plumbness of wells. 103-D-1522.

FIGURE 16-5.—Template for measuring deviation of a well. PX-D-34588.

16-8. Bibliography.—

[1] "Well Drilling Manual," Speed Star Division of Koehring Co., Enid, Okla., 1967.

[2] "Well Drilling," Operations, Department of the Army Technical Manual 5-297, Washington, D.C., September 1965.

[3] "Ground Water and Wells," UOP Johnson Division, St. Paul, Minn., 1966.

[4] Bennison, E., "Ground-water, Its Development, Uses, and Conservation," Edward E. Johnson Company, St. Paul, Minn. 1947.

[5] Gibson, U. P., and Singer, R. D., "Small Wells Manual," U.S. Agency for International Development, Washington, D.C., January 1969.

[6] Gordon, R. W., "Water Well Drilling with Cable Tools," Bucyrus-Erie Company, South Milwaukee, Wis., 1958.

[7] "Water Well Drillers Beginning Training Manual," National Water Well Association, Columbus, Ohio, 1971.

[8] Campbell, M. D., and Lehr, J. H., "Water Well Technology," McGraw-Hill, New York, 1973.

[9] "AWWA Standard for Deep Wells," AWWA A100-66, American Water Works Association, New York, 1967.

Chapter XVII

WATER WELL DEVELOPMENT

17–1. Purpose of Well Development.—The primary purpose of well development or stimulation is to obtain maximum production efficiency from the well. Incidental benefits are stabilization of the structure, minimization of sand pumping, and the improvement of corrosion and encrustation conditions. Development also removes the mud cake from the face of the hole and breaks down the compacted annulus about the hole caused by drilling. Fines are removed from the pack and the aquifer, thus increasing the porosity and the permeability of the pack and aquifer. Water is made to surge back and forth through the screen, pack, and aquifer and to flow into the well at higher velocities than during pumping at design rates. Material which is brought to stability under high development velocities and surging will remain stable under velocities present during normal pumping operations.

Proper and careful development will improve the performance of most wells. Well development is not expensive in view of the benefits derived and only under unusual circumstances or improper methods will it cause harm.

Depending upon the circumstances, a number of methods and supplemental chemicals may be used in developing a well. Some of the common methods and the conditions for which they are used are described in the following sections.

17–2. Development of Wells in Unconsolidated Aquifers.—(a) *Overpumping.*—Pumping a well at a discharge rate considerably higher than design capacity is often the only well development procedure used. However, except in thin, relatively uniformed grained, permeable aquifers, this method alone is not recommended. The pump is normally set above the top of the screen; hence, development is primarily concentrated in the upper one-quarter of one-half the screen length. With the water moving in one direction only, stable bridging of the sand grains occurs so long as pumping continues. When pumping is stopped, the water in the column pipe drops back into the well causing a reverse flow which destroys the bridging. When the well is again pumped, sand will enter the well until stable bridging is reestablished. A well so developed may pump sand for several minutes each time the pump is started. This may continue for months or even years but may eventually clear up.

(b) *Rawhiding (Pumping and Surging)*.—The arrangement for rawhiding is similar to that for overpumping. However, the pump must not be equipped with either a rachet or other device that would prevent reverse rotation of the pump or a check valve. The well is pumped in steps, for example at ¼, ½, 1, 1½, and 2 times the design capacity. At the beginning of each step the well is pumped until the discharge is relatively sand free. The power is then shut off and the water in the column pipe is allowed to surge back into the well to break up bridging. The well may be surged one or more additional times by operating the pump until water is discharged at the surface and then stopping the pump. The pump is then operated again at the same rate repeating the surging cycle whenever the discharge clears. The rate of discharge is then increased and the same procedure followed at each of the higher rates, with the final rate being at the maximum capacity of the pump or well. Rawhiding is definitely superior to simple overpumping, but when used alone will usually result in development of only the upper portion of the screened aquifer. Rawhiding is recommended as a finishing procedure following initial development by any of the methods described in the following subsections (c), (d), and (e) of this section.

During final development by rawhiding, the amount of sand discharged by the well is measured when pumping is resumed after each cycle of surging. The initial discharge on resumption of pumping is usually almost sand free. Within a few seconds or minutes, depending upon the rate of discharge and the depth of the well, the sand will increase to a maximum. This condition will usually persist for a short period and then the amount of sand will begin to decrease until the discharge is practically sand free. At this time the well should be surged again.

The approximate concentration of sand being discharged can be estimated by looking through the discharge stream. The sand will be concentrated at the bottom of the stream where it issues from a discharge pipe with free discharge. It will look like a dark gray or brown layer. If an orifice is attached to the end of the pipe, the sand will appear as a dark vein in the center of the jet. The orifice should always be removed to avoid sand cutting its edge during rawhiding.

The time of maximum concentration of sand can be judged closely by observing discharge flow at the beginning of each rate of discharge. A sample is taken when the sand discharge is maximum.

Sand traps are available which will permit relatively accurate determination of sand content of the discharge, but they are expensive, heavy pieces of equipment. An Imhoff cone is commonly used to catch samples (see fig. 17–1). The cone should be held firmly with both hands, and the outside lip of the cone slipped into the bottom

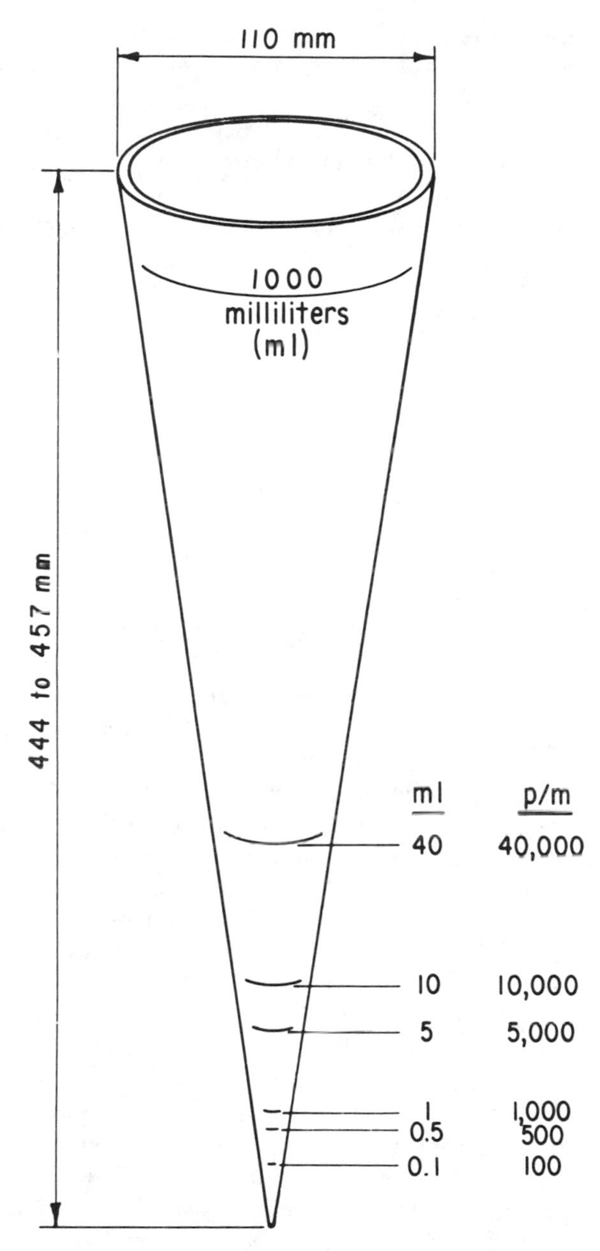

FIGURE 17–1.—Imhoff cone used in determination of sand content in pump discharge. 103–D–1523.

of the discharge stream to the center of the sand concentration. The cone fills in a fraction of a second and the entire procedure must be done rapidly.

The cone is then set in a holder to permit the contents to settle for a few minutes, and then the sand content by volume is estimated. The smallest division on a cone is 0.1 ml (milliliters). About one-tenth of the smallest division on the scale is approximately 10 p/m by volume or 20 p/m by weight. For estimating purposes, multiply the volume by 2 to get weight. Acceptable sand content for various purposes is as follows:

(1) Municipal, domestic, and industrial supply—0.01 ml or 20 p/m by weight
(2) Sprinkler irrigation—0.025 ml or 50 p/m by weight
(3) Other irrigation (furrow, flooding, etc.)—0.075 ml or 150 p/m by weight

Since the sample is taken during the period of highest sand concentration in the discharge, the estimated sand content is probably somewhat high and on the safe side.

Rawhiding, pumping, and sampling should be continued at the maximum discharge rate until the desired sand content is reached.

Imhoff cones are made in two styles. One has a somewhat rounded bottom while the other has a more pointed bottom. The model with the pointed bottom is preferable for estimating small volumes of material. Most Imhoff cones are made of glass and the breakage frequency is sometimes high, particularly when sampling high capacity wells. Recently, a plastic model has been produced which is less likely to break, easier to clean, and less expensive, but unfortunately, it has the rounded rather than the pointed bottom [1,2,5].[1]

(c) *Surge Block Development.*—The surge block is one of the oldest and most effective methods of well development. Such blocks are particularly applicable for use with a cable tool rig, and often such a rig equipped with a surge block is used to develop a well drilled by other methods. Solid, vented, and spring-loaded surge blocks are used. The solid and vented blocks consist of a body block 1 to 2 inches smaller in diameter than the well screen, and fitted with as many as four ¼- to ½-inch-thick disks of belting, rubber, or other tough material having a diameter the same as the inside diameter of the screen in which they will be used. Most surge blocks are made by well drilling contractors.

The solid surge block has a solid body, whereas the vented one has a number of holes drilled through the body parallel to the axis. The

[1] Numbers in brackets refer to items in the bibliography, section 17–5.

top of the body is fitted with rubber or similar flap valve which seals the holes on the upward stroke and permits water to move through them on the downstroke. Figure 17–2 shows a design for a vented surge block for use in an 8-inch-diameter well screen. The same general design can be used for screens from 4 to 12 inches in diameter. Figure 17–3 shows a design for a larger diameter spring-loaded vented surge block. Similar designs can be used for solid surge blocks by eliminating the vents and flap valves.

As the block is moved up and down in the screen, the solid surge blade imparts a surging action to the water which is about equal in both directions. The gentler downstroke of the vented surge block causes only sufficient backwash to break up any bridging which may occur and the stronger upstroke pulls in the sand grains freed by the destruction of the bridging. The solid surge block is usually most effective in dirty sands containing large percentages of clay, silt, and organic matter, while the vented surge block is best in cleaner sands.

The spring-loaded surge block may be vented or solid, is more effective than the other two blocks, and offers some advantage in avoiding sand or locking in the screen.

The surge block is attached to the bottom of a drill stem of adequate weight to ensure a fairly rapid downstroke under the action of gravity.

Before surging is begun, the well should be bailed clean and the surge block cable marked to identify the bottom of the well and top of the screen.

Surging should be started above the screen to bring in the initial flow of sand, thus minimizing the hazard of sand locking the block in the screen. Following this, surging is continued at the bottom of the screen using the longest stroke and slowest rate of which the rig is capable. The surge block is raised slowly through the screen as surging progresses until all the screen has been surged. The procedure is then repeated using a faster stroke. Several passes should be made until the maximum stroke rate is attained at which the line can be maintained in tension on the downstroke. The drill stem should not be permitted to fall out of plumb because of the hazard of falling against and possibly damaging the screen.

After each upward pass of surging, the block should be lowered to the bottom of the well to check the accumulation of material. When the material accumulated at the bottom begins to encroach upon the screen, the block should be pulled and the hole bailed clean. The rate of accumulation should be recorded to provide data on the progress of development.

Each time the surge block is removed from the well, the disks should be measured and if they have worn so that their diameter is

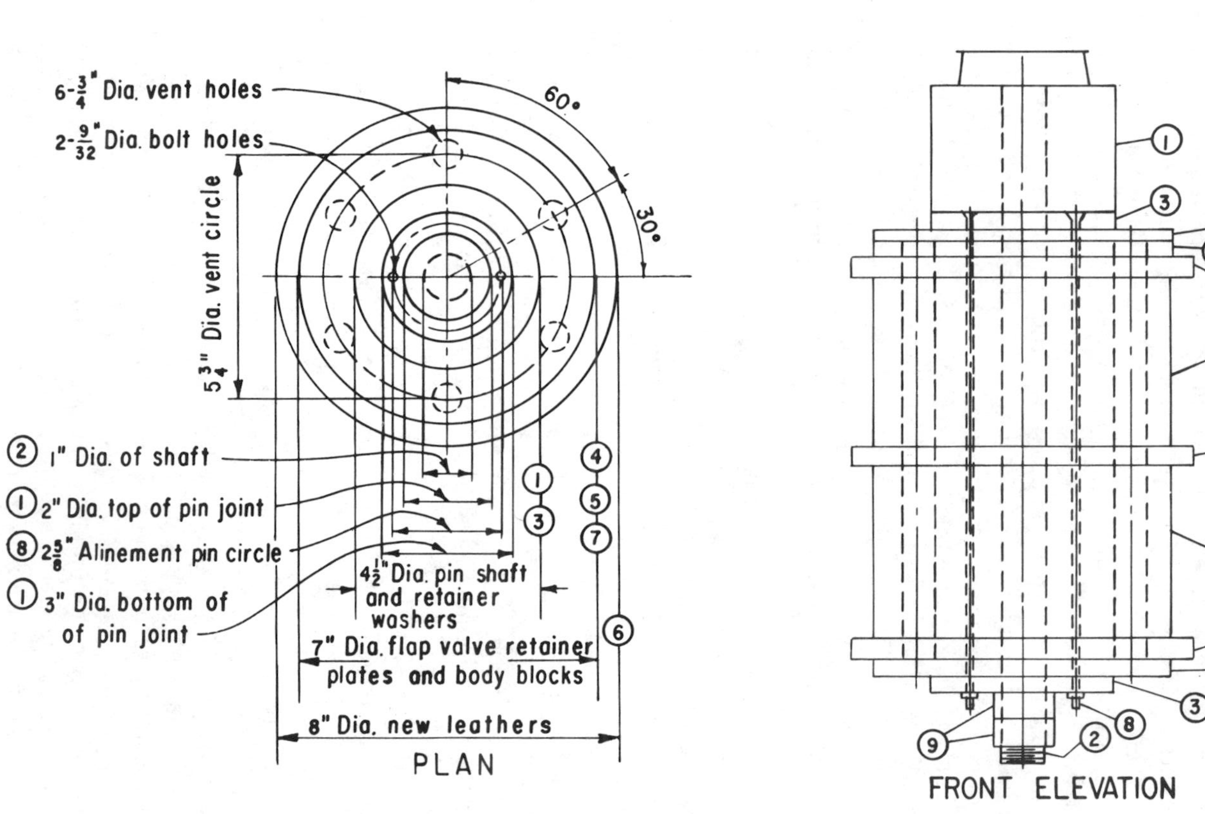

FIGURE 17-2.—Design of a vented surge block for an 8-inch well screen. (Sheet 1 of 2.) 103-D-1524-1.

① API pin joint, collar tapped and threaded for a 1-inch-diameter shaft.

② One-inch-diameter shaft threaded the required length on each end.

③ Two $4\frac{1}{2}$-inch o.d. steel retainer washers $\frac{3}{8}$-inch thick with 1-inch dia. center hole. Two $\frac{9}{32}$-inch dia. alinement holes drilled 180° apart on a $2\frac{5}{8}$-inch dia. circle about the center. Alinement holes for tap washer countersunk for flathead bolts.

④ Flap valve, $\frac{1}{4}$-inch thick flexible belting, 7-inch o.d., 1-inch dia. center hole, two $\frac{9}{32}$-inch alinement holes 180° apart on a $2\frac{5}{8}$-dia. circle about the center.

⑤ Two steel retainer plates, $\frac{3}{8}$-inch thick, 7-inch o.d., 1-inch center hole. Two $\frac{9}{32}$-inch dia. alin-ement holes 180° apart drilled on a $2\frac{5}{8}$-inch dia. circle about the center. Six $\frac{3}{4}$-inch dia. vent holes 60° apart drilled on a $5\frac{3}{4}$-inch dia. circle about the center.

⑥ Three "leathers", $\frac{1}{2}$-inch thick belting, 8-inch o.d., 1-inch center hole. Two $\frac{9}{32}$-inch dia. alinement holes 180° apart drilled on a $2\frac{5}{8}$-inch dia. circle about the center. Six $\frac{3}{4}$-inch dia. vent holes 60° apart drilled on a $5\frac{3}{4}$-inch dia. circle about the center.

⑦ Body blocks of hardwood, marine plywood, etc. with 7-inch o.d., 4-inches thick, 1-inch center hole, two $\frac{9}{32}$-inch alinement holes 180° apart drilled on a $2\frac{5}{8}$-inch dia. circle about the center. Six $\frac{3}{4}$-inch dia. vent holes 60° apart drilled on a $5\frac{3}{4}$-inch dia. circle about the center.

⑧ Two flathead bolts, $\frac{1}{4}$-inch dia., 11-inches long with nuts.

⑨ Heavy nut and locknut to fit threads on 1-inch shaft, or a castelated nut with cotter pin.

NOTE: For a solid surge block, the same materials and design may be used but omitting the flap valve ④ and the six $\frac{3}{4}$-inch dia. vent holes specified in ⑦.

FIGURE 17-2.—Design of a vented surge block for an 8-inch well screen. (Sheet 2 of 2.) 103-D-1524-2.

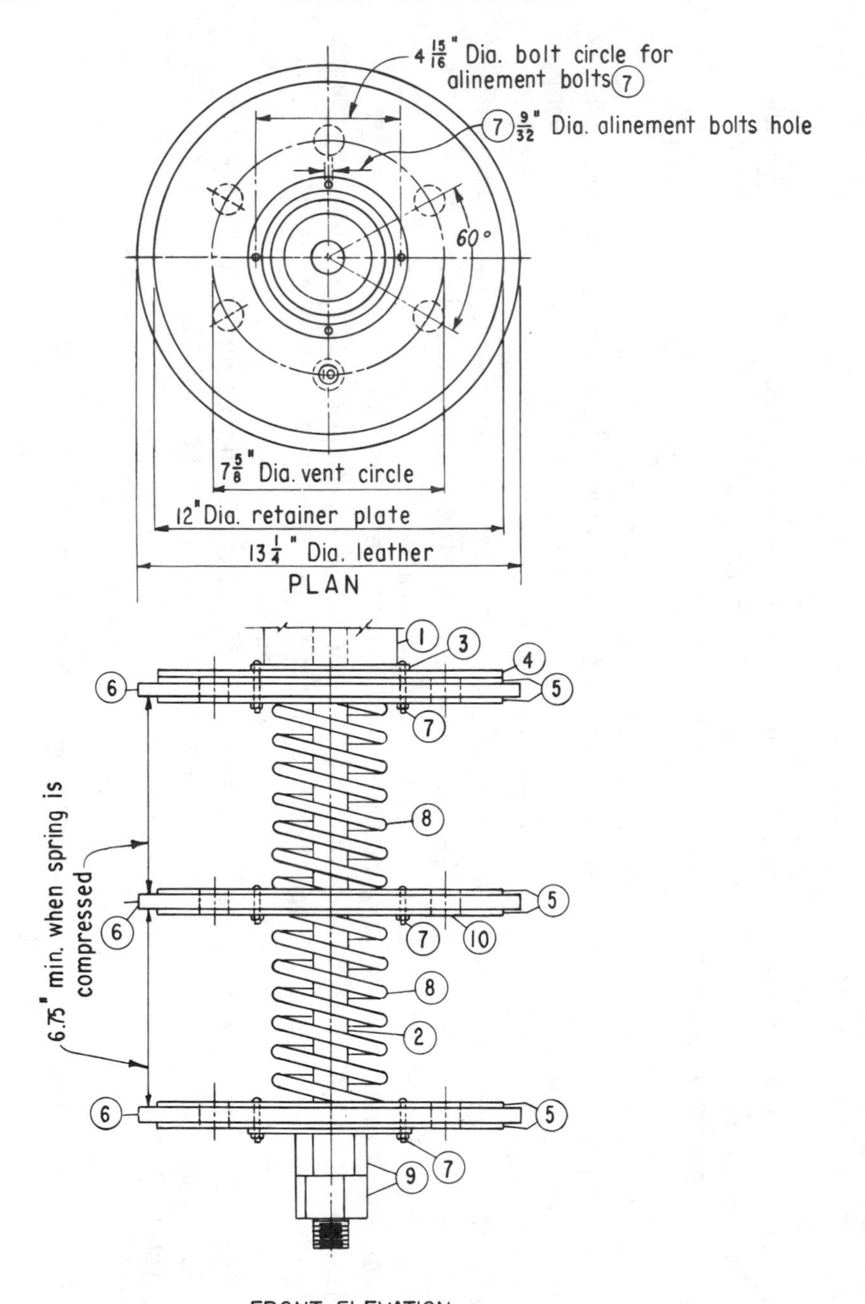

FIGURE 17-3.—Design of a spring-loaded, vented surge block. (Sheet 1 of 2.) 103-D-1525-1.

① API Pin joint, collar tapped and threaded for shaft.

② One $\frac{3}{16}$- or $\frac{1}{2}$-inch shaft of required length, threaded at both ends.

③ Two $5\frac{1}{2}$-inch dia. retaining washers, $\frac{3}{8}$-inch thick, center hole drilled to fit shaft. Four $\frac{9}{32}$-inch dia. alinement holes 90° apart drilled on a $4\frac{15}{16}$-inch dia. circle about center.

④ Flap valve, $\frac{1}{4}$ to $\frac{1}{2}$-inch thick flexible belting or impregnated fabric, 12-inch o.d., center hole drilled to fit shaft. Four $\frac{9}{32}$-inch alinement holes 90° apart drilled on a $4\frac{15}{16}$-inch dia. circle about center. (Desirable that at least two spare flap valves be available at the rig.).

⑤ Six retainer plates, $\frac{3}{8}$ inch steel, 12-inch o.d., center hole drilled to fit shaft. Four $\frac{9}{32}$-inch alinement holes 90° apart drilled on a $4\frac{15}{16}$-inch dia. circle about center. Six 1-inch dia. vent holes 60° apart drilled on a $7\frac{5}{8}$-inch dia. circle about center.

⑥ Three "leathers", $\frac{1}{4}$- to $\frac{1}{2}$-inch thick belting, $13\frac{1}{4}$-inch o.d., center hole drilled to fit shaft. Four $\frac{9}{32}$-inch alinement holes 90° apart drilled on a $4\frac{15}{16}$-inch dia. circle about center. Six 1-inch dia. vent holes 60° apart drilled on a $7\frac{5}{8}$-inch dia. circle about center.

⑦ Alinement bolts. 4-$2\frac{3}{4}$" x $\frac{1}{4}$", 4-$2\frac{1}{4}$" x $\frac{1}{4}$" and 4-$1\frac{3}{4}$" x $\frac{1}{4}$", with lock washers and nuts.

⑧ Two coiled steel compression springs with squared ends, max. 4-inch o.d., $\frac{1}{2}$-inch or heavier spring stock, installed under load. (Automobile coil springs have been satisfactory).

⑨ Heavy nut and locknut threaded to fit shaft or castellated nut with cotter pin lock.

FIGURE 17-3.—Design of a spring-loaded, vented surge block. (Sheet 2 of 2) 103-D-1525-2.

three-fourths of an inch smaller than the inside diameter of the screen, they should be replaced.

Surging is often done with a flap valve bailer. The action is similar to a vented surge block; if the difference between the diameter of the bail and inside diameter of the pipe is an inch or less, it is almost as effective as a surge block. If the difference in diameters is greater than an inch, use of a bailer for surging is relatively ineffective. Smaller diameter bailers are often built up with wrappings of burlap, clamp-on rings, etc., to fit the inside diameter of the pipe more closely.

The time required to properly surge a well depends upon the character of the aquifer material and its apparent response to development. As a guide, if the thickness of material accumulated at the bottom of the well is less than one-third of the inside diameter of the screen during one-half hour of surging on each 20 feet of screen, development with the surge block is probably adequate. On completion of surging, the well should be bailed clean to the bottom. Final development by rawhiding should precede testing of the well.

Development by a surge block should be started at the slowest possible rate and be increased as the well develops. A too vigorous initial development, particularly in clayey formations, can damage a well rather than improve it [1,2,3,4].

Swabbing is a violent method which is not recommended for initial well development because it may pack fines tightly around a screen. However, swabbing of wells which do not respond to normal surging or chemical treatment is frequently effective in breaking down a heavy wall cake or particularly tight annulus. Once these are broken down, normal development is usually possible. Swabbing should be used only as a last resort and never used in wells cased or screened with thin-walled pipe, California stovepipe, plastic, or similar materials since these lighter casings are likely to collapse. Swabbing is done using a surge block which is lowered to the bottom of the screen and then pulled up as rapidly as possible until suction is broken or the block comes out of the casing.

(d) *Development with Air*—Development of wells using compressed air is an effective method that requires considerable equipment and skill on the part of the operator. Figure 17–4 shows typical installations for development of wells by air. Two methods, backwashing and surging, are generally used [2]:

(1) In the backwashing method, water is alternately pumped from the well by airlift and then forced through the screen and into the water-bearing formation by compressed air introduced through a tight seal at the top of the casing. The three-way valve (fig. 17–4(A)) is turned to deliver air down the air line which pumps water out of the well through the discharge pipe.

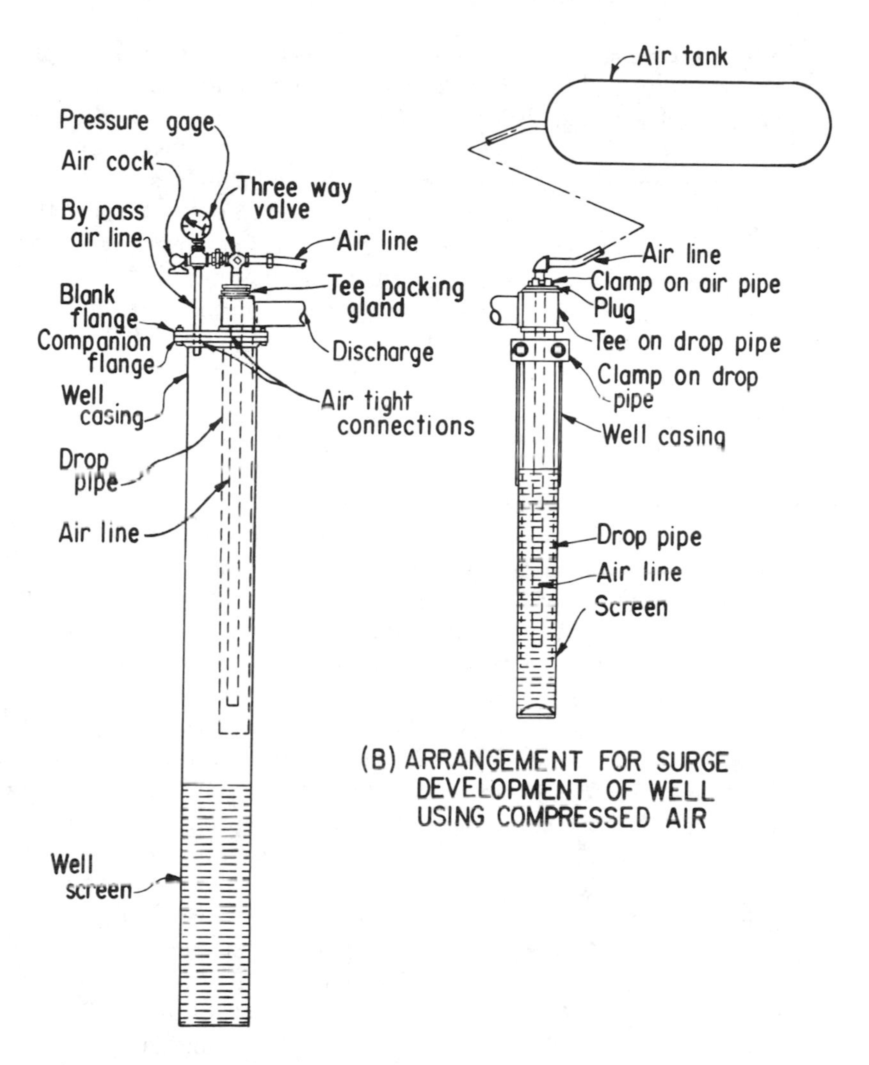

FIGURE 17-4.—Typical installation for development of a well using compressed air. 103-D-1526.

The bottom of the discharge pipe is usually set 1 to 2 feet above the top of the screen. When the discharged water becomes clear, the supply of air is cut off and the air cock is opened. The water in the well is allowed to return to the static level, which can be determined by listening to the escape of air through the air cock as the water rises in the casing.

When air ceases to flow from the air cock, the water no longer rises in the well. The air cock is then closed and the three-way valve turned to direct the air supply down the bypass into the well above the static water level. This forces the water back through the aquifer, agitating it and breaking down bridges of sand grains. When the water surface in the well is lowered to the bottom of the drop pipe, the air escapes through the drop pipe and the water is not depressed to any greater depth, thus avoiding air logging of the formation. When water reaches the bottom of the drop pipe, air can be heard escaping from the discharge pipe and the pressure gage will show no increase in pressure with time. At this point, the supply of air is cut off and the air cock is opened to allow the water to return to its static level. Then the three-way valve is turned and the air supply again directed down the air line to pump the well. This procedure is repeated until the well is thoroughly developed. The well should be sounded and, if necessary, bailed to remove any sand which may have accumulated on the bottom of the hole during the development operation. This method is not very effective except on short lengths of screen and open hole because development is limited to the upper part of longer screens or open hole.

(2) Figure 17–4(B) shows the arrangement and equipment necessary for development with air by the surging method. It should be noted that in this method the drop pipe is placed inside the screen. For the best operation, the drop pipe should have a submergence of at least 60 percent; that is, 60 percent of the total length of the drop pipe should be beneath the water surface when the well is pumping. However, a good operator can do an acceptable job of pumping with as little as 35 percent submergence. Development consists of a combination of surging and pumping. The sudden release of large volumes of air produces a strong surge within the well and pumping is accomplished as with an ordinary airlift pump.

At the start of development, the drop pipe is lowered to within about 2 feet of the bottom of the screen and the air line is lowered within it, so that its lower end is about 1 foot above the bottom of the drop pipe. Air is turned on and the well pumped until the water discharged is free of sand. The air tank is then pumped to maximum pressure while the air line is lowered a foot below the bottom of the drop pipe. When the air tank is full, the quick-opening valve is thrown open allowing the air in the tank to rush with great force into the well. This causes a brief but forceful surge of the water. The air line is immediately pulled back so that its lower end is again about 1 foot above the bottom of the drop pipe. Pumping by airlift is then resumed until the water is again free of sand and the cycle repeated until little or no sand is evident on pumping immediately after surging. The drop pipe is then raised about two screen diameters and the cycles of surging and pumping repeated until the entire length of the screen has been developed. The drop pipe and air line should then be lowered to the bottom of the well and an effort made to pump out any sand which may have accumulated at the bottom of the screen. If this is not successful, it may be necessary to use a sand pump in the well to remove any accumulation of sand.

Under certain conditions, the foregoing procedures should be used with caution because of the potential hazard of air binding of the aquifer. A safer procedure is to maintain the air pipe several feet up within the drop pipe and accomplish the surging by the falling column of water within the drop pipe.

In deep wells with a considerable depth of water, the use of compressed air may be limited by the volume and pressure capacities of available compressors. For example, a compressor with a 150 pounds per square inch pressure rating would be counterbalanced by a depth of water of about 349 feet. For best results, about 5.61 cubic feet per second of free air is required for each cubic foot of water pumped. When surging and pumping with air, the discharge should be directed into a fairly large tank in which the sand discharged can be collected so that the degree of effectiveness and progress of development can be determined.

When developing by backwashing or surging with air, an arrangement should be made which will permit periodic measurement of the volume of sand accumulating in the bottom of the well. Should sand begin to encroach on the screen, development should be stopped and the well bailed or pumped clean before resuming the operation. On completion of development, the well should be bailed or pumped clean [2].

Figure 17–5 shows an effective arrangement for developing wells with long screen sections by surging with air. The double packer is set at the base of the eductor pipe and all surging and pumping is done within the double packer. The double packer maximizes the effectiveness by confining the action to a specific length of the screen.

(e) *Hydraulic Jetting.*—Hydraulic jetting is most effective in open rock holes and in wells having cage-type wire-wound screen and some types of louvre screen. The jetting tool consists of a head with two or more $^{3}/_{16}$- to ½-inch jet nozzles equally spaced in a plane about the circumference (see fig. 17–6). The head is attached to the bottom of a string of 1¼-inch or larger pipe that is connected through a swivel

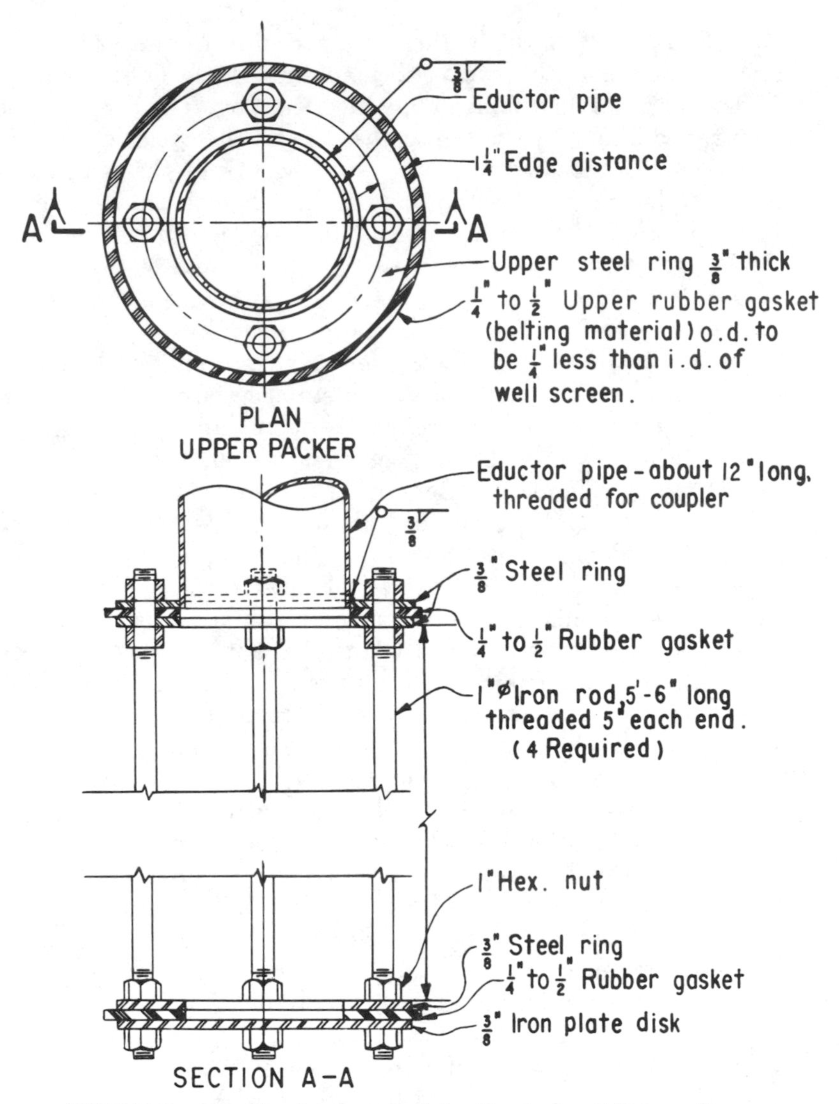

FIGURE 17–5.—Double packer air development assembly. 103–D–1527.

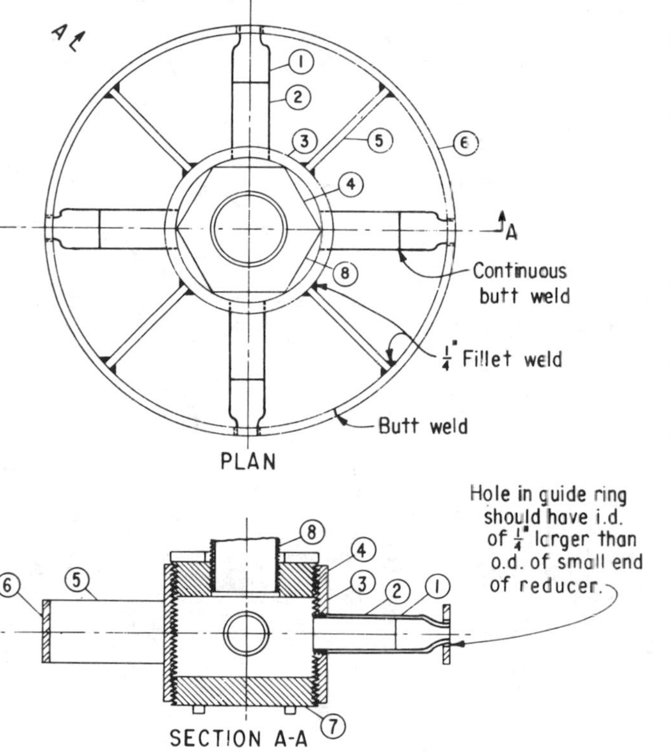

(1) Concentric reducer, 1" x $\frac{1}{4}$" or 1" x $\frac{3}{8}$", steel butt welding or similar, black iron (extra strong), two req'd. for 4-inch coupling and 2 or 4 req'd. for 6-inch couping.

(2) Line pipe T and C, 1-inch grade A API 5L, black iron (extra strong), two req'd. for 4-inch coupling and 2 or 4 required for 6-inch coupling. Length should be such that small end of reducer (1) is flush with o.d. of guide ring (6).

(3) Standard coupling, 4.5-inches long for 4-inch coupling and 4.78-inches long for 6-inch coupling, black iron (extra strong).

(4) Hexagon bushing (outside or inside), double tapped, 4" x 2" for 4-inch coupling, 6" x 3" for 6-inch coupling, black iron (extra strong).

(5) Bar 2" x $\frac{1}{4}$", strap steel, 4 req'd., length depends on i.d. of screen.

(6) Guide ring, 2" x $\frac{1}{4}$", strap steel, o.d. of guide ring should be 1-inch smaller than i.d. of screen.

(7) Bar plug, cast iron.

(8) Line pipe T and C, black iron (extra strong), length should be such as to permit easy removal when swivel is at max. height, sufficient pipe should be available to reach from casing collar to bottom of screen.

NOTE: Above material is for use with a 4 or 6-inch nominal coupling.

FIGURE 17-6.—Jetting tool for well development. 103-D-1528.

TABLE 17-1.—*Approximate jet velocity and discharge per nozzle*

Size of nozzle orifice, inches	Effective pressures lb/in²							
	100		150		200		250	
	Exit velocity, ft/s	Discharge, gal/min	Exit velocity, ft/s	Discharge, gal/min	Exit velocity, ft/s	Discharge, gal/min	Exit velocity, ft/s	Discharge, gal/min
3/16	120	9	150	12	170	13	190	15
1/4	120	16	150	21	170	23	190	26
3/8	120	36	150	46	170	53	190	59
1/2	120	66	150	82	170	93	190	104

and hose to a high pressure, high capacity pump. Water is pumped down the pipe in sufficient quantity and at sufficient pressure to give a nozzle velocity of 150 feet per second or more. Table 17-1 indicates the discharge rate for various exit velocities of different size nozzles and the required pressures in the jet head. Pumping pressures may have to be somewhat higher, depending on the number of nozzles, required volume, and the sizes and arrangement of the plumbing.

For small diameter wells of 3 inches and less, an efficient jetting tool can be made by attaching a coupling and plug to the bottom of the air line. Two or four 3/16 -inch-diameter holes are drilled through the coupling to serve as nozzles.

The jet head is turned at 1 revolution per minute or less, but should not be maintained for more than 2 minutes at a given setting. The tool is successively raised a distance equal to about one-half the screen diameter after completion of jetting at each setting until the entire surface of the screen or the open hole has been developed. During jetting, the jet head should always be rotated. The sand-bearing water returning to the well is picked up by the jet stream and if a jet impinges on one spot or a circumference for only a few minutes, it may erode a hole in the screen or even cut it.

The jetting tool has been effective in removing stubborn mud cakes from some holes and in opening up formations of dirty sand which have been plugged by too rapid and vigorous development by surging. The jet is particularly effective in developing gravel-packed wells.

If possible, water should be pumped from the well during development by jetting. Ideally, the pump discharge should exceed the jet discharge by 1.5 to 2 times. This practice removes the fines as they are washed into the screen and keeps ground water flowing into the wall, thus avoiding the buildup of positive head in the well which acts to force the fines back into the formation. An airlift pump is usually used for this purpose.

The pump should discharge into a large tank which permits an appraisal of the effectiveness of the jetting on the basis of the materials collected in the bottom of the tank. The tank also permits recirculation of the water to the jet head which is advantageous when chemical additives are used [2].

17-3. Development of Wells in Hard Rock.—Open hole wells drilled in hard rock supposedly do not benefit from development, but experience has shown this to be in error. In consolidated granular materials, a mud cake forms and fines are forced into the walls of the hole by the drilling operation. In fractured and jointed rocks where water yields depend upon the interception by the well bore of water-filled cracks or solution openings, such openings are frequently sealed by much the same action as well as by mud invasion. Practically all the methods used in developing screened wells can be used effectively in open-hole hard-rock wells. Under some circumstances, however, some additional practices may also be effective.

Wells in carbonate rock are often developed by the addition of sulfamic acid or muriatic acid which attack the carbonate rock and enlarges existing and creates new openings. When the acid is spent it is pumped to waste and the well treated with polyphosphates and surging or jetting. Under some circumstances, shooting a well in limestone with dynamite or other explosives using 50- to 100-pound charges of 60 percent gelatine or equivalent every 5 feet to fracture the rock has been effective. There is a definite risk associated with all of these procedures and they should be planned and carried out only under experienced and informed direction using adequate equipment and safeguards.

Wells in sandstone drilled with cable tools or down-the-hole tools should be developed using polyphosphates and vigorous surging. Wells drilled with rotary rigs are sometimes under-reamed about one-half inch using plain water as a drilling fluid after which the well is bailed clean. The well is further developed using polyphosphates and strong surging. In very competent, cemented sandstones, shooting with 5- to 10-pound charges of 50 percent gelatine at 5-foot intervals along the hole to enlarge it, or shooting with a few heavier charges to break and fracture the rock are sometimes effective.

Regardless of the method of drilling, wells in basalt and crystalline rocks should be developed using polyphosphates and jetting, vigorous surging, or both. Spot shooting with 50 to 100 pounds of 50 percent gelatine of selected portions of the hole is sometimes effective in increasing yields.

Hydraulic fracturing has been of limited effectiveness in increasing yields of sedimentary, crystalline, and volcanic rocks. Inflatable packers on a pipe leading to the surface are used to isolate 5- to 10-foot lengths of the hole. The pipe and isolated section are filled with water and pump pressure is applied to fracture the rock. Continued pumping may result in another buildup in pressure and additional fracturing. Sand fracturing, a refinement of the method, consists of pumping selected sizes and types of sand into the fractures to prop them and keep them open. Some wells have showed an increase in yield of as much as 200 percent as a result of fracturing but in all cases the initial yield was small, from less than 1 to possibly 3 gallons per minute [2].

17-4. Chemicals Used in Well Development.—Numerous chemicals are used to aid in well development. The most common are probably the polyphosphates: sodium tripolyphosphate ($Na_5P_3O_{10}$), sodium pyrophosphate ($Na_4P_2O_7$), tetra sodium pyrophosphate ($Na_1P_2O_7$), and sodium hexametaphosphate $(NaPO_2)_6$. These compounds act as deflocculants and dispersants of clays and other fine grained materials and permit the mud cake on the wall of a hole and the clay fractions in the aquifer formation to be more readily removed by the development. They are also chelating agents available for some of the heavy minerals and nearly all of them are mixtures of a polyphosphate and minor amounts of other compounds such as wetting agents, sterilants, and chelating compounds.

A mixture usually used by the Bureau of Reclamation is: 16 pounds of sodium polyphosphate, 4 pounds of sodium carbonate, and 1 quart of 5.25 percent sodium hypochlorite for 100 gallons of water in the well.

The action of the polyphosphates is enhanced by including a wetting agent in the mixture such as 1 pound of Pluronic F-68 or equivalent per 100 gallons of water when developing dirty formations. Use of the wetting agent is not recommended in formations consisting of thin interbedded sand and clay layers.

When polyphosphates or commercially produced development com pounds are not available, the common household phosphate-base detergents are a usable but somewhat expensive substitute. Most o them have the disadvantage of containing a foaming agent and whe the well is pumped, excessive foaming may result [2].

17-5. Bibliography.—

[1] "Well Drilling," Operations, Department of the Army Technic Manual 5-297, Washington, D.C., September 1965.

[2] "Ground Water and Wells," UOP Johnson Division, St. Pa Minn., 1966.

[3] "Water Well Drillers Beginning Training Manual," National Water Well Association, Columbus, Ohio, 1971.
[4] Campbell, M. D., and Lehr, J. H., "Water Well Technology," McGraw-Hill, New York, 1973.
[5] Gibson, U. P., and Singer, R. D., "Small Wells Manual," U.S. Agency for International Development, Washington, D.C., January 1969.

Chapter XVIII

WELL STERILIZATION

18-1. Purpose of Well Sterilization.—Many States and local political subdivisions require the sterilization of domestic and municipal water supply wells to assure the absence of pathogenic bacteria. All wells regardless of the water use should be sterilized on completion to prevent or retard the growth of corrosion or incrustation fostering organisms. Many of these organisms are not harmful, but they can accelerate and aggravate corrosion and incrustation problems and reduce the life of a well. Although sterilization may not always eliminate such problems, it is a worthwhile and relatively inexpensive precautionary measure.

18-2. Chlorination.—Sterilization is usually accomplished by introducing chlorine, or a compound yielding chlorine, into the water in the well and the immediate aquifer surrounding the well. Chlorine gas may be used, but the safest and usually most readily available materials to furnish chlorine for field operations are calcium hypochlorite ($Ca(ClO)_2$) or sodium hypochlorite ($NaClO$) and chlorinated lime. Calcium hypochlorite is available in granular or tablet form and contains about 70 percent available chlorine by weight. Sodium hypochlorite is available commercially in aqueous solutions ranging from about 3 to 15 percent chlorine. Commercially available chloride of lime is not a pure compound and does not have a definite formula but usually contains about 23 percent available chlorine.

Calcium hypochlorite is probably the least costly and most convenient material to use for sterilization. However, if the calcium in the ground water plus that added in the hypochlorite solution exceeds about 300 p/m, a precipitate of calcium hydroxide may form which may reduce the permeability of the aquifer adjacent to the well. If the sterilization solution is mixed to give 1,000 p/m available chlorine, about 280 p/m calcium would be present which, when combined with the calcium already present in the natural water, could result in precipitation of calcium hydroxide. Since the percentage of calcium is much higher in chlorinated lime, the danger of such a similar occurrence is even greater. Consequently, sodium hypochlorite is recommended for use in well sterilization. Numerous household bleach solutions containing sodium hypochlorite are commonly available. These solutions usually contain about 3 to 5.25 percent available chlorine. Commercial solutions which may be purchased from chemical supply houses contain 15 to 20 percent chlorine. Thus, in using sodium hypochlorite, the chlorine content should always be determined if not shown on the container. Also, because sodium hypochlorite deteriorates with age, the freshness and concentration should be considered in its purchase and use.

The trade percentage may be converted to p/m (parts per million) of chlorine by the following equation:

$$p/m = (\text{Trade percent})\ (10{,}000)$$

Thus, a 5.25 percent solution would be equivalent to approximately 52,500 p/m chlorine.

For well sterilization for pathogens, usually 50 to 100 p/m available chlorine and a contact time of from 30 minutes to 2 hours are specified. Many organisms, however, such as sulfate-reducing and filamentous iron bacteria, require 400 p/m or more available chlorine and contact times up to 24 hours for an effective kill [1,2,3,4].[1]

Wells may contain oil and other organic materials which combine with and neutralize the effectiveness of the chlorine. In addition, an unknown amount of dilution takes place in the well. Therefore, to assure an adequate concentration of chlorine in the well, the volume of water in the screen and casing should be estimated and sufficient chlorine added to yield a chlorine concentration of about 1,000 p/m.

The solution should be thoroughly mixed in the well by surging with a bailer or other similar tool from the bottom of the well to the water surface or by surging with the pump. The solution should remain in the well for at least 6 hours during which time the well should be surged at about 2-hour intervals.

In wells containing considerable oil or organic material in the water, or in which the aquifer contains considerable organic matter, the solution should be tested for residual chlorine after each cycle of surging and mixing. If the residual chlorine falls below the desired concentration, additional compounds should be added to bring it up to the desired concentration.

If the pump is in the well, the well should be pumped at the expiration of the contact time and the discharge diverted to flow back into the well to thoroughly flush the inside of the casing and the column pipe, and gravel pack if present, for at least 30 minutes. The well should then be pumped to waste until there is little or no odor or taste of chlorine in the discharge [4].

To add calcium hypochlorite or chlorinated lime, the required amount is first determined from the volume of water in the well (see

---

[1] Numbers in brackets refer to items in the bibliography, section 18–5.

TABLE 18–1.—*Volume of water in well per foot of depth*

Nominal casing size, inches	Schedule No.	Volume, gallons per foot of depth
4	40	0. 66
5	40	1. 04
6	40	1. 50
8	30	2. 66
10	30	4. 19
12	30	5. 80
14	30	7. 16
16	30	9. 49
18	30	11. 96
20	30	14. 73
22	30	17. 99
24	30	21. 58

table 18–1). This amount of the compound is then placed in a bail or similar tool through which water can flow. The tool is then raised and lowered between the bottom of the well and the water level until the material is completely dissolved. While this is an effective method, it may require considerable time.

A more rapid method involves dissolving the chlorine compound in clear water using 1 gallon of water per pound of compound. Depending upon temperature and the quality of the water, all the compound may not go into solution, but if there is some solid material remaining, it can be broken up and stirred into suspension at the time the solution is poured into the well. It will readily dissolve in the well water.

If sodium hypochlorite is used, the solution may be poured into the well as received [2].

The amount of various additives to use to obtain about 1,000 p/m of chlorine in a well could be estimated as in the following example:

Known: 425-foot-deep well, 16-inch casing from plus 1 foot to 300 feet, 14-inch casing and screen from 300 to 425 feet, and the static water level at 190 feet.

Using table 18–1, find the volume of water in the well:

190 to 300 ft: 110 ft of 16-inch casing at 9.49 gal/ft=1,044 gal

300 to 425 ft: 125 ft of 14-inch casing at 7.16 gal/ft=895 gal

Total volume of water=1,044+895=1,939 gal.

Using 70 percent calcium hypochlorite:

$$\text{Wt(lb)} = (\text{volume of water})(\text{wt of water})\left(\frac{\text{concentration desired}}{\text{concentration of sterilant}}\right)$$

$$\text{Wt}=(1{,}939)(8.33\ \text{lb/gal})\left(\frac{0.001}{0.70}\right)=23\ \text{lb}$$

Using 23 percent chlorinated lime:

$$\text{Wt}=(1{,}939)\ (8.33)\left(\frac{0.001}{0.23}\right)=70\ \text{lb}$$

Using 5.25 percent sodium hypochlorite:

$$\text{Vol.}=(1{,}939)(8.33)\left(\frac{0.001}{0.0525}\right)=308\ \text{lb}/8.33=37\ \text{gal}$$

Table 18–2 shows chlorine compounds and water required to give various concentrations of chlorine.

TABLE 18–2.—*Chlorine compounds and water required to give various concentrations of chlorine*

Chlorine, p/m	Volume of water per gallon of 5.25 percent sodium hypochlorite, (gal)	Volume of water per gallon of 10 percent sodium hypochlorite, (gal)	Weight of calcium hypochlorite for 70 percent available chlorine per 1,000 gallons of water	
			(oz)	(lb)
10	5, 250	10, 000	1. 91	0. 12
20	2, 625	5, 000	3. 82	0. 24
30	1, 750	3, 300	5. 73	0. 36
40	1, 315	2, 500	7. 64	0. 48
50	1, 050	2, 000	9. 55	0. 60
60	875	1, 650	11. 46	0. 72
70	750	1, 420	13. 37	0. 84
80	660	1, 250	15. 28	0. 96
90	585	1, 110	17. 19	1. 1
100	525	1, 000	19. 10	1. 2

Sterilization of a well, except possibly for some pathogens, is seldom 100 percent effective. Organisms may be covered with incrustation or corrosion products or lodged in crevices not readily penetrated by the sterilizing solution. While most such organisms may be destroyed, the few remaining continue to multiply and, depending upon conditions, periodic sterilization may be required at intervals to control them.

In some instances, continuous chlorination may be required to control mineral incrustation in addition to organisms. When this is necessary, chlorine gas should be discharged continuously through a suitable pipe to the bottom of the well. Where pathogenic contamination is present, chlorine should be injected into the pump discharge pipe and sufficient storage capacity should be provided to permit adequate contact time.

Sterilization of a new well is best deferred until installation of the permanent pump unless excessive delay is incurred. The required amount of sterilant should be added to the well just prior to installing the pump. After the pump is installed, but before it is bolted down permanently, the pump may be used to surge the well periodically to increase the effectiveness of the solution and flush the casing, pump column, and gravel pack, where present.

18-3. Other Sterilants.—A number of other sterilants are equal or superior to chlorine or chlorine compounds for controlling certain organisms. Such sterilants are usually more expensive and less readily available than chlorine and some are too toxic for use in potable water supply wells. Some of those which are suitable in potable water supplies are:

- A mixture of a polyphosphates detergent and anthium dioxide (chlorine dioxide) for control of filamentous algae.
- Cocomines and cocodiamines for sulfate-reducing bacteria.
- Quaternary ammonium chloride compounds for general use.

Most of these compounds are sold under proprietary names. The manufacturers should be consulted regarding recommended concentrations to use, contact time, and other factors.

Other sterilants not recommended for use in water supply wells but which might be used in waste disposal or similar wells are:

- Copper sulfate.
- Formaldehyde.
- Some mercury compounds.

18-4. Sterilization of Gravel Pack.—When wells are constructed using a gravel pack or formation stabilizer, sterilization of the pack at the time it is installed in the well is recommended. A fairly common and acceptable practice in most cases is to mix 1 pound of calcium hypochlorite in each cubic yard of gravel as it is installed in the well.

Another method is to pour one of the chlorine compound solutions down the tremie pipes with the gravel. Recommended amounts per cubic yard of gravel are as follows:

84,000 p/m calcium hypochlorite solution (1 lb per gallon of water), 3 quarts per cubic yard

27,000 p/m chlorinated lime solution (1 lb per gallon of water), 2.5 gallons per cubic yard

52,500 p/m sodium hypochlorite solution (1 gallon), 5 quarts per cubic yard.

When sterilization is completed, the well should be sealed and pumped to waste until there is no odor or taste of chlorine in the discharge. The waste solution may require treatment or other special disposition to minimize ecological effects.

18-5. Bibliography.—

[1] "Well Drilling Manual," Speedstar Division of Koehring Co., Enid, Okla., 1967.

[2] "Well Drilling," Operations, Department of the Army TM 5-247, Washington, D.C., September 1965.

[3] "Ground Water and Wells," UOP Johnson Division, St. Paul, Minn., 1966.

[4] Campbell, M. D., and Lehr, J. H., "Water Well Technology," McGraw-Hill, New York, 1973.

«Chapter XIX

VERTICAL TURBINE PUMPS

19–1. Turbine Pump Principles.—The line shaft vertical turbine pump is the most suitable pump for ground-water applications, especially for moderate to large discharge rates. Improved materials and design, combined with increased efficiency, have greatly broadened the field of vertical turbine pump application. There are very few ground-water pumping problems that cannot be solved efficiently by using the vertical turbine pump.

Pump selection varies with the type and temperature of the fluid being pumped. The following discussion is based on the pumping of water at temperatures in the range of 40° to 80°F, which includes most ground-water applications.

The capacity, pressure (head), efficiency, and power requirements of a vertical turbine pump depend on the diameter and design of the impeller (runner) and bowl (stage) and on the rate of rotation [2,4, 6].[1] In most applications, selecting the required pump is based on choosing impellers of the correct diameter and design which operate at the rotation speed affording the greatest efficiency. The basic principles upon which vertical turbine pump characteristics are determined will be discussed briefly to show the effect of different requirements on design.

The pressure developed by a vertical turbine pump is a function of the peripheral velocity of the impeller, which in turn is a function of the impeller diameter and rate of rotation. The pressure may be expressed in pounds per square inch but is usually expressed in feet of water. As the revolving impeller imparts kinetic energy to the water, directional vanes in the bowl surrounding the impeller convert this energy to pressure and guide the fluid vertically so that the flow becomes axial with the pump shaft. A turbine pump impeller and bowl of a certain design at a given speed adds energy to water, expressed in feet of head, and this head is constant for a given capacity, disregarding friction loss and viscosity.

19–2. Turbine Pump Operating Characteristics.—Pump performance characteristics, when determined by tests in the manufacturer's laboratory and plotted on a graph, furnish an understandable picture of the performance of a particular impeller design [2,4,6]. This graph, called a performance curve, is the key to selecting the type of impeller

---

[1] Numbers in brackets refer to items in the bibliography, section 19–18.

to suit the pumping requirements. The performance curves for two 12-inch pumps, shown on figure 19–1, were plotted for a constant speed of 1,760 revolutions per minute. The difference in the performance curves for the two pumps of the same size and rate of rotation is due to the differences in impeller and bowl design. On figure 19–1, head-capacity curves vary from the shutoff head on the left of the graph to the maximum capacity to the right. (The discharge of a turbine or a centrifugal pump can be completely stopped by closing the discharge valve without injury to the pump, which is not true of a positive displacement pump. However, both the pump and motor may overheat.) In general, a performance curve is based on the operating characteristics of one stage and shows the relationships between the head in feet and capacity in gallons per minute, horsepower, and efficiency of such a stage. The efficiency curve is derived from the comparison of theoretical water horsepower (whp) to actual input horsepower (bhp) in the test laboratory. Thus, whp/bhp times 100 equals the pump efficiency. The horsepower curve represents the brake horsepower delivered to the shaft at various head capacities as determined by laboratory tests.

In the examples that follow, the relationships of capacity, head, and horsepower, as shown on figure 19–1, are valid provided that the rate of rotation is constant at 1,760 revolutions per minute. If the rate of rotation is changed, the capacity, head, and horsepower will change as follows: capacity varies directly as the rate, head varies as the square of the rate, and horsepower varies as the cube of the rate. These relationships can be expressed symbolically by the equations:

$$\frac{N_1}{N_2}=\frac{Q_1}{Q_2}=\frac{H_1{}^2}{H_2{}^2}=\frac{bhp_1{}^3}{bhp_2{}^3} \tag{1}$$

where:

Q=capacity (or discharge),
H=total head,
bhp=brake horsepower, and
N=rate of rotation.

Theoretically, the efficiency does not change materially. However, in field service the efficiency may vary slightly depending on mechanical losses and other factors. With a variable-speed driver, the rotation speed of a vertical turbine pump may usually be varied as much as 20 percent below the design value without serious loss of efficiency. To illustrate the effect of speed on pump characteristics, calculate the change in head, capacity, and horsepower on the pump shown by the dashed lines on figure 19–1 when the rotation speed is reduced from 1,760 to 1,400 revolutions per minute. From the curves,

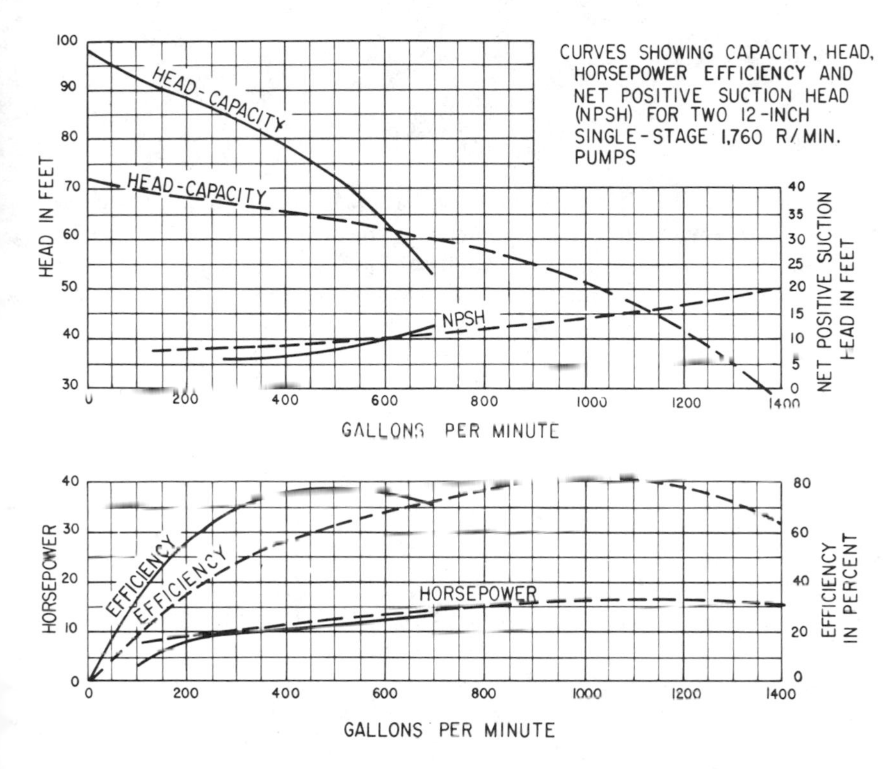

FIGURE 19-1.—Laboratory performance curves for two 12-inch single stage deep well turbine pumps. 103-D-1529.

which are based on the higher speed, near the point of maximum efficiency:

Capacity = 1,000 gal/min
Head = 52 feet
Horsepower = 17

For operation at 1,400 r/min:

Capacity varies directly as the speed,

$$(1{,}000)\left(\frac{1{,}400}{1{,}760}\right) = 795 \text{ gal/min.}$$

Head varies as the square of the speed,

$$(52)\left(\frac{1{,}400}{1{,}760}\right)^2 = 33 \text{ ft.}$$

Horsepower varies as the cube of the speed,

$$(17)\left(\frac{1{,}400}{1{,}760}\right)^3 = 8.5 \text{ hp.}$$

The capacity, head, and horsepower relationships for a particular impeller and bowl operating at constant speed can also be varied by changing the diameter of the impeller. This is known as trimming.

Actually, when the impeller of a pump is trimmed, the design relationships are changed and in reality a new design results. The diameter of an impeller should not be changed in the field without first consulting the pump manufacturer to determine the effects of trimming.

A steep head-capacity characteristic curve and a flat efficiency curve are desirable for deep well turbine pumps as the water level in the well may vary considerably during the life of the pump. As the water level declines in the well, the head increases, but the capacity decreases only slightly in proportion. Also, with this type impeller, the brake horsepower curve is almost flat, with the power input highest at highest pump efficiency. This nonoverloading power characteristic is desirable for protection of the motor from overload. Also, a steep head-capacity curve indicates that this type of impeller is suitable for service where the capacity is varied slightly by operation of a valve in the discharge line [2,4,6].

19-3. Turbine Pump Construction Features.—Turbine pumps were originally designed for use in drilled wells; thus, the nominal diameter of the impellers and bowls is designated by standard well casing sizes [1,3,4,5,6]. The pump size designation of 4, 6, 8 inches, etc., indicates the smallest inside diameter of a well casing of standard size (sec. 11-3(b)) into which the pump can be installed. For clearance, the outside diameter of the pump bowl is usually designed a fraction of an inch smaller than the inside diameter of the casing. In practice; however, more clearance than this is usually specified. A minimum clearance of 1 inch around the pump bowls (casing diameter of 2 inches larger than the pump diameter) is recommended. More clearance may be required for large pumps or very deep settings.

Where several stages (bowls) are assembled in series on a common shaft, they constitute a multistage pump. The head produced is directly proportional to the number of stages. For example, if the head requirement is 140 feet at 500 gallons per minute and one bowl and impeller develop 72 feet of head, then two stages will furnish the required head. As the velocity is converted into pressure in one stage and guided to the next stage; the same velocity is sustained, additional pressure energy is added, and the required horsepower is increased by each additional stage.

Deep well turbine pumps have the line shaft lubricated by oil or water. In oil-lubricated pumps, the line shaft and bearings are enclosed in a tube into which oil is dripped (while the pump is operating) from a suitable oil reservoir mounted on the pump base or discharge head. In a water-lubricated pump, the tube enclosing the shaft and bearings is omitted and the water flowing up the pump column acts as the lubricant. The bearings in an oil-lubricated pump are usually bronze, while those in the water lubricated pump are made of special types of rubber. Oil-lubricated pumps are generally used when the depth to water is 50 feet or more and water-lubricated pumps are generally used when the depth to water is less than 50 feet. Water-lubricated pumps can be used where the depth to water is more than 50 feet, but they must be equipped with a means of prelubricating the bearings before the pump is started. Use of oil-lubricated pumps usually results in leakage of oil into and contamination of the well.

Stepdown tests are made after completion and development of a well to determine the pumping lift for various discharges. These data must be known to select the correct pump for the well (sec. 5-15).

19-4. Discharge Heads.——Discharge heads are components which convey the pump discharge from the vertical column pipe to the horizontal discharge, usually above ground. Such discharge heads are mounted above and attached to the column pipe and below the drive motor or drive gears. Discharge heads are made of cast iron or fabricated steel, depending on pressure heads and other factors. Most are standardized to National Electrical Manufacturers' Association (NEMA) standard dimensions and will permit the use of a matching NEMA standard, vertical, hollow shaft motor; a right-angle gear drive; a belt drive; or a combination drive. Discharge head nomenclature is based on nominal drive base and column and discharge pipe dimensions. For example, a 150-hp motor may have a base diameter of either 16½ or 20 inches. The column and discharge pipes for a 1,000-gallon per minute pump, that would reasonably be used with such a motor, are usually 8-inch diameter. The discharge head for such an installation would be designated as a 1608 or 2008 or 8 by 8 by 16 or 8 by 8 by 20. Discharge head sizes larger than 1004 or 4 by 4 by 10 are usually equipped with flanges for the attachment of the column and discharge pipes. Smaller heads are threaded. Discharge heads are also available for solid shaft motors, but these are seldom used on deep well turbine pumps and may be made to a different standard.

Discharge heads are made for either surface or underground discharge. Surface discharge heads are selected on the basis of drive base diameter, discharge pipe size, and column pipe and shaft size.

Below ground discharge heads are selected on the basis of the drive base diameter, shaft and stuffing box size, and column and discharge pipe sizes. Most of these heads are designed for a single type of drive, but combination heads are available which permit changes from vertical hollow-shaft motors to right-angle gear or belt drive. This permits use of standby power in the event of an electrical power outage.

Manufacturers' catalogs of pumping equipment usually contain instructions on the selection of discharge heads for various pump sizes and horsepower requirements [1,5,6].

19–5. Submergence and Net Positive Suction Heads.—Many deep well turbine pumps will operate efficiently with a suction lift. Bowls are often equipped with a length of suction pipe as a factor of safety in anticipation that pumping water levels may fall below the bottom of the bowls, or in the belief that a savings is being made by reducing the length of column and shaft required. Neither of these reasons is defended from a practical standpoint.

It is always advisable for the pump bowls in a well to be submerged when the pump is operating. Submergence avoids corrosion problems resulting from the bowls being alternately submerged and exposed to the atmosphere and permits elimination of the suction pipe. In addition, adequate submergence precludes the formation of vortices with resultant air entrainment and a decrease of available net positive suction head below that required—conditions which could cause cavitation, corrosion, and inefficient operation of the pump [6].

If conditions are analyzed adequately, an estimate can be made of the probable lowest pumping water level in a well, and a pump setting can be selected which will keep the top of the bowls below the pumping level for several years. The pump should be set at a depth such that the top of the bowls is 5 feet or more below the estimated lowest water elevation when the well is pumping. Some pumps may require deeper settings because of net positive suction head requirements.

In any pumping system, the pressures in the liquid being pumped vary from location to location in the system. In a centrifugal type, such as a deep well turbine pump, pressures are greatest in the discharge end since the liquid has already passed through the impeller where pressure was added. Pressures are low in the suction end and diminish to their lowest value as the liquid enters the eye of the impeller. There must be sufficient pressure at the eye of the impeller so that the liquid being pumped will not flash into vapor and reduce the efficiency of the pump or cause cavitation.

The term, net positive suction head (NPSH), is used to describe the amount of pressure required to prevent vaporization of the liquid and is a characteristic of each particular pump. The NPSH is a function of the distance from the eye of the impeller to the water level in the well while pumping. Thus, a requirement for a higher value of NPSH is satisfied by lowering the pump in the well. It should be noted that the required NPSH is the minimum value for satisfactory operation of the pump, but that there is no maximum value. Additional pump submergence does no harm; neither does it increase the efficiency of the installation. The value of the NPSH required for a particular pump can be determined by a test during which the head, capacity, efficiency, and horsepower relationships are established. The NPSH is a characteristic that should be furnished by the pump manufacturer. Because the NPSH varies with the rate of pumping, it is furnished as a performance curve, as are the other pump characteristics. The following equation is used to determine the bowl setting that satisfies NPSH requirements:

$$H_z = \text{NPSHR} - B_p + H_f + V_p \tag{2}$$

where:

H_z = distance from the eye of the lowest impeller to the surface of the water in the well while pumping, in feet (the water level must be at least this high); a positive value indicates submergence of the eye of the impeller and a negative value indicates a suction.

NPSHR = required net positive suction head for the particular pump, in feet (obtained from the pump manufacturer).

B_p = normal barometric pressure for the elevation of the installation, in feet (see table 19–1).

H_f = head lost in friction through the suction piping, in feet [5,6].

V_p = vapor pressure of the water at its temperature in the well, in feet (see table 19–2) [2,4,6].

19–6. Estimating Projected Pumping Levels.—Projections of anticipated future ground-water levels are often difficult and unreliable because of poorly known sources and magnitudes of factors which influence such levels. However, reasonably reliable estimates within the range of pump operation can be made. Seasonal and longtime increases or decreases in static ground-water levels occur because of seasonal variations in precipitation and longtime changes in the climatic cycle. Longtime declines due to withdrawals imposed on natural fluctuations may reflect normal aquifer development. An

TABLE 19-1.—*Normal barometric pressure at various elevations*

(In feet of water)

Altitude, feet	Barometric pressure, in feet H_2O	Altitude, feet	Barometric pressure, in feet H_2O
0	34. 0	7, 500	25. 7
500	33. 4	8, 000	25. 2
1, 000	32. 8	8, 500	24. 8
1, 500	32. 2	9, 000	24. 3
2, 000	31. 6	9, 500	23. 8
2, 500	31. 0	10, 000	23. 4
3, 000	30. 5	10, 500	22. 8
3, 500	29. 9	11, 000	22. 4
4, 000	29. 4	11, 500	21. 9
4, 500	28. 8	12, 000	21. 4
5, 000	28. 3	12, 500	21. 0
5, 500	27. 8	13, 000	20. 6
6, 000	27. 3	13, 500	20. 2
6, 500	26. 7	14, 000	19. 8
7, 000	26. 2		

TABLE 19-2.—*Vapor pressure of water at various temperatures*

(In feet of water)

°F	Vapor pressure, in feet H_2O	°F	Vapor pressure, in feet H_2O	°F	Vapor pressure, in feet H_2O
40	0. 28	54	0. 48	68	0. 79
41	. 29	55	. 50	69	. 81
42	. 30	56	. 51	70	. 84
43	. 32	57	. 53	71	. 86
44	. 33	58	. 55	72	. 90
45	. 34	59	. 57	73	. 93
46	. 35	60	. 59	74	. 96
47	. 37	61	. 62	75	. 99
48	. 38	62	. 64	76	1. 02
49	. 40	63	. 66	77	1. 06
50	. 41	64	. 68	78	1. 10
51	. 43	65	. 71	79	1. 14
52	. 44	66	. 73	80	1. 17
53	. 46	67	. 75		

extended period of pumping from any well is accompanied by a continuous, slow decline at a constantly diminishing rate unless recharge balances withdrawals. Interference from existing or future wells, installation of recharge facilities, or a change in boundary conditions may cause a change in pumping levels. Deterioration of a well due to corrosion and incrustation may cause a significant decrease in the specific capacity. All these factors should be considered in estimating probable maximum and minimum pumping levels.

Projections of static water levels in conjunction with analysis of results of pumping tests and the probable pumping schedule are made to determine the required pump characteristics and setting. The drawdown for a given discharge over any period of pumping under conditions prevailing at the time of the test can be approximated by extending the straight-line portion of a semilog time-drawdown plot (secs. 5-14 and 5-15). This estimated drawdown is adjusted to compensate for the projected decrease or increase in saturated thickness or the static water level present at the time of the tests (sec. 2-1). A judgment estimate of the possible influence of well interference is added to the adjusted estimated drawdown to obtain a value of possible maximum and minimum pumping elevations over the projected future. This analysis gives an estimate of the minimum and maximum pumping levels for a given minimum discharge for a specific pumping schedule.

19-7. Selection of Pump Bowl and Impeller.—If the required discharge Q and head H are known, the pump bowl unit can be selected from a manufacturer's pump performance or H/Q curves. To obtain the lowest initial cost and most economical operation of the pump, the rate of rotation should be kept as high as possible without sacrificing efficiency. Except for unusually low or high capacities, this usually results in the selection of a pump rotating at approximately 1,760 revolutions per minute. The nominal diameter of the pump bowl is given on each pump H/Q curve sheet. This indicates the smallest inside diameter of well casing in which the unit can be installed. However, the inside diameter of the pump chamber casing should be at least 2 inches larger than the nominal size of the bowl for most pumps and in excess of 2 inches for large pumps and deep settings.

An H/Q curve shows the capacity and lift of a single bowl. Additional bowls or stages must be added to increase the head or lift, but such additions will not increase the discharge. However, the power requirement is approximately proportional to the number of stages used. Two or more pumps of different diameters usually can be found which will meet the head-capacity requirements.

The smaller diameter pumps will usually require more stages and will have lower efficiency than the larger pumps. The pump should be selected which will more closely approach the required capacity at the maximum head with the highest efficiencies over the estimated range of heads and the smallest number of bowls. In general, pumps having a steep head-capacity curve are most suitable for water wells.

19-8. Analyses of Basic Data on Well and Pump Performance.—To amplify the previous discussion, a summary of the procedures and methods in a hypothetical situation is given in this section.

(a) In a developed area, a review of existing well performance, hydrographs, and logs of wells in the vicinity prior to drilling the pilot hole indicated that the desired yield could be obtained from a well within the following range of conditions:

(1) Depth to static water level	245 to 285 feet
(2) Annual fluctuation in static water level	6 to 8 feet
(3) Probable drawdown for 900 gal/min	25 to 40 feet
(4) Average annual water level decline during a 6-year trend.	5 feet per year
(5) Depth of existing wells	400 to 500 feet
(6) Thickness of saturated aquifer	+190 to 250 feet
(7) Average age of wells	15 years
(8) Problems from incrustation or corrosion	little or none reported

(b) The pilot hole showed the following conditions at the well site:

(1) Depth to static water level	254 feet
(2) Thickness of saturated aquifer	226 feet
(3) Depth to bottom of aquifer	480 feet

(4) Mechanical analyses of the aquifer materials and study of the log showed an adequate thickness of materials was present opposite which a 0.050-inch slot screen (No. 50) could be set to furnish the desired yield. In an undeveloped area, the pilot hole information probably would be supplemented by a pumping test to determine aquifer characteristics.

(5) Chemical analysis of the water indicated total dissolved solids content of 300 p/m, pH of 7.2, and favorable Ryzner

and Langlier corrosion and incrustation indices.

(c) Basic well design:

(1) Required minimum yield	900 gal/min
(2) Minimum pump bowl nominal diameter	10 to 12 inches
(3) Minimum pump chamber diameter	12 to 16 inches
(4) Pump chamber depth:	
a. Present static water level (b)(1)	254 feet
b. Possible maximum drawdown (a)(3).	40 feet
c. Decline in static water level in 20 years (a)(4).	100 feet
d. Decline in pumping level in 20 years due to 1 percent annual deterioration of well (judgment estimate).	8 feet
e. Overlap between screen assembly and pump chamber (standard for this type of well).	10 feet
f. Estimated total depth of pump chamber casing required: Sum of a. through e. above (254+40+100+8+10)	412 feet

Since the aquifer thickness (b)(2) is 226 feet and maximum desirable drawdown is 65 percent, the maximum drawdown is: (226)(0.65) =147 feet.

The maximum desirable pumping level would then be:

(b)(1)+147=254+147=401 feet

The maximum pump chamber depth is:

401 + 10-foot overlap = 411 feet, or 401 feet of usable pump chamber depth.

(5) Screen assembly:

a. 10-inch telescoping screen (recommended diameter for 900 gal/min from table 11–7 in sec. 11–4) with 0.050-inch (No. 50) slots has about 125 to 130 in² of open

area per linear foot (table 11–9, sec. 11–4)
(127.5) (0.31) = 39 gal/min per foot for 0.1 ft/s entrance velocity (to estimate gal/min per foot at 0.1 ft/s entrance velocity, multiply square inches of open area per linear foot of screen by 0.31 or square feet of open area by 45)
900/39 = 23 feet of screen (minimum), use 30 feet.

b. As dictated by aquifer conditions, set two 15-foot-long screen sections separated by a 39-foot-long flush tube section between depths of 411 and 480 feet.

c. 10-foot section of flush tube overlap.

d. Sump: 10 feet of blank flush tube extension on bottom of screen with closed bail bottom or other seal.

(6) Total well depth:

a. Pump chamber depth (c)(4)f------ 411 feet

b. Casing and screen assembly below pump chamber including 10-foot sump:

Sum of (c)(5)a. through d.
(30+39+10+10)------- 89 feet

Total depth of well:
Sum of a.+b. above less 10-foot overlap--------- 490 feet

(7) Estimated pump requirements:

a. Q=900 gal/min (discharge)

b. Drawdown (at end of 5 years)
(c)(4)b.+(a)(4)+(c)(4)d.
(40+25+2)---------------- 67 feet

c. Drawdown (at end of 20 years)
(c)(4)b.+(c)(4)c.+(c)(4)d.
(40+100+8)--------------- 148 feet

d. Pump lift (at end of 5 years)
(b)(1)+(c)(8)b.
(254+67)----------------- 321 feet

e. Initial bowl setting for first 5 years (c)(7)d. rounded to standard column lengths	325 feet
(8) Estimated probable pump head losses:	
a. Length of 8-inch column with 1½-inch shaft	325 feet
b. Column loss at 900 gal/min	10.4 feet
c. Discharge head loss	0.3 feet
d. Total pump head loss b.+c.	10.7 feet
(9) Estimated surface losses:	
a. Elevation of bottom of storage tank	50 feet
b. Elevation of shutoff elevation of storage tank	56 feet
c. Effective length 8-inch pipe and fittings	104 feet
d. Pipe head loss	3 feet
e. Maximum surface head requirement b.+d.	59 feet
(10) Estimated total head: (c)(7)d.+(c)(8)d.+(c)(9)e. (321+10.7+59)	391 feet
(11) Probable pump (from manufacturer's data):	
Bowl diameter	12-inch nominal
Head per stage	80.5 feet
Number of stages	5
Horsepower per stage	22
Bowl efficiency	82 percent

(12) For 12-inch nominal bowls use 16-inch casing (clearance between 14-inch, 0.375-inch wall casing (i.d.=13.25-inch) and bowls (o.d.=11.5-inch) would be inadequate)

(13) Final well design:

a. Casing: 16-inch casing from +1 to depth of 411 feet.

b. Screen and casing assembly: 30 feet of 10-inch by 0.050-inch slot screen and 59 feet of 10-inch casing from 401 feet to 490 feet.

(d) Results of step and 72-hour production tests on completion of well and development:

(1) Elevation of well head . . . . 5,011 feet

(2) Static water level start of test (low water period) . . . . 256 feet

(3) Thickness of aquifer . . . . 224 feet

(4) Temperature of water . . . . 54°F

(5) Step test: 3 steps at 387, 701, and 1,001 gal/min, each step run 4 hours.

(6) Pump schedule may call for 30-day continuous pumping. Projection of drawdown through 900 gal/min parallel to plot of first step indicates 22-foot drawdown in 30 days. Projection of 72-hour pumping test drawdown to 30 days indicates 24 feet.

(e) Refinement of pump requirements for first 5 years:

(1) Static water level at end of 5 years:

(d)(2)+(a)(4)=256+25=281 feet

(2) Thickness of aquifer in 5 years.

(b)(3)−(e)(1)=480−281=199 feet

(3) Drawdown for 900 gal/min for 30 days at end of 5 years:

$$\text{(d)(6) times } \frac{(d)(3)}{(e)(2)} = (24)\left(\frac{224}{199}\right) = 27 \text{ feet}$$

(4) Pump lift at end of 5 years—no deterioration:

(e)(1)+(e)(3)=281+27=308 feet

(5) Pump lift at end of 5 years and 1 percent a year well deterioration:

(e)(1)+(e)(3) times 1.05=281 +(27)(1.05)=309 feet.

(f) Estimate of pump and well performance for first 5 years:

(1) Present water conditions:

	Minimum pump lift
a. Static water level (d)(2)	256 feet
b. Aquifer thickness (d)(3)	224 feet
c. Drawdown at 900 gal/min (d)(6)	24 feet
d. Pump lift a. + c.	280 feet

(2) Low water conditions in 5 years:

a. Low static water level (e)(1)	281 feet
b. Thickness of aquifer (e)(2)	199 feet

c. Drawdown 900 gal/min, 30 days:
(e)(3) .... 27 feet
d. Pump lift (e)(4) .... 308 feet
+ 1 percent pump deterioration for 5 years .... 309 feet

(3) bhp=gal/min times total head in feet divided by 3,960 times efficiency of the pumping unit:
Efficiency of pumping unit is equal to the product of the bowl and motor efficiency:
(0.82) (0.90)=0.74, use 75 percent

$$\text{bhp}=\frac{(900)(379)}{(3{,}960)(0.75)}=115 \text{ hp}$$

(4) Shaft loss hp=4
(5) Total hp=119, use 125-hp motor.

(g) NPSH (Net Positive Suction Head) required:
(1) Net positive suction head required at 900 gal/min. 15 feet
(2) Vapor pressure of water at 50°F .... 0.4 feet
(3) Barometric pressure at 5,000 feet elevation. 28.2 feet
(4) Available net positive suction head (3)+(2). 28.6 feet
(5) Excessive net positive suction head (4)—(1). 13.6 feet

Theoretically this pump could operate with a 13-foot plus suction lift but for other reasons it is preferable that the bowls be submerged. The 325-foot bowl setting originally estimated will be satisfactory.

The well with the above pump would perform satisfactorily for 5 years or more if the projections regarding aquifer thickness, etc., are realized. Eventually an additional bowl would have to be added, the bowl setting increased, and a larger motor installed. If the decrease in aquifer thickness continued, at some still later date, the yield would have to be reduced and a second well drilled if the minimum yield requirements were to be met. Before any such changes were made, a step test of the well would be desirable, followed by rehabilitation if necessary, and a subsequent step test before the new pump is specified.

When the basic pump bowls have been selected, the above data and the charts and tables in the manufacturer's technical manual permit estimates of additional values for use in the preparation of designs and specification of components.

19-9. Additional Factors in Pumping Equipment Design.—The hypothetical situation in section 19-8 covered selection of the basic pumping equipment for the particular application. Additional data on complete design include diameter of the column pipe, diameter of the drive shaft, discharge head size and type, lubrication, power selection, and type of drive. Most of these items have been standardized by the industry and methods of determining the required components are included in the catalogs of pump curves and equipment issued by the various pump manufacturers.

Most large pump installations use weatherproof electric motors and control equipment. However, a pumphouse may be necessary under certain conditions. To facilitate pulling the pump, a roof hatch located over the well or a removable roof should be provided in the design of a pumphouse.

A pit installation below the ground may be advantageous in some instances; however, such installations are particularly susceptible to flooding and are prohibited by some State regulations.

Pumps operating in corrosive waters may require use of corrosion-resistant metals. The general conditions governing the use of such metals are covered in the discussion on corrosion and incrustation (ch. XIV). Specific solutions for a particular case should be discussed with corrosion specialists and the pump manufacturer.

The type of power unit selected usually depends on the availability and cost of fuel. If electric service is available within a reasonable distance, an electric motor is generally preferred because of lower first cost, lower maintenance cost, and its reliability without regular servicing and periodic attendance. If electric service is not available, an engine fueled with gasoline, diesel fuel, natural gas, or liquid petroleum gas usually must be chosen. Such an engine can either be belted or geared to the pump and can be fitted with many different appurtenances. Small pumps can be powered by windmills where wind conditions are favorable.

19-10. Other Vertical Shaft Driven Pumps.—(a) *Submersible Pumps.*—The term "submersible" is usually applied to turbine pumps in which the motors are close-coupled beneath the bowls of the pumping unit and both are installed under water. This type of construction eliminates the surface motor, long drive shaft, shaft bearings, and lubrication system of the conventional turbine pump.

The motors of submersible pumps are cooled by the water flowing vertically past the motor to the pump intake. This cooling system permits a different motor design than is possible with air-cooled motors. The submerged motors are usually much longer and of smaller diameter than surface motors of the same horsepower and rate of rotation.

To avoid high head losses in flow through the annular space and into the intake, the pump chamber should be large enough so that velocity of flow in the annular space should not exceed 5 feet per second and should preferably be nearer 1 foot per second. A minimum velocity of 1 foot per second is needed to assure adequate cooling of the motor.

Head losses due to high velocities in restricted annular spaces may result in a reduction of the available NPSH at the pump. This may be compensated for by increasing the submergence of the unit below the pumping water level. Where use of large-capacity submersible pumps is contemplated, the manufacturer should be consulted regarding desirable submergence, pump-chamber diameter, and length of the pump and motor assembly.

From the previous discussion, it is obvious that use of large-capacity submersible pumps necessitates larger and deeper pump chambers which increase the well cost. In addition, the initial cost and installation charges of submersible pumps are greater than those of conventional deep well turbine pumps. Proposed use of large-capacity submersible pumps in lieu of conventional deep well turbines should be carefully analyzed. The small increase in efficiency and savings in bearing and shaft costs, as compared with the higher initial costs of both pump and well, will usually show the submersible pump to be less economical overall than the conventional deep well turbine installation. Despite these adverse factors, submersible pumps do offer several distinct advantages including: (1) noiseless, below-ground operation; (2) a factory installed and sealed lubrication system which eliminates oil contamination of the well; (3) simple installation without special tools; and (4) the well and pump are subject to complete burial in flood-prone areas.

Submersible pumps are especially useful for high-head, low-capacity applications such as domestic water supply. With the exception of the factors discussed above, the selection of submersible pumps is identical to that of conventional deep well turbines.

(b) *Lemoineau-Type Pump.*—The Lemoineau-type pump is an especially designed positive displacement pump available in both surface-mounted and submersible models. The most widely used pumping element consists of a hard-surfaced, corrosion-resistant, helical-contoured metal rotor which revolves inside a tough, abrasion-resistant, double-helical-contoured, flexible rubber stator. At the prescribed speed of rotation, discharge is practically constant regardless of the lift, although the horsepower requirement increases with increased lift. Conversely, since this is a positive displacement pump, discharge varies almost directly with the speed. The power unit and the column pipe and shaft above the pumping unit are similar to those of a water-lubricated turbine pump of the same capacity, and features of the submersible type are similar to a submersible turbine pump. The design of the pump results in high resistance to electrolytic corrosion and damage by sediment-laden water. The Lemoineau-type pump is especially useful for pumping sediment-laden water at high heads but low capacities.

19–11. Measuring Pump Performance.—Cost of energy is one of the principal expenses incurred in the operation of pumps. Therefore, pumps should be monitored to ensure that they are operating at or near peak efficiency. Three factors must be measured to check pump efficiency: (1) total head, (2) input horsepower, and (3) quantity of water pumped. When internal combustion engines are used, it is also necessary to determine the rotation speed. These measurements must be taken simultaneously when the flow, head, and speed are steady. Other flow rates must determine if the whole performance curve of the pump is desired.

19–12. Estimating Total Pumping Head.—The total dynamic head against which the pump is operating includes the vertical distance from the water level in the well while pumping to the center of the free-flowing discharge, plus all losses in the line between the point of entry of the water and the point of discharge. If the discharge is maintained under pressure, the pressure required at the pump head to operate the system is added to the lift and line losses to obtain the total head.

Losses in pipe and fittings can be obtained from a hydraulics handbook and pump-column losses from pump manufacturer catalogs or from Standards of the Hydraulic Institute [1,5].

19–13. Estimating Horsepower Input.—A convenient method of measuring the power input to electric motors without interrupting their operation is with a hook-on voltammeter. Usually there is enough slack in the wires in the motor starter box to permit reading each phase. The method given here pertains to three-phase circuits but can be adapted to others. The power input is obtained by dividing the average current of the three phases by the full load amperes as stamped on the nameplate of the motor. For example, if the average current in the three phases for an 1,800 revolutions per minute, 200-volt, 30-horsepower motor, with a full load current of 75 amperes, is 50 amperes, then 50 divided by 75 equals 67 percent of the full load current, and the power input is 67 percent of the rated horsepower, or 20 horsepower. The voltammeter is not only convenient for determining that the motor nameplate voltage is maintained, but it also will reveal any serious unbalance between the three phases. This method of power measurement should usually yield results accurate to within about 3 percent [6].

Another method of determining power input is with the aid of the watt-hour meter. A watt-hour meter installed on the pump control panel can be used. The procedure is simply to count the number of revolutions of the meter disk for a time interval (3 minutes is usually enough), during which time water discharge measurements are also taken. The electrical input to the motor is given by the formula,

$$\text{hp input} = \frac{3{,}600\,RK}{746t}$$

where:

R=number of revolutions of the disk in time t,
K=meter constant taken from meter nameplate, and
t=time in seconds for R revolutions.

If current and potential transformers are used, the meter constant must be multiplied by the current transformer ratio, the potential transformer ratio, or the product of both, and the computations would then be made as follows:

$$\text{hp input} = \frac{3{,}600\,RKM}{746t}$$

where:

M=transformer ratio.

Unless one is experienced in the operation and testing of large, high-voltage motors, a qualified industrial electrician should be consulted prior to testing for pump performance.

Where an internal-combustion engine is used as the prime mover for a pump, the input horsepower can be calculated by methods described in various mechanical engineering handbooks or manufacturers' catalogs.

19–14. Measuring Pump Discharge.—Several means of measuring the discharge of a pump are available, but for freely discharging pumps, a weir or orifice plate is widely used and each is adaptable to most field situations. Tables and information on weirs, as well as on some other measuring devices, are available in the Bureau's *Water Measurement Manual* [7] and orifice plates have been previously described in section 9–9. Where a closed system is involved, there are several types of flowmeters which can be used.

19–15. Measuring Pump Efficiency.—With measurements of total head, input horsepower, and quantity of water pumped, the efficiency of the installation, expressed as a decimal, may be determined from the following formula:

$$\text{Plant efficiency} = \frac{Q \text{ (gal/min) times total head (ft)}}{3{,}960 \text{ times input horsepower}}$$

The pump efficiency may be determined by dividing the plant efficiency by the efficiency of the electric motor or of the engine and drive mechanism:

$$\text{Pump efficincy} = \frac{\text{plant efficiency}}{\text{motor efficiency}}$$

The efficiency of an electric motor is usually between 90 and 95 percent, depending on size and type, but an exact value can be obtained from the information furnished by the manufacturer for the particular motor. The efficiency of an internal-combustion engine is more difficult to obtain because it changes as wear occurs. Plant efficiency (sometimes called wire-to-water efficiency) should be determined, at least annually, as a means of checking wear or changes in pumping conditions. In some areas where power costs are high and pumps are operated a large part of the year, plant efficiency should be checked every 2 months [2,6].

19–16. Selection of Electric Motors.—The designer should consult an electrical specialist for advice and assistance on selecting electric motors. However, the following summary of motor characteristics is included as a guide.

Electric motors are usually selected according to NEMA (National Electrical Manufacturer's Association) standards including its definitions of enclosures and cooling methods.

Dripproof motors are built for a 40°C ambient temperature rise. These motors are satisfactory where equipment is installed within a shelter.

Splashproof motors are built to tolerate a 50°C ambient temperature rise. Precipitation coming to the motor at angles less than 100° from the vertical cannot enter the motor. These motors are satisfactory for use in the open where rain, snow, and wind velocities are not excessive.

Weather-protected motors are made with provision for ambient temperature rise of either 40°C type I or 50°C type II. The Type I motor has the ventilation openings so constructed as to minimize the entrance of rain, snow, or airborne particles into the motor. Most are so constructed as to prevent the insertion of a rod three-fourths inch in diameter through the ventilation openings. These motors are suitable for installation in the open but screening of the ventilation openings is mandatory. They are used in relatively unprotected locations where extreme adverse weather conditions exist, that is, areas where hurricanes, repetitious storms, snow, extreme heat, and abundant rain are prevalent.

The number phases, frequencies, and voltage of the motor are usually established in advance by the power service available.

The motor should be selected to deliver the estimated maximum power required by the pump without overloading but with consideration of the service factor and desirable insulation.

Thrust bearings are usually built into the motor and vary in type of construction consistent with the magnitude of the thrust expected. Total thrust consists of the weight of the rotating elements of the pump, the weight of the column of water, and the hydraulic thrust developed by the pump. Most pump manufacturers' catalogs furnish thrust and bearing data.

Pump motors should be equipped with nonreverse protection. This usually consists of a releasing coupling which disengages the motor when the pump is stopped for cause, such as a power failure. The coupling allows the pump drive shaft to spin in reverse as water drains from the column pipe without driving the motor. This eliminates the possibility of the motor turning in reverse or of snapping the drive shaft in the event the power outage is only momentary.

Supply line limitations often limit the amount of inrush power required as a motor is started. If the supply line permits an inrush of 600 percent of the full load current, the most economical control is across-the-line starting. However, if limitations preclude the 600 percent, a reduce-voltage starter should be used.

An electrical specialist should be consulted on all aspects of selection, installation, and operation of electrical pumping equipment.

In summary, in selecting an electric motor the following factors should be considered:

- Power required by the pump and service factor of the motor.
- Compatibility of design rotation rates of pump and motor.
- Use of shelters or protected motors.
- Adequate thrust bearing capacity.
- Self-releasing couplings or other nonreverse protection.
- Compatibility of pump discharge head, column pipe, and motor dimensions.

- Inclusion of thrust horsepower loss in wire-to-warter efficiency.
- Inrush limitations and need of reduced voltage starting.

19-17. Selection of Internal-Combustion Engines.—Selection of an internal-combustion engine as a source of power for pumps is more complex than selection of an electric motor. Internal-combustion engine horsepower ratings are usually given without consideration for power consumed by accessories and are rated for sea level operation. The developed horsepower decreases with increase in altitude. The maximum developed horsepower is usually rated at a given revolutions per minute and varies with different manufacturers so sheave ratios for belt drives and gear ratios for gear drives must be selected to give compatibility of pump and motor speeds. Most internal-combustion engines undergo up to 25 percent reduction in developed horsepower if used continuously as compared to intermittent use.

When an engine is adapted to the use of natural gas or other similar fuel, the Btu rating of the fuel is also a factor in estimating the developed horsepower. Engine manufacturers can furnish data which will permit estimates of the horsepower and speed developed by their engines at various altitudes, Btu content of fuel, and other factors.

19-18. Bibliography.—

[1] "American National Standard for Deep Well Vertical Turbine Pumps, Line Shaft and Submersible Types," AWWA E101-71, No. 45101, American Water Works Association, New York, 1971.

[2] Fabrin, A. O., "The Answers to Your Questions About Layne Vertical Turbine Pumps," Layne Bowler, Inc., Memphis, Tenn. 1954.

[3] "Hydraulic Handbook," third edition, Fairbanks Morse and Co., Kansas City, Mo., 1959.

[4] Moore, A. W., and Sens, H. (editors), "The Vertical Turbine Pump by Johnston," Johnston Pump Co., Pasadena, Calif., 1954.

[5] "Standards of the Hydraulic Institute," tenth edition, Hydraulic Institute, New York, 1955.

[6] "Turbine Pump Facts," Vertical Turbine Pump Association, Pasadena, Calif., 1962.

[7] "Water Measurement Manual," second edition, Bureau of Reclamation, 1967.

«Chapter XX

WELL AND PUMP COST FACTORS, OPERATION, AND MAINTENANCE

20-1. Well Construction Costs.—Until recently, standards in the water well drilling industry were largely determined by local custom. Accordingly, many design and construction practices were questionable. The situation was further complicated by the geographic concentrations of drilling contractors, many part time contractors, reluctance of contractors to move more than 40 or 50 miles from their base of operations, diverse geologic and hydrologic conditions, diverse drilling methods, seasonal operations, and other factors. Well construction costs that developed under such conditions tended to be erratic and unpredictable. Although these problems have not been eliminated entirely, the industry has been stabilized by several factors including: (1) enactment by many States of minimum well construction standards; (2) development of more efficient and versatile equipment; (3) training of contractors in good engineering and business practices; and (4) the organization of State, regional, and national drillers associations.

Well construction costs show a marked seasonal variation in much of the country. Costs are usually highest in the early spring and lowest from early fall to midwinter. Move-in costs for small wells are usually relatively low, but such costs can be extremely variable for larger, more complex jobs. This variation on larger jobs may be attributable to the unbalancing of bids to obtain operating funds early in the operation.

The foregoing practices and other factors have precluded the establishment of a meaningful well construction cost index similar to those made available to the general construction industry by various engineering publications and reporting services. Consequently, cost estimates for large wells involve consideration of local costs, the proximity and availability of competent contractors, and seasonal factors. For smaller wells, 8 inches or less in diameter to depths of as much as 300 feet, the common practice is to estimate from $1.50 to $2 per inch diameter per foot of depth (1973-74 base), plus casing costs, depending upon the availability of local drillers and season of the year.

20-2. Pump Costs.—The vertical turbine pump is practically standard equipment for water wells of moderate to large capacity. Manufacturers have essentially standardized all motors, motor controls, pump discharge heads, and column assembly features so that they are generally interchangeable for pumps of a given size, capacity, and rotation rate.

A wide variety of off-the-shelf pumping units of standard construction are available and some manufacturers offer off-the-shelf units for use in corrosive environments.

Most manufacturers publish manuals and catalogs describing their products which contain performance curves showing head-capacity relationships, bowl efficiencies, horsepower requirements, pump speeds, and net positive suction head requirements. Similar publications are available from motor, electrical control, valve, and flowmeter manufacturers. A review of the available literature will usually permit preparation of specifications which will permit bidding on a competitive basis.

Small wells usually are equipped with jet, lift, or small submersible pumps. Local distributors can usually furnish literature on capacities, costs, etc., for the preparation of specifications and cost estimates.

20-3. Operation and Maintenance Responsibilities.—In the E&R (Engineering and Research) Center of the Bureau of Reclamation at Denver, Colorado, the design, construction, development, and testing of wells; selection of test and other temporary type pumps; and recommendations regarding the lift, capacity, and possibly special design features of permanent pumps, are usually the responsibility of regional or E&R Center ground-water specialists. Permanent pump selection, installation, and design of discharge facilities, controls, and housing are usually the responsibility of mechanical design personnel. Close coordination between these two technical groups assures that certain features will be included in the discharge and distribution facilities to permit proper monitoring and maintenance of the well and pump. The more important of these features include: (1) provision of an outlet in the discharge system to permit diversion from the system during future test pumping and water sample collection; (2) a permanent throttling valve on the discharge; (3) a permanent air line with valve and gage for water level reading; (4) access into the pump chamber casing which can also be used to measure water levels to permit backup water leveling reading by tape or electric probe; and (5) ready access to the well to pull the pump and maintain the well.

A regional or E&R Center specialist may supervise the initial test of a newly completed well, but subsequent tests and operation and maintenance are normally project responsibilities unless cooperation or assistance from the E&R Center is specifically requested.

20-4. Operation and Maintenance Basic Records.—Most Bureau water storage, conveyance, and control structures are subject to periodic inspection and testing. However, wells and pumps are often neglected. The nature of the deterioration which occurs in a well may not be readily discernible during operation and may not be recognized until the well fails. Since the greater portion of both well and pump are located beneath the ground surface, some of the neglect may be due to an "out-of-sight, out-of-mind" attitude.

The deterioration which occurs in a well usually develops slowly to a critical point and then accelerates rapidly to failure. If the deterioration can be recognized before the progression reaches the critical point, rehabilitation may be possible. If the progress of deterioration is neglected too long, the potential of successful rehabilitation is materially decreased.

It can generally be assumed for Bureau installations that: (1) the well was carefully designed, constructed, developed, and tested after completion to permit the determination of specific capacity and related characteristics of the well and the quality of the water; (2) the pump was selected to discharge the minimum acceptable volume of water at the estimated maximum probable pump lift and within an acceptable range of efficiencies; (3) the pump was installed in the well and tested for conformance to the specifications; and (4) the pump and motor have received the service and maintenance recommended by the supplier.

Every well upon completion should be tested for and results recorded of sand content of the discharge, drawdown of various yields (step drawdown), drawdown at design capacity, plumbness and alinement, and chemical and biological suitability of the water. In addition, an as-built construction diagram of the well, formation log, and mechanical analyses of aquifer and gravel pack (if used) samples should be a part of the record. After the permanent pump is installed and adjusted, it should be tested for wire-to-water efficiency (actual discharge of pump compared to theoretical discharge considering amount of energy used), shut-in head, and conformance to the performance curves furnished by the manufacturer. These data, together with a copy of the well and pump specifications, should be included in the record. These are basic data to which the results of subsequent tests of the entire installation will be compared so that pump and well conditions can be evaluated and the need for rehabilitation or other maintenance determined.

20-5. Routine Measurements, Tests, and Observations on Large Capacity Wells.—Irrigation and other large capacity wells are often operated seasonally. Initially, the static level of each well should be measured a week or two before the pumping season begins. Shortly after the start of the pumping season, each well, after having operated continuously for at least 8 hours, should be measured for drawdown, discharge, and sand content of the discharge, and the measurements recorded. In multiwell fields, each well should be tested individually and drawdowns in the adjacent wells should be measured during the test.

At intervals of several months during the pumping season, the static water level in each well should be measured and recorded after the well has been shut down for 12 hours or more. Also, the discharge and drawdown should be measured and recorded after 8 or more hours of operation.

After the end of the pumping season each well should be sounded, if possible, to determine the total depth. If sand or other material has accumulated in the bottom of the well to a level where it has encroached on the screen, or may encroach on the screen during the following pumping season, the pump should be pulled and inspected and the well bailed clean before any other tests or measurements are made. On completion of this work the following measurements and tests should be made: static water level, a step test at about the same rates and for the same period of time as was made when the well was initially completed, closed-in head, wire-to-water efficiency, sand content of the discharge 5 minutes and 30 minutes after pumping is started, and water samples taken for chemical and possibly bacterial analysis.

The static level in each well also should be measured each year about midway between the end of the pumping season and the initiation of the following pumping season. These measurements and tests should be made for each well at approximately the same date each year. Continuous hydrographs should be plotted of the static levels, pumping levels, and specific capacities of each well. The tests should be analyzed and the results compared with those of the initial tests made when the well was completed.

During routine lubrication and servicing of each installation, the following should be observed and recorded:

- any increase in sand content of the discharge
- decrease in discharge
- excessive heating of the motor
- excessive oil consumption
- excessive vibration
- sounds possibly attributable to cavitation
- cracking or uneven settlement of the pump pad or foundation

- settlement or cracking of the ground
- ground surface gradient around the well.

20–6. Interpretation of Observed or Measured Changes in Well Performance or Conditions.—(a) A decrease in specific capacity without a proportional decline in the static water level may indicate blockage of the screen by accumulated sediment in the bottom of the well, blockage of the screen or gravel pack by encrustation, or collapse of casing or screen.

Should the specific capacity during a step test show a decline of 10 percent or more from the original step test at a given discharge, the well should be surveyed with a dolly or bailer (sec. 16–7) to determine the location and extent of possible contributing conditions. If collapse appears to be the problem, the well should be surveyed (sec. 8–3) to determine the location and nature of the collapse. If collapse is not the problem, the inside of the well should be scraped and the sediment that was subsequently bailed from the bottom should be examined to determine the chemical composition, nature, and extent of the encrusting material as a basis for a plan of rehabilitation.

(b) An increase in sand content of the discharge, particularly if it is associated with a measurable accumulation of sand in the bottom of the well, may indicate enlargement of slot sizes by corrosion; settlement of gravel pack beneath a bridge leaving an unpacked zone opposite a screened section; a break in the casing or screen, usually at a joint; or failure of a packer seal. Mechanical and mineralogical examination of a sample bailed from the bottom of the hole and comparison with the original description of the aquifer and gravel pack materials made during construction of the well may give some indication of the nature of the difficulty. If the material is noticeably smaller in grain size than the grain size of any aquifer screened in the well, or if the material contains the full range of sizes of the gravel pack, there is probably a break in the casing or screen. If all the material is smaller than the screen slot sizes, it is probably a bridge. If the above interpretations of grain size and distribution are not applicable, the problem may be due to enlargement of a slot size by corrosion. If the problem is apparently due to bridging, it can frequently be corrected by redevelopment while pouring water down the gravel refill tremies and the addition of gravel pack material. The other problems usually require a photographic survey to be made of the well to more clearly assess the problem. Decisions can then be made concerning the practicability of rehabilitation and the procedures to be followed.

(c) Settlement or cratering of the land surface around a well, the development on the ground surface of small drainage channels toward the well, and cracking and settlement of pump pads and foundations are all indicative of settlement of the well structure. In some areas, the problem may be associated with land subsidence due to excessive pumping of the aquifers. Usually, however, the problem is related to poor well design, construction, or development, and results from excessive pumping of sand. In many instances the sand pumping is complicated by collapse of casing or screen, bridging of gravel packs, and similar deterioration. When such conditions are encountered and as a basis for rehabilitation, the well should be taken out of service, sounded for depth, and surveyed photographically to determine whether any structural damage has occurred. If the well cannot be shut down because of the need for water, the casing should be temporarily supported by welding heavy I-beams to it (sec. 11–2).

The foregoing problems are related primarily to the well and are the most commonly encountered. Many of them may occur because of conditions that were not considered in the original well design; others are caused by inadequate investigation prior to construction or the attempt to standardize on a particular well design. In any event, any failure or deterioration should be thoroughly investigated and reported along with the rehabilitation program used and the success of such a program. These data should be made a permanent part of the well record and used as a guide in the design and construction of wells drilled in the area in the future.

(d) Decline in pump discharge and head may be due to deterioration of the pump or simultaneous deterioration of both the well and the pump. A common occurrence is a decrease in shut-in head and significant decrease in discharge without a corresponding decline in static water level and specific capacity. Such an occurrence is usually due to one of the following conditions: (1) improper adjustment of the impeller due to wear or other causes, (2) a hole in the column pipe, or (3) erosion or corrosion of the impeller or bowls. The latter condition is usually associated with considerable vibration when the pump is running. If the condition cannot be corrected by adjusting the impellers, the pump should be pulled and repaired or replaced. The cause of the problem should be thoroughly investigated and made a part of the permanent well and pump record.

(e) Excessive vibration of the pump may result from imbalance of the impeller or from the pump being installed in a crooked well.

A pump which makes a crackling noise similar to gravel being thrown on a tin roof is probably experiencing cavitation of the impellers. This is particuarly true if the discharge is surging and irregular and contains considerable air. The condition usually results from a decline in the static water level or reduced well capacity because of encrustation or accumulation of sand in the screen. Either of these conditions results in excessive drawdown for the pump and a decline below that required in the available net positive suction head. If the condition is due to a decline in the static water level, it can usually be corrected by lowering the bowls. In severe cases, it may be necessary to add additional stages and a larger motor in addition to lowering the bowls. The well should also be checked for possible encrustation of the screen or other causes of reduced efficiency.

(f) Excessive heating of the motor is occasionally encountered and is usually associated with an overload condition and the consumption of excessive electrical energy. Such heating may be caused by a poorly adjusted impeller which is dragging on the bowls, too tight a packing gland, improper or unbalanced voltage, poor electrical connections, or improper sizing of the motor.

Occasionally, an inadequate discharge will be associated with trash that has lodged in the bowls or blockage of the impellers or bowl channels by products of corrosion and encrustation. Correction entails pulling the pump for repair. These conditions may also be reflected in overheating of the motor. Where overheating is encountered, the installation should be first checked by an electrician to determine whether the trouble is in the power system or in the pump, rather than in the well.

(g) Occasionally, a noticeable increase in oil consumption is encountered in oil-lubricated pumps. The excessive consumption may be due to a hole in the wall of the oil tubing or excessive wear on a packing gland in the tubing. These conditions can result in a decrease in differential pressure in the oil tubing and loss of oil into the well. The first condition can result in inflow of water into the tubing and formation of an emulsion of water and oil. The emulsion lacks adequate lubricating qualities and can result in excessive wear or burning out of the bearings. The escape of oil into the well can result in the accumulation of oil floating on the water surface in the casing. With adequate pump submergence, this latter condition may not cause serious trouble, but if drawdown increases due to a decline in water table or deterioration of the well, oil may be drawn into the pump, causing impairment of water quality. In addition, the presence of oil may contaminate the ground water and preclude accurate measurement of static water and pumping water levels.

(h) Small capacity wells usually have discharges of less than 125 gallons per minute. Casing and screen used for such wells commonly are 6 inches in diameter or smaller, and materials used in their construction are relatively light in weight. While the observations and measurements outlined above for large capacity pumps and wells are equally applicable, they are usually difficult to justify economically.

Pumps may be shaft-driven vertical or submersible turbines, ejector, cylinder, or suction types of various kinds. The cost of the well construction is usually minor compared to a large capacity well. In many instances, when such a well fails it may be less expensive to replace it than attempt rehabilitation. While the continued observation and periodic inspection and testing of such wells can seldom be justified economically, they should be checked at least once a year for discharge, drawdown, specific capacity, sand content of discharge, effective depth, and static and pumping levels. Many manufacturers' handbooks give methods of testing and evaluating the condition of their pumps. Such literature should be consulted and, where practical, the recommendations applied.

«Chapter XXI

WELL REHABILITATION

21-1. Initial Investigations.—Chapter XX discusses well deterioration and methods of diagnosing such deterioration. This chapter discusses methods, equipment, and materials commonly used in well rehabilitation. Water well rehabilitation includes the repair of wells which have failed because of collapse, broken casing and screen, or other similar major damage, and the treatment of wells which have begun to pump sand, have experienced a change in water quality, or have shown a marked decrease in discharge and efficiency because of incrustation, corrosion, or other factors which tend to reduce the intake area of the screen or permeability of the adjacent aquifer. Normally, well rehabilitation does not include deepening or other major changes in the well structure necessitated by declining water levels, a need for increased discharge, or other similar factors.

In some instances of well deterioration, rehabilitation may be impracticable, so construction of a replacement well may be necessary. Furthermore, a major problem in well rehabilitation may be in determining the exact nature of the deterioration since the screen and other components most likely to deteriorate are not subject to direct visual inspection or testing. Accordingly, well rehabilitation usually involves the risk of further damaging a well or destroying its usefulness. However, the element of risk can be reduced by adequate investigations and planning prior to undertaking the work.

Data should include:

(a) Original design and construction (as-built conditions)

- (1) When drilled
- (2) Method of drilling
- (3) Materials log
- (4) Geophysical logs
- (5) Casing log
  - a. Length and diameter
  - b. Wall thickness
  - c. Type and location of joints
- (6) Screen or perforated casing description
  - a. Type and material
  - b. Length and diameter
  - c. Slot size
  - d. Depth of settings
  - e. Type and location of joints

(7) Grout or seals
  a. Type and composition
  b. How placed

(b) Mechanical analysis of aquifer material and gravel pack
(c) Relation of aquifer material or gravel pack to screen slot opening
(d) Method and completeness of development
(e) Original pump test results: step and constant yield tests, sand content of discharge
(f) Ground-water hydrographs of area
(g) Quality of water determinations
(h) Summary of historical performance and operation
(i) Summary of needed maintenance and rehabilitation.

21-2. Sand Pumping.—Most wells pump sand to some degree. However, proper design and adequate development usually can limit sand pumping to an acceptable concentration. Excessive sand pumping by a well is accompanied by numerous undesirable side effects. Pump bowls and impellers may be eroded by sand and necessitate frequent replacement. Not all sand entering a well is pumped out with the discharge. A certain amount, usually the larger grained portion, settles to the bottom of the well where it may encroach on the screen and reduce well efficiency. This results in increased drawdowns, increased entrance velocity, and perhaps accelerated corrosion and incrustation.

The sand in the discharge may collect in pipelines and channels, thereby reducing their carrying capacity and necessitating periodic cleaning. Furthermore, sufficient sand may enter a well to create fairly large cavities in the aquifer around the screen. As a consequence of collapse of such cavities, the casing or screen may be broken or deformed and upward caving to the ground surface may occur with resultant subsidence at the surface and damage to the entire installation. Where sprinkler systems are directly supplied from a well, excessive sand may block the pipelines and erode the orifices in the sprinkler heads.

If sand pumping is due to a broken screen or casing or faulty packer, the location and nature of the break can usually be determined by sounding and verified by a photographic survey of the well. In a few instances, the break may consist only of a parting of the casing or screen at a joint with no offset of the axis. This is the easiest type of break to repair. In most cases, the break will be associated with displacement of the axis and possibly deformation of the casing and screen on one or both sides of the break. The correct procedure is to run a hydraulic or mechanical casing swage into the well to round out and, if possible, realine the casing or screen. In some instances, the casing and screen may be so far out of line that it is impossible to realine them without causing buckling of the casing elsewhere. In this circumstance, there is no fully satisfactory solution, although the well possibly may be modified by insertion of a liner to produce sand-free water at a lower discharge and specific capacity. Where approximate or complete realinement is possible, the well can be repaired by inserting a liner through the break and anchoring the liner in place either by using a hydraulic swage or by cementing it in place. This usually will result in reduction in the yield or specific capacity and may require the use of a different pump because of the reduced inside diameter.

If a broken or defective seal is involved, several possibilities may exist. It is almost impossible to remove a swaged lead seal without pulling the casing or screen. One solution is to telescope 10 feet or more of liner with neoprene rubber seals sized to both the smaller and larger casings into the smaller casing. Another solution is to swage a liner into the smaller casing with the end extending about 3 feet above the original lead packer and then fill the annular space with a neat cement grout.

If the problem is due either to localized enlargement of screen slots or to a hole in the casing or screen as a result of corrosion, a liner may sometimes be swaged in place opposite the corroded section. The liner in all cases should be of the same material as the casing or screen within which it is placed.

In some instances, because of poor initial slot selection or enlargement of slot sizes over the greater length of a screen because of corrosion, insertion of a liner is impractical. If telescoping construction has been used, the screen may be pulled and a new screen with smaller slots or casing made of more corrosion-resistant materials installed. This is impossible, however, where single string design has been used. Nevertheless, rehabilitation has been made on single string designs by ripping the original screen to increase the open area and then telescoping a smaller diameter screen inside of it. This may be used as a temporary expedient, but is not recommended as a permanent repair. Well efficiency is reduced and will deteriorate rapidly although several years' service may be obtained from the well. In some instances, stainless steel screen has been installed inside low carbon steel. This is not recommended. The same material should be used as in the original screen; otherwise, aggravated corrosion of the original screen and blockage of the inserted screen by incrusting corrosion products are almost certain to result.

Where sand pumping has resulted from settlement and bridging of a gravel pack, the best procedure is to vigorously redevelop the well while injecting large quantities of water into the pack from the surface. This will usually cause the bridge to collapse, thereby reestablishing the integrity of the pack.

Additional material should then be added to replenish the pack. In older wells, the bridging may be a result of local cementation of the pack and the foregoing procedure may be ineffective. In such a case, the well should be acidized and then another attempt made to cause collapse of the bridge.

Where casing settlement has occurred, the structure should first be supported by welding parallel I-beams of adequate strength and length to opposite sides of the casing collar at the ground surface (sec. 11–2).

The well should then be surged vigorously while water is injected into the gravel pack or, if there is no pack, water should be applied to the caved area about the top of the well. Selected gravel pack material should be added to the pack or to the caved area about the well, as required. This will fill existing caverns and cause bridging to collapse and ensure a stable condition before further rehabilitation is undertaken. Excessive subsurface movement during this work may aggravate the situation or cause additional casing failure which may make further work impractical. During work of this type, care should be taken to assure that rapid settlement does not endanger workmen or equipment.

Collapse from excessive hydraulic differential head may result from overpumping a well which has inadequate entrance area in the perforations or screen or from loss of intake area because of incrustation. This is seldom a problem where casing with an adequate diameter to wall thickness ratio has been used. Necessary corrective measures may include swaging the affected component to full diameter or reducing pumping.

21–3. Decline in Discharge.—A decline in discharge and an increase in drawdown are usually caused by a decline in the static water level, the installation of additional nearby wells that have overlapping areas of influence, an accumulation of sediment on the bottom of a well sufficient to cover a significant part of the screen, collapse of the screen, or incrustation of the screen and gravel pack.

Where the decline in yield is the result of a decline in the static water level or interference from other wells, it may be possible to correct the situation by merely lowering the pump bowl and if necessary, adding additional bowls and a larger motor. If regular measurements are made of static and pumping levels in the well, the cause is usually apparent.

If decline in yield is due to loss of screen length resulting from accumulation of sand over part of the screen, this can be determined by sounding the well. The solution is to bail the well clean. However, such an accumulation is usually indicative of other problems. The discharge should be tested for sand content and, if too high, the investigations outlined in section 20–6(b) made.

If collapse of the casing or screen is suspected, lowering a bailer or dolly down the hole on a cable will usually show the approximate location of the trouble (sec. 16–7). If collapse is indicated, a television or photographic survey should be made to determine the nature of the damage and the possibility of repair. If the casing or screen is not broken, the trouble may be corrected by using a hydraulic or mechanical casing swage. If there is a break, however, a liner should be installed as described in section 20–6.

21–4. Shooting with Explosives and Acidizing.—If all other possibilities are eliminated, the problem of reduced yield is usually one of incrustation of the screen or pack. The first step in rehabilitation is to scrape the inside of the screen with a steel disk on a drill stem or rod to break loose some of the incrusting materials which will settle to the bottom of the well. These scrapings should be examined to determine the nature and chemical composition of the incrustation. If it consists primarily of calcium, magnesium and iron carbonates, or iron hydroxides, rehabilitation using sulfamic or hydrochloric acid may be possible.

If iron and manganese compounds constitute over 20 percent of the material, other than included sands, corrosion should be suspected as a contributing factor. If the molecular ratio of $Fe(OH)_3$ (ferric hydroxide) to FeS (ferrous sulfide) is 3 to 1, sulfate-reducing bacteria are probably a contributing factor.

After the nature of the incrustation has been determined, reexamination of the television or photographic survey of the well is recommended to assess the extent and location of concentrated zones of incrustation.

If the incrustation does not appear to be heavy, and the condition of the screen is believed to be good, a single string of 50 grains per foot of Prima Cord, cut to the length of each screen section, can be fired opposite each section. In some instances, two shots may be advisable, but in no case should more than one string be fired at a time. Shooting will crack and break the incrustation and cause it to be more readily attacked by the subsequent acid treatment. The practice usually results in some incrustation being broken out of the screen and settling to the bottom of the well. This should be removed by bailing before acidizing.

Shooting should not be attempted in any well that has pumped appreciable sand or where there is evidence that the casing or screen might not be fully supported by earth materials.

Sonar Jet cleaning, a patented process, consists of shooting a series of small explosive charges in a well with a slight delay between the detonation of each successive charge. This process has not been extensively used by the Bureau, and reports of its effectiveness are variable. Where other procedures have been poor or ineffective, Sonar Jet cleaning should be considered. Whether the well is shot with Prima Cord or by Sonar Jet, it may require acidization subsequent to shooting.

When complete rehabilitation is impractical, shooting alone will often give a temporary improvement of well performance. This may be done by lifting the pump head from the base and moving it to one side. The Prima Cord is lowered alongside the column pipe into the well and detonated opposite the screen or screens and the well developed by placing the pump back in position and rawhiding. The above procedure is a temporary expedient at best and should be followed by a more complete program as soon as conditions permit.

For successful well acidizing, the acids must be strong and the products of the reaction soluble. The more commonly used acids are muriatic or hydrochloric (HCl), sulfuric (H_2SO_4), and sulfamic (amino sulfamic) (H_2NSO_3H).

Where iron or manganese constitute a significant part of the incrustation and the pH of the acid solution reaches about three, the iron and manganese compounds form insoluble precipitates which settle out. Under these conditions, chelating agents should be used to keep the iron and manganese compounds in solution so that they may be readily pumped from the well. Commonly used chelating agents are:

Citric Acid $(COOH)CH_2\ C(OH)\ (COOH)\ CH_2\ COOH$
Phosphoric Acid H_3PO_4
Tartaric Acid $HOOC\ (CHOH)\ COOH$
Rochelle Salt $KN_aC_4H_4O_6$
Glycolic Acid $(HOCH)_2\ COOH$

Usual amounts of chelating agents used are:

1 pound of agent to 15 pounds of sulfamic acid powder
2 pounds to each gallon of 15% HCl
4 pounds to each gallon of H_2SO_4

Sulfuric acid is seldom used in acidizing wells because the reaction of sulfuric acid with calcium carbonate forms calcium sulfate (gypsum) which is relatively insoluble and difficult to remove from the well.

In addition, even when inhibited, sulfuric acid is aggressive to most metals, particularly copper alloys. It should be used only as a last resort when two or more treatments with less active acids have been unsuccessful and the only other alternative is the construction of a new well.

Muriatic or hydrochloric acid was for years the most commonly used acidizing agent and is still popular. However, hydrochloric acid should not be used even in inhibited form on wells equipped with type 304 or 308 stainless steel screens, casing, or other components because it causes stress corrosion cracking of these alloys. The damage caused by the acid may not show up for some time after treatment of the well. However, use of hydrochloric acid would probably be safe with type 316 or 321 stainless steel.

Muriatic acid is available commercially in three strengths, but that most commonly used for well acidizing is 18° Baume or 27.92 percent hydrochloric acid. The acid is usually used full strength. The volume of water within each screen section is estimated, and 2 to 2½ times as much acid as water is placed in the well through a plastic or black iron pipe opposite each screen. One-half pound of diethylthiourea or a similar inhibitor is used per 100 gallons of acid as well as chelating agents if required. The acid is normally left in the well for 4 to 6 hours; the well is then surged with a surge block for 15 to 20 minutes at about 1-hour intervals, after which the solution is pumped or bailed from the well and wasted.

Hydrochloric acid is dangerous to use unless handled by experienced personnel and special equipment. Such work should be contracted to specialized well-servicing firms.

Sulfamic acid is being used increasingly for well treatment. It is more costly than hydrochloric acid but is much more convenient, safer to use, and more easily shipped and stored. It is not as aggressive as hydrochloric acid and more time is required for an equivalent treatment. If the work is done by force account or a local contractor, sulfamic acid is usually less expensive than treatment with hydrochloric acid.

The product of the reaction of sulfamic acid with calcium carbonate is calcium sulfamate which is highly soluble and readily pumped from the well. If iron compounds make up a considerable part of the incrustation, a chelating agent should be used. While not highly aggressive, sulfamic acid should not be used on copper alloy screens and other components without an inhibitor.

The solubility of sulfamic acid is as follows:

Water temperature, °F	*Solubility, lb/gal*
32	1. 38
41	1. 45
50	1. 54
59	1. 66
68	1. 79

Sulfamic acid in the amount of 5.5 pounds per gallon of water would be equivalent in reaction potential to a gallon of 18° B (27.97 percent) hydrochloric acid and 3.0 pounds to a gallon of 15 percent hydrochloric acid. It is obvious that such sulfamic acid concentrations cannot be obtained as true solutions; however, a slurry can be mixed and pumped into the well. The mix proportions in 100 gallons of water include 300 pounds of sulfamic acid, 20 pounds of citric acid, 17 pounds of diethylthiourea, 3.5 pounds of pluronic F 68 or L 62, and 150 pounds of sodium chloride. The chemicals are dissolved and suspended in water at the surface in a volume equivalent to the volume of water within the well casing and screen. The slurry is pumped or poured into the well through a black iron or plastic pipe which initially extends to the bottom of the well. The pipe is raised in 5- or 10-foot stages as sufficient solution is added to displace an equivalent volume of water in the well.

The solution is left in the well from 12 to 24 hours during which time it is surged for 15 to 20 minutes at hourly intervals. When the solution in the well, when tested with litmus or similar paper, shows a pH of between 6 and 7, the acid may be considered exhausted. The solution should then be pumped from the well and the well tested for yield and drawdown. If a marked improvement is apparent, the well should then be redeveloped, sterilized, tested as a new well, and put back into production. If the improvement on initial testing after acidizing is relatively slight, another treatment should be made.

When acidizing a well, a tank of concentrated sodium bicarbonate solution should be available to permit neutralizing the acid in event of an accident. In addition, workmen should wear protective rubber shoes, clothing, gloves, hood, and goggles. Until all components are mixed in water, a filter respirator should also be worn. Special equipment may be required such as mixing tanks and piping fabricated of black iron, plastic, or wood.

Incrustation may consist primarily of silica, clay particles, and other materials resistant to normal acid treatment. Under such circumstances, successful acidizing usually entails the use of hydrofluoric and similar strong acid. In many instances, because of the specialized nature of the treatment, the cost of such treatment may approach or exceed that of a new well. In some instances, where telescoping screen construction has been used, removal and cleaning of the screen above ground may be preferable to acidization.

21–5. Chlorine Treatment.—Where screen blockage is caused by slime-forming organisms, chlorine gas may be an effective treatment agent. Chlorine gas is dangerous to use without experienced personnel and adequate equipment. Usually a 100- to 150-pound pressure cylinder of chlorine gas is used. The cylinder is mounted on a scale to permit checking on the rate of feed. The gas is fed through a plastic or black iron pipe which extends to approximately the bottom of the well. The bottom of the pipe should be centered with an appropriate device so that the released gas does not impinge directly on the casing or screen, and an approved feeder should be employed to avoid back sucking. The cylinder is opened slowly, one full counter-clockwise turn of the valve. Rate of discharge of the cylinder should not exceed 40 pounds per 24 hours.

When the cylinder is exhausted, the chlorine in the well can be neutralized by adding sodium hydroxide or calcium hydroxide to the water prior to pumping it to waste.

Hypochlorite solutions are cheaper, more convenient, and safer to use than gas but generally are less effective (sec. 18–2).

Sufficient hypochlorite should be added to the water in the well to give an estimated 1,000 p/m chlorine content. The hypochlorite is poured or pumped into the well, then thoroughly mixed and diffused by surging for about 30 minutes. The solution is left in the well for about 6 hours during which it is surged for 15 to 20 minutes at hourly intervals, then pumped or bailed to waste. Following this, the well can usually be adequately redeveloped by rawhiding. This treatment can usually be carried out without pulling the pump.

21–6. Rehabilitation of Rock Wells.—The previous discussion has been primarily applicable to cased and screened wells in unconsolidated materials. The fractures and other voids in uncased rock wells may become clogged and sealed by deposition products similar to incrustation of a well screen.

Hydrochloric acid of 18° B strength is commonly used full strength to treat rock wells. A volume of acid equal to about 2.5 times the volume of water in the well is pumped or poured into the well through a plastic or black iron pipe extending to the bottom of the well. The pipe is raised as acid displaces the water in the well. If the water level in the well is within the casing, an inhibitor should be used. The acid is permitted to remain in the well for at least 6 hours, surged for 15 or 20 minutes at hourly intervals, and then pumped to waste.

In some open hole rock wells, shooting has been more effective than acid treatment and at times a combination treatment has been used, with acidization following shooting.

Ten-pound shots of 50 to 60 percent dynamite are used at 5-foot intervals within the open hole. Shots should not be fired within 10 feet of a shale formation or within 50 feet of the bottom of the casing. Shots are fired separately beginning at the bottom of the open hole. After shooting is completed, the well should be bailed clean and developed.

In very hard rock, up to 100 pounds of explosive have been used in shots 10 to 12 feet apart. The amount of power to use and the spacing is a matter of judgment based on experience. One or more test shots are advisable when operating in unfamiliar rocks.

Nitramine has also been used instead of dynamite. While more expensive than dynamite, it is much safer and easier to handle. One can of nitramine is equivalent to 1.6 pounds of 50 percent or 1 pound of 60 percent dynamite.

On completion of acidizing or shooting work, the well should be thoroughly developed in the same manner as a new well.

APPENDIX

International System (SI metric)/U.S. Customary Conversion Tables

Length

To convert from	*To*	*Multiply by*
angstrom units	nanometers (nm)	0.1
	micrometers (μm)	1.0×10^{-4}
	millimeters (mm)	1.0×10^{-7}
	meters (m)	1.0×10^{-10}
	mils	$3.937\ 01\times10^{-6}$
	inches (in)	$3.937\ 01\times10^{-9}$
micrometers	millimeters	1.0×10^{-3}
	meters	1.0×10^{-6}
	angstrom units (A)	1.0×10^{4}
	mils	0.039 37
	inches	$3.937\ 01\times10^{-5}$
millimeters	micrometers	1.0×10^{3}
	centimeters (cm)	0.1
	meters	1.0×10^{-3}
	mils	39.370 08
	inches	0.039 37
	feet (ft)	$3.280\ 84\times10^{-3}$
centimeters	millimeters	10.0
	meters	0.01
	mils	0.3937×10^{3}
	inches	0.3937
	feet	0.032 81
inches	millimeters	25.40
	meters	0.0254
	mils	1.0×10^{3}
	feet	0.083 33
feet	millimeters	304.8
	meters	0.3048
	inches	12.0
	yards (yd)	0.333 33

To convert from	*To*	*Multiply by*
yards	meters	0.9144
	inches	36.0
	feet	3.0
meters	millimeters	1.0×10^{3}
	kilometers (km)	1.0×10^{-3}
	inches	39.370 08
	yards	1.093 61
	miles (mi)	$6.213\ 71 \times 10^{-4}$
kilometers	meters	1.0×10^{3}
	feet	$3.280\ 84 \times 10^{3}$
	miles	0.621 37
miles	meters	$1.609\ 34 \times 10^{3}$
	kilometers	1.609 34
	feet	5280.0
	yards	1760.0
nautical miles (nmi)	kilometers	1.8520
	miles	1.1508

Area

To convert from	*To*	*Multiply by*
square millimeters	square centimeters (cm^2)	0.01
	square inches (in^2)	1.550×10^{-3}
square centimeters	square millimeters (mm^2)	100.0
	square meters (m^2)	1.0×10^{-4}
	square inches	0.1550
	square feet (ft^2)	$1.076\ 39 \times 10^{-3}$
square inches	square millimeters	645.16
	square centimeters	6.4516
	square meters	6.4516×10^{-4}
	square feet	69.444×10^{-4}
square feet	square meters	0.0929
	hectares (ha)	9.2903×10^{-6}
	square inches	144.0
	acres	$2.295\ 68 \times 10^{-5}$
square yards	square meters	0.836 13
	hectares	8.3613×10^{-5}
	square feet	9.0
	acres	$2.066\ 12 \times 10^{-4}$
square meters	hectares	1.0×10^{-4}
	square feet	10.763 91
	acres	2.471×10^{-4}
	square yards (yd^2)	1.195 99

To convert from	*To*	*Multiply by*
acres	square meters	4046.8564
	hectares	0.404 69
	square feet	4.356×10^{4}
hectares	square meters	1.0×10^{4}
	acres	2.471
square kilometers	square meters	1.0×10^{6}
	hectares	100.0
	square feet	107.6391×10^{5}
	acres	247.105 38
	square miles (mi²)	0.3861
square miles	square meters	$258.998\ 81\times10^{4}$
	hectares	258.998 81
	square kilometers (km²)	2.589 99
	square feet	$2.787\ 81\times10^{7}$
	acres	640.0

Volume—Capacity

To convert from	*To*	*Multiply by*
cubic millimeters	cubic centimeters (cm³)	1.0×10^{-3}
	liters (l)	1.0×10^{-6}
	cubic inches (in³)	$61.023\ 74\times10^{-6}$
cubic centimeters	liters	1.0×10^{-3}
	milliliters (ml)	1.0
	cubic inches	$61.023\ 74\times10^{-3}$
	fluid ounces (fl. oz)	33.814×10^{-3}
milliliters	liters	1.0×10^{-3}
	cubic centimeters	1.0
cubic inches	milliliters	16.387 06
	cubic feet (ft³)	$57.870\ 37\times10^{-5}$
liters	cubic meters	1.0×10^{-3}
	cubic feet	0.035 31
	gallons	0.264 17
	fluid ounces	33.814
gallons	liters	3.785 41
	cubic meters	$3.785\ 41\times10^{-3}$
	fluid ounces	128.0
	cubic feet	0.133 68
cubic feet	liters	28.316 85
	cubic meters (m³)	$28.316\ 85\times10^{-3}$
	cubic dekameters (dam³)	$28.316\ 85\times10^{-6}$
	cubic inches	1728.0
	cubic yards (yd³)	$37.037\ 04\times10^{-3}$
	gallons (gal)	7.480 52
	acre-feet (acre-ft)	$22.956\ 84\times10^{-6}$

To convert from	*To*	*Multiply by*
cubic miles	cubic dekameters	$4.168\ 18 \times 10^6$
	cubic kilometers (km^3)	4.168 18
	acre-feet	3.3792×10^6
cubic yards	cubic meters	0.764 55
	cubic feet	27.0
cubic meters	liters	1.0×10^3
	cubic dekameters	1.0×10^{-3}
	gallons	264.1721
	cubic feet	35.314 67
	cubic yards	1.307 95
	acre-feet	8.107×10^{-4}
acre-feet	cubic meters	1233.482
	cubic dekameters	1.233 48
	cubic feet	43.560×10^3
	gallons	325.8514×10^3
cubic dekameters	cubic meters	1.0×10^3
	cubic feet	$35.314\ 67 \times 10^3$
	acre-feet	0.810 71
	gallons	$26.417\ 21 \times 10^4$
cubic kilometers	cubic dekameters	1.0×10^6
	acre-feet	$0.810\ 71 \times 10^6$
	cubic miles (mi^3)	0.239 91

Temperature

degrees Celsius (°C)	t_c
kelvin (K)	t_k
degrees Fahrenheit (°F)	t_f
degrees rankine (R)	t_r

$t_c = (t_f - 32)/1.8$
$\quad = t_k - 273.15$

$t_k = t_c + 273.15$
$\quad = (t_f + 459.67)/1.8$
$\quad = t_r/1.8$

$t_f = t_c/1.8 + 32$

$t_r = 1.8\ t_k$
$\quad = 1.8\ t_c + 491.68$

Acceleration

To convert from	*To*	*Multiply by*
feet per second squared	meters per second squared (m/s^2)	0.3048
	G's	0.031 08
meters per second squared	feet per second squared (ft/s^2)	3.280 84
	G's	0.101 97

To convert from	*To*	*Multiply by*
G's (standard gravitational acceleration)	meters per second squared	9.806 65
	feet per second squared	32.174 05

Velocity

To convert from	*To*	*Multiply by*
feet per second	meters per second (m/s)	0.3048
	kilometers per hour (km/h)	1.097 28
	miles per hour (mi/h)	0.681 82
meters per second	kilometers per hour	3.60
	feet per second (ft/s)	3.280 84
	miles per hour	2.236 94
kilometers per hour	meters per second	0.277 78
	feet per second	0.911 34
	miles per hour	0.621 47
miles per hour	kilometers per hour	1.609 34
	meters per second	0.447 04
	feet per second	1.466 67
feet per year (ft/yr)	millimeters per second (mm/s)	$9.665\ 14 \times 10^{-6}$

Force

To convert from	*To*	*Multiply by*
pounds	newtons (N)	4.4482
kilograms	newtons	9.806 65
	pounds (lb)	2.2046
newtons	pounds	0.224 81
dynes	newtons	1.0×10^{-5}

Mass

To convert from	*To*	*Multiply by*
grams	kilograms (kg)	1.0×10^{-3}
	ounces (avdp)	0.035 27
ounces (avdp)	grams (g)	28.349 52
	kilograms	0.028 35
	pounds (avdp)	0.0625
pounds (avdp)	kilograms	0.453 59
	ounces (avdp)	16.00
kilograms	kilograms (force)-second squared per meter ($kgf \cdot s^2/m$)	0.101 97
	pounds (avdp)	2.204 62
	slugs	0.068 52

To convert from	*To*	*Multiply by*
slugs	kilograms	14.5939
short tons	kilograms	907.1847
	metric tons (t)	0.907 18
	pounds (avdp)	2000.0
metric tons (tonne or megagram)	kilograms	1.0×10^{3}
	pounds (avdp)	$2.204\ 62\times10^{3}$
	short tons	1.102 31
long tons	kilograms	1016.047
	metric tons	1.016 05
	pounds (avdp)	2240.0
	short tons	1.120

Volume Per Unit Time Flow

To convert from	*To*	*Multiply by*
cubic feet per second	liters per second (l/s)	28.316 85
	cubic meters per second (m^3/s)	0.028 32
	cubic dekameters per day (dam^3/d)	2.446 57
	gallons per minute (gal/min)	448.831 17
	acre-feet per day (acre-ft/d)	1.983 47
	cubic feet per minute (ft^3/min)	60.0
gallons per minute	cubic meters per second	0.631×10^{-4}
	liters per second	0.0631
	cubic dekameters per day	5.451×10^{-3}
	cubic feet per second (ft^3/s)	2.228×10^{-3}
	acre-feet per day	4.4192×10^{-3}
acre-feet per day	cubic meters per second	0.014 28
	cubic dekameters per day	1.233 48
	cubic feet per second	0.504 17
cubic dekameters per day	cubic meters per second	0.011 57
	cubic feet per second	0.408 74
	acre-feet per day	0.810 71

Viscosity

To convert from	*To*	*Multiply by*
centipoise	pascal-second (Pa·s)	1.0×10^{-3}
	poise	0.01
	pound per foot-hour (lb/ft·h)	2.419 09
	pound per foot-second (lb/ft·s)	$6.719\ 69\times10^{-4}$
	slug per foot-second (slug/ft·s)	$2.088\ 54\times10^{-5}$
pascal-second	centipoise	1000.0
	pound per foot-hour	$2.419\ 09\times10^{3}$
	pound per foot-second	0.671 97
	slug per foot-second	20.8854×10^{-3}

To convert from	*To*	*Multiply by*
pound per foot-hour	pascal-second	$4.133\ 79 \times 10^{-4}$
	pound per foot-second	$2.777\ 78 \times 10^{-4}$
	centipoise	0.413 38
pound per foot-second	pascal-second	1.488 16
	slug per foot-second	31.0809×10^{-3}
	centipoise	$1.488\ 16 \times 10^{3}$
centistokes	square meters per second (m^2/s)	1.0×10^{-6}
	square feet per second (ft^2/s)	$10.763\ 91 \times 10^{-6}$
	stokes	0.01
square feet per second	square meters per second	9.2903×10^{-2}
	centistokes	9.2903×10^{4}
stokes	square meters per second	1.0×10^{-4}
rhe	1 per pascal-second (1/Pa·s)	10.0

Force Per Unit Area
Pressure Stress

To convert from	*To*	*Multiply by*
pounds per square inch	kilopascals (kPa)	6.894 76
	[1] meters-head	0.703 09
	[2] mm of Hg	51.7151
	[1] feet of water	2.3067
	pounds per square foot (lb/ft^2)	144.0
	std. atmospheres	68.046×10^{-3}
pounds per square foot	kilopascals	0.047 88
	[1] meters-head	4.8826×10^{-3}
	[2] mm of Hg	0.359 13
	[1] feet of water	16.0189×10^{-3}
	pounds per square inch	6.9444×10^{-3}
	std. atmospheres	$0.472\ 54 \times 10^{-3}$
short tons per square foot	kilopascals	95.760 52
	pounds per square inch (lb/in^2)	13.888 89
[1] meters-head	kilopascals	9.806 36
	[2] mm of Hg	73.554
	[1] feet of water	3.280 84
	pounds per square inch	1.422 29
	pounds per square foot	204.81
[1] feet of water	kilopascals	2.988 98
	[1] meters-head	0.3048
	[2] mm of Hg	22.4193
	[2] inches of Hg	0.882 65
	pounds per square inch	0.433 51
	pounds per square foot	62.4261

[1] Column of H_2O (water) measured at 4 °C.
[2] Column of Hg (mercury) measured at 0 °C.

To convert from	*To*	*Multiply by*
kilopascals	newtons per square meter (N/m^2)	1.0×10^3
	[2]mm of Hg	7.500 64
	[1]meters-head	0.101 97
	[2]inches of Hg	0.2953
	pounds per square foot	20.8854
	pounds per square inch	0.145 04
	std. atmospheres	9.8692×10^{-3}
kilograms (f) per square meter	kilopascals	$9.806\ 65 \times 10^{-3}$
	[2]mm of Hg	73.556×10^{-3}
	pounds per square inch	1.4223×10^{-3}
millibars (mbar)	kilopascals	0.10
bars	kilopascals	100.0
std. atmospheres	kilopascals	101.325
	[2]mm of Hg	760.0
	pounds per square inch	14.70
	[1]feet of water	33.90

Mass Per Unit Volume

Density and Mass Capacity

To convert from	*To*	*Multiply by*
pounds per cubic foot	kilogram per cubic meter (kg/m^3)	16.018 46
	slugs per cubic foot ($slug/ft^3$)	0.031 08
	pounds per gallon (lb/gal)	0.133 68
pounds per gallon	kilograms per cubic meter (kg/m^3)	119.8264
	slugs per cubic foot	0.2325
pounds per cubic yard	kilograms per cubic meter	0.593 28
	pounds per cubic foot (lb/ft^3)	0.037 04
grams per cubic centimeter	kilograms per cubic meter	1.0×10^3
	pounds per cubic yard	1.6856×10^3
ounces per gallon (oz/gal)	grams per liter (g/l)	7.489 15
	kilograms per cubic meter	7.489 15
kilograms per cubic meter	grams per cubic centimeter (g/cm^3)	1.0×10^{-3}
	metric tons per cubic meter (t/m^3)	1.0×10^{-3}
	pounds per cubic foot (lb/ft^3)	62.4297×10^{-3}
	pounds per gallon	8.3454×10^{-3}
	pounds per cubic yard	1.685 56

[1] Column of H_2O (water) measured at 4 °C.
[2] Column of Hg (mercury) measured at 0 °C.

To convert from	*To*	*Multiply by*
long tons per cubic yard	kilograms per cubic meter	1328.939
ounces per cubic inch (oz/in^3)	kilograms per cubic meter	1729.994
slugs per cubic foot	kilograms per cubic meter	515.3788

Volume Per Unit Area Per Unit Time [1] Hydraulic Conductivity (Permeability)

To convert from	*To*	*Multiply by*
cubic feet per square foot per day	cubic meters per square meter per day ($m^3/(m^2\cdot d)$)	0.3048
	cubic feet per square foot per minute ($ft^3/(ft^2\cdot min)$)	0.6944×10^{-3}
	liters per square meter per day ($l/m^2\cdot d$))	304.8
	gallons per square foot per day ($gal/(ft^2\cdot d)$)	7.480 52
	cubic millimeters per square millimeter per day ($mm^3/(mm^2\cdot d)$)	304.8
	cubic millimeters per square millimeter per hour ($mm^3/(mm^2\cdot h)$)	25.4
	cubic inches per square inch per hour ($in^3/(in^2\cdot h)$)	0.5
gallons per square foot per day	cubic meters per square meter per day ($m^3/(m^2\cdot d)$)	40.7458×10^{-3}
	liters per square meter per day ($l/(m^2\cdot d)$)	40.7458
	cubic feet per square foot per day ($ft^3/(ft^2\cdot d)$)	0.133 68

Volume Per Cross Sectional Area Per Unit Time [1] Transmissivity

To convert from	*To*	*Multiply by*
cubic feet per foot per day ($ft^3/(ft\cdot d)$)	cubic meters per meter per day ($m^3/(m\cdot d)$)	0. 0929
	gallons per foot per day ($gal/(ft\cdot d)$)	7. 480 52
	liters per meter per day ($l/(m\cdot d)$)	92. 903
gallons per foot per day	cubic meters per meter per day ($m^3/(m\cdot d)$)	0. 012 42
	cubic feet per foot per day ($ft^3/(ft\cdot d)$)	0. 133 68

[1] Many of these units can be dimensionally simplified. For example, $m^3/(m\cdot d)$ can also be written m^2/d.

INDEX

☆ U.S. GOVERNMENT PRINTING OFFICE: 1981 O—778–098